Benchmark Papers
in Geology

Series Editor: Rhodes W. Fairbridge
Columbia University

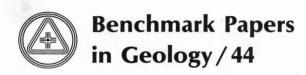

**Benchmark Papers
in Geology / 44**

A BENCHMARK ® Books Series

MINERAL DEPOSITS, CONTINENTAL DRIFT AND PLATE TECTONICS

Edited by

J. B. WRIGHT

The Open University

Dowden, Hutchinson & Ross, Inc.

STROUDSBURG, PENNSYLVANIA

Copyright © 1977 by **Dowden, Hutchinson & Ross, Inc.**
Benchmark Papers in Geology, Volume 44
Library of Congress Catalog Card Number: 77-1602
ISBN: 0-87933-290-5

79 78 77   1 2 3 4 5
Manufactured in the United States of America.

LIBRARY OF CONGRESS CATALOGING IN PUBLICATION DATA
Main entry under title:
Mineral deposits and Continental Drift
   (Benchmark papers in geology ; 44)
   Includes bibliographies and indexes.
   1. Continental drift—Addresses, essays, lectures. 2. Mineralogy—
Addresses, essays, lectures. 3. Plate tectonics—Addresses, essays, lectures.
I. Wright, J. B.
QE511.5.M56   553'.1   77-1602
ISBN 0-87933-290-5

Distributed by
ACADEMIC PRESS
A Subsidiary of Harcourt Brace Jovanovich, Publishers

SERIES EDITOR'S FOREWORD

The philosophy behind the "Benchmark Papers in Geology" is one of collection, sifting, and rediffusion. Scientific literature today is so vast, so dispersed, and, in the case of old papers, so inaccessible for readers not in the immediate neighborhood of major libraries that much valuable information has been ignored by default. It has become just so difficult, or so time consuming, to search out the key papers in any basic area of research that one can hardly blame a busy man for skimping on some of his "homework."

This series of volumes has been devised, therefore, to make a practical contribution to this critical problem. The geologist, perhaps even more than any other scientist, often suffers from twin difficulties—isolation from central library resources and immensely diffused sources of material. New colleges and industrial libraries simply cannot afford to purchase complete runs of all the world's earth science literature. Specialists simply cannot locate reprints or copies of all their principal reference materials. So it is that we are now making a concerted effort to gather into single volumes the critical material needed to reconstruct the background of any and every major topic of our discipline.

We are interpreting "geology" in its broadest sense: the fundamental science of the planet Earth, its materials, its history, and its dynamics. Because of training and experience in "earthy" materials, we also take in astrogeology, the corresponding aspect of the planetary sciences. Besides the classical core disciplines such as mineralogy, petrology, structure, geomorphology, paleontology, and stratigraphy, we embrace the newer fields of geophysics and geochemistry, applied also to oceanography, geochronology, and paleoecology. We recognize the work of the mining geologists, the petroleum goelogists, the hydrologists, the engineering and environmental geologists. Each specialist needs his working library. We are endeavoring to make his task a little easier.

Each volume in the series contains an Introduction prepared by a specialist (the volume editor)—a "state of the art" opening or a summary of the object and content of the volume. The articles, usually some twenty to fifty reproduced either in their entirety or in significant extracts, are selected in an attempt to cover the field, from the key papers of the last century to fairly recent work. Where the original works are in foreign languages, we have endeavored to locate or commission translations. Geologists, because of their global subject, are often acutely aware of the oneness of our world. The selections cannot, therefore, be restricted to any one country, and whenever possible an attempt is made to scan the world literature.

To each article, or group of kindred articles, some sort of "highlight commentary" is usually supplied by the volume editor. This commentary should serve to bring that article into historical perspective and to emphasize its particular role in the growth of the field. References, or citations, wherever possible, will be reproduced in their entirety—for by this means the observant reader can assess the background material available to that particular author, or, if he wishes, he, too, can double check the earlier sources.

A "benchmark," in surveyor's terminology, is an established point on the ground, recorded on our maps. It is usually anything that is a vantage point, from a modest hill to a mountain peak. From the historical viewpoint, these benchmarks are the bricks of our scientific edifice.

RHODES W. FAIRBRIDGE

PREFACE

The literature for this volume has been grouped into two somewhat unequal parts. Part I attempts to provide an historical perspective, with articles summarizing earlier ideas on continental drift and its relation to mineral deposits. When mentioned together continental drift and the formation of mineral deposits were treated essentially as two independent processes: continental drift merely broke up pre-existing mineral provinces, and in this context only the Mesozoic disruption of supercontinents and subsequent movement of the fragments were considered. Part I also includes a brief examination of ideas on the development of metallogenic provinces and the source of ore metals.

Part II contains a selection of the necessarily recent literature that emphasizes the causal links between plate tectonics and formation of mineral deposits, metallic as well as nonmetallic, both in a general way and with reference to specific plate tectonic or geographic settings. The influence of plate tectonic concepts on ideas about the source of ore metals is also examined.

Plate tectonic concepts are not accepted by all earth scientists and dissenting views are also given a place in this compilation.

The nouns *metallogeny* and *metallogenesis* and the adjectives *metallogenic* and *metallogenetic* seem to be used interchangeably in the literature and the commentaries on the articles in this volume are no exception. These words are close enough in appearance and meaning for such usage to present no problem, except to the dedicated purist.

Some authors receive considerable coverage in terms of numbers of articles. This is inevitable where the field is still new, and the number of authors relatively small. In fifty years time, an Earth Sciences Citation Index might tell us which of these papers, if any, represented true milestones (benchmarks). As it is, while many papers appear to handle the same subject matter, closer inspection shows that there are contrasted points of detail or emphasis that are worth recording at this early stage.

There still remains a large number of papers that have not been included, and the References on pages 381–387 list them. Therefore this volume should also serve as a fairly comprehensive bibliography for the literature on the relationship between mineral deposits and plate tectonics (see also Kasbeer, 1973).

ACKNOWLEDGMENT

I would like to thank Eileen Thompson for her help in typing and keeping track of successive drafts of this publication.

J. B. WRIGHT

CONTENTS

Contents

Contents

CONTENTS BY AUTHOR

Part I

HISTORICAL PERSPECTIVE

HISTORICAL BACKGROUND TO CONTINENTAL DRIFT AND PLATE TECTONICS

The idea that continents can move relative to one another has been with us since at least the mid-nineteenth century and possibly since the seventeenth, although this remains a matter of debate (Carozzi, 1970; Rupke, 1970). Antonio Snider (1858) can probably be credited with the first published illustration of continental drift (Figure 1).

In the twentieth century Taylor (1910), Wegener (1929), and du

Figure 1 Maps published by A. Snider in 1855 to illustrate his conception of continental drift. The map on the left represents his reassembly of the continents for late Carboniferous times. Source: Holmes, 1965, Fig. 860.

Toit (1937) were chiefly responsible for collating the evidence and formulating the hypothesis of continental drift in scientific terms. However, even the wealth of supporting data presented first by Wegener and then by du Toit failed to gain general acceptance for the theory until well into the 1950s. It had the support of some distinguished advocates, nonetheless, such as King (1953), Vening Meinesz (1964) and Holmes (1931, 1944, Fig. 262), whose descriptions of mantle convection currents as a driving mechanism are remarkably similar to current ideas. But even Holmes (1953) was not always a totally convinced supporter.

The failure of the continental drift theory to achieve respectability was due partly to the incomplete nature of the evidence (e.g. Longwell, 1944), but principally to the contemporary state of knowledge about the structure and properties of the earth's outer layers. Geophysicists could not reconcile their data, either with Wegener's picture of sialic continents "ploughing" through a simatic "sea" and pushing up mountain ranges in front of them, or with mantle convection currents. So continental drift remained "impossible" dismissed by the more forthright as little more than a fairy tale (Willis, 1944).

In the decades following the mid-1950s, the situation reversed quite suddenly. Paleomagnetic evidence in favor of relative continental movements was accumulating rapidly. New information about the crust and upper mantle meant that convection currents were no longer "impossible." The resulting "revolution in the earth sciences" is chronicled in many books and articles (e.g. Blackett et al., 1965; Kanamori et al., 1967; Gass et al., 1972; Hallam, 1973; Cox, 1973). Earth scientists have become equipped with a unifying global theory of sea-floor spreading and plate tectonics, within which most geological processes can be accommodated.

The present picture shows the outer part of the solid earth—the lithosphere—as a carapace or mosaic of interlocking plates (Figure 2) that continually change shape in response to processes taking place at their margins. Since the continents are an integral part of such plates, changes in plate shape cause the continents to change their relative positions.

Plate margins can be classified into three types on the basis of the processes taking place there.

1. Constructive or accreting margins are the ocean ridge crests, where new oceanic lithosphere is continually generated by rise of material from the upper mantle, which is hotter and of lower density than elsewhere, so that ocean ridges are regions of relatively high heat flow and shallow focus earthquakes (10–20 km). Plates move away from each other at constructive margins, at rates of between 1 and 8 cm per year (Figure 2).

2. Destructive or consuming or convergent margins are recognized

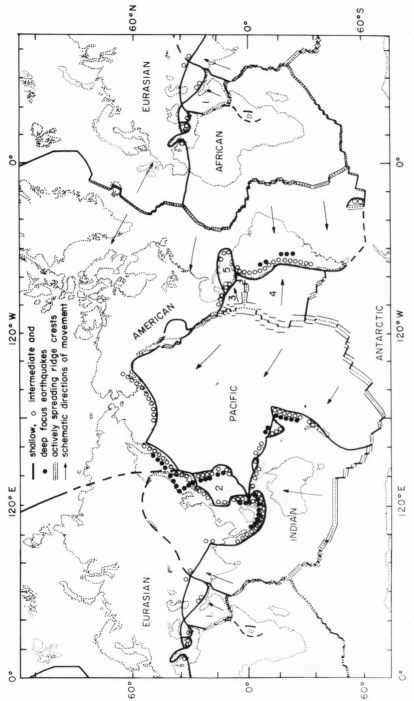

Figure 2 Summary of the seismicity of the earth, and distribution of lithospheric plates. The six major lithospheric plates bounded by active ridge crests, transform faults, trench systems and zones of compression are named. The following minor plates are numbered: (1) Arabian; (2) Phillipine; (3) Cocos; (4) Nasca; (5) Caribbean. Spreading rates at ridge crests are indicated schematically and vary from 1 cm per year per ridge flank in the vicinity of Iceland to 8 cm per year per ridge flank in the equatorial Pacific Ocean. Source: Professor F. J. Vine.

3

at ocean trenches near island arcs and orogenic continental margins (e.g. South America), where lithosphere is underthrust or *subducted* and resorbed into the mantle along an inclined plane identified by a zone of earthquake foci deepening to a maximum of 700 km—the Benioff Zone. Plates move toward each other at destructive margins and if two continental masses meet (collide) at such a margin they become sutured together, with a mountain range between them—continental crust is not dense enough to be taken down into the mantle. Such sutures may be marked by ophiolite complexes, which are interpreted as slices of oceanic lithosphere overthrust or obducted onto the crust.

3. Passive or transform margins are giant fault planes along which the plates move past each other, involving lateral movements measurable in hundreds of kilometers.

Transform faults also occur as cross-fractures at both constructive and destructive plate margins and probably represent accommodation to distortional stresses resulting from changes in the direction or rate of sea floor spreading or subduction (Wilson, 1965). They are typically a feature of oceanic lithosphere, but many transcurrent faults of continental regions are now recognized as landward extensions of major transform fractures. Two well-known examples, which in this case happen also to be plate boundaries (Figure 1), are the San Andreas system of western North America, and the New Zealand Alpine fault.

The lithosphere is 100–150 km thick, and consists largely of upper mantle peridotite, overlain by a thin (c. 5 km) skin of gabbroic and basaltic crust (sima) beneath the oceans and a much thicker (c. 35 km) granitic crust (sial) in continental areas.

Lithospheric plates can move at rates of a few cm per year because the underlying asthenosphere or low velocity layer (so-called because earthquake waves are slowed down and attenuated as they pass through it), which is 100–200 km thick, is believed to be a zone of partial melting and hence of lower shear strength. Recognition of the asthenosphere as a zone of relatively low friction upon which the lithosphere can move, eliminated one of the major obstacles that prevented acceptance of Wegener's ideas. The driving force for plate movements is less satisfactorily established, but some form of convective circulation is the most generally accepted.

The validity of the new global tectonics is now accepted by the great majority of earth scientists, despite a spirited rearguard action by some well-known names (e.g. Beloussov, 1970; Jeffreys, 1970a, p. 450–459 and 1970b; Meyerhoff et al., 1972a, b). Even among the majority who accept the principle, however, dissenting voices may be heard. As some of the papers in this volume will show, not everyone is agreed that continental drift and plate tectonics have anything at all to do with the formation of mineral deposits.

Editor's Comments
on Papers 1, 2, and 3

MINERAL DEPOSITS AS EVIDENCE FOR CONTINENTAL DRIFT

As long as continental drift remained merely a hypothesis, rejected by most earth scientists because it was impossible according to contemporary geophysical knowledge, the main concern of its adherents was naturally to accumulate further evidence in its favor. Little attempt was made to establish causal relationships with geological processes such as magmatism, regional metamorphism, development of major sedimentary basins, or accumulation of mineral deposits—even the connection with mountain building was treated in a rather superficial way. For example:

> Thus the East Indies went to the north and are still moving north, causing the damming or urging up of the great mass we call the Himalayan Range; just as the earlier drift of South America is puckering up the great American backbone (Haddock, 1936, p. 68).

So it is hardly surprising that the only references to mineral deposits in the early continental drift literature occur almost incidentally, when distribution patterns were used merely as evidence to test the hypothesis.

Paper 1 by Hall is an example of how the distribution of Carboniferous coals could be used to support continental drift theory. It is a review article, based on more up-to-date information than was available to Wegener and du Toit.

Climate obviously exerts an important control over the formation

5

of coal and most other mineral deposits of sedimentary origin. The global distribution of such deposits in the geological column could thus reveal the extent to which climatic belts have changed with time. Huene (Paper 2) suggests how this could provide a test of the continental drift hypothesis; the idea was independently applied again much later, as we shall see in subsequent papers.

The internal earth processes responsible for magmatic and hydro-thermal ore deposits were less well understood, a situation that still persists today. In consequence, they were rarely used in the continental drift argument. For example, du Toit (1937, Fig. 18) refers only in passing to the important metal deposits in the Hercynian orogenic belt, extending from Europe to Appalachia across what is now the Atlantic Ocean. He evidently regarded metallogenesis as being related to volcanism and orogeny, but not to the causal mechanism of continental drift.

Dauvillier and Henry (Paper 3) are the only ones who refer specifically to ore deposits of this type in relation to continental drift, that could be found in the available literature. It is an interesting reflection upon the prevailing climate of opinion that these authors could present a case so much flimsier than the theory it was intended to demolish!

1

Copyright © 1950 by Mining Journal Limited

Reprinted from *Mining Mag.* **82**:201–203, 204–207, 208, 210 (1950)

The Coal of Gondwanaland

T. C. F. Hall

Introduction

The recent issue by the Fourth Empire Mining and Metallurgical Congress of two pamphlets dealing with Commonwealth coals,[1] and the previous publication in the MAGAZINE of articles descriptive of individual occurrences, have drawn attention to the immense resources of Permian or Permo-Carboniferous coal that occur in Africa, India and Australia. For the Commonwealth coals, on the contrary, evidently owe their origin to the accumulation of vegetable matter in continental inland basins under cold climatic conditions.

The southern coals are distributed over a vast area, being found alike in South, Central and East Africa, Madagascar, Peninsular India, Australia, Tasmania, South America and Antarctica. They occur, in other words, within the confines of the lands assigned to the hypothetical ancient conti-

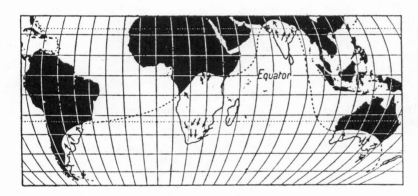

Fig. 1.[1]—Map Showing the Distribution of the Late-Carboniferous Glaciation of Gondwanaland (Holmes) :

Arrows indicate direction of ice movements.

countries alone the estimated reserves amount to not much less than 150,000 million tons, a figure that takes no account of the resources of the Belgian Congo, South America and other extra-Commonwealth countries where similar coals occur.

These coals, which, with the exception of those of India, are confined to the Southern Hemisphere, are distinct from those contained in the Coal Measures of the Northern Hemisphere, found in North America, Europe and northern Asia, and differ from them both in age and mode of origin. The two were formed in widely different circumstances. Thus, while the northern coals, which are the earlier, are to be regarded as having originated in tropical lagoonal or deltaic swamps, the later, southern

nent to which Suess gave the name " Gondwanaland." [2]

The seams form part of a group of dominantly terrestrial formations, which range in age from late-Carboniferous to Rhaetic and which, despite their widespread distribution, show remarkable lithological uniformity throughout their extent, consisting in the main of shales, mudstones and sandstones. These strata make up the *Gondwana System*, so called after the Indian locality where they were first discriminated and studied.

At the base of the succession occur consolidated boulder-clays (tillites) and fluvioglacial boulder-beds, notable examples of which are the " Dwyka Tillite " of South

---

[1] A list of works to which reference is made is given at the end of article.

[1] Reproduced from " Principles of Geology," by courtesy of the author and publishers.

[2] Now generally shortened to *Gondwana*.

Africa, the " Talchir Boulder-bed " of India, and the " Itararé Boulder-bed " of southern Brazil. These morainic deposits, which may attain a thickness of 1,000 ft. or more, encompass a great part of the area occupied by the Gondwana rocks and are remnants of a formerly continuous girdle of such deposits, now broken by intervention of the ocean, removal by erosion, or concealment beneath later formations. The extent of their distribution is shown in Fig. 1. They reveal, together with striated rock surfaces, erratic blocks, rock-basins, and other records of ice action, that during late-Carboniferous time much of Gondwana was ice capped. The Gondwana System thus rests in great part upon an ice-eroded platform of ancient rocks.

The coal occurs at a number of horizons but is most abundant towards the base of the System and the principal deposits are of Permian, especially Lower Permian, age. Those of Triassic and Rhaetic ages are relatively unimportant and the seams are generally narrow and of poor quality. Lower Permian coals occur in the Union of South Africa, Northern and Southern Rhodesia, Nyasaland, Mozambique, Tanganyika, Belgian Congo, Madagascar, India, Australia, Tasmania and Brazil. Those of Upper Permian age are less abundant, but occur in South Africa, India, Australia, Argentina and Antarctica. The deposits differ generally from those of the Northern coalfields in that they tend to consist of single or compound seams rather than of a number of superposed ones, but the seams are often of relatively great thickness.

Working of the deposits has been done principally in Natal (Newcastle, etc.), Transvaal (Witbank, Middleburg), Orange Free State (Vereeniging), Southern Rhodesia (Wankie), Belgian Congo (Lukuga), India (Bihar, Orissa, Central Provinces), New South Wales (Newcastle, Greta, Aberdare, etc.), southern Brazil and Argentina. Most of the coal is of the bituminous variety, much of it with good coking properties, but it varies considerably in quality from place to place.

Coal Formation

Coal formation in Gondwana began with the disappearance of the ice-cap and generally the lowermost coal-bearing strata are immediately underlain by tillite or rest directly upon the ice-eroded platform of older rocks. These strata are of early Permian or Permo-Carboniferous age for, apart from

some local recurrence, as shown by the presence of Permian tillites in Australia and South America, the glaciation of Gondwana, on the whole, came to an end at the close of the Carboniferous Period.

In its retreat the ice-sheet left behind it ice-excavated basins and dammed-up glacier valleys, which became the sites of lakes, swamps and marshes supporting the new vegetation that sprang up to replace that destroyed by the glaciation. This vegetation, which is characteristically Gondwanian and distinctive from any of equivalent age found elsewhere, is the *Glossopteris Flora*, so called after one of its commonest forms, a genus of fern almost peculiar to the region, being found otherwise only in Russia and China. Another prominent, though more restricted, genus is *Gangamopteris*. This flora, whose remains are a constant feature of the Permian coal-bearing strata, is believed to have originated in Carboniferous time somewhere on the fringe of Gondwana, possibly in Patagonia, outside the area of glaciation, whence, with the retreat of the ice, it spread rapidly over the continent. It was fully established by mid-Permian time, some genera having migrated beyond the confines of Gondwana into Central Asia (Fig. 5). The rapid spread and uniformity of this vegetation testify to the absence of obstacles, such as mountain ranges, to its migration, and to a general sameness of climatic and other physiographical conditions obtaining in Gondwana. *Glossopteris* survived in a few places into Triassic time, when it was overtaken by a new flora and died out. In the fewness of its genera and species the *Glossopteris Flora* shows a strong contrast to the contemporary flora of the Northern Hemisphere.

The Gondwana coal can be regarded, therefore, as having originated in sub-polar peat bogs, and was thus formed under conditions very different from those attending the formation of the Carboniferous coal of the Northern Hemisphere. The conditions may be paralleled, however, with those obtaining in that Hemisphere at the close of the Pleistocene Ice Age.

The contrasted modes of origin of the two coals may seem the more remarkable when their present distribution is considered. They appear, as it were, to have exchanged environments, for while the Gondwana seams now lie very largely within the tropics the northern coal measures occur at high latitudes. Such distribution evidently implies a considerable change in the relative positions of the continental areas and the earth's climatic belts, a state of affairs that could be accomplished by a shifting of the polar axis or by movement of the continents themselves. Before considering this question it will be as well to look at Gondwana.

Gondwana

The uniformity of the Gondwana System throughout its distribution, the lithological correspondence of the different exposures, the constant recurrence of basal glacial deposits overlain by coal-bearing strata with a *Glossopteris Flora*, and the extraordinarily similar sequence of events recorded by the System wherever it occurs, these and other factors testify to the collective origin of these dominantly terrestrial formations, despite their widespread distribution.

To account for these resemblances Suess identified the distribution of the System with the one-time existence of a great continent (Gondwanaland) incorporating the sites of South America, Africa, India and Australasia, together with much of the area now occupied by the South Atlantic and Indian Oceans.[1] To the north of this continent lay a shallow sea, the *Tethys*, dividing it from the northern lands.

But, although a belief in the collective origin of the Gondwana formations is fully supported by the evidence, there are, nevertheless, strong reasons for doubting the existence of so vast a continent as that envisaged by Suess. One such reason is the difficulty of explaining its disappearance, or rather of those parts of it now occupied by ocean, for the supposition that they foundered is beset by baffling problems and is rejected by those who maintain that the subsidence to ocean depths of continental masses is contrary to the principle of isostasy.

This difficulty is overcome by advocates of "Continental Drift," who regard Gondwanaland as having consisted of an assemblage of the existing continents, which have since drifted apart to their present positions.[2]

[1] The existence of such a continent had been previously suggested by M. Neumayr and also by W. T. Blanford, but it was Suess who elaborated and substantiated the idea and who named the continent Gondwanaland.

[2] A full account of the Theory of Continental Drift will be found in A. L. Du Toit's *Our Wandering Continents*. Other references are given at the conclusion of this article.

This assemblage is illustrated in the accompanying figures. The idea of displacement also removes the difficulty of explaining the identity of terrestrial formations occurring in places so far apart as are India, Africa, South America, Australia and Antarctica since, instead of having to be regarded as remnants of formations that once spread over millions of square miles, they can be looked upon as detached fragments that were formerly contiguous. Such an enormous expanse of stratigraphical continuity and lithological persistency as is implied in the restoration of Gondwana in terms of present-day geography is highly improbable. Furthermore, the contemporaneous glaciation of so vast an area is very problematical. Such conformity becomes possible, however, when the continents are assembled (Fig. 4). This glaciation provides, in fact, perhaps the most compelling reason for regarding Gondwana as a continental unit, and it will be appropriate, therefore, to consider it in some detail, more especially as it paved the way for coal formation and the distribution of the coal-bearing strata is conterminous with its extent.

[*Editor's Note:* Material has been omitted at this point.]

Continental Drift

All such postulations are unnecessary with recourse to the " Displacement Hypothesis " and the assumption that Gondwana, whose existence can hardly be doubted, consisted of a grouping of the continents over which its formations are now scattered.

According to Wegener the world's land area in Carboniferous time was comprised in two large continents—a northern one, or *Laurasia*, and a southern one, or *Gondwana*, the former being an assemblage of what are now known as Europe, Asia (for the most part), North America and Greenland; the latter of the present-day South America, Africa, India and south-east Asia, Australasia and Antarctica, with Africa as the core. Between the two lay the *Tethys*, communicating east and west with the one great ocean, the primeval Pacific, by which the lands were completely enveloped. This conception of Carboniferous geography is illustrated in Fig. 2, based on one of Wegener's famous maps.

An arresting feature of this map, of particular significance in our inquiry, is the disposal of the lands with respect to the earth's poles and equator, this being such that the southern part of Laurasia occupies the equatorial belt while much of Gondwana lies within the Antarctic Circle, with the South Pole adjacent to the south-east corner of Africa. Thus regions of North America, Europe and Asia that now have a temperate climate become tropical, whereas portions of Africa, India and Australasia that now have a tropical or sub-tropical climate become polar. This arrangement is fully in accord with the claims of lithological and palæontological evidence shown by the rocks of the period. Indeed, it is difficult to interpret this evidence except in terms of continental displacement and explain the presence of glacial deposits among the Gondwana formations and of the tropically-formed Carboniferous coal measures in a temperate region.

Du Toit's grouping of the continents to form Gondwana, shown in Fig. 3, differs from that of Wegener in several respects, especially as regards the position assigned to Antarctica, but the overall picture remains the same.

The conception of continental displacement thus makes possible a reconstruction of Gondwana in harmony with its records. It removes its glacial deposits from the tropics, a position in which, it is true, glaciers may originate at a sufficient elevation, but one wholly unsuited to the formation of a great continental ice-sheet, and it brings the glaciated regions into a compass consistent with their contemporaneous glaciation and comparable in size with that glaciated in the Northern Hemisphere during the Pleistocene Period (Fig. 4). On the other hand, the coal-bearing region of the Northern Hemisphere falls into an environment suitable for the existence of luxuriant tropical forests and coastal swamps, while of similar, perhaps even of greater, significance is the affording of an equally favourable environment to the Carboniferous laterites and bauxites, deposits of undoubted tropical origin. As Holmes says : " The Carboniferous climatic girdles fall consistently into their appropriate places." Moreover, a " wandering of the poles," so highly questionable as an actuality, becomes relative.

This reconstruction, therefore, places Gondwana in Carboniferous time, before its northerly, equatorward, drift carried it into warmer climes, astride the South Pole, with its components, except Antarctica, away to the south of their present sites and for the most part in an ice-bound environment. The

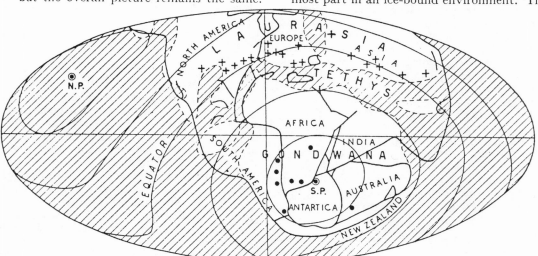

Fig. 2.—Upper Carboniferous World Geography

(after Köppen and Wegener) : Unshaded areas, land ; shaded areas, ocean and epicontinental seas ;
+ Laurasian coal deposits ; • *Glossopteris Flora.*

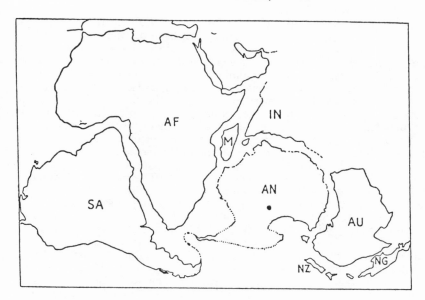

Fig. 3.[1]—
Reconstruction
of Gondwana
(Du Toit).

[1] Reproduced from "The Geology of South Africa," by courtesy of the publishers.

Fig. 4.[1]—Map Showing Distribution of Late-Carboniferous Glaciation of Gondwanaland

(Continents re-assembled, though not so closely, as interpreted by Wegener) (Holmes).

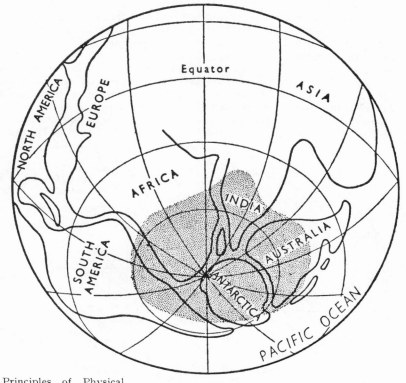

[1] Reproduced from "Principles of Physical Geology" by courtesy of the author and publishers.

extent of its glaciation thus conforms to that of a polar ice-cap.

Whatever view is taken as to its make-up, however, Gondwana, after the retreat of the ice and during the time of principal coal formation—that is, during the Permian Period—is to be pictured as very largely an ice-worn, low-lying, marshy region, cut off from the ocean by a mountain range, the *Gondwanides*, which was rising in the south and converting it into an area of interior drainage, with great lakes and swamps occupying ice-excavated hollows and depressions due to warping. The climate had in general ameliorated and was probably humid.

Wegener's reconstruction for this Period is shown in Fig. 5. A comparison of this with Fig. 2 shows the spread of the *Glossopteris Flora*, with migration into Laurasia, a land connexion with Laurasia, *via* Persia (due to the Hercynian orogeny), and a northward displacement of the continents.

Geological History

The Permian Period was the time of maximum coal formation. Thereafter it declined and seams are fewer and less widely distributed among the later strata, while such as there are are on the whole of comparatively minor value. This falling off may be correlated with a change of climate, for at the close of the Permian Period Gondwana entered on a phase of aridity, and during Triassic time was in great part desertic, as evidenced by the occurrence of æolian and current-bedded sandstones, " red beds " and wind-faceted pebbles among the formations. This aridity has been attributed to interception by the Gondwanides of the moisture-laden winds from the southern ocean, since its culmination corresponded with that of the orogeny, but perhaps also, in consequence of northerly drift, Gondwana then lay astride the desert climatic belt. During this time it was the haunt of a host of reptiles, whose abundant remains have furnished an unparalleled record of the evolution, step by step, of these vertebrates. Cotylosaurians, Dinocephalians, Anomodonts and Theriodonts dominated the land, and among the last-named were those specializing along the line of descent that culminated in the mammals. Of particularly significant interest among Gondwana reptiles is the small aquatic known as *Mesosaurus*, which inhabited the deltas during Permo-Carboniferous time. Remains of this little creature, which was only about 18 in. in length, have been found exclusively in South Africa and Brazil, and its appearance in these widely-separated places is highly suggestive of their former closer connexion, for it is most improbable, to say the least, that an inhabitant of fresh or brackish water, as *Mesosaurus* evidently was, could have migrated across the width of ocean (some 4,000 miles) that now intervenes.

The Dinosaurs, those bizarre reptiles whose remains have been found so plentifully in the Northern Hemisphere, are not so well represented among Gondwana fossils, but this is not surprising, for during the closing stages of the Triassic Period and the opening

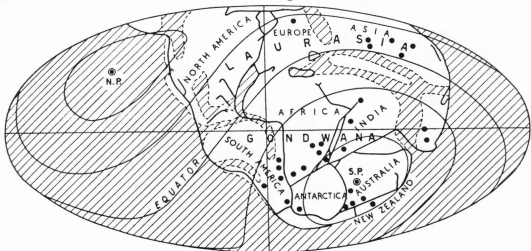

Fig. 5.—Permian World Geography

(after Köppen and Wegener): Unshaded areas, land; shaded areas, ocean and epicontinental seas; ● *Glossopteris Flora*.

of the Jurassic, when these creatures were expanding, much of Gondwana was being overwhelmed by great outpourings of lava. Subsequently depression set in and its margins and basins were invaded by the Jurassic and Cretaceous seas.

Finally, if we accept the Drift Hypothesis, Gondwana broke up, probably in late Cretaceous or early Tertiary time, and as a great single continent passed from the geological scene after having been in existence since Pre-Cambrian time, a duration of possibly not less than 1,000 million years. With the drift of its fragments to lower latitudes much of it entered the tropical belt, and those who now swelter in the heat of Africa or India may perhaps find consolation in the thought that there was a time, not so remote geologically, when in those places they would have had to contend with frost-bite !

[*Editor's Note:* Material has been omitted at this point.]

The Theory of Continental Drift

For a full account of this theory, known also as the " Displacement Hypothesis," readers should consult A. L. Du Toit's " Our Wandering Continents."

[*Editor's Note:* Material has been omitted at this point.]

A collection of papers by various authors, for and against the theory, was issued by the American Association of Petroleum Geologists in 1928 (" Theory of Continental Drift : A Symposium ") and in the following year readers of the MAGAZINE were given a " Review of the Hypothesis " by Dr. Arthur Holmes (vol. xl, pp. 205–209, 286–288, 340–347). The same author gives an up-to-date summary of the theory in his " Principles of Physical Geology " (Chap. XXI), of which the writer of the present article has fully availed himself.

Here we have been concerned with the theory only in so far as it has a bearing on the nature and distribution of the Gondwana formations and it is unnecessary to enter into the arguments for and against it. Suffice it to say in passing that its opponents are concerned chiefly with the difficulty of accounting for the mechanical force necessary to engineer the horizontal movement of great continental masses and maintain that forces acting from outside the earth—such as, tidal friction, as advocated by Wegener and others—are insufficient or inapplicable.

This hardly seems to be a sufficient reason for discarding the theory, however, since the difficulty is one that may well be overcome in time with increase of knowledge concerning the forces that operate within the earth. The happening of the Pleistocene Great Ice Age was not disputed because of the problem of explaining its cause ; the evidence in its favour was accepted as conclusive. It is equally permissible to accept the evidence of continental drift despite the problem of accounting for the phenomenon. It is possible, as has been suggested, that sub-crustal convection currents may provide the motive force. In any case, against the difficulty of accounting for the operation of continental drift must be set an abundance of evidence that can be adduced in its favour—such as, the tectonic, stratigraphical, and palæontological similarities of now widely separated areas.

It is true that these similarities have been otherwise explained, but the explanations offered, such as the occurrence of former land bridges and the subsidence to great depths of large blocks of sial, present at least as great, if not greater, problems for solution. One of the great merits of the Drift Hypothesis is the light it throws upon the seeming vagary of past climatic changes. It certainly provides, as we have seen, a plausible explanation of the spread of the late-Carboniferous glaciation over much of Africa, Peninsular India, Australia and South America, and of the occurrence in all those far-flung areas of similarly formed deposits of coal, co-extensive with the remains of a peculiar land flora.

References

CARLOW, C. A. : " Coal Resources of the British Commonwealth." Fourth Empire Mining and Metallurgical Congress. Paper No. F.1. London, 1949.

FOOT, R. : " The Characteristics of the Empire Coals and their Bearing on the Industrial Potential of the Empire." Fourth Empire Mining and Metallurgical Congress. Paper No. F.2. London, 1949.

HOLMES, A. : " Principles of Physical Geology." (Chap. XXI), London, 1944.

DU TOIT, A. L. : " Our Wandering Continents," London and Edinburgh, 1937.

DU TOIT, A. L. : " The Geology of South Africa," 2nd Ed. London and Edinburgh, 1939.

KOPPEN, W., and WEGENER, A. : " Die Klimate der geologischen Vorzeit." Berlin, 1924.

2

Copyright © 1939 by the American Journal of Science

Reprinted from *Am. Jour. Sci.* **237**(6):439 (1939)

A POSSIBLE METHOD FOR THE PROOF OR DISPROOF OF THE WEGENER THEORY

WEGENER'S theory of continental displacement was proposed seventeen years ago. During this time its influence on research has been stimulating and fruitful, but nevertheless its validity is still much discussed and thoroughly convincing evidence in favor of it is still missing. Some scientists have proceeded as if the theory had been fully proved, but apparently a satisfactory solution of the question will not be easy. A method should be sought which, after it has been applied carefully and systematically, will lead to a solution that will be unequivocal. Such a method is here proposed. The carrying out of it, however, would require long and detailed work for which the writer has not the time, as his research is along another line.

The method would be to fix climatic zones for every possible stratigraphical level in every continent, if possible without gap, in order to see how they coincide. It ought to be demonstrated whether, in every continent and in the different ages, they parallel those of to-day, or whether they form angles with the direction of the present climatic zones; and whether they continue as normal rings around the globe or whether the zones of one continent form angles with those of the next continent. Possibly, then, the continents will form a "Wegener-configuration" if shifted—on paper—so that the climatic zones continue unbroken around the globe, and possibly they will not.

The criterion for the most reliable recognition of the climatic zones seems to the writer to be the distribution of gypsum. Already Köppen and Wegener in "Die Klimate der geologischen Vorzeit," 1924, p. 39 ff., have referred to gypsum in connection with salt, desert sandstones, and red beds.

The prosecution of this study would demand much work over a considerable number of years. Probably it would pay to have a number of geologists from all continents and many countries unite for joint work. At the end, an organization of geologists would be desirable, to prepare detailed transparent charts of the globe for every stratigraphic level, to be put on the present geodetic net of the globe, so that the eventual shifting of the continents or the wandering of the poles could be directly seen.

FRIEDRICH VON HUENE.

3

A DECISIVE ARGUMENT AGAINST
THE THEORY OF CONTINENTAL DRIFT

A. Dauvillier and P. Henry

This translation was prepared expressly for this Benchmark volume by J. B. Wright from "Un argument décisif contre la théorie des translations continentales," in C. R. Acad. Sci. Paris 221 (24): 757–758 (1939).

Geologists are divided by two major and fundamentally opposed doctrines: the theory of the permanence of ocean basins on the one hand and that of continental displacements on the other. The latter, developed by Taylor and A. Wegener in particular, provides a simple interpretation of many paleo-geographical and paleoclimatic phenomena that still has many adherents.

To the geophysicist the hypothesis of continental drift is unacceptable for several reasons. But if it could be shown that some fundamental feature on the Earth's surface has persisted unchanged throughout geological time, then the theory would be effectively demolished.

J. E. Spurr[1] showed in 1923 that silver deposits in the Americas show a remarkable great circle alignment, oriented NW–SE and stretching from the Blue Mountains on the west coast of North America, down to the River Plate in South America (Argentina). This alignment is over 10,000 km long and crosses the Pacific west of Panama at depths in excess of 4,000 m, as well as traversing mountain ranges and regions of contrasted geological age and structure.

Although there is no surface manifestation of its existence (sic), the lineament defines the site of a deep straight fracture in the Earth's crust, which has given rise to emanations responsible for the silver mineralization. This fracture must be geologically very ancient and appears to be a radial fissure, produced by crustal cooling in the earliest stages of the Earth's history.

It is similar to fractures still visible on the Moon and to the Martian canals and provides an additional argument in support of an hypothesis advanced by one of us.[2] This attributes the principal surface features of the Earth (ocean basins, continental drifts, mountain chains) to a mechanism identical to that responsible for the surface relief of the Moon. The persistence of such a lineament throughout geological time is quite incompatible with any kind of continental drift, however limited, and conclusively proves the permanence of the Earth's principal surface features.

---

1. J.E. Spurr. The Ore Magmas. 2 vol., 915 pp, New York, 1923.
2. A. Dauvillier, Comptes Rendues, 207, 1938, p. 452.

Editor's Comments
on Papers 4 Through 8

CONTINENTAL DRIFT AND CORRELATION OF MINERAL PROVINCES

Once continental drift had become generally accepted in the 1960s, its relationship to mineral deposits began to receive serious consideration. Early attempts to define this relationship consisted of reassembling continental fragments into their predrift configurations, to correlate mineral deposits across what are now wide expanses of ocean. Such correlations were believed to be of potential interest to exploration geologists, for where mineral-rich regions were truncated by coastlines their continuation could perhaps be sought in other parts of the continental "jigsaw puzzle."

A feature common to all the papers in this section is that mineral deposits are regarded as little more than passengers on drifting continental rafts. There is little attempt to relate metallogenesis to the mechanism of drift: for the most part, these are simply exercises in correlation.

This approach remains quite valid and is still occasionally used (e.g. Reid, 1974) but it can perforce only be applied in relation to the Mesozoic breakup of Gondwanaland and Laurasia (or of Pangea)—earlier drift phases are not susceptible to this kind of treatment.

Paper 4 is Schuiling's now classic paper on the distribution of tin belts around the Atlantic. Its appearance in the literature was almost immediately followed by Petrascheck's more general survey (Paper 5). Both papers also consider the source of ore metals in provinces that had been split up by drift movements, a theme to which we return later. It should be remembered that Paper 5 appeared in 1968 and some of the detail is sketchy and may appear dated and even incorrect in the light of present knowledge. As with other papers in this section, its importance lies in the contribution it makes to our understanding of how ore deposits are (and have been) distributed in space and time over the earth as a whole—and this in turn provides further information enabling us to reach more reliable conclusions about the origins of metal-bearing fluids.

In Paper 6, Petrascheck continues his general theme with particular reference to Greenland. In this case, however, only the map and summary have been excerpted.

The Indo-Australian region is examined in more detail in Paper 7A. Here we find the first reference (page 13) to the probability that Mesozoic drift was not a unique event, and that some generally similar process has occurred at intervals as far back as Precambrian time, though Crawford evidently did not consider these movements to have been on the same scale as in the Mesozoic.

Bosazza's discussion of this paper (Paper 7B) is especially interesting for it cites one of the few examples where the location of previously unknown mineral deposits was correctly identified on the basis of a continental drift reconstruction. It must be conceded, however, that in view of the debate over the predrift position of Madagascar, this could have been fortuitous (e.g. Smith and Hallam, 1970; Flower and Strong, 1969; Wright and McCurry, 1970; Green, 1972).

Crawford's reply (Paper 7C) heralds some revisions of his earlier views and is instructive because it shows how quite plausible reconstructions can be made untenable by new information—and how difficult it is to fit the "jigsaw" together in many places. A quite different arrangement of India and Australia, based largely on more detailed paleomagnetic evidence, is illustrated in Figure 3, which was kindly provided by

17

Professor Crawford and has already appeared in print (Crawford, 1974; compare the map in Paper 7A).

It harks back to a reconstruction by Du Toit (1937), and shows that a link between the Kolar and Kalgoorlie gold fields (cf. Paper 5) cannot be sustained; nor can another previous suggestion by Crawford (1973), that Tibet was the source of alluvial diamonds found in southern Thailand. In a written communication (1974) on this subject, Professor Crawford says: "the diamonds in the Pilbara district of northwest Australia may have come from sources in the Tarim Basin" (see Figure 3).

Correlation of mineral provinces between Antarctica and other Gondwana continents is attempted by Runnells in Paper 8.

After reading the papers in this section it could be argued that correlating fragmented mineral provinces across ocean basins is, after all, of limited use to the exploration geologist, despite claims to the contrary—the reason being that for many places, more than one 'fit' is possible, so that correlations are profoundly affected by the configuration chosen. In some cases, moreover, mineral deposits seem to be treated as little more than additional pieces of evidence to refine continental reconstructions.

There may also be other important geological variables to consider. A recent attempt to attribute a West African origin to alluvial diamonds in the Guianan shield, for example (Reid, 1974), has attracted critiscism, not because the continental fit is in doubt, but mainly because of discrepancies in the distribution and size of diamonds (Schönberger and de Roever, 1974), and in the age and distribution of West African kimberlites (Hastings, 1974).

Metallogenic Provinces

It is clear from the papers in this section that correlation of mineral deposits between separated continental fragments is only possible because some parts of the crust are much richer in economic minerals than others.

The uneven distribution of metal ores was recognized almost as soon as men started to mine them thousands of years ago. The concept of the metallogenic province, in which metal concentrations are seen as the products of geological processes related in space and time, began to evolve around the end of the nineteenth century.

There is naturally a voluminous literature dealing with the classification and origin of metallogenic provinces in relation to their geological and tectonic setting. Useful reviews on this subject can be found in Turneaure (1955), Petrascheck (1965), and Dunham (1973).

A recurring theme in metallogenic studies is the concept of deep

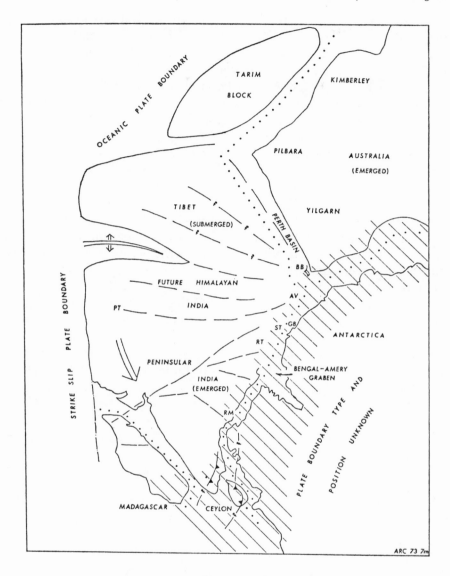

Figure 3 Part of Gondwanaland during Permian to late Jurassic times, as envisaged by Crawford. Abbreviations in capital letters refer to volcanics of various types and ages. Source: Crawford, 1974, Fig. 3.

and possibly long-lived fracture systems acting as loci of mineralization and controlling the distribution of mineral deposits. Metallogenic hypotheses based upon continental drift or plate tectonic mechanisms tend to incorporate this concept rather than exclude it, as subsequent papers will show.

An integral part of any metallogenic theory concerns the origin of ore metals. Do metals originate in the crust, the mantle, or both, depending upon circumstances? If fracture systems do control ore deposition, how deep do they extend?

Both Schuiling (Paper 4) and Petrascheck (Paper 5) have made the point that some metallogenic provinces are blessed with repeated mineralization. (For other examples, see Clifford, 1966; Landwehr, 1968; Wright, 1970; Turneaure, 1971, Goossens, 1972).

At least some of these provinces are known to have changed position between successive mineralizing epochs, which suggests that the source of mineralization lies somewhere within the moving lithosphere plate.

Schuiling (Paper 4) postulates so-called geochemical culminations in the crust; while Petrascheck (Paper 5) concludes that most ore deposits are the result of crustal recycling, except perhaps those in rocks of great antiquity and ultrabasic character, where the metals such as Cu, Ni, Cr, Pt could be of mantle origin (see also Petrascheck, 1969).

Ideas on the subject have naturally been influenced by the acceptance of continental drift and by the development of plate tectonic concepts. Although the structure and composition of the earth's outer layers are now better understood than ever before, the ultimate source of ore metals remains a matter for continuing debate, as we shall see in later sections.

4

Reprinted from *Econ. Geol.* **62**:540-550 (1967)

TIN BELTS ON THE CONTINENTS AROUND THE ATLANTIC OCEAN

R. D. SCHUILING

ABSTRACT

Tin is inhomogeneously distributed on the continents, in relatively narrow, continent-sized belts. On a reconstruction of the continents around the Atlantic Ocean before continental drift, the belts extend unbroken from one continent to another. As the ages of the tin mineralizations vary within the belts, an argument can be derived that the source of the tin and its associated elements must be in the crust. It is further speculated that concentrations of workable tin deposits occur in the intersection of orogenic belts with zones of primitive enrichment of tin. The ultimate cause of such primary geochemical culminations may well lie back in the early history of the earth.

INTRODUCTION

A METALLOGENETIC province may be defined as an area characterized by a conspicuous concentration of a certain element or group of elements, as compared with other areas. In such provinces the element, or elements, are commonly deposited by more than one geological process and at different times. The concept of a metallogenetic province implies the existence of large-scale chemical inhomogeneities in that part of the crust or the mantle from which the ore-deposits ultimately were derived. As tin seems to be an element that shows a strongly inhomogeneous distribution, it seemed worth while to plot all the known economic and uneconomic occurrences of this element in North and South America, Africa and Europe. The investigation was restricted to these continents, because it was hoped that the data might, at the same time, have some bearing on the theory of continental drift. Furthermore a review of this kind might be of some help in planning of future prospecting for tin deposits. The map (Fig. 1) provides the factual basis for the subsequent discussion.

Acknowledgments.—Valuable information was obtained from the Directors of the Geological Surveys of Algiers, Angola, Argentina, Cameroon, Congo-Brazzaville, Congo-Leopoldville, Dahomey, Ecuador, Liberia, Libya, Malawi, Mauretania, Peru, Senegal, Sierra Leone, Spanish Sahara, Tanzania and Zambia, for which the writer wishes to express his thanks. The writer is further indebted for information and assistance to Sir Edward Bullard, Cambridge; Dr. G. C. Brouwer, formerly of the Bureau de Recherches Géologiques et Minières, French Guyana; Professor Dj. Guimaraes and Dr. J. B. Kloosterman, Brazil; Dr. Lepersonne of the Musée Royal de l'Afrique Centrale at Tervuuren, Belgium; and Dr. Tagini of the Organisation S.O.D.E.M.I., Ivory Coast. Professor W. C. Burnham of the Pennsylvania State University kindly read the manuscript, and suggested a number of improvements; Professor J. Kalliokoski of Princeton University is thanked for his stimulating interest in the subject. A grant of the Billiton Mining Company as well as the stimulating interest of its Chief Geologist, Dr. G. L. Krol, is gratefully acknowledged. The Netherlands Organization for the Advancement of Pure Science granted the author a NATO Research Fellowship at Princeton University, where most of this work was carried out.

CONSTRUCTION OF MAP

The heading "Tin" in the Annoted Bibliography of ECONOMIC GEOLOGY, 1928–1963, provided a first set of references to tin-occurrences, which was expanded by a search for further data, both in older and in very recent literature. The final list of references from which the map (Fig. 1) was compiled contains well over 500 titles, to which new literature is still being added. From these I have selected a shorter list of references that contains most of the data used in the construction of the map. The map contains also much unpublished information, which was obtained on request from the Directors of many Geological Surveys, as well as a number of individuals, all mentioned in the "acknowledgments."

Compared to the amount of information presented on some metallogenetic maps (age, size, type of deposit), on the maps presented herein no distinction is made on the basis of size of deposits. The only distinction made is between those deposits that have been or are in production, and uneconomic occurrences. By differentiating between productive and unproductive locations I feel that some qualitative measure of tin concentration is introduced.

Some objection might be raised against including placer deposits in a map

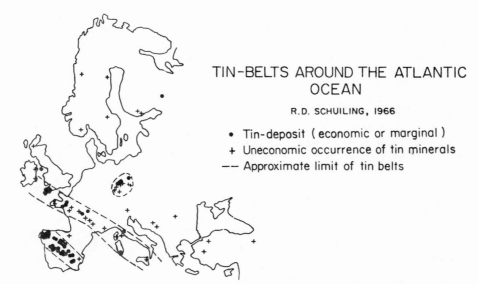

TIN-BELTS AROUND THE ATLANTIC OCEAN

R.D. SCHUILING, 1966

• Tin-deposit (economic or marginal)
+ Uneconomic occurrence of tin minerals
— Approximate limit of tin belts

FIG. 1, A–D. A, Europe; B, Africa; C, South America; D, North America.

Fig. 1B.

which purports to give information on the primary distribution of tin. However, as it is common experience that placer deposits of tin are never, or rarely, located far from their primary source, it is felt that on the scale of the map this introduces no ambiguities.

TIN BELTS

From an inspection of Figure 1 it is obvious that tin occurrences show an unevenly distribution over the continents. They are concentrated in elongated

23

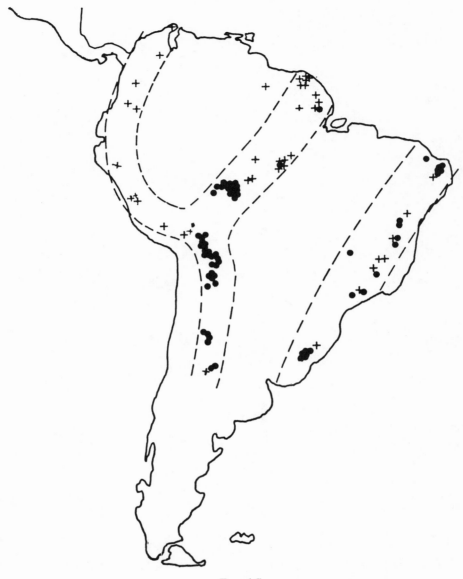

Fig. 1C.

zones that can be refered to as "tin belts." Although in detail one may have a different opinion on where to draw the boundaries of these belts (and on this scale details may be matters of several hundred kilometers!), it seems certain that the overall distribution can be characterized as "belt-like." It may well be that some clusters of tin occurrences that have not been dis-

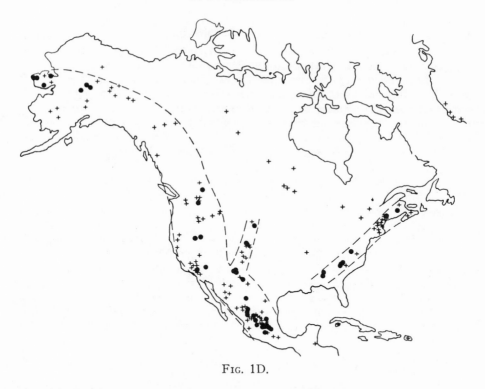

Fig. 1D.

tinguished as belts, such as those in Canada and Central Africa, will eventually form parts of belts when more occurrences are found. The following tin belts are distinguished:

1. *The Andean belt,* richest in Bolivia, can be traced from Argentina into Bolivia, Peru and probably Colombia. The age of the deposits ranges from Mesozoic to Middle Tertiary, although in Argentina the belt contains some small Precambrian occurrences. The deposits are related genetically to Mesozoic intrusives or Tertiary extrusives.

2. *The East Brazilian belt* extends from Rio Grande do Sul in a northeasterly direction through the coastal states of Brazil up to Ceará and Rio Grande del Norte, where it terminates at the Atlantic Ocean. Many of the deposits are in pegmatites, associated with beryl and lithium silicates. As far as is known, most or all of the deposits are Precambrian and a number of them in this belt have been dated at 750 m.y.

3. *The Rondônia-Guyana belt.* This area is still largely unknown, with the tin fields of Rondônia having been discovered only in recent years, and some of the occurrences to the northeast of Rondônia discovered as recently as 1965. A large part of the connection between Rondônia and Guyana is obscured by the Tertiary of the Amazon basin. An age of 940 m.y. is reported from Rondônia (29), whereas the ages of the granites and pegmatites associated with tin mineralization in French Guyana are around 2,100 m.y.

4. *The Rocky Mountain belt* extends from Mexico through California, New Mexico, Nevada, Washington and British Columbia to Alaska. It seems to be the counterpart of the Andean belt in South America, as it contains deposits both related to Tertiary volcanics and to Mesozoic intrusives.

5. *A small belt* may branch off the Rocky Mountain belt in the U. S., going into Colorado and South Dakota (Black Hills). Although this belt is based on rather few occurrences, it is interesting to note that it would correspond to Burnham's (3) Eastern belt, defined on the basis of high trace-element content (especially tin) in chalcopyrite and to a lesser extent sphalerite. In New Mexico and South Dakota some pegmatitic deposits of Precambrian age are included in this belt.

6. *The Appalachian belt* extends from Alabama to New Brunswick and Nova Scotia. Its southern end disappears under the Mississippi delta and the Gulf of Mexico, and it may be buried in and around New Jersey under a Mesozoic cover. It contains both Precambrian, mainly pegmatitic deposits, and Paleozoic occurrences which are in some cases characterized by greisen.

7. *The Central African belt* extending from Natal through Swaziland, Transvaal, Southern Rhodesia into Eastern Congo and Uganda, contains pegmatites, lodes and pipes, which are definitely Precambrian (around 2,100 m.y.; in Rwanda-Burundi and Katanga between 870 and 1,000 m.y.). It is tempting to speculate that the recently discovered Precambrian tin deposits of the Eastern Desert of Egypt and those to the north of Kharoum are part of this same belt.

8. *The South West Africa-Nigeria belt* starts at the very tip of South Africa, where tin was mined from the Kuils river, near Capetown. Most of the numerous tin deposits in South West Africa, are related to "Younger Granites" of late or post-Karroo age (Jurassic?), but others are related to Precambrian granites and pegmatites. The belt forms a relatively narrow strip along the Atlantic Ocean in Angola, the Congos, Gabon and Cameroon, striking almost due north into Nigeria. Extensive tin mineralization, both Precambrian and Jurassic occur in Nigeria, and analogous mineralization has been discovered recently in former French Nigeria (Air and Zinder massives). In southern Algeria extensive tin mineralization occurs in the Hoggar Massif. The belt is probably continuous with some tin occurrences in eastern Algeria, through the still poorly prospected Sahara, where the lack of water prevents prospecting by panning. The age of the younger granites is dated as Jurassic in Nigeria and as Eocene in Cameroon.

9. *A rather poorly defined belt* extends from Liberia to Morocco. In its southern part tin mineralization occurs mainly in Precambrian pegmatites, but in Morocco some occurrences seem to be related to Hercynian intrusives.

10. *The Iberian belt* strikes through Spain and northern Portugal. The tin-tungsten mineralization is related to late-Hercynian granites.

11. *The Armorican belt*, in the western part of the Massif Central, extends through Brittany into Cornwall. All deposits are related to Hercynian granites and consist of veins, and disseminated cassiterite in greisen. If, as suggested in Figure 1 this belt includes also the Tuscany and Elba occurrences, then part of the belt is of Tertiary age.

26

12. *The Erzgebirge province.* Here the predominant deposits are greisen-type, with associated lithium-micas, related to Hercynian granites. It cannot be ascertained whether this province is part of a larger tin belt.

Roughly 90% of the occurrences, including all but one of the past or present producers fall into the above belts that occupy less than half the land surface under consideration. A calculation shows that the average density of tin-occurrences within belts is more than 20 times the average outside the belts. The actual number of occurrences and deposits within the belts is even higher than indicated, because it is necessary to represent closely spaced deposits in tin-fields by only one point, whereas occurrences outside the belt are commonly single, isolated localities representing mineralogical curiosities in some well-studied deposit of other minerals. The lithological map of the Republic of Ruanda for example shows over 100 tin localities which are represented by only 5 points on Figure 1. This is one of the reasons why a method of contouring to establish the outlines of the belt would be rather unsatisfactory. A count of the number of occurrences and deposits shows the following:

452 points

54 points

One could argue that the belts represent mountainous areas where discovery of deposits is made easier by the work of erosion. This is only part of the explanation of the belt-like patterns, as tin deposits are absent from areas that are rich in other mineral deposits, and as the belts themselves include large segments covered by younger sediments or extrusives. Some examples of this have been noted above, and several examples could be given to show that the belts are not simply the more favorably exposed parts of the continents under consideration. It must be recognized, however, that the belts cannot be traced on an absolutely objective basis, because our sampling of the continents is still very uneven. The discovery of two tin occurrences in Antarctica is more significant, with possible ties to Tasmania and Eastern Australia, than the discovery of two tin occurrences in Western Europe. It is interesting to note that the accompanying maps would have shown more blank space only a few years ago; some of the major discoveries in the last decades include the Rondônia tin fields, the Eastern Egypt occurrences, the stanniferous granites in northern Nigeria, the cassiterite find north of Khartoum and the tin deposit at Mount Pleasant, New Brunswick.

TIN BELTS AND CONTINENTAL DRIFT

The foregoing section deals mainly with the directly observable distribution of tin occurrences and points to a possible practical application of the observed distribution pattern in prospecting. In this section the observed distribution of these continent-sized phenomena is considered in their relationship as evidence for continental drift. The tin belts of Figure 1 are replotted on Bullard's reconstruction of the continents before drifting (2). As can be seen from this Figure 2, there are some remarkable coincidences.

a. The East Brazilian belt seems to be the direct counterpart of the Southwest African-Nigeria belt.

b. The Rondônia-Guyana belt seems to have its extension in the Liberia-Morocco belt on the African side.

c. The Appalachian belt seems to join the Armorican belt.

d. The Late-Hercynian Iberian belt joins more closely the late-Hercynian Armorican belt, if the Gulf of Biscay is closed.

e. The Iberian belt seems to be the continuation of the Southwest Africa-Nigeria belt.

f. The Andean and Rocky Mountain belts are continuous and tin-free; Central America is missing.

g. The group of occurrences in southwest Greenland forms the continuation of the Armorican and the Iberian belt.

The remarkable pattern of belts in Figure 2 alone cannot constitute convincing evidence for continental drift, but it is safe to say that the concept of tin concentration in belts and the theory of continental drift derive some strength from each other. If we assume for the moment that continental drift is supported sufficiently by independent evidence of paleomagnetism, paleoclimatology, and other lines of geological reasoning to be used as a working hypothesis, let us see where this leads in connection with the tin belts.

Continental drift is usually considered to have taken place either from the Permian to the Tertiary, or, alternatively, is still taking place. If the ages of the tin deposits in the Southwest Africa-Nigeria belt are considered, a number are Precambrian (pre-drift), an important part are Jurassic (during drift) and some are Eocene (after drift or at a later stage during drift). This is interpreted to mean that throughout this time the source of the tin must have been carried along by the moving continents. Hence this source must be located in the crust, or in that part of the upper mantle which adhered to the crust during drifting. Otherwise if the source of tin is some linear zone in the mantle, mineralization would have taken place in successively displaced belts as the crust slid over this source of tin. It seems reasonable to suppose that the idea of a crustal source holds also for the commonly associated elements W, Ta, Nb, Be, Li, F.

CAUSES OF TIN-CONCENTRATIONS

There is another point that merits some speculation. It is rather obvious that in a general way several of the tin belts follow orogenic trends (Andes, Rocky Mountains, Appalachians, Armorican belt). It is only in much smaller segments of those belts, however, that large concentrations of occurrences and workable tin deposits are found. There may be some indication on Figure 2 that for economic concentrations to occur, a combination of a geochemical culmination and an "event" is necessary. A geochemical culmination is considered to be a continent-sized, lower-crustal belt along which a particular element has been enriched relative to its normal abundance. An "event" is any geological process, by means of which this enriched material is brought up, or further concentrated as for example by the development of granite intrusions

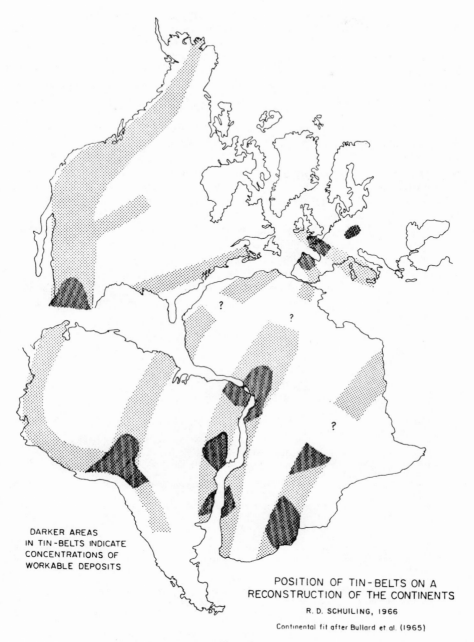

DARKER AREAS
IN TIN-BELTS INDICATE
CONCENTRATIONS OF
WORKABLE DEPOSITS

POSITION OF TIN-BELTS ON A
RECONSTRUCTION OF THE CONTINENTS

R. D. SCHUILING, 1966

Continental fit after Bullard et al. (1965)

Fig. 2.

accompanying an orogeny. Each of these factors alone is not sufficient to produce workable deposits. A geochemical culmination is passive, whereas certain events like pegmatite formation, granite intrusion, or volcanic extrusions, may under favorable conditions produce some tin minerals, even when starting with materials originally poor in tin. Such occurrences will rarely exceed the status of mineralogical curiosities. Only if the appropriate process acts on already enriched starting material will economic concentrations of tin minerals result.

These concepts find some support in the data in Figure 2. The rich Bolivian tin deposits are found where the Rondônia-Guyana belt strikes into the Andes, whereas elsewhere in the Andes the same apparent geological processes produced only insignificant concentrations of tin minerals. The same relationship may hold true for Mexico, where the Appalachian culmination joins the Rocky Mountain belt. Another example might be Cornwall, where the Iberian and the Armorican belt come together.

Clearly, with only 12 belts and provinces, and with even fewer intersections, arguments like these can never be completely convincing. The idea of an interaction between geological event and geochemical culmination does seem logical, and it is consistent with the small amount of available data. The significant point is that the ultimate cause of the geochemical inhomogeneities may be due to some process accompanying the earliest formation of the sialic crust.

VENING MEINESZ LABORATORY OF GEOPHYSICS AND GEOCHEMISTRY,
UTRECHT,
April 24, 1967

REFERENCES

1. Brouwer, G. C., 1962, Sur la métallogénie du Bassin du Maroni: 3me Congrès Géologique des Caraïbes, Jamaica.
2. Bullard, E., Everett, J. E., and Smith, A. G., 1965, The fit of the Continents around the Atlantic Ocean: Trans. Roy. Phil. Soc., v. 258, p. 41–51.
3. Burnham, C. W., 1959, Metallogenic provinces of the Southwestern United States and Northern Mexico: New Mexico Inst. Mines and Techn., Bull. 59, p. 1–76.
4. Chauris, L., 1965, Les Minéralisations pneumatolytiques du Massif Armoricain: Mém. B.R.G.M., v. 31, p. 1–217.
5. Dadet, P., 1966, Essai de réinterprétation stratigraphique des formations précambriennes du Mayombe congolais: Bull. B.R.G.M., v. 2, p. 77–91.
6. Davies, K. A., 1947, The geology and mineral deposits of Uganda: Imp. Inst. Min. Res. Dept. Bull. XLV, no. 2, v. 161–180.
7. Davies, D. N., 1964, The tin deposits of Swaziland: in Some ore deposits in Southern Africa, v. 2, p. 535–542.
8. De Kun, N., 1965, The mineral resources of Africa: Elseviers Publ. Co., Amsterdam-London-New York, 740 pp.
9. Dennis, J. G., 1959, Notes on some cassiterite-bearing pegmatites near Brandberg, South West Africa: ECON. GEOL., v. 54, p. 1115–1121.
10. Drysdall, A. R., 1963, The tin belt of the southern province: North. Rhodesia Geol. Surv. Dept., Econ. Unit, Rept. 1, 35 pp.
11. Fick, L. J., 1960, The geology of the tin pegmatites at Kamativi, S. Rhodesia: Geol. Mijnb. Nwe. Ser., v. 39, p. 472–491.
12. Foshag, W. F., and Fries, C., Jr., 1943, Tin deposits of Mexico: U.S.G.S. Bull. 935C, p. 99–176.
13. Gazel, J., Lasserre, M., Limasset, J. C., and Vachette, M., 1963, Ages absolus des massifs granitiques et de la minéralisation en étain du Cameroun Central: C. R. Acad. Sc., Paris, v. 256, p. 2875–2878.

14. Geologic Map of S.W. Africa, 1963, Dept. of Mines, S.W. Africa.
15. Guimarães, D., and Dutra, C. V., 1964, Contribuçao à petrografia e geoquimica de jazida estanifera de Ipameri-Goiás: Div. Fom, Prod. Min. Av., v. 86, p. 13–39.
16. Harris, J. F., 1961, Summary of the geology of Tanganyika, Pt. IV: Econ. Geol. Geol. Surv. Dept. Tanganyika, Mem. 1, 43 pp.
17. Haughton, S. H., Ed., 1940, The mineral resources of the Union of South Africa: 3d ed., Geol. Surv. S. Africa, 544.
18. Hinze, C., 1915, Hanbuch der Mineralogie: Bd. 1, Abt. 2, Leipzig.
19. Johnston, W. D., Jr., 1945, Beryl-tantalite pegmatites of northeastern Brazil: Geol. Soc. America Bull., v. 56, p. 1015–1070.
20. Jouravsky, G., and Destombes, J., 1961, Les différents types de minéralisations du domaine de l'Anti-Atlas: Mines et Géol., Bull. trim. Dir. Min. et Géologie du Maroc., v. 13, p. 19–58.
21. Killeen, P. L., and Newman, W. L., 1965, Tin in the United States: Min. Invest. Res. Map MR-44, U.S.G.S., p. 1–9.
22. Kloosterman, J. B., 1967 (in press), A tin province of the Nigerian type in Southern Amazonia: Proc. Int. Techn. Conf. Tin, London.
23. Marić, L., 1954, Magmatismus und Alkalimetasomatose im jugoslawischen Raum (auch mit Hinblick auf ostserbischen Scheelit- und mittelbosnischen Kassiieritvorkommen): N. JhB. Min. Abh., v. 87, H. 1, p. 1–32.
24. Montagne, D. G., 1964, An interesting pegmatite deposit in northeastern Surinam: Geol. Mijnb. Nwe. Ser., v. 43, p. 360–374.
25. Moussu, R., 1965, Les gisements d'étain et wolfram de Saxe et Bohème: Bull. B.R.G.M., v. 4, p. 1–68.
26. Mulligan, R., 1966, Geology of Canadian tin occurrences: Paper 64-54, Geol. Surv. Canada, 22 pp.
27. Neves Ferrão, C. A., 1965, Occorrencias Minerais: Direcçao Provincial dos Serviços de Geologia e Minas, p. 1–59.
28. Pennsylvania State College, Div. of Min. Economics, 1953, Material Survey, Tin, 774 pp.
28a. Petersen, G., 1960, Sobre Condoriguina y otros depósitos de estano en el Peru: Bol. Soc. Nac. Min. Petr. No. 72, p. 36–44.
29. Priem, H. N. A., Boelrijk, N. A. I. M., Hebeda, E. H., Verschure, R. H., and Bon, E. H., 1966, Isotopic age of tin granites in Rondônia, N.W. Brazil: Geol. Mijnb. Nwe. Ser., v. 45, p. 191–192.
30. Ramirez y Ramirez, E., 1953, Proyecto de investigación y estudio de los yacimientos wolframo-estanníferos de España: Inst. Geol. y Min., Notas y Com., v. 31, p. 123–161.
31. Raulais, M., 1946, La série granitique ultime de l'Aïr au Niger et sa minéralisation stannifère: C. R. Acad. Sc. Paris, v. 223, p. 96–98.
32. Rwanda, 1963, Carte lithologique du Rwanda: Serv. Géol. Rép. Rwandaise.
33. Soen, Oen Ing, 1958, The geology, petrology and ore deposits of the Visean region, northern Portugal: Meded. Geol. Inst. Amsterdam.
34. Turneaure, F. S., 1960, A comparative study of major ore deposits of Central Bolivia: Econ. Geol., v. 55, p. 217–254 and 574–606.
35. Vokes, F. M., 1960, Contributions to the mineralogy of Norway, no. 7: Cassiterite in the Bleikvassli ores: Norsk Geol. Tidsskr., v. 40, p. 193–201.
36. Wokittel, R., 1960, Recursos minerales de Colombia: Serv. Geol. Nac., 10, 393 pp.

5

CONTINENTAL DRIFT AND ORE PROVINCES

W. E. Petrascheck

This abridged translation was prepared expressly for this Bench-mark volume by J. B. Wright from "Kontinental-verschiebung und Erzprovinzen," in Mineral. Deposita *3:56–65 (1968). English Summary and Figures 2, 3, and 4 are reprinted from the original publication. Copyright © 1968 by Springer-Verlag.*

ORE PROVINCES OF GONDWANA

The picture of a metallogenetic unity is comparatively convincing when Africa is put together with South America (Figure 2). The <u>auriferous quartz veins</u> of El Callao, cutting early Algonkian greenschists in the Guianese highlands, can be matched with those found at the boundary between green-schists and Birrimian rocks in Ghana. Placer deposits derived from these veins characterize the somewhat younger Minas Series in Guiana (Grabert, 1962) and the Tarkwaian Series in Ghana and Ivory Coast (Karpoff, 1952; Krenkel, 1957). Further south, the late-Algonkian gold province of Montevideo can be correlated with that of Walvis Bay in southwest Africa.

A tin-tungsten belt stretches some 5,000 km from southern Brazil to the Borborema Plateau on the west, and from southwest Africa to Nigeria and perhaps as far as In Toumine, deep in the Sahara, on the east. In both north-eastern Brazil and west Africa, tungsten deposits are associated with so-called Younger Granites (Ahlfeld, 1958; Lombard, 1966), although it should be noted that the dominant ore mineral is scheelite in Brazil, wolfram in Nigeria. The distribution of niobium-tantalum and beryl deposits, which are commonly associated with alkali-rich intrusives and pegmatites, helps to strengthen the metallogenic link. The Nb-Ta-Sn ores of Gabon may correspond to those in western Minas Gerais, while beryl-bearing pegmatites with columbite, apatite, uraninite and sulphides are found in both the pampas regions of Cordoba in Argentina and the uplands of southwest Africa.

There appears to be little correspondence between the relatively rare occurrence of <u>base metals</u> (Cu, Pb, Zn) along the Atlantic coasts of these continents, although the vanadium ores found with lead and copper deposits in the San Luis and Cordoba regions of Argentina could perhaps be matched with the well-known deposits of Tsumeb in southwest Africa.

There may be a more definite link between the Variscan Sn-W mineralization associated with rich Bi, Pb, Zn, Ag sulphides, which stretches from Queensland to Tasmania, and the similar polymetallic ores of Variscan age in the Argentinian cordillera (Stoll, 1965). Ahlfeld (1958) noted that the tin sulphide "pipes" of Herberton in Queensland are comparable with the tin "pipes" of Bolivia. The suggestion that these two elongate ore provinces are part of a Variscan metallogenic belt is supported by Runcorn's (1965) reconstruction, in which mid-Mesozoic Australia lay 160° to the right of its present position, bringing this Variscan belt into an alignment--interrupted by Antarctica--with the South American cordillera (Figure 3).

The Pilbara goldfield of western Australia which can be related to Middle Algonkian granites, and is associated with zinc-tantalum pegmatites, recalls similar deposits in southern and central Africa. The mid-Algonkian gold ores of Kolar in southern India could perhaps be linked to South Africa through those of Western Australia (Bendigo-Kalgoorlie-Mount Morgan)* via a gap in east Antarctica, to form a circum-polar gold belt (Figure 3).

The Arabian subcontinent provides a northward extension of the meridional gold belt which runs the length of eastern Africa, from the Witwatersrand along the western border of Ethiopia, thence north of Asmara across the Red Sea, and embracing the gold occurrences east of Jidda and north of Medina.

The remaining Gondwana continents show much less convincing metallogenic links. Similar types of ore deposits can be identified, but clearly defined belts are lacking, and the problems are made more difficult by lack of knowledge about what underlies the Antarctic ice sheet.

The Hatches Creek <u>quartz-wolfram</u> ores with associated bismuth, in the Precambrian of western Australia, can be compared to the wolfram occurrences of the pampas plateau in Argentina where bismuth and molybdenum are also found.

ORE PROVINCES OF LAURASIA

It is difficult to define connections among the ore deposits of Precambrian shield regions in the giant North American-Asiatic continental plate, which remained intact till early Tertiary times. Deposits complementing well-known occurrences in Fennoscandia or Canada, for example, such as the Kiruna iron ores or Lake Superior copper, are not found elsewhere. Some general similarities can be discerned, however: the titanomagnetite deposits of Allard Lake (Canada) and southern Sweden are analogous, while the Sudbury nickel deposits are similar to those of southern Sweden and Finland, although the latter are more deformed. Greenland offers no connecting link across northern parts of Laurasia, nor is there much in common between the sedimentary ore deposits of the Siberian and North American platforms.

The Paleozoic ore provinces along the North Atlantic margins are more promising; sulphide ores in the Ordovician tuffs and marine sediments of the northern Appalachian belt in Newfoundland and Quebec, resemble sulphides in the Norwegian Caledonides. The Variscan wolfram-gold deposits of the southern Appalachians in Carolina, possibly related to late Paleozoic granites (Turneaure, 1955), may be linked to similar deposits in the Variscan belt of western Europe. The metasomatic (hydrothermal replacement) Pb-Zn ores of Tennessee could be matched with those of Santander and the Aachen region-- except that the Tennessee deposits are in Ordovician limestones, those of northern Spain and Aachen in Carboniferous limestones.

---

*Editor's Note: This is geographically incorrect; only Kalgoorlie is in (south) western Australia. Bendigo is in southeastern Australia, and Mount Morgan is in the northeast.

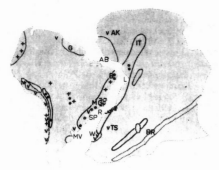

Abb. 2. Die metallogenetische Verbindung Afrikas
und Südamerikas. (G Guayana, B Borborema Plateau, R Rio de Janeiro, MG Minas Geraes, SP San
Paolo, MV Montevideo, AK Akjuvit, IT In Tou-
nine, TS Tsumeb, W Walfischbay, BR Broken
Hill)

Figure 2.   Metallogenic links between Africa
and South America.   Lagerst. =
Lagerstätten = Ore deposits.

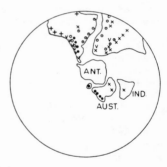

Abb. 3. Die metallogenetischen Gürtel der Gond-
wana-Kontinente im mittleren Mesozoikum

Figure 3.   Mid–Mesozoic metallogenic
belts of Gondwanaland.

THE CIRCUM-PACIFIC PROVINCE

The picture here is much clearer and more easily understood in terms of continental drift. In South America, according to a summary by Stoll (1965), the Chilean copper belt (with Cu, Ag, Au, Pb, W, and Hg), related to late Jurassic and early Tertiary (Laramide) magmatism, lies along the western front of the Cordillera, which was itself mineralized in Paleozoic times. Younger Tertiary ore deposits are found along this belt in Chile, Argentina, and Bolivia. Further north in the Andes, mineralization is characterized by significant amounts of Sn, W, Ag, Sb (in Bolivia) and of Au in Colombia. Schneider-Scherbina (1962) has shown that the Sn-W-Ag deposits in Bolivia resulted from remobilization of older Jurassic tin-tungsten ores.

Upper Jurassic (Nevadan) and Tertiary magmatism and associated mineralization characterize the Pacific coastal ranges of North America. Granodioritic and andesitic magmas were the main carriers of Au-Ag, Cu, W, and other metals.

In Alaska there is a late Mesozoic gold province (Juneau), as well as tin (Jucco), and there is tungsten in British Columbia. The Alaskan gold province can be convincingly traced into the Siberian gold province of similar age, to which belong the deposits of Kolyma, Allakhyum, and Darasun on the Amur River (Figure 4). The Sn-W belt which is weakly developed in North America becomes much more prominent in Asia, extending from the Ege-Khaya region in northeast Siberia, along the border regions between Siberia and Manchuria and across southern China to Tavoy, Thailand and Banka. It is noteworthy that the Sn-W belt of eastern Asia closely follows the western side of the gold belt and thus lies closer to the Siberian shield, while the Sn-W belt of South America follows the eastern side of the gold belt, and similarly lies closer to the Brazilian shield. In both continents, however, the Mesozoic Sn-W ore deposits are merely the remobilization products of earlier Sn-W mineralization, as shown by Ahlfeld (1958) and Stoll (1965) for South America and by Smirnov (1959) for the Asian segment of the Soviet Union. Smirnov (1966) has also shown that the metallogenic belts are concentrically disposed about the ancient shield blocks.

Abb. 4. Die metallogenetischen Gürtel der Nord-
kontinente im mittleren Mesozoikum ∖ Mesoz.-
tertiär. Pb-Zn-Cu; ▼ Paläozoisches Pb-Zn-Cu;
+ Mesozoisches Au; O Mesozoisches Sn-W

Figure 4. Mid-Mesozoic metallogenic belts
of the northern continents.

35

THE TETHYAN PROVINCE

(*Editor's Note: After criticizing the postulated "Operation Tethys Twist" involving a few thousand kilometers of dextral movement between Laurasia and Gondwanaland, which has been proposed on paleomagnetic grounds (van Hilten, 1964), the author concedes that some shearing movements must have occurred.*)

It is now generally known that the Mediterranean ophiolite belt includes ultrabasic and basic rocks ranging in age from late Paleozoic to Eocene, as shown by de Wijkersloot (1942) in Asia Minor, and by the author in parts of the eastern Mediterranean. Such repeated injection of ultrabasic and basic magmas can only have originated from subcrustal depths, not along a smooth fracture plane, but rather from a broad irregular shear zone extending down and "ploughing into" the mantle, to give rise to the numerous chromite deposits and perhaps also the copper ores of, for example, Ergani Maden and Cyprus, which would have resulted from mixing of basic magma with sheared fragments from the lower crust.

Continued mixing of primitive magmas with anatectic melts from the sialic crust in the Mesozoic and Tertiary orogenies yielded large volumes of andesite and granodiorite, especially in the eastern Mediterranean and Indonesian mountains, and the variable mineralization which accompanied it produced deposits of iron as well as the base and noble metals. The northern flank of the Alpine Mediterranean orogenic belt is characterized by Cu, Fe, and Au, the southern by Pb, Sn and Sb (Petrascheck, 1963), which is consistent with the postulate that the northern and southern sides of this great shear zone belonged to different crustal segments, with contrasted metal contents.

(*Editor's Note: The final part of the paper deals with the provenance of ore metals, and the author has adequately summarized his opinions on this topic in his English summary, which is reproduced here.*)

Summary

This paper deals with an attempt of checking the position of the great ore provinces in the frame of the theory of continental drift. Contemporaneous with and independent from the studies of the present author D. SCHUILING made a comparison of African and South American Sn-S-provinces, the result of which was shown in a map presented by W. NIEUWENKAMP in a slide during a lecture in Mainz in March 1967[5]).

The picture of a metallogenetic unity is comparatively convincing when Africa is put together with South America. The gold province of Guiana corresponds to the province of the Ivory-, and the Goldcoast, both rich in placers derived from quartz-veins in Algonkian schists. The SW-African and the S-Brasilian goldfields too seem to be related. The apparent connection of the Brasilian Sn-W-province with the Sn-W-provinces of Nigeria, Congo and farther north into the Sahara results in a 5000 km long Sn-W-belt. Minor metals like niobium, tantalum and beryllium are found in corresponding places in the two continents.

The community of the Gondwanian ore provinces becomes much less clear when we consider the other southern continents. Only deposits of similar age and similar metal association can be recognized. The connection is veiled by the ice covering the Antarctics. A possible junction to a belt might exist between the Hercynian Sn-W-province in Eastern Australia and Argentinia, according to the relative position of these landmasses in early Mesozoic time.

The Laurasian shields show surprisingly few analogies, and some very characteristic types of deposits in Canada are lacking in Fennoscandia and vice-versa. Greenland does not help as a connecting link. Somewhat better are the transatlantic relationships between the Palaeozoic ore provinces in Europe — Caledonian and Hercynian — and those of the Appalachians. But again the deposits of the sedimentary cover of the North American and the Siberian platforms do not show any striking similarity.

The great circumpacific orogenic belt incorporates a metallogenetic belt of the same age of formation. It is characterized by the abundance of base metals (Cu, Pb, Zn, Sb), but these are very often connected with the metals of the Precambrian shields, such as W, Sn, Au. In South America, as well as in the Transbaikalian provinces of the Soviet Union, the Palaeozoic and Mesozoic Sn-W belts are situated closer to the respective Precambrian platforms than the gold- and basemetal belts.

At last, the Tethys belt shows a similar petrologic and metallogenetic assemblage as the Pacific belt, but in addition plenty of ultrabasic rocks with chromite deposits (SE-Europa, Turkey, Iran, Indonesia). This belt was considered recently as the orogenetic result of an anticlockwise rotation of Laurasia versus Gondwana of about 50° (VAN HILTEN, 1964, VAN BEMMELEN 1966). Important transcurrent faults in the Mediterranean area, like the seismic line of Northern Anatolia, are used as arguments for a yet active rotation. In the present authors' opinion, neither palaeogeographic nor structural evidence is yet sufficient for the assumption

[5] This synopsis, dealing with the tin belts on both sides of the Atlantic Ocean, recently appeared in Econ. Geology Vol. 62/4, 1967. D. SCHUILING came to very similar conclusions as the author.

of a transversal displacement of such an extent since late Palaeozoic time. Some features, e. g., the Hercynian Mauretanides or the facies of the Upper Cretaceous are not in agreement with an opposite position of NW-Africa and Asia Minor — nor are old metallogenetic provinces fitted together by this "Operation Tethys Twist". On the contrary, the submerged sialic block in the Western Mediterranean Sea probably contained the magmatic center of the Tertiary base metal mineralization, found along the Spanish and the North-African coastal areas; and this would prove an unchanged relative position of the two continents since Miocene time.

However, tremendous shearing movements along the Tethys belt are fully recognized by the author. The repeated intrusions and extrusions of ultrabasic magmas with rich chromite ores in Palaeozoic, Mesozoic and Eocene times are probably the result of this "deep ploughing" into the substratum. Later and partly already contemporaneously the many ore deposits of various metals, related to intermediate and acid magmas were formed.

Here we touch the problem of the provenance of the ore metals. Sn, W, Au, U are frequent in the sialic crust of the ancient shields. To these lithophile or granitophile elements an increasing amount of the chalcophile base metals was added later. C. I. SULLIVAN has stated already in 1948 that this afflux was caused by the basic magma which was induced to the crust during the later orogenic epochs.

But nevertheless, the primary basaltic magma of the substratum which is derived from the mantle, cannot be accepted as the source of the chalcophile base metals. A. RITTMANN has stressed this point mentioning that Hawaiian basaltic lavas do not even contain traces of base metals. Not only the oceanites, but also the continental tholeiites are sterile. Thus we may assume that the ophiolitic magma of the initial magmatism in the orthogeosynclines which has produced so many and various ore deposits, is already the result of hybridization by anatectic magmas derived from the basis of the geosynclines.

The metals, therefore, exhaled by the submarine diabasic, porphyritic and even more acidic volcanos, came from the crust. Only the siderophile metals such as chromium, platinum, nickel have a mantle origin. The progressive hybridization during geosynclinal subsidence and subsequent orogenesis produced the andesites and granodiorites, associated with chalcophile and granitophile elements. The initial basic magma of the substratum has figured as a kind of collector for the chalcophile metals.

Even the dubious stratiform deposits in the ancient platform may have been formed by a diffusion of the primary volatiles of the basaltic substratum, which have leached and redeposited metals from the crust — an assumed process called transsudation by A. RITTMANN. We have the impression, yet to be checked critically with the argument, that this metal diffusion was more active during the periods, when the continents did not move. H. BACKLUND was the first to state that the earth's crust is autarchic as far as the metal supply is concerned. But he considered mainly granitization within limited areas.

If we accept a continental drift, then the well-known persistence of certain metals in certain regions during long periods of the earths' history may be taken as another argument for the origin of the metals in the wandering crust and not in the mantle below.

Literatur

AHLFELD, F.: Zinn und Wolfram. In: Die Metallischen Rohstoffe etc., Bd. 11, Stuttgart: Enke, 212 p., 1958.

Amt für Bodenforschung Hannover, Bericht über die Lagerstätten Südamerikas. Hannover 1957 (hektographiert).

BACKLUND, H.: Zum Werdegang der Erze. Geol. Rundschau 32, 60—66 (1941).

BRINKMANN, R.: Geotektonische Gliederung von Westanatolien. Geol. Mh. 10, 603—618 (1966).

DAVIDSON, C. F.: Some genetic relationship between ore deposits and evaporites. Trans. Inst. Mining Met., Sect. B 75, 216—226 (1966).

GRABERT, H.: Zum Bau des Brasilianischen Schildes. Geol. Rundschau 52, 292—317 (1962).

KARPOFF, R.: Sur quelques conglomerats antecambriens de la Côte d'Ivoir. Congr. Geol. Intern., Compt. Rend., 19., Algiers 20, 129—140 (1952).

KETIN, I.: Über die tektonisch-mechanischen Folgerungen aus den großen anatolischen Erdbeben des letzten Dezenniums. Geol. Rundscnau 34, 77—83 (1948).

KRENKEL, E.: Geologie und Bodenschätze Afrikas. 2. Aufl., Leipzig: Geest & Portig 1957.

LOMBARD, L.: Répartition lineamentaire de quelques metaux en Afrique. Chronique Mines 357, 319—328 (1966).

MARTIN, H.: The hypothesis of continental drift in the light of recent advances of the geological knowledge in Brazil and SW-Afrika. Geol. Soc. S-Africa 64, 1—47 (1961).

PAVONI, N.: Die nordanatolische Horizontalverschiebung. Geol. Rundschau 51, 122—139 (1961).

PETRASCHECK, W. E.: Intrusiver und extrusiver Peridotitmagmatismus in alpinotypen Bereich. Geol. Rundschau 48, 205—217 (1959).

—, Die alpin-mediterane Metallogenese. Geol. Rundschau 53, 376—389 (1963).

— Typical features of metallogenetic provinces. Econ. Geol. 60, 1620—1634 (1965).

RAGUIN, E.: Sur l'ampleur de la Metallogenie regenerée. Congr. Geol. Intern. Algiers 1952, Sect. XII, Fasc. XII, 105—115 (1954).

RITTMANN, A.: Über die Herkunft der vulkanischen Energie und die Entstehung des Sials. Geol. Rundschau 30, 52—60 (1939).

— Volcanos and their activity. New York: Wiley, 305 p., 1961.

RUNCORN, S. K.: In: Symposium on Continental Drift. Phil. Trans. Roy. Soc. London 1088, 328 p. (1965).

SCHNEIDER-SCHERBINA, A.: Über metallogenetische Epochen und den hybriden Charakter der sog. Zinn-Silber-Formation. Geol. Jahrb. 81, 157—170 (1962).

SCHNEIDERHÖHN, H.: Genetische Lagerstättengliederung auf geotektonischer Grundlage. Neues Jahrb. Mineral. Mh., Abt. a., 47—89 (1952).

— Die Erzlagerstätten der Erde. Bd. II: Die Pegmatite. Stuttgart: Fischer, 1961.

SCHUILING, R. D.: Tin belts on the continents around the Atlantic Ocean. Econ. Geol. 62, 540—550 (1967).

SMIRNOW, W. I.: Versuch der metallogenetischen Rayonierung des Gebietes der UdSSR. (russ.). Mitt. Akad. Wiss. UdSSR, Geol. Ser. 4, 3—21 (1959).

— Besonderheiten der Metallogenese des NW-Teiles der pazifischen Provinz. (russ.) Vestnik Moskov. Univ., Ser. IV, 5, 3—11 (1966).

SOUGHY, J.: Westafrican fold belt. Bull. Geol. Soc. Am. 73, 871—876 (1962).

STOLL, W. C.: Metallogenetic Provinces of South Amerika. Mining Magaz. (London) 112, 22—33 and 90—99 (1965).

SULLIVAN, C. J.: Ore and granitization. Econ. Geol. 43, 471—498 (1948).

TURNEAURE, F. S.: Metallogenetic provinces and epochs. Econ. Geol. 55, 38—98 (1955).

VAN BEMMELEN, R. W.: On mega-undations, a new model for the earths evolution. Tectonophysics 3, 83—127 (1966).

VAN HILTEN, D.: Evaluation of some geotectonic hypotheses by paleomagnetism. Tectonophysics 2, 3—72 (1964).

WEGENER, A.: Die Entstehung der Kontinente und Ozeane. Petermanns Mitt. 58, 185, 235, 305 (1912).

WIJKERSLOOTH, P.: Die Chromerzprovinzen der Türkei und des Balkan und ihr Verhalten zur Großtektonik dieser Länder. M. T. A. Bulletin, Ankara 1/26, 54—75 (1942).

Received November 3 / December 8, 1967

Prof. Dr. W. E. PETRASCHECK
Institut für Geologie und Lagerstättenlehre,
Montanistische Hochschule, A-8700 Leoben,
Österreich

6

THE ORE POTENTIAL OF GREENLAND IN THE LIGHT OF CONTINENTAL DRIFT

W. E. Petrascheck

This English summary was prepared expressly for this Bench-mark volume by J. B. Wright from "Die Erahöffigkeit Grönlands im Lichte der Kontinentalverschiebung," in Erzmetall *24(6): 257–306 (1972). Figure 1 is reprinted from p. 259 of the original publication.*

SUMMARY

Restoration of the north Atlantic land masses to their pre-late Mesozoic continental drift positions provides a fundamental insight into the ore potential of Greenland.

Iron-titanium and uranium ores are known in the ancient shield region, analogous to those of the Grenville Series and Labrador coast of eastern Canada, and others may be expected. Pegmatites merit special attention.

The Caledonides of eastern Greenland, from Scorsby Sound to Clavering Island, are not very promising, for the metamorphic and migmatitic complex is mainly derived from clastic young proterozoic sediments, which for the most part are equivalent to the metal-poor Torridonian of Scotland and the Norwegian Sparagmite Formation. In any case, this coastal segment of Greenland formerly lay opposite the ore-poor gneiss-migmatite province of western Norway. Better prospects could be found north of the 74th parallel, corresponding to the ore-rich Norwegian Caledonides.

The Tertiary alkali plutonic complexes, of which those west of Mesters Vig carry a major molybdenum deposit as well as some galena and barytes occurrences, could repay further prospecting in other parts of the Greenland coast.

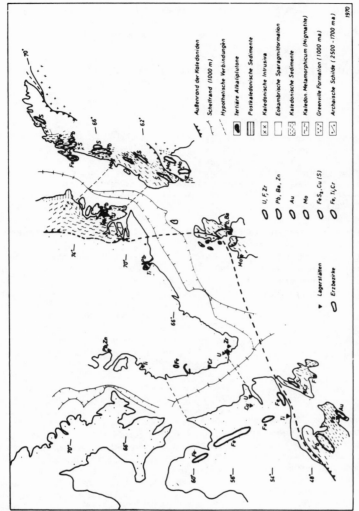

Bild 1. Skizze erzreicher und erzarmer Bereiche der ursprünglich an Grönland angrenzenden Gebiete Norwegens, Schottlands und Ost-Kanadas mit ihren möglichen Fortsetzungen an den Küsten des heutigen Grönland

Figure 1. Sketch map showing ore-rich and ore-poor provinces in Norway, Scotland and eastern Canada, which formerly adjoined Greenland, with with inferred projections to the present Greenland coast. Latitudes have present-day values. (Note: Ausenrand der Kaledonian = Caledonide margin; Schelfrand = Continental shelf edge; Hypothetische Verbindungen = hypothetical correlation; Lagerstätten = ore deposits; Erzbezirke = ore districts (minor provinces).)

41

7A

Reprinted from *Econ. Geol.* **65**:11–16 (1970)

Continental Drift and Un-Continental Thinking

ARTHUR RAYMOND CRAWFORD

Abstract

A synoptic approach to earth history is now made much easier as a result of acceptance of the theory of continental drift following additional evidence mainly from paleomagnetic and marine geophysical and geological work.

The value of this approach for mineral search is illustrated by examples chosen from India, Ceylon and Australia. Many aspects of such search necessitate recognition of the earlier close association of geological features now widely separated geographically. Continents can no longer be studied effectively in isolation.

MOST exploration geologists possibly regard the theory of continental drift as not very relevant to their daily work. This paper is written to point out that the adoption of the theory might often usefully influence prospecting. The examples are chosen from India, Ceylon and Australia, areas I feel most competent to discuss.

Earlier studies have been made by Schuiling (1967) and Petrascheck (1968). The former restricted himself to tin belts around the Atlantic and the latter adopted a general approach to ore provinces and their world distribution.

Continental Drift and a Synoptic View of Earth History

Continental drift (or displacement) is now respectable. The immense amount of new knowledge of the ocean floors, together with much paleomagnetic and geochronological work on the continents, give results which cannot be explained in terms of fixed continents. It has made necessary an entirely new approach to earth history, which can now be viewed synoptically. In fact it must be so viewed.

As Wilson (1968) has said, geology is in revolution. For years we geologists have been regarded by our scientific colleagues, not without justice, as people who merely observe and describe, and perhaps crudely interpret. Our interpretations have been local or regional, rarely even continental. They have not been truly geological in the sense of developing a theory of the whole earth. In Australia, as elsewhere, most geologists look at the rocks of a district, attempt to work out the sequence of their deposition or intrusion, and describe commonly in more detail the economically interesting mineral deposits that occur in them—usually, it is often alleged by mining men, after some prospector has already found them. They interpret the geological history in written or tabular form and provide a geological map. If this is a good map—geologically and cartographically—it is itself an interpretation of the history.[1]

Later this kind of work is synthesized to outline the geological history of larger regions, and still later, of continents. Valuable pioneer attempts at such syntheses have been made for example in Australia. We were fortunate to have Edgeworth David's fine "Explanatory Notes . . . " as early as 1932. It was not until 1950 that the second important continental study appeared, a hugely extended David, in large part the work of Browne, though we had to wait till 1960 for a revised map. A subsequent work, very much briefer, and more restricted in scope (though it covers New Zealand but, oddly, not New Guinea) is the recent volume by Brown, Campbell and Crook (1968). All these are of great value. For India, the equivalent is the latest edition of Krishnan (1968). Wadia's book (1957) is now sadly out of date on the Precambrian but we have for India an excellent account by Pichamuthu (1968) though his account of Ceylon in the same volume is less up to date. The equivalent of David and Browne is the monumental work of Pascoe (1950, 1959, 1964), which regrettably lacks not only an index, but an adequate number of illustrations (and is as abominably badly bound as any book one can buy).

There has been no satisfactory attempt this century to view synoptically the whole earth. The only successful attempt was by the Austrian geologist Suess. Though at the time (1885–1909) geological exploration was very incomplete, his achievement was remarkable. The useful works of Umbgrove (1947) and Kummel (1961) are not in the same category. Today the usual comment made when such a synthesis is suggested is that the local and regional data are now far too voluminous, and that no one person, nor even a team, can really do the

[1] Very often it is the only one available, because of the lamentable delays in publishing texts which accompany the maps.

job. But it is not a classical Suess-type synthesis that we need.

Our view of the earth is changing. The earth is a dynamic body. We now believe that the continents are not fixed and that the ocean basins in their present form are young features. The most fundamental division of the earth's crust is not into land and water, nor even into continents and their shelves on the one hand and true oceans on the other, but into large plates, the boundaries of which are suggested by earthquake epicentres. The boundaries appear as mid-ocean ridges, young fold mountain belts, deep ocean trenches, or long strike-slip or transform faults.

The story of how this concept developed is sufficiently important to warrant a brief summary.[2] It is a consequence largely of paleomagnetic work followed by the discovery by marine geologists of the Atlantic mid-ocean ridge and recognition of Iceland as the product of local lava eruption along it. The concept of sea-floor spreading, developed from early ideas of Arthur Holmes by Hess at Princeton, was confirmed when Tuzo Wilson at Toronto noticed that the isotopic dating of the volcanic rocks of Atlantic Ocean islands showed that the islands increase in age away from the ridge. Vine and Matthews at Cambridge then noticed that the volcanic rocks which emerge from the mid-ocean ridges, pushing the blocks apart as they spread out, produce a magnetic zebra pattern which is easily mapped from the air. The stripes can be matched with dated reversals of the geomagnetic field; the timing of these has been praticable because volcanic rocks retain the imprint of the earth's magnetic field at the time they crystallize. Morgan and Le Pichon have correlated this timed sea-floor spreading with the patterns of continental plates, and have worked out, for example, that in the southern Atlantic the spreading is going on at the rate of about a centimetre and a half a year. As Vine states, the history of the ocean basins is frozen in the oceanic crust.

It now appears that the Atlantic Ocean is widening by creation of new crust along its mid-ocean ridge. Thus the east Atlantic and most or all of Eurasia moves away slowly from the west Atlantic and North America.

Our task now is to reconstruct the history of the earth in the light of these new concepts. The detail of this new world division has yet to be worked out and it certainly presents us with new problems. But great strides have been made, for example by Le Pichon and Heirtzler (1968).

[2] Rather than give numerous references, I refer the reader to the review paper by J. T. Wilson, Static or Mobile Earth: The Current Scientific Revolution, *Proc. Amer. Phil. Soc.* 112, 309-320 (1968).

Australia, India and Ceylon

Australia forms one plate with peninsular India and Ceylon. The Pacific Ocean in the geological sense begins east of New Caledonia and New Zealand, though the recent earthquakes in both peninsular India (at Koyna near Poona) and Western Australia (at Meckering near Perth) tell us that even the older parts of the major blocks, which are themselves part of the plates, are not completely rigid. The most active earthquake area nearest to Australia is the axial zone of New Guinea. It seems probable that oceanic crust is there being thrust over continental material. By contrast, the really active zone in India is at the foot of and along the Himalaya, where continental crust is being thrust under continental crust.

This Australian plate developed by growth of oceanic crust between Indo-Ceylon and Australia, by sea-floor spreading from ridges older than those now active in the Indian Ocean, the relics of which lie south-west of Java. The history of the Indian Ocean is exceedingly complicated. Its present tripartite division into an African, a southern and a north-eastern part is fairly recent in origin. Paleomagnetic work done at the Australian National University by Irving, Briden and McElhinny shows that Australia was previously farther south and attached to Antarctica, from which it separated only about 80 m.y. ago. India and Ceylon were attached also and this relationship with Australia lasted rather longer. The movement away from Australia of the Indian platform block was so vigorous that the highest land mountains on the earth, the Himalaya, were formed when it drove under Tethyan crustal material to amalgamate with a proto-Asia. The floor of the Indian Ocean is remarkable in having enormous scars aligned north-south, associated with this movement. The present movement seems to be north-eastward, because of the more recent and vigorous activity of the Carlsberg Ridge. There is no satisfactory evidence for, and much evidence against, the rotation of India from Arabia as postulated by Carey (1958); and such rotation is not needed to explain the height of the Tibetan Plateau.

Thus the geological history of Australia is intimately bound up with that of peninsular India and Ceylon, and also Antarctica and indeed Madagascar. The work of Compston, Arriens, Turek and Bofinger at the Australian National University (Compston and Arriens 1968) on the dating of the development of the ancient rock pattern of Western Australia permits comparison with the dating of the pattern in Ceylon and India (Crawford, 1969) and that reveals close similarities. The picture is a good deal more complicated than any revealed by Atlantic comparisons, because of the complexity of the break-up. But it can be argued that the Cape Leeuwin-Naturaliste peninsula of Western Australia is really a piece of Ceylon left behind. We still need to know more about the Australian relationship with Antarctica, a less revealing terrain. Sproll and Dietz (1969) have demonstrated that a morphological fit is satisfactory except for a small overlap for which one of their suggested explanations (involving some young volcanism) is, I believe, valid.

Although all this is of the very greatest intrinsic interest, it also has an immediate significance for those engaged in mineral search. We must think along new lines. As Australia was for much of its geological history attached to Ceylon, to peninsular India and to Antarctica, our whole concept of the development of its geology has to be rethought. In fact, for the time before separation, the terms "Australia" and "India" are not only meaningless but misleading. We must remember that in much of the past, "Australia" had very different continental limits. Thus a simple accretionary model for the growth of an Australian continent generally eastwards, fashionable some years ago, either upon one or more nuclei in what is now Western Australia, is no longer acceptable. Instead, we need to recognize the existence of an "Indo-Australia" in which there were at least six such nuclei of rocks over 2,500 m.y. old. At least one more existed in East Antarctica, which seems likely to have been closely associated with "Indo-Australia" for a very long period. The six are the nuclei of Pilbara, Yilgarn, central Ceylon, the South Deccan, Bundelkhand, and Singhbhum. Kerala may have been a separate nucleus, or perhaps attached to the South Deccan. There is some evidence that in the early stages Pilbara and Singhbhum were joined. I do not wish to discuss here the complex problem of what Hurley (1968b) has called "pre-drift drift" except to say that as drift is clearly a continuous process, the term itself is perhaps not a very happy one. I believe that no major displacement of western Australian cratons occurred relative to Indian cratons for a very long period. I equally believe that minor displacement very definitely occurred and that this "jostling" went on for a comparably very long period and affected some parts of all the area during most of the Precambrian and much of the Phanerozoic.

We should not be simple-minded about drift. Although there is much evidence of the preservation of geological patterns without significant distortion in any platform block as it moves about, the flow structures suggested by, for example, the vast mountain systems of Central Asia surely suggest also that deformation of such patterns eventually takes

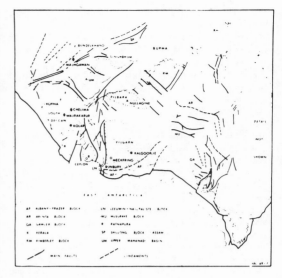

place. It would therefore be naive to believe that in an area such as Western Australia the present relative positions of the Pilbara and Yilgarn cratons are exactly those of the past.

The cratonic nuclei were perhaps at the earliest stage an immense single block. The areas between what are now cratons were for some reason less stable and rifts formed. In the deep rifts these were predominantly graywackes, pillow lavas and tuffaceous marine sediments. These rifts were later translated into horsts after burial metamorphism and regional metamorphism as the cratons re-approached. We now see them either as belts of largely granulitic rock—where the horsts are deeply eroded—or as belts of shallow-sea sediments, largely unfolded. Whatever happened in detail, one thing is certain. The erosion of the old rocks and the deposition of the succeeding sequences of middle and upper Precambrian and Phanerozoic rocks was mostly unrelated to the present continental boundaries. Though the forms of these are sometimes apparently influenced by the existing geology, it may be rather that deep-seated and persistent zones of weakness strongly influenced the delineation of areas of deposition and metamorphism, lines and points of extrusion and lines of fracture. There is much evidence put forward by many workers (in Africa especially) to suggest that a world-wide pattern of such zones of weakness exists. It was along sections of this pattern that the most recent break-up has taken place. It is interesting that the concept of long, narrow and long-lived features of this type has been developed independently in the Soviet Union (e.g., Peive

1956, Nalivkin 1963) but there no attempt seems yet to have been made to relate these "aulacogenes" (graben-like trenches, from the Greek *aulax*, channel, furrow) to a continental drift model.

Relevance to Mineral Search

Such concepts of geological development are very relevant to the problems of the origin of some metalliferous ores, to the distribution of gem minerals and, particularly in Phanerozoic rocks, to the occurrence of oil source and reservoir rocks in this region.

We can consider two aspects: comparison of mineral deposits known to occur in both Australian and Indo-Ceylon (for Antarctica is too poorly known yet) and the question of the possible existence in the one continent of mineral deposits known in economic quantity only in the other.

In the first case iron ores and gold are of particular interest. I shall not discuss the former, which is now the expert field of Dr. A. F. Trendall,[3] except to say that a close study of the Singhbhum iron-ore area of India in respect also of its *non-ferruginous economic* mineral deposits might be very fruitful for Western Australia. Concerning gold, the age of the gold mineralization at Kolar in Mysore is the same as that at Kalgoorlie. Although the Kolar area has not itself yielded good samples dateable by Rb-Sr, the lavas of the lower part of the Dharwar System in which the gold occurs give a total-rock age of 2,345 ± 60 m.y. These may, as suggested by Sreenivas and Srinivasan (1968) have been effused during folding of the System. The concentration of gold mineralization at Kalgoorlie is also apparently unrelated directly to the deposition of the greenstone sequence but is located in a fracture pattern dated at 2,420 ± 30 m.y. (Dr. A. Turek, personal communication). The pattern is partly occupied by large basic dikes independently dated at the same age. Although many parts of the world have gold of this age, it is my belief that at the time of formation these two very rich areas were probably much closer to each other than even a pre-drift assembly suggests. The South Deccan craton and the Yilgarn craton appear to have formed by separation, with a rift developing between. Horsts later developed as the cratons re-approached. Later more rifting took place along slightly different trends. It was along these later rifts that the narrow straits of Phanerozoic time existed. The continental break-up at the end of the Mesozoic shows the influence of both trends and the influence of the older one can still be seen in Australia in the Yandanooka-Cape Riche seismic zone on which the Meckering

[3] A major **study** of the Brockman Iron Formation of Western **Australia** by Dr. Trendall will shortly be published by the Government Printer, Perth.

earthquake epicentre lies (Doyle, Everingham and Sutton, 1968). It is therefore important, in studying gold occurrences throughout Western Australia, to recognize the probable original contiguity, and not just proximity, of what are now very widely separated cratons.

Ceylon is rich in graphite and in precious gems. The most valuable graphite deposits are epigenetic and follow the strike of the Highland Series rocks, granulites of charnockitic type which are at least 2,100 m.y. old. (Crawford and Oliver, 1969). In that Ceylon in any reconstruction lay close to what is now southwestern Australia, important graphite deposits may be expected to occur in similar rocks in the latter area. Graphite is certainly known, but total production so far has been only 153 tons. Ceylon production for the single year 1965 was 8,740 tons. Similarly, because the gems of Ceylon are mostly from the Ratnapura area in the southwest, it might be expected that comparably rich occurrences could be found in rocks of similar type and age in Western Australia. These are probably in the Albany-Frazer block and the Cape Leeuwin-Naturaliste peninsula. Ceylon is also especially characterized by lanthanide minerals. Monazite and thorianite are common in the southwest of Ceylon. It is difficult to believe that substantial occurrences in Western Australia are restricted to those near Bunbury.

In India, diamonds have been produced in the south for centuries, and they are still being mined in the north at Majhgawan near Panna in Madyha Pradesh. The age of the Majhgawan pipe is 1,140 m.y. (Crawford & Compston, 1969a) and it lies on what was the north bounding fault of an old rift. The primary sources in the south are the pipes and dikes of kimberlitic affinities near Wajrakarur in the Ananatapur District, and at Chelima in the Kurnool District of Andhra Pradesh. The former is of the same age as Majhgawan and the latter slightly older. The secondary sources include the basal Banganapalli Conglomerate of the Kurnool System (about 800 m.y.). Apart from the possibility that diamond-bearing pipes of similar type (not necessarily of the same age) may occur in northwestern Australia, and conglomerates in that area of similar age to the Banganapalli Conglomerate, or younger, could in theory be diamondiferous. The two continents should be thought of as one in considering areas of erosion feeding sedimentary basins of that period. It should be remembered that the distribution of diamond occurrences in India is far wider than that of known pipes. Apart from the deposits in the Vindhyan System proper, related quite definitely to the pipe at Majhgawan (Sinor 1930) and to other pipes of probably the same age cutting older rocks in Bundelkhand, there are diamonds in the Upper Mahanadi Basin in young Precambrian rocks, which are unrelated to any known sources nearby (King 1885). These may well have been eroded and the products deposited in what is now Australia. It is equally possible that the Upper Mahanadi diamonds were derived *from* Australia. Diamonds have indeed been found in northwestern Australia, but so far at only one locality, Nullagine. They have usually been very small. The possibility of more widespread occurrences should be studied.

Diamoniferous diatremes are closely related to carbonatites. The latter have been recognized in India since 1963 (Deans and Powell 1968) but in Australia the first was found in 1968 (Crohn and Gellatly 1968). In each continent their occurrence is associated with rift networks which were once connected. These can be used as a clue in tracking down the location of carbonatites yet undiscovered. I believe, for example, that they will be found in the northern Flinders Ranges of South Australia.

Any continental margin fit of Western Australia to peninsular India implies apparent apposition of the Kimberley area to Burma. In fact the fit suggested by Hurley (1968a) and others involve much overlap. This is not such a problem as first appears. Apart from the likelihood of relative movement between much of what is now Burma and Assam, southern Burma west of the Salween is geologically young and need not have been in existence in its present form at the time of separation. Work by Gellatly, Derrick and Plumb (1968) has shown that paleocurrent directions in the Kimberley Basin suggest a northern provenance. Apart from the immediate relevance of all this to continental drift, it should be borne in mind that the Precambrian of central and northeastern Burma is rich in metalliferous ores and gemstones.

The question of oil search is equally important but I shall not discuss this enormous subject. Ahmad (1960) has made a pioneer study comparing Phanerozoic formations in India and Australia. This needs to be repeated in detail with the benefit of recent work. Clearly, acceptance of continental drift implies recognition of late Phanerozoic break-up of sedimentary basins which transgress present continental boundaries, and it is better to study the whole rather than a fragment.

In conclusion I mention two more items of interest. First, I noticed in south-east India the "duricrusting" of the upper surface of the Kurnool System rocks, and I wondered whether this superficial and climatic effect so common in Australia had developed while the Indian peninsula was still in the southern hemisphere. Second, we should realize that the river system of peninsular India is very old and must in part surely be related to that which once existed in Western Australia. A wholly new aspect is thus given to the study of Indo-Australian geomorphology.

These are only some of the problems we ought to be attempting to study. We need to look harder. To look effectively we need also to extend the sweep of our vision.

Acknowledgments

I thank Drs. J. Tuzo Wilson and D. C. Tozer for critical comment on the draft. This paper was written following much geochronological work on the Precambrian rocks of India, Pakistan and Ceylon, carried out with the help of numerous geologists in those countries and made possible by the generosity of the Australian National University.

ERINDALE COLLEGE,
    UNIVERSITY OF TORONTO,
    CANADA,
        July 15; October 6, 1969

REFERENCES

Ahmad, F., 1960, A brief comparative study of the geological formations of western Australia and peninsular India and its bearing on the drift hypothesis: Rec. Geol. Soc. India, v. 86, p. 621–636.

Brown, D. A., Campbell, K. S. W., and Crook, K. A. W., 1968, The Geological Evolution of Australia and New Zealand: London, Pergamon Press.

Carey, S. W., 1958, A tectonic approach to continental drift *in* Continental drift, a symposium: Geology Department, University of Tasmania, Hobart.

Compston, W., and Arriens, P. A., 1968, The Precambrian geochronology of Australia: Canad. Jl. Earth Sci., v. 5, p. 561–583.

Crawford, A. R., 1969, India, Ceylon and Pakistan: New age data and comparisons with Australia: Nature, Lond., v. 223, pp. 380–384.

——, and Compston, W., 1969, The age of the Vindhyan System: Quart. Jl. geol. Soc. Lond. (in press).

——, and Oliver, R. L., 1969, The precambrian geochronology of Ceylon: Spec. Pubs. Geol. Soc. Aust. No. 2, p. 283–306.

Crohn, P. W., and Gellatly, D. C., 1968, Probable carbonatites in the Strangways Range area, Central Australia: Rec. No. 1968/114, Bur. Min. Res. Geol. Geophys., Canberra.

David, T. W. E., 1950, (W. R. Browne, Ed.) The Geology of the Commonwealth of Australia. 3 vs.: London, Edward Arnold.

Deans, T., and Powell, J. L., 1968, Trace elements and strontium isotopes in carbonatites, fluorites and limestones from India and Pakistan: Nature, Lond., v. 218, p. 750–752.

Doyle, H., Everingham, I. B., and Sutton, D. J., 1968, Seismology of the Australian Continent: J. geol. Soc. Aust., v. 15, p. 295–312.

Gellatly, D. C., Derrick, G. M., and Plumb, K. A., 1968, Proterozoic palaeocurrent directions in the Kimberley region, northwestern Australia: Rec. No. 1968/141, Bur. Min. Res. Geol. Geophys., Canberra.

Hurley, P. M., 1968a, The confirmation of continental drift: Scient. Amer. v. 218, p. 52–64.

——, 1968b, No pre-drift drift (?): Ann. Progress Rep. for 1968, Dept. of Geology & Geophys., Mass. Inst. Tech., Boston, Mass.

King, W., 1885, Sketch of the progress of geological work in the Chattisgarh Division of the Central Provinces: Rec. geol. Surv. India v. 18, p. 169–200.

Krishnan, M. S., 1968, Geology of India and Burma, 5th Edn., Madras, Higginbotham's.

Kummel, B., 1961, History of the Earth: An Introduction to Historical Geology. San Francisco, Freeman.

Le Pichon, X., and Heirtzler, J. R., 1968, Magnetic anomalies in the Indian Ocean and sea-floor spreading: J. Geophys. Res., v. 73, p. 2101–2117.

Nalivkin, V. D., 1963, Graben-like trenches in the east of the Russian platform: Sovyetskaya Geologiya, v. 1, p. 40–52 [in Russian].

Pascoe, E. H., 1950, 1959, 1964, A Manual of the Geology of India and Burma, 3 vols.: Delhi, Govt. of India Press.

Peive, A. V., 1956, The general description, the classification, and the spatial distribution of depth-faults: Izv. Akad. Nauk SSSR, Geol. Series, v. 1, p. 90–105 [Pt. I] and v. 3, p. 57–71 [Pt. II]. [In Russian].

Petrascheck, W. E., 1968, Kontinentalverschiebung und Erzprovinzen: Mineral. Deposita (Berl.), v. 3, p. 56–65.

Pichamuthu, C. S., 1968, The Precambrian of India, in: Rankama, K. (Ed.) The Precambrian, Vol. III: London, Interscience.

Schuiling, R. D., 1967, Tin Belts on the continents around the Atlantic Ocean: Econ. Geol., v. 62, p. 540–550.

Sinor, K. P., 1930, The Diamond Mines of Panna State: Bombay, Taraporevala.

Sreenivas, B. L., and Srinivasan, R., 1968, Tectonic significance of pillow lavas, Mysore, India: Geol. Soc. India Bull., v. 5, p. 56–58.

Sproll, W. P., and Dietz, R. S., 1969, Morphological continental drift fit of Australia and Antarctica: Nature, Lond., v. 222, p. 345–348.

Suess, E., 1904, 1906, 1908, 1909, 1924, The Face of the Earth. 5 vols.: Oxford, Clarendon Press.

Umbgrove, J. H. F., 1947, The Pulse of the Earth: The Hague, Martinus Nijhoff.

Wadia, D. N., 1957, Geology of India, 3rd Edn.: London, Macmillan.

Wilson, J. T., 1968, A revolution in earth science: Canad. Jl. Min. Metal., v. 61, p. 185–192.

7B

Reprinted from *Econ. Geol.* **65**:892 (1970)

CONTINENTAL DRIFT AND UN–CONTINENTAL THINKING

Sir: The paper by Arthur Crawford (ECON. GEOL., v. 65, pp. 11–16) deals with an area in which I have had an interest, namely the continental margins of Central East Africa and Madagascar. In 1956 as a result of the postulation by Wellington (1955) that Madagascar had drifted from Africa not from the position indicated by du Toit much to the north, but from the Zambesi area, I predicted that there ought to occur in Southern Malawi the rare uranium-niobium-tantalum minerals such as brannerite and betafite.

These were found early in 1957 in the Tambane area and along the Mwanza Fault, reported by Bosazza (1959) and the detailed geology by Cooper and Bloomfield (1961). At that stage I accepted Continental Drift not only as an hypothesis but as a principle. King (1953) in fact has considered that there is a necessity for continental drift.

The reconstruction by Wellington merits greater and more careful study than it has been accorded, and du Toit (1937), not quoted by Crawford, is still worthy of reading. Indeed the carbonatite-kimberlite (Crockett and Mason, 1968) conception is a very useful working hypothesis in the search for kimberlites, possibly diamondiferous in the rift valley margins of Malawi, and in the Zambesi-Shire confluence area. Kersantite-lamprophyres are associated with kimberlites in the Livingstonia and Fort Johnstone areas of Malawi. Basalts containing melilite perovskite appear to be closely related as well.

V. L. BOSAZZA

39 BARKLY ROAD, PARKTOWN, JOHANNESBURG.
May 7, 1970

REFERENCES

Wellington, J. H., 1955, Southern Africa. A geographical Study: Cambridge Univ. Press, Vol. 1, 528 pp. See Chapter 15, 460–473.

Bosazza, V. L., 1959, Radioactive minerals in Southern Nyasaland: Mining Mag. Lond., v. 101, p. 49–55.

Cooper, W. G. C., and Bloomfield, K., 1961, The Geology of Tamani-Salambidwe: Ministry Londs Surv., Geol. Surv., Nyasaland (Malawi), Bull. No. 13, 63 pp.

King, L. C., 1953, A necessity for continental drift: Bull. Amer. Ass. Petrol. Geol., v. 37(9), p. 2163–2177.

du Toit, A. L., 1937, Our Wandering Continents. An Hypothesis of Continental Drifting: Edin., Oliver and Boyd, 366 pp.

Crockett, R. N., and Mason, R., 1968, Foci of mantle disturbance in southern Africa and their economic significance: ECON. GEOL., v. 63, p.532–540.

7C

Reprinted from *Econ. Geol.* **66**:499 (1971)

CONTINENTAL DRIFT AND UN-CONTINENTAL THINKING—A REPLY

Sir: I appreciate the comments of Dr. V. L. Bosazza (ECON. GEOL., v. 65, p. 892) on my paper (ECON. GEOL., v. 65, pp. 11–16). I had seen the reconstruction by Wellington, but it is interesting to note the success in mineral discovery following predictions based upon it.

Geology progresses so fast now that papers rapidly go out of date. This seems a good opportunity to state that the latest paleomagnetic data (McElhinny 1970a, b; McElhinny and Luck 1970) compel us to revise our thinking about Gondwanaland and its manner of break-up. It is no longer possible to postulate *for the Phanerozoic* any contiguity of the eastern continental margin of India with the western continental margin of Australia. The assembly proposed by du Toit (1937)), whom Dr. Bosazza so rightly admires, fits the data better.

Such an assembly does not rule out an earlier possible association of "India" with "western Australia," as Gondwanaland probably formed fairly late in the Precambrian. Whatever the truth is, we need to think in terms of recurrent drift and of different continent/ocean relationships at different times, and especially not allow ourselves to be led astray by present-day nomenclature. I have developed this in a paper entitled "Gondwanaland and the Growth of India" submitted in January 1971 to the Journal of the Geological Society of India.

A. RAYMOND CRAWFORD
DEPARTMENT OF GEOPHYSICS AND GEOCHEMISTRY,
AUSTRALIAN NATIONAL UNIVERSITY,
CANBERRA 2600, AUSTRALIA,
February 2, 1971

REFERENCES

du Toit, A. L., 1937, Our Wandering Continents, An Hypothesis of Continental Drifting: Oliver and Boyd, Edin., 366 pp.

McElhinny, M. W., 1970a, Palaeomagnetism of the Cambrian purple sandstone from the salt range, West Pakistan: Earth, Planet, Sci. Lett., v. 8, pp. 149–156.

——, 1970b, The formation of the Indian Ocean: Nature, Lond., v. 228, pp. 977–979.

McElhinny, M. W., and Luck, G., 1970, Paleomagnetism and Gondwanaland: Science, v. 168, pp. 830–832.

8

Reprinted from *Earth Planet. Sci. Lett.* **8**:400–402 (1970)

CONTINENTAL DRIFT AND ECONOMIC MINERALS IN ANTARCTICA

Donald D.RUNNELLS

*Department of Geological Sciences, University of Colorado,
Boulder, Colorado 80302, U.S.A.*

Received 19 May 1970
Revised version received 5 June 1970

At present there are no deposits of minerals known in Antarctica that can be economically exploited. Minor amounts of copper, iron, chromium, and major beds of coal are known, but these cannot presently support mining operations. Projections of known belts and areas of economic mineralization into Antarctica from other segments of Gondawanaland show that a potential exists for discovering deposits of nickel, copper, gold, diamonds, and other high value commodities. Reconstruction of pre-drift positions of Africa and Australia adjacent to Antarctica makes it possible to pinpoint the most promising areas for intial exploration. Although exploration for deposits in Antarctica is not now economically attractive, the continent does offer potential for the future.

1. Introduction

As the number of easily-discovered deposits of valuable minerals dwindles, and the demand for mineral commodities spirals upward, we must turn to regions of the world and concepts of exploration that have not been tested. A few workers have considered the role of continental drift in the exploration for new deposits of minerals. Schuiling [1] investigated the distribution of tin deposits in South America and Europe and found good correspondence between similar provinces on opposite sides of the Atlantic. Petrascheck [2] showed that certain other metallogenic provinces can also be correlated across the present oceans, including a tungsten-tin province that may extend in pre-drift positions along the eastern coast of Australia through Antarctica and into western Argentina. Recently, Crawford [3] drew attention to specific areas for prospecting in India, Ceylon, and Australia, based on projections of known belts of mineralization among the continents as they were situated prior to the breakup of Gondawanaland. Ravich [4] and Potter [5] mention the possibility that mineral deposits similar to those of other Gondawana continents may exist in Antarctica, and Swan [6] suggests that deposits like those of Australia may be present in Antarctica because of the geologic similarities between the two areas.

Good summaries of the geology of Antarctica are available in Adie [7] and Hadley [8].

Little is known about occurrences of valuable minerals in Antarctica. Mueller [9] described trivial occurrences of copper minerals in veins with barite, calcite, and other minerals on Greenwich Island, and Anderson [10] reports that copper, iron, and chromium minerals occur in the mafic complex of the Dufek Massif. Major deposits of coal are known [5] in the Transantarctic Mountains.

2. Targets for exploration in Antarctica

Projection of belts of mineralization known from the other Gondawana continents offers loci for exploration in Antarctica.

Dietz and Sproll [11] and Smith and Hallam [12] demonstrate a remarkably good fit between the outline of the southeastern coast of Africa and the portion of the coast of East Antarctica from the Weddell Sea to the Princess Martha Coast. This fit, together with the corresponding geologic [12] and sea-floor features, is good evidence that these two continents were continuous prior to the breakup of Gondawanaland in Mesozoic time. Such a reconstruction affords the basis for searching for deposits of minerals in Antarctica similar to those of South Africa and Rhodesia. In particular, the postulated east-west Lesotho structural trend [13] of South Africa, with its deposits of nickel and clusters of diamond-bearing kimberlite pipes, may extend beneath the Weddell Sea and Filchner Ice Shelf into East Antarctica at about 42° W. Longitude. Exceptional among the rare outcrops of rock in this area is the Dufek Massif (82°40′ S. Lat., 50—54° W. Long). The Dufek Massif is a mafic igneous complex with near horizontal layering, exposed over an area about 30 miles long and 2 to 10 miles wide [10]. Anderson [10] reports that iron, copper, and chromium minerals are present in the complex. It is interesting to note that the large Insizwa sill, with its layered mafic rocks and magmatic deposits of nickel and copper, occurs in eastern South Africa along the same postulated Lesotho trend [13]. Although the diamond-bearing kimberlite pipes of South Africa may be as young as Cretaceous [14], the same fundamental zones of weakness along which the pipes were intruded in South Africa [13] may have permitted similar activity in Antarctica, regardless of the exact time of breakup of Gondawanaland.

Another major zone of mineralization that may project into Antarctica from Africa is the northeasterly Lompopo trend [13] of Rhodesia and northern South Africa. The Lompopo trend is a well-defined belt of crustal weakness that has endured since Precambrian time, and Crockett and Mason [13] emphasize the genetic relationship between this zone and the numerous magmatic nickel deposits of Rhodesia. The Lompopo trend, if continuous, should project onto the Princess Martha Coast of East Antarctica at about 0° Longitude. Outcrops of rock are fairly abundant in this region [15] and should lend themselves to investigation.

The Antarctic Peninsula is usually correlated with the Andes Mountains of South America in reconstructions of Gondawanaland, although Dietz and Sproll [11] postulate the independent development of these two orogenic belts. Regardless of the details of reconstruction, the strong geologic similarities between the two areas suggest that deposits of base and precious metals like those of the Andes may be present along the Antarctic Peninsula.

Sproll and Dietz [16] demonstrate an excellent fit between the Wilkes Land sector of East Antarctica and the southwestern coast of Australia. This reconstruction places Tasmania adjacent to the northern margin of Victoria Land. It is then tempting to project the Paleozoic rocks of the Tasman geosynclinal belt of eastern Australia into Hamiltons's [17] Middle Paleozoic geosynclinal belt east of the Transantarctic Mountains in Victoria Land. However, Hamilton [17] points out that the non-volcanic Paleozoic section of Victoria Land cannot be correlated directly with the volcanic-rich rocks of similar age in the Tasman geosynclinal belt. He suggests that the Australian analogue of the rocks of Victoria Land is buried beneath the Tertiary cover of the Murray Basin in southeastern South Australia. Although not mentioned by Hamilton, his suggestion is supported by exposures of a thick sequence of Ordovician, Silurian, and Devonian non-volcanic clastic sediments [18] in central Victoria in Australia. The major gold deposits of Ballarat, Bendigo, and Mount Alexander occur in these sediments [19], and similar deposits might therefore be present in Antarctica at the northeastern corner of Victoria Land.

The reconstruction by Sproll and Dietz [16] and work by Hamilton [17] indicate that the Adelaide geosynclinal belt of south-central Australia projects into the Transantartic Mountains. These strata in Australia host significant deposits of base and precious metals of early Paleozoic age [20].

The Precambrian shield of southwestern Australia contains deposits of gold, banded iron formation, nickel, and pegmatites. Similar mineral deposits should be present in the shield area of East Antarctica, inland from the Knox Coast. There are a few exposures of rock along the Knox Coast [15] that could be examined for mineralization.

Finally, Hamilton [17] suggests that by analogy with the other provinces, the central part of West Antartica should corresponds geologically to eastern Australia, although the area in Antarctica is too poorly known to permit verification. If his suggestion proves to be valid, it would encourage a search in the scattered interior mountains of West Antarctica for analogues of the rich deposits of base and precious metals that occur in the Tasman geosynclinal strata of eastern Australia.

References

[1] R.D.Schuiling, Tin belts on continents around the Atlantic Ocean, Econ. Geol. 62 (1967) 540.

[2] W.E.Petrascheck, Kontinentalverschiebung and Erzprovinzen, Mineralium Deposita 3 (1968) 56.

[3] A.R.Crawford, Continental drift and un-continental
thinking, Econ. Geol. 65 (1970) 11.

[4] M.G.Ravich, Antarctica Commission Reports 1960
(Transl. for National Science Foundation by Israel Program for Scientific Translations) (1960) 20.

[5] N.Potter, Economic potential of the Antarctic, Antarctic
Jour. of the U.S. 4 (1969) 61.

[6] R.A.Swan, Australia in the Antarctic (Cambridge Univ.
Press, London, 1962) p. 369.

[7] R.J.Adie (ed.), Antarctic Geology (North-Holland Publ.
Co., Amsterdam, 1964) 758 pp.

[8] J.B.Hadley (ed.), Geology and paleontology of the Antarctic, Antarctic Research Series 6 (Amer. Geophys.
Union, 1965) 281 pp.

[9] G.Mueller, Some notes on mineralization in Antarctica,
in: R.J.Adie (ed.), Antarctic Geology (North-Holland
Publ. Co., Amsterdam, 1964) 393.

[10] J.J.Anderson, Bedrock geology of Antarctica: a sum-
mary of exploration 1831–1962, in: J.B.Hadley (ed.),
Geology and paleontology of the Antarctic, Antarctic
Research Series 6 (Amer. Geophys. Union, 1965) 34.

[11] R.S.Dietz and W.P.Sproll, Fit between Africa and Antarctica: a continental drift reconstruction, Science 167
(1970) 1612.

[12] A.G.Smith and A.Hallam, The fit of the southern continents, Nature 225 (1970) 139.

[13] R.N.Crockett and R.Mason, Foci of mantle disturbance
in southern Africa and their geologic significance, Econ.
Geol. 63 (1968) 532.

[14] A.L.DuToit, Geology of South Africa, 3rd ed. (Oliver
and Boyd, Edinburgh, 1954) 611 pp.

[15] C.R.Bentley, R.L.Cameron, C.Bull, K.Kojima and A.J.
Gow, Physical characteristics of the Antarctic ice sheet,
American Geographical Soc. Antarctic Map Folio Series,
Folio 2 (1964) Plate 3.

[16] W.P.Sproll and R.S.Dietz, Morphological continental drift
fit of Australia and Antarctica, Nature 222 (1969) 345.

[17] W.Hamilton, Tectonics of Antarctica, Tectonophysics 4
(1967) 555.

[18] O.P.Singleton, Geology and mineralization of Victoria,
in: J.McAndrew (ed.) Geology of Australian Ore Deposits, Eighth Commonwealth Mining and Metallurgical
Congress Vol. 1 (Australasian Inst. Mining and Metall.,
Melbourne, 1965) 440.

[19] J.McAndrew, Gold deposits of Victoria, in: J.McAndrew
(ed.) Geology of Australian Ore Deposits, Eighth Commonwealth Mining and Metallurgical Congress, Vol. 1
(Australasian Inst. Mining and Metall., Melbourne, 1965)
450.

[20] E.S.Hills, Tectonic setting on Australian ore deposits, in:
J.McAndrew (ed.) Geology of Australian Ore Deposits,
Eighth Commonwealth Mining and Metallurgical Congress
Vol. 1 (Australasian Inst. Mining and Metall., Melbourne,
1965) 3.

Editor's Comments
on Papers 9 Through 12

METALLOGENESIS RELATED TO CONTINENTAL BREAK-UP

Acceptance of continental drift entailed recognizing the global ridge-rift system (Figure 2, p. 3) as the site of upwelling convection currents that fragmented continents and moved them over the earth's surface. The metal-rich sediments and hot brines already known from the Red Sea and Salton Sea rifts suggested that convection leading to continental fragmentation could also involve mineralization.

However, the proximity of continental crust is not essential for mineralization related to crustal rifting. This was demonstrated in the pioneer study by Boström and Peterson (Paper 9) on metal enrichment in East Pacific Rise sediments—a nice example of scientific serendipity. They concluded that the metals were precipitated from hydrothermal solutions of juvenile origin, thus providing a more convenient source for the metals in manganese nodules and some deep sea muds, than direct precipitation of land-derived metals from solution in sea water (e.g. Mero, 1965). Their work also suggested that the metalliferous potential of oceanic crust was greater than had previously been supposed by many authorities, a topic to which we return later.

The uniformitarian thesis that mineralization processes in ancient rift valleys resembled those in the present-day Red Sea, was developed by Kanasewich in Paper 10. Most of his paper is concerned with identifying an ancient rift system in Canada (see his Figure 2). The analogy lacks some precision, for there is no continental crust beneath the median zone of the Red Sea (where the mineralized deeps are); and the metals there are no longer generally considered to be of direct mantle origin.

Russell (Paper 11) probably made the first major attempt to link the stresses accompanying continental breakup with development or reactivation of mineralizing fractures. Russell proposed that the rich base metal ores in the Carboniferous limestone of Ireland came from hydrothermal solutions controlled by a north-south fissure system related to opening of the North Atlantic. This paper aroused much discussion and criticism (Russell, 1969).

As in Paper 10, comparisons between the Irish Carboniferous and present-day Red Sea and Salton Sea areas are not particularly apt for the tectonic settings are rather different, and in any case the hydrothermal fluids could have originated from connate water in the Carboniferous sediments. Moreover, continental breakup in the North Atlantic did not begin until at least Jurassic times, perhaps not until the late Cretaceous. Russell's proposed fractures could just as well have been related to plate movements associated with the Hercynian orogeny, since they must have been of Carboniferous age. Another important criticism is that there is no surface manifestation of the postulated fissures. Indeed, Morrisey et al. (1971) found the only obvious correlation between mineralization and fracturing to be the occurrence of ore deposits near east-northeasterly trending normal faults bordering pre-Carboniferous inliers. After further discussing the age range and possible mechanisms of mineralization, they concluded: "The evidence is at present too equivocal to indicate a genetic model which would be specifically useful in pinpointing targest for mineral exploration."

Nonetheless, Russell (1973) has further developed his Irish study, taking account of these and other criticisms, and extended it to Scotland, where he identifies other north-south mineralizing geofractures, also of Carboniferous age.

A similar attempt to relate the Jurassic tin mineralization of Nigeria to reactivation of an ancient lineament system early in the separation of Africa and South America (Wright, 1970), can be criticised on similar grounds: there is no obvious surface manifestation of these lineaments, so their existence remains speculative until more information is available.

On the other hand, Vokes (1973) has convincingly shown that base metal sulphide mineralization in the Oslo region is clearly related to

Permian magmatism, and there is much better evidence than in Ireland (or Nigeria), that both are controlled by deep fracture systems. Vokes attributes the fractures to stresses accompanying the breakup of Laurasia, which may already have begun in late Palaeozoic times, although Hercynian movements could presumably also have had some effect.

Kutina (1972) has also correlated the distribution of ore deposits in several areas with rift structures, particularly where different fracture trends intersect. However, he makes no explicit reference to continental drift movements being in any way related to the development of fracture systems, although this is not unusual in such studies (compare, for example, Crockett and Mason, 1968; James, 1972).

Rift Valleys, Ocean Ridges and the Source of Ore Metals

Although the list of metals in the sediments and waters of the East Pacific Rise and other ocean ridges has been extended (Paper 9) and new occurrences continue to be found (e.g. Fisher and Boström, 1969; Boström and Fisher, 1969; Carr et al., 1974; Anderson and Halunen, 1974), metal concentrations are much higher in sediments and brines of the Red Sea and Salton Sea (e.g. Degens and Ross, 1969; White, 1968), areas that lie respectively close to and upon continental crust. Oxygen isotope studies suggest that the water in the brines is meteoric and it is now generally considered (e.g. Tooms, 1970) that the salts and metals are largely derived from surrounding igneous rocks and sediments, including evaporites which reach a thickness of 4 km along the Red Sea margin.

An origin of this general type is proposed by Grant for the lead-zinc mineralization of the Benue trough in Nigeria (Paper 12), and in a more recent paper, Degens and Kulbicki (1973) reached very similar conclusions about the origin of hydrothermal metal deposits in lakes of the Western Rift.

If mineralized sediments develop in the earliest stages of continental rifting and separation, it follows that metal-rich muds might be found at the base of sequences deposited along the continental margins of opening oceans such as the Atlantic (cf. Schoell, 1975). According to Bullard (1974), such sediments have been identified in drill cores.

Where there is continental crust in the vicinity, the relative contribution of ore metals from crust and upper mantle remains a matter for debate, as we shall see in Part II. By contrast, only a mantle source is possible in truly oceanic environments although the contribution is not quite as direct as originally envisaged by Boström and Peterson (Paper 9).

Metals in the hydrothermal solutions emanating from ocean-ridge crests are now believed to have been leached by sea water circulating through newly generated hot basaltic crust. In other words, the metals have a two-stage origin, as summarised by Spooner and Fyfe (1973, p. 295) for example: "Circulation of hot brines in geothermal systems . . . is . . . an important means of transporting a variety of elements from the oceanic crust into sea water, and since the . . . basalts are initially derived by mantle anatexis (partial melting), a significant stage in transport from the upper mantle to the hydrosphere."

Ridge (1973) has examined some of the constraints on the behavior of such geothermal brines, notably salinity, vapour pressure, temperature, and depth of sea water. With the application of plate tectonic concepts to metallogenesis (Part II), there is now little doubt that many important metal ore deposits owe their origin to hydrothermal processes along the chain of constructive or accreting plate margins which constitute the global ridge-rift system (Figure 2, p. 3). But which metals? Could concentrations of "granitic" elements be found along accreting margins in a truly oceanic setting, e.g. the present Mid-Atlantic ridge? Sparce base-metal mineralization in Icelandic granophyres comprises chalcopyrite, galena, and sphalerite, with minor molybdenum (Jankovic, 1972), all of which must be of mantle origin, since there is no continental crust beneath Iceland.

Of greater interest are dredged fragments of oceanic crustal rocks from the Carlsberg Ridge and the Azores segment of the Mid-Atlantic Ridge, enriched in elements such as Sn, Hg, Ag, Pb, Sn, Be, and B, as well as the more typically "oceanic" elements, Cu, Zn, V, Cr and Co (Dmitriev et al., 1971). The mineralized rocks have an aspect consistent with hydrothermal (fumarolic) alteration, and the source of the metals must lie in the upper mantle.

It follows that if elements such as tin can demonstrably be derived from the mantle in a truly oceanic situation, there is no a priori reason why they should not do so in a continental crust setting. To which it may be replied that *most* of the tin and other granitophile metals in the mantle were transferred to the continental crust well before the end of Precambrian time, and have been simply recycled into new deposits ever since. What emerges at oceanic sites such as the Azores is merely a small residue of no quantitative importance. The debate continues in Part II.

9

PRECIPITATES FROM HYDROTHERMAL EXHALATIONS ON THE EAST PACIFIC RISE

K. BOSTROM AND M. N. A. PETERSON

ABSTRACT

On the very crest of the East Pacific Rise, in equatorial latitudes—particularly 12° to 16° South, the sediments are enriched in Fe, Mn, Cu, Cr, Ni, and Pb. The correlation of these areas of enrichment to areas of high heat flow is marked. It is believed that these precipitates are caused by ascending solutions of deep-seated origin, which are probably related to magmatic processes at depth. The Rise is considered to be a zone of exhalation from the mantle of the earth, and these emanations could serve as the original enrichment in certain ore-forming processes. Ba and Sr are enriched on the flanks of the Rise.

INTRODUCTION

VOLCANIC processes on the continents commonly release volatile emanations like H_2O, CO_2, H_2S, H_2, HF and various mineralizing aqueous solutions. Zies (17) and Rubey (12), among others, have estimated the amounts of subaerial volcanic emanations that are released into the atmosphere and hydrosphere. One may infer from Rubey (12) that the quantities of emanations released by submarine volcanic activity are of the same order of magnitude. In addition, many magmatic events are not manifested as surficial volcanism. The possibility that precipitates that originate from hydrothermal solutions might be accumulating in deep-sea sediments was suggested a long time ago by von Gümbel (14). The idea continues to attract attention: very recently Richards (11) has favored submarine exhalations as the source for the Broken Hill banded iron formations. Descriptions of solutions from hydrothermal wells near the Salton Sea, California (16) have added recent impetus to this general trend of thought.

THE EAST PACIFIC RISE SEDIMENTS

On the very crest of the East Pacific Rise in equatorial latitudes—6° North and 12°–16° South (where expedition RisePac of Scripps Institution crossed the Rise with two traverses), the sediments are different from most other sediments of the Pacific Ocean. They contain an abundant, brown, X-ray amorphous, metal oxide precipitate. Rather than being a coating or impregnation of other mineral grains, as is common for the normal red-brown sediments of the deep ocean, it is a separate, very finely disseminated or loosely aggregated material. It is possible to stabilize a suspension of this material that does not settle out for many months. Our attention was first attracted to the possibility that this material could be a precipitate related to hydrothermal activity by the observation that, during the treatment to remove calcium carbonate from the sediment, this very fine precipitate would not settle out after having been dispersed. Figure 1 shows a series of samples taken across the Rise (the southern traverse of RisePac Expedition—section AA of Fig. 1). The supernatant liquid of only those samples on the very crest of the Rise is colored a bright orange-brown by this finely dispersed material that did not settle from suspension after several weeks. This same effect, although not as marked, was clearly present for the northern traverse of RisePac expedition, which crossed the Rise at about 6° North Latitude. Where abundant, this precipitate stains the normal light, pure globigerina ooze a dark brown.

Fig. 1. Series of samples taken during southern traverse of RisePac expedition showing very finely disseminated metal oxide precipitate in suspension. Only those samples very near the crest of the Rise show this effect.

We therefore decided to study the distributions of various elements that could serve as tracers of hydrothermal processes.

The ideal tracer for this purpose should have a comparatively short residence time so that the element would precipitate as an aureole around the area of emanation. Too short a "residence time" would lead to deposition just around, or even deep inside, the orifice, and would make it difficult or impossible to locate regions of hydrothermal activity. Too long residence times would lead to mixing and transport over long distances and thus, not produce easily interpreted distribution patterns. The major constituents of sea water, of course, have very long residence times and are poor tracers of hydrothermal activity unless they were to occur in unusually large concentrations.

The locations of the analyzed sediment cores are shown in Figure 2. All

samples were collected during RisePac expedition.   Samples were taken from
the upper several centimeters of the core.   Two analytical procedures were
used:

1) One set of samples, each consisting of 0.15 gm dry, ground unwashed
sediment, was digested in boiling HCl, diluted and analyzed with a Perkin
Elmer 303 Atomic Absorption spectrophotometer.   All samples of the north-
ern traverse of RisePac and that part of the southern traverse designated AA
(Fig. 2) were analyzed in this way for Au, Ba, Co, Cr, Cu, Fe, Mn, Ni,
Pb and Sr.   Carbonate analyses were of this same mixed sample.

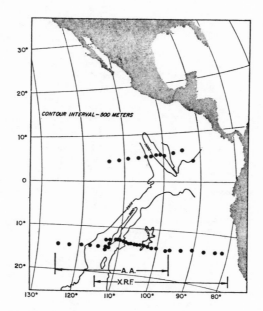

FIG. 2.   Map of the East Pacific, showing the location of
analyzed sediment samples.

2) The samples from the southern traverse designed XRF were pressed
into pellets and analyzed on a Norelco X-ray fluorescence unit for Ba, Ca, Fe,
Mn, Si, and Sr.   The range of samples studied by XRF was purposely
extended more to the east than those studied by method 1) for reasons
given below.

Results of the analyses are in Figures 3, 4, and 5.   Heat flow data and
water depths for these two traverses across the Rise are also shown (15).

There is a marked enrichment of Fe, Mn, Cu, Cr, Pb, and Ni in sediments
from the very crest of the Rise; in addition, a clear co-variation exists between
these elements.   There is also, however, especially on the southern traverse,
a very pronounced maximum in the concentration of these elements to the

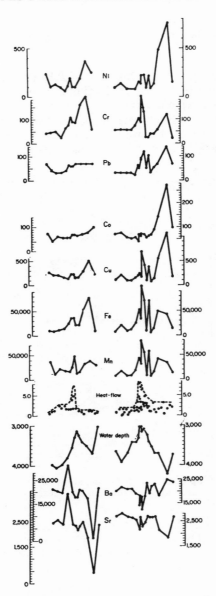

Fig. 3. Results of chemical analyses by atomic absorption. Analyses of samples from the northern traverse of the RisePac expedition are given to the left; to the right are given analyses of samples from the southern traverse of the RisePac expedition. Concentrations in ppm, water depths in m, heat flow in 10^{-6} cal cm^{-2} sec^{-1}.

east of the Rise, closely coinciding with a depression in the ocean floor between the Rise and a parallel ridge to the east.   Because of the deep water, very little carbonate is present in these sediments of the depression.   In order more clearly to determine the normal concentration values for Fe and Mn, and most abundant elements in the oxide precipitate, these elements were determined by X-ray fluorescence for samples farther to the east of the Rise. As can be seen, Figure 5, the high Fe and Mn values for the sediments of the depression mentioned above are not typical for most of the sediments on the sides of the Rise.   On the contrary, it is evident that, calculated on a carbonate free basis, the total concentrations of Fe and Mn reach their highest concentrations on the very crest of the Rise (Figs. 4, 5).

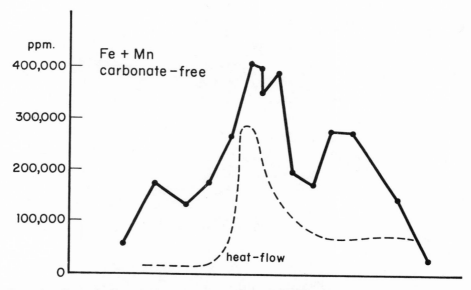

Fig. 4. Co-variation of heat flow and Fe and Mn (combined) on a carbonate free basis.   Data from the southern traverse.   Concentrations in ppm.

Ba and Sr are enriched on the flanks of the Rise, but not on its crest— a distribution pattern clearly different from that of Fe, Mn, and associated elements.   A clear co-variation exists for Ba and Sr over the crest and flanks of the Rise; this co-variation is probably due to the fact that Ba and Sr occur as a celestobarite.   Farther to the east, however, the distribution of Sr is that more commonly found in pelagic sediments, that is, Sr co-varies with Ca, presumably in a carbonate phase.   In the depression Ba is probably present both in barite and a zeolite; both these minerals have been observed in sediments from this area (Bonatti, pers. comm.).   Au showed no enrich- ment across the crest of the Rise.   Summing up the evidence, Fe, Mn, Cu, Cr, Ni, and Pb are present in important quantities only on the crest of the

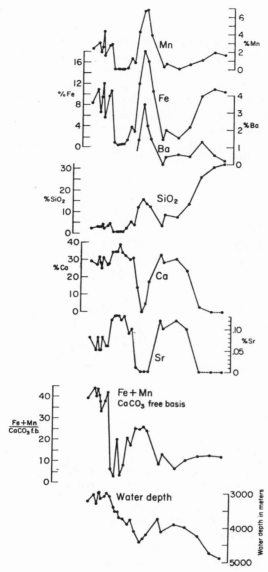

Fig. 5. Results of chemical analysis by X-ray fluorescence. Location of analyzed samples is indicated by the symbol XRF on Figure 1. Concentrations in weight percent.

Rise; on a carbonate free basis, these sediments may exceed 60 percent, by weight, metal oxide precipitate. Similar results for Fe and Mn have been reported by Skornyakova (13) to be due to hydrothermal activity. This is the only explanation that seems reasonable also to the present authors.

DISCUSSION

The crest of the East Pacific Rise appears to be a more or less persistent zone of magmatic activity. High heat flow anomalies on the East Pacific Rise (15) suggest that some magmatic event may exist under its crest (4, 6). This conclusion is supported by the evidence for volcanic activity that has been observed along the trend of the East Pacific Rise (9, 10). Silica cementation (1) and barite distribution in sediments in the vicinity of the Rise has also been suggested as due to hydrothermal processes (2). The hot springs of the western United States and the high heat flow areas of the top of the East Pacific Rise appear to be part of the same main geological feature (7). The presence of the Rise appears to be related to rifting of the Gulf of California, where the Rise extends under the continent; similar relationships are to be found in the Red Sea, in which hot pools of metalliferous brines have recently been discovered (8).

It is believed that these precipitates are caused by ascending solutions, of deep-seated origin, which are probably related to magmatic processes at depth. It is futile to speculate on the exact nature of the magmatic events, whether interflow, ascending magma chamber, rift filling or other. Drake and Girdler (5) interpret the data of the Red Sea rift to indicate the presence of a long, narrow igneous intrusive at shallow depth extending along the middle of the rift. Judging from the suite of elements that debouch at the surface, it would appear likely that the magma type would be basic. These ascending solutions, which would be expected to be acidic and reducing, would transport these elements upward, and precipitate them upon reaching alkaline, oxidizing environments associated with sea water and ocean bottom conditions. Such solutions would also surely mobilize elements such as Mn while traversing deeper sections of the sediment (3).

The major Rise-Rift systems of the world are exceedingly large features, and the reasons for their existence must most likely lie within the mantle of the earth. Whatever else the Rise-Rift systems may be, they appear also to be zones of emanation from the upper mantle. Additional evidence for this statement may be the anomalously high quantities of He that have been found in deep waters close to the crest of the East Pacific Rise (Bieri, Koide, and Goldberg, pers. comm.). The well known fact that high heat flow values are found on the crests of oceanic rises indeed suggests that the same deep seated processes that cause the formation of rises and their high heat flows also trigger the formation of ascending mineralizing solutions. It seems likely that we are seeing here an example of a process that could serve as the basic enrichment mechanism and source for the elements in some ore formations. If such metals in hydrothermal solution could find their way into the proper depositional situation and become deposited there, and later perhaps reworked in various ways or metamorphosed, these later processes would leave their imprint on the geologic history of the deposit, and leave in doubt the original concentration mechanism.

ACKNOWLEDGMENTS

We thank R. Bieri and E. Bonatti for valuable comments. Carbonate analyses were made by Miss B. Meinke in the laboratory of Dr. E. D. Goldberg. Mr. E. Jarosevich kindly helped us with the X-ray fluorescence analyses. Support by NSF grants GP-489, GP-5112 and the ONR is gratefully acknowledged.

Scripps Institution of Oceanography,
    UCSD, La Jolla, California,
    August 16, 1966

REFERENCES

1. Arrhenius, G., 1952, Sediment cores from the east Pacific: Reports of the Swedish deep sea expedition, 1947–48, Vol. V., Goteborg.
2. Arrhenius, G., and Bonatti, E., 1965, Neptunism and volcanism in the ocean: in Progress in Oceanography, v. 3.
3. Bostrom, K. (in prep.), The problems of excess manganese in pelagic sediments.
4. Bullard, E. C., 1963, The flow of heat through the floor of the ocean: in The Sea, ed. M. N. Hill, Interscience Publishers, J. Wiley and Sons, N. Y.
5. Drake, C. L., and Girdler, R. W., 1964, A geophysical study of the Red Sea: Geophys. Jour. Royal Astro. Soc., v. 8, p. 473.
6. McBirney, A. R., 1963, Conductivity variations and terrestrial heat-flow distribution: J. Geophys. Research, v. 68, p. 6323–6329.
7. Menard, H. W., 1964, Marine Geology of the Pacific: McGraw-Hill, N. Y.
8. Miller, A. R., Densmore, C. D., Dogens, E. T., Hathaway, J. C., Manheim, F. T., McFarlin, P. F., Pocklington, R., and Jokela, A., 1966, Hot brines and recent iron deposits in deeps of the Red Sea: Geoch. et Cosmochim. Acta, v. 30, p. 341–359.
9. Peterson, M. N. A., and Goldberg, E. D., 1962, Feldspar distributions in South Pacific pelagic sediments: J. Geophys. Research, v. 67, p. 3477–3492.
10. Peterson, M. N. A., and Griffin, J., 1964, Volcanism and clay minerals in the southeastern Pacific: J. Marine Research, v. 22, p. 13–21.
11. Richards, S. M., 1966, The banded iron formations at Broken Hill, Australia, and their relationship to the lead-zinc ore bodies: Econ. Geol., v. 61, p. 257–274.
12. Rubey, W. W., 1951, Geologic history of sea water: Geol. Soc. America Bull. v. 62, p. 1111.
13. Skornyakova, I. S., 1964, Dispersed iron and manganese in Pacific Ocean sediments: Lithology and Mineral Resources, v. 5, p. 3–20 (In Russian). Internat. Geol. Rev., v. 7, p. 2161–2174 (In English).
14. von Gümbel, G., 1878, Ueber die im Stillen Ocean auf dem meeresgrunde vorkommenden Manganknollen, Sitz. Berichte d. K. Bayerischen Akademie d. Wissenschaften Munchen, *Matem-Physik Klasse*, p. 189–209.
15. von Herzen, R. P., and Uyeda, S., 1963, Heat flow through the eastern Pacific Ocean floor: J. Geophysical Research, v. 68, p. 4219–4250.
16. White, D. E., Anderson, E. T., and Grubbs, D. K., 1963, Geothermal brine well: mile-deep drill hole may tap ore-bearing magmatic water and rocks undergoing metamorphism: Science, v. 139, p. 919–922.
17. Zies, E. G., 1929, The valley of ten thousand smokes. II The acid gases contributed to the sea during volcanic activity: Nat. Geogr. Soc. Contrib. Tech. Papers, Katmai Series 1, v. 4, p. 61–79.

10

Precambrian Rift: Genesis of Strata-Bound Ore Deposits

E. R. Kanasewich

Until recently (*1*), reliable seismic reflections from the lower crust were difficult to obtain; one relied almost entirely on refraction seismology, which produces useful broad-scale regional data but has no great resolving power. Modern instrumentation, improved field techniques, magnetic recording, and digital processing now obtain detailed and reliable information on crustal structure by reflection seismology (*2*).

Several deep reflecting horizons appear to be continuous, except for zones of fracturing, over at least 100 km in southern Alberta. The most prominent event is from the top of an intermediate layer, locally called the Riel discontinuity, at which the velocity increases abruptly from 6.5 to 7.2 km/sec; this may be equivalent to the Conrad discontinuity in Europe. The marker maintains so much character across zones of faulting that the probability of miscorrelation is small. Dips of up to 20 deg, present on many records, necessitated migration of all reflections so that

the horizons might be displayed in their true subsurface positions on the migrated section (Fig. 1). The events are plotted in terms of two-way vertical-travel time. The depth scale varies slightly along the section as the average velocity changes. Weak reflection *M* correlates with the Mohorovičić discontinuity from reversed-refraction profiles.

The Moho changes in depth from 47 to 38 km while the Riel discontinuity changes from 37 to 26 km. An in-line reversed-refraction profile along the strike of the deepest part of the graben confirms the great thickness of the crust in this region, while a broadside refraction profile from a shot point 470 km to the west substantiates the structural variation and the faulting perpendicular to the strike. The amount of structural relief is great but not excessive when one remembers that from surface evidence alone Florensov (*3*) found 5 to 7 km of crustal displacement from the deepest part of the rift to the highest nearby mountain at Lake Baikal.

The seismic structure coincides with a notable linear anomaly in the gravitational and magnetic fields; the area (Figs. 2 and 3) has yielded more than 1800 gravity and ground-magnetometer readings, with a reasonable distribution except in the main range of the Rocky Mountains. With a few notable exceptions the rift was apparently filled with low-density, nonmagnetic, Precambrian sediments. Model calculations, based on a relation between seismic velocities and rock densities, have been made (Fig. 4); agreement with the seismic structure is excellent.

The postulated trace of the rift, determined from geophysical evidence, is superimposed in Fig. 2 together with the locations of gravity high and low anomalies, exploratory wells drilled to basement, reflection and refraction lines, and mining developments on the Precambrian Belt Series. The gravity and magnetic trends may be used to trace the remains of the rift along 450 km; it passes at right angles to the strike of

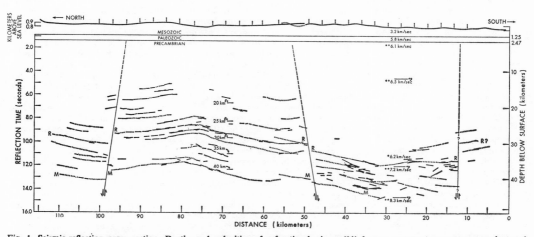

Fig. 1. Seismic-reflection cross section. Depths and velocities of refracting horizons (**) fom an east–west survey are superimposed at the positions of intersection. The *6.2-km/sec velocity is an average vertical velocity between the top of the Paleozoic and the Riel (*R*) discontinuity as obtained from reflection data (*2*). The profile is located in Fig. 2.

the more recent Rocky Mountain system. Ages determined (4) from samples from wells into the Precambrian crystalline basement in Alberta make it clear that the rift cuts across Churchill province which is 1.5 to 2.0 × 10⁹ years old. Before deposition of the Cambrian clastics the rift must have been filled with 11 km of sediments in a narrow sea extending from the ancestor of the Pacific Ocean through Idaho, British Columbia, and Alberta into Saskatchewan. Precambrian sediments are exposed over the western section of the proposed rift zone in unusual thicknesses; the Purcell and Belt series vary from a minimum of 11 km near Kimberley to about 15 km in northern Idaho (5). A small section of the rift probably remained as a depositional basin near Cranbrook into the Middle Cambrian (6). Thrusting during formation of the Rocky Mountains may have moved some of the rocks of the Belt and the Paleozoic eastward from their original positions, but this possibility does not affect my correlation with ore deposition, because any movement must have taken place largely along the strike of the magnetic-anomaly trends in British Columbia.

There are two broad areas in which positive gravity anomalies intrude within the confines of the postulated rift valley, both of which are accounted for by available data: the first, southwest of Cranbrook, British Columbia, is associated with the Moyie Intrusion, a group of thick sills of dense basaltic lava of Precambrian age; the second is over a prominent gravity feature called the Princess High, northeast of Brooks, Alberta. A wildcat well drilled to this feature (6) indicates that it was an island that protruded from the Middle Cambrian seas; probably it is the remnant of a basic volcano and was eventually covered by Upper Cambrian clastics.

Because Sullivan (at Kimberley) and Coeur d'Alene ore bodies appear to lie within the Precambrian rift zone, it is natural to investigate a possible relation between formation of these ore bodies and igneous activity associated with the development of a rift. I shall briefly review the evidence. The Sullivan ore body is an example of a strata-bound or conformable deposit (7). A hydrothermal theory, according to which thermal waters transport the metals from a magma at depth, has been postulated by some geologists, while others favor a syngenetic origin in which deposition of sulfides is contemporaneous with sedimentation.

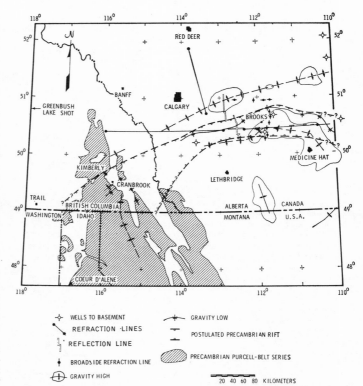

Fig. 2. The postulated trace of the Precambrian rift as determined by geophysical evidence. Reflection and refraction profiles and prominent Bouguer gravity anomalies are located, together with Precambrian outcrops in the area.

Other deposits in this category are the lead-zinc mines in Australia (Mount Isa, Broken Hill, Captain's Flat, Read Roseberry, Hall's Peak), Canada (Bathurst, Buchans, Lake Geneva, Yukon Treadwell, Manitouwadge), and Austria (Bleiberg), and the copper ores in Rhodesia. The lead from the strataform deposits often has isotope ratios that are very uniform within each district, and are related on a worldwide basis by a simple physical model, with ages correlating with an early stage in the development of crustal material in the area (8). A single average growth curve has been shown to fit all lead-isotope ratios of ordinary leads (8) within 1 percent. It is argued (9) that only leads not having traversed crustal rocks are most likely to satisfy the requirements for an ordinary lead, and Stanton suggested that many lead-zinc-copper ores are of syngenetic origin, deriving from volcanoes, and were concentrated in nearshore sediments around islands in an active island arc (10).

The syngenetic hypothesis, with biogenic reduction in an island-arc environment, has been criticized by many geologists more familiar with individual mines. Sulfur-isotope measurements on minerals from these deposits (11) suggest a more variable mechanism, and furthermore the model lead ages are inconsistent with the ages of the rocks in the general areas of many of these deposits (12). Despite the difficulties with the postulated geological environment and the mechanism for concentrating the metal, the evidence of a unique growth curve for ordinary leads has increased as more-precise mass spectrometry has become available (8, 12). A deep source is necessary for the worldwide uniformity, over a long period of time, of the parent uranium and thorium whose daughter products are lead.

The Red Sea is the site of an active rift zone as Africa and the Arabian peninsula break up and drift apart. In the deepest part of the Red Sea rift

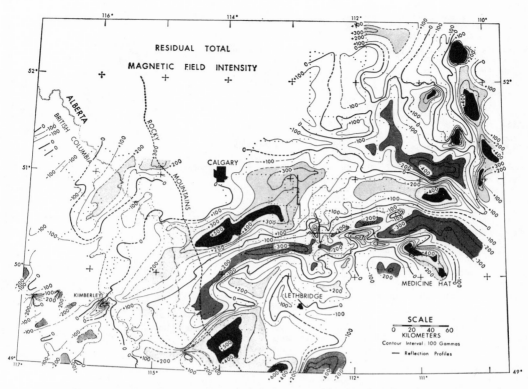

Fig. 3. Residual total-magnetic-field map for southern Alberta and southeastern British Columbia. This is formed by removal of the first six spherical harmonics from the raw data by use of a program from Lamont; essentially a slightly curved surface (due mainly to Earth's dipole field) is thus removed without distortion of magnetic anomalies due to geologic features.

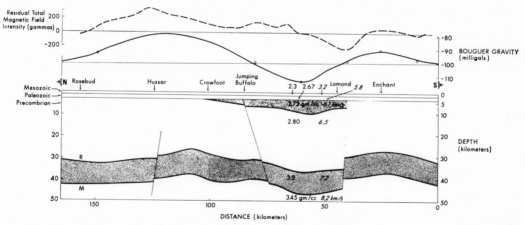

Fig. 4. Generalized crustal section across southern Alberta. Dashed curve, residual total-magnetic-field intensity. Solid curve, Bouguer gravity profile; the circles show the fit of the proposed crustal model depicted at the bottom.

valley have been found (*13, 14*) three pools of hot acidic brines with heavy-metal concentrations as high as 50,000 times those in normal ocean waters. Red Sea water is normally 20°C at *p*H 8.5; the deep brines at 2000 m may be as hot as 56°C at *p*H 4. Under the hot pools is a sedimentary layer consisting mainly of metal oxides and sulfides that is 10 to 100 m thick (*14*). The pools appear to be large natural chemical laboratories in which metals that have migrated from the upper mantle, through faults, are concentrated and precipitated, from colloidal suspension, on the sea floor. The Coeur d'Alene mining district was used (*14*) to show that the value of ore was probably greater in the Red Sea sediments.

The graben structure of and a similar origin for Coeur d'Alene have been discussed (*15*), as have the subsequent diagenesis, metasomatism, and metamorphism (*16*). Areas similar to the Red Sea pools have been found over the East Pacific Rise and its extensions into continental North America (*17*).

I suggest that the strata-form deposits at Kimberley originated within the bounds of the Precambrian rift that crosses western Canada. The method of ore concentration may have been similar to that now occurring in the Red Sea hot brines. Lead isotopes from the Sullivan body (*18*) yield an age of 1340 ± 50 million years for the time of mineralization. Model lead ages of 1340 to 1500 million years have been calculated by me from older measurements (*19*) from nearby in British Columbia and from the Coeur d'Alene district. A syngenetic origin is supported by a rubidium-strontium isochron age of 1315 ± 35 million years (*20*) from rocks from the lower part of the

Belt Series in Montana; the intercept of the isochron has an initial $Sr^{87}:Sr^{86}$ ratio of 0.7075, which is consistent with a subsialic or mantle origin for the strontium (*21*).

Thus there is evidence that strata-form deposits, and possibly some having characteristics of a hydrothermal origin, were formed in active rift valleys or in areas under so much tension that fractures extended into the mantle to form channels for mineralizing solutions. The lead-isotope ratios for such deposits may be interpreted with a simple single-stage model (*11*). Measurements (*22*) on a sample from the Red Sea brine shows that the model lead age is zero as it should be. However, if the crust is very old and thick, some modification of the lead-isotope ratios may be anticipated because of contamination with lead formed in a sialic environment having highly variable lead, uranium, and thorium ratios. If rifting and the formation of hot pools were under a deep sea, the sulfur-isotope ratio would probably be close to the meteoritic value, since only chemical fractionation is involved. Under shallow seas, biogenic activity may be important, and the sulfur ratios may show large amounts of fractionation.

Perhaps the most significant finding is that the seismic-reflection method may be used to discover and map an ancient rift zone. In the search for oil it is used indirectly to locate suitable structural or stratigraphic traps. If major ore deposits lie within the confines of rift zones, seismology may be successful in mining exploration also.

References

1. J. S. Steinhart and R. P. Meyer, *Carnegie Inst. Wash. Publ. 622* (1961).
2. E. R. Kanasewich and G. L. Cumming, *J.*
Geophys. Res. **70**, 3441 (1965); R. M. Clowes, E. R. Kanasewich, G. L. Cumming, *Geophysics*, in press.
3. N. A. Florensov, *Geol. Surv. Can. Paper 66-14* (1966), p. 173.
4. R. A. Burwash, H. Baadsgaard, Z. E. Peterman, *J. Geophys. Res.* **67**, 1617 (1962); E. R. Kanasewich, *Nature* **208**, 1275 (1965).
5. G. B. Leech, *Geol. Surv. Can. Paper 62-13* (1962); R. G. Yates, G. E. Becraft, A. B. Campbell, R. C. Pearson, in *Tectonic History and Mineral Deposits of the Western Cordillera*, Special Volume 8 of Canadian Inst. of Mining and Metallurgy (1966), p. 47; S. W. Hobbs, A. B. Griggs, R. E. Wallace, A. B. Campbell, *U.S. Geol. Surv. Profess. Paper 478* (1965).
6. H. van Hees and F. K. North, in *Geological History of Western Canada* (Alberta Soc. of Petroleum Geologists, Calgary, 1965), chap. 3, p. 22.
7. A. C. Freeze in *Tectonic History and Mineral Deposits of the Western Cordillera* (Can. Inst. Mining Met. spec. vol. 8, 1966), p. 263.
8. C. B. Collins, R. D. Russell, R. M. Farquhar, *Can. J. Phys.* **31**, 402 (1953); R. G. Ostic, R. D. Russell, R. L. Stanton, *Can. J. Earth Sci.* **4**, 245 (1967).
9. R. L. Stanton and R. D. Russell, *Econ. Geol.* **54**, 588 (1959).
10. R. L. Stanton, *Can. Mining Met. Bull.* **63**, 22 (1960).
11. H. R. Krouse, A. Sasaki, E. R. Kanasewich, "Sulfur-isotope results from Captain's Flat, Cobar and Bathurst," unpublished; L. J. Lawrence and T. A. Rafter, *Econ. Geol.* **57**, 217 (1962); P. J. Solomon, *ibid.* **60**, 737 (1965).
12. E. R. Kanasewich, *Radiometric Dating for Geologists*, R. M. Farquhar and E. R. Hamilton, Eds. (Wiley, New York, in press).
13. A. R. Miller, *Nature* **203**, 590 (1964); H. Charnock, *ibid.*, p. 591; J. C. Swallow and J. Crease, *ibid.* **205**, 165 (1965); J. M. Hunt, E. W. Hays, E. T. Degens, D. A. Ross, *Science* **156**, 514 (1967); D. A. Ross and J. M. Hunt, *Nature* **213**, 687 (1967).
14. E. T. Degens and D. A. Ross, *Oceanus* **13**, 24 (1967).
15. R. A. Anderson, *Econ. Geol.* **62**, 1092 (1967).
16. B. W. Robinson and R. G. J. Strens, *Nature* **217**, 535 (1968).
17. K. Bostrom and M. N. A. Peterson, *Econ. Geol.* **61**, 1258 (1966); B. R. Doe, C. E. Hedge, D. E. White, *ibid.*, p. 462.
18. A. J. Sinclair, in *Tectonic History and Mineral Deposits of the Western Cordillera*, Special Volume 8 of Canadian Inst. of Mining and Metallurgy (1966), p. 249.
19. R. M. Farquhar and G. L. Cumming, *Roy. Soc. Can. Trans.* **48**, 9 (1954); A. Long, A. J. Silverman, J. L. Kulp, *Econ. Geol.* **55**, 645 (1960).
20. J. D. Obradovich and Z. E. Peterman, in *Proc. Conf. Geochronology of Precambrian Stratified Rocks Univ. of Alberta* (1967), p. 75.
21. H. W. Fairbairn, P. M. Hurley, W. H. Pinson, *J. Geophys. Res.* **69**, 4889 (1964).
22. M. H. Delavaux, B. R. Doe, G. F. Brown, *Earth Planetary Sci. Letters* **3**, 139 (1967).

11

Copyright © 1968 by The Institution of Mining and Metallurgy

Reprinted from *Trans. Instn Min. Metall.* 77:B117–B128 (1968)

Structural controls of base metal mineralization in Ireland in relation to continental drift

M. J. Russell B.Sc., Stud.I.M.M.
Department of Geology, University of Durham

551.241 : 553.43 /.44(417)

Synopsis

A hypothesis is presented that suggests that both the genesis and the siting of the newly discovered base metal deposits in the Carboniferous rocks of Ireland are controlled by the intersection of north–south upper mantle fissures with east–west to northeast–southwest faults of Caledonian trend. It is proposed that the intrusion of magma at these intersections gave rise to a convective or a partial convective system within the pore waters which leached metals from the Lower Palaeozoic geosynclinal sediments and precipitated them in the overlying Carboniferous rocks or on the Carboniferous sea floor. The formation of the north–south upper mantle fissures parallel to the continental margin is believed to relate to continental rifting in Devonian and Carboniferous times. A similar origin is ascribed to the comparable base metal deposits occurring in the Lower Carboniferous rocks of the Maritime Provinces of Canada.

Concentrated investigation of these intersections may result in further ore discoveries.

Four large base metal deposits have been discovered in Ireland in the past seven years—at Tynagh, Silvermines, Gortdrum and Riofinex at Keel.[2–4a] These discoveries occur in a variety of Lower Carboniferous lithologies. Host rocks comprise sandstones, limestone shales, muddy limestones, dolomites and a carbonate mud bank complex. These low-grade high-tonnage ores contrast with most of the previously worked small sulphide veins elsewhere in Ireland,[5] but show similarities with the strata-bound deposit at Abbeytown mine, County Sligo. The latter has been known for over 100 years and consists of sphalerite and galena disseminations in a Lower Carboniferous calcareous sandstone.[6] The mineral assemblages are typically those of low-temperature hydrothermal origin, but it is important to note the absence of fluorite in association with the ores. Four of the five large-tonnage, low-grade sulphide deposits— Abbeytown, Tynagh, Silvermines and Gortdrum—are distributed along an approximately north–south line.

Known sulphide ore deposits are widely separated, the minimum distance between any two being 18 miles. The extensive occurrence of Lower Carboniferous rocks in the Central Plain presents possibilities for further discoveries, but the boulder clay and peat cover necessarily limits exploration

The juxtaposition of Carboniferous Limestone with older rocks along east–west trending faults is significant in connexion with disseminated sulphide deposits in Ireland. Exploration companies have concentrated their investigations in such areas. Pereira[7] suggested that northeast–southwest sub-mantle fissures govern the distribution of ore deposits in Ireland, and also stressed the importance of the sedimentary environment in Lower Carboniferous times. A hydrothermal origin has been suggested by Derry and co-workers[2]

Manuscript received by the Institution of Mining and Metallurgy on 5 July 1968. Paper published on 10 August, 1968.

for the Tynagh deposit, which is associated with an east–west fault. Their theory postulated shallow replacement and early infilling of organic structures in a mud bank complex; Schultz,[8] however, favoured a normal hydrothermal origin of later date.

A sedimentary-exhalative theory was proposed by Gordon-Smith[4b] to explain the stratiform Silvermines zinc–lead orebody.

Several points from these theories have been incorporated in a theory of genesis that is capable of explaining the north–south distribution of the ore deposits. This proposes that deposition of minerals in Carboniferous rocks is controlled by the initiation of north–south fissures extending from the upper mantle, and their intersection with faults of Caledonian trend. Geosynclinal deposits underlying the Carboniferous are believed to have provided metals for these deposits. The initiation of continental drift contemporaneously with the formation of structures parallel to the continental margin adds support to the hypothesis.

Geological environment

Structurally, Ireland can be visualized as a large rhomb, with north–south and east–northeast trending sides, situated close to the continental margin. Gravity surveys have shown positive anomalies of the order of 20 mgal associated with the margins of the island.

The important mineralization occurs in the Lower Carboniferous rocks of the Central Plain of Ireland—an area of about 16 000 square miles.

The oldest rocks outcrop in the Ox Mountains and in Donegal in the northwest. They are metamorphic schists and gneisses similar to the Moinian of Scotland. Overlying these rocks are metamorphosed Dalradian geosynclinal sediments which also outcrop in the north and west. Lithologically, these sediments include black pelitic schists with lesser amounts of quartzite, limestone and metadolerite sills and lavas.

Lower Palaeozoic geosynclinal rocks outcrop in the east and west of Ireland and in the inliers of the Central Plain. They include shales, mudstones, silts and greywackes, together with andesitic and acid extrusives and minor intrusives. These rocks are folded, but have suffered only minor metamorphism. Some of the argillaceous rocks have developed slaty cleavage parallel to the axes of folding. The general trend of the axes of folding is sub-parallel to that of the two older groups of rocks; this direction is northeast–southwest in the east, changing to approximately east–west in the west. These Lower Palaeozoic rocks presumably extend under the Central Plain; Murphy[9] suggested a vertical thickness of the order of 10 000 ft for them. Caledonian granite intrusions outcrop in the east, north and west of Ireland.

Old Red Sandstone overlies the Lower Palaeozoic

rocks with strong unconformity. The thickness is not known in the Central Plain, but is less than 1000 ft on the anticlinal inliers. In the south of Ireland the Upper Old Red Sandstone is often enriched in chalcopyrite and bornite and this is most marked in the Kiltorcan sandstone and conglomerates. Andesitic volcanic activity is

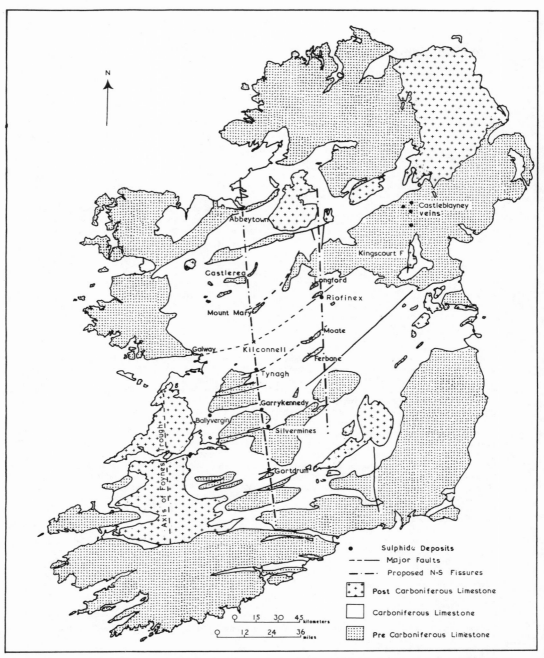

Fig. 1 Geology of Ireland (from the Geological Survey Map of Ireland[44]) with the postulated north–south fissures and faults of Caledonian trend. (Fault running northeast from Garrykennedy, from Murphy[41])

evident in the north and the south, and is similar to the Devonian vulcanicity of Scotland. Charlesworth[10] noted a large number of north–south trending felsite dykes in the northern half of the country.

Carboniferous rocks of shelf-sea facies overlie the Old Red Sandstone conformably, or with slight unconformity, in the southern part of the Central Plain. They consist of limestone, shales and sandstones passing up into bedded muddy limestones and dolomites overlain by a calcareous mud bank complex of upper Tournaisian and lower Viséan age. The Lower Carboniferous Sea transgressed slowly upon the land to the north in upper Tournaisian times and finally covered the northern half of the plain by the early Viséan age. Here the Carboniferous sandstones and limestones of C_1 or C_2S_1 age lie unconformably on Old Red Sandstone and Lower

Palaeozoic rocks and, in the northwest, on the Precambrian metamorphic rocks. In the central and eastern districts of this area the diachronous Waulsortian mud bank complex[11] overlies lagoonal limestones of lower Viséan age.

Tuffaceous deposits have been recorded in the Lower Carboniferous rocks of some of the mine areas and may be present throughout the central and southern regions of the plain. The most important lavas and sills occur at Pallas Green, County Limerick, and the feeding vents outcrop to the south along the Gortdrum Fault. The early flows are trachytes, trachyandesites, trachybasalts and olivine basalts of C_2S_1 age. The later igneous rocks of D_1 age also include tuffs and vent agglomerates. These igneous rocks are similar to those of the same age seen in the Midland Valley of Scotland.

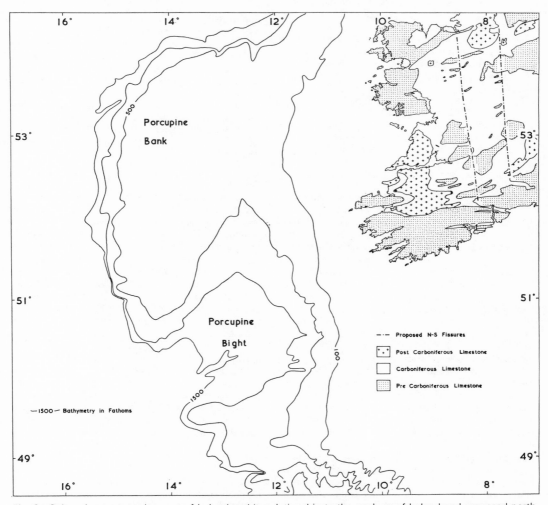

Fig. 2 Submarine topography west of Ireland and its relationship to the geology of Ireland and proposed north–south fissures. (Bathymetry from National Institute of Oceanography Maps, Area 16)

It is possible that east–west faults associated with the base metal deposits were active in Lower Carboniferous times; in support of this Derry and co-workers[2] described some thinning of Lower Carboniferous beds towards the Tynagh fault. There is no stratigraphic evidence to support a theory that the inliers in the Central Plain were large upstanding islands in the Lower Carboniferous Sea, but there may have been some minor emergence. Major movements along the east–west faults were of a later date.

Upper Carboniferous shales, grits and coals occur in County Antrim, County Kilkenny and in the Foynes area to the west. The succession is 3500 ft thick at Foynes and has been deposited in a north–south trending trough (Fig. 1).

Most of the structures in Ireland have previously been related to Caledonian and Armorican earth movements. The Armorican disturbance is envisaged as giving rise to east–west structures in the south and reactivating structures parallel and normal to the Caledonian trends elsewhere. These movements do not adequately explain the many north–south features which include faults, dykes, the predominant joint direction and the Foynes trough.

To the west of Ireland certain major submarine features are also predominantly north–south (Fig. 2): (1) the western edge of the Porcupine Bank; (2) a gravity survey by Gray and Stacey[12] revealed a steep gradient trending north–south associated with the western margin of Porcupine Bank; and (3) the strike of the maximum gravity anomaly off the west coast of Ireland is N 8° W.[12]

The theory of continental drift is now accepted by many geologists. If the validity of this process is assumed, these north–south structures take on a new significance. From the fit of North America and Europe proposed by Fitch[13] the continuation of the Caledonian system is apparent. Westoll[14] believed that the first fractures in the formation of the present North Atlantic Ocean developed between Lower and Middle Old Red Sandstone times. Evidence for this fracturing is provided by the sinistral transcurrent movement of 100 km along the Great Glen Fault, which can be dated as post Lower and pre Middle Old Red Sandstone by stratigraphic means. Such a movement depended upon the opening of a rhombo-chasm, or large rift, between Norway and Greenland. As additional evidence Westoll[14] cited the occurrence of an *Ostracoderm* fauna in a unique development of non-marine sediments of Middle and Upper Devonian age, deposited along the line of the fracture system. This fauna is similar to that of Spitzbergen, but dissimilar to that of the rest of the British Isles. The 30° sinistral rotation of Newfoundland Island during the late Devonian,[15] together with transcurrent movement along the Cabot fault system of similar age,[16] would suggest that there was major movement on both sides of the Atlantic at this time. The splitting of the continents in the North Atlantic region took advantage of Caledonian trends, except in the area west of the British Isles. In this area there appears to have been a crossover from the northwest to the southeast side of the Caledonian orogenic belt (Fig. 3). From the trend of the continental margin west of Ireland the split was apparently north–south. The rifting may have been governed by structures normal to the Caledonian trend, which appears to have swung round to an east–west direction west of Ireland. The split may be the southerly extension of the Greenland–Newfoundland break along the trend of the Ketilides of southwest Greenland[17] (Fig. 3). Structures parallel to this major split were formed during the later Devonian and Carboniferous periods—possibly in response to east–west tension.

Description of base metal deposits

Abbeytown

At Abbeytown mine sphalerite, galena and pyrite are disseminated in, and replace, a calcareous sandstone 24 ft thick.[6] They also partially replace the limestone above and below this sandstone. The limestones are dolomitized in the vicinity of the ore. The sandstone may be equivalent to the Mullaghmore Sandstone of S_2 age[18] which outcrops to the northeast. The ore deposit lies between two east–west faults which have brought the Carboniferous Limestone against schists of Moinian type. These same schists presumably underlie the limestone.

Tynagh

At Tynagh mine fine-grained pyrite, galena, sphalerite and some chalcopyrite, often intimately mixed with barytes, replace and infill a Waulsortian mud bank complex of C_1 to C_2S_1 age. The complex lies adjacent to an east–west fault, with a throw of 1200 ft to the north which brings Old Red Sandstone against the ore deposit. A bedded iron formation,[19] associated with tuffs, lies to the north of the sulphide deposit at the same stratigraphic horizon. Schultz[19] suggested that this iron deposit was derived by intensive chemical weathering of lower Dinantian sediments that had been exposed on newly emergent land. Derry and co-workers,[2] however, regarded the ore deposit as a precipitate from metal-bearing solutions, possibly arising along the east–west fault and seeping into an area of recent organic growth of the mud bank complex. The silica, manganese and much of the iron, however, remained in solution and were deposited in a protected basin off the reef. The green tuff bands interbedded with the iron ore point to simultaneous local volcanic activity. The banded and colloform texture of the sulphide suggest either penecontemporaneous deposition or early diagenetic replacement with or within the mud bank complex. The thinning of some of the Carboniferous formations towards the fault led Derry and co-workers[2] to conclude that movement continued throughout ore deposition.

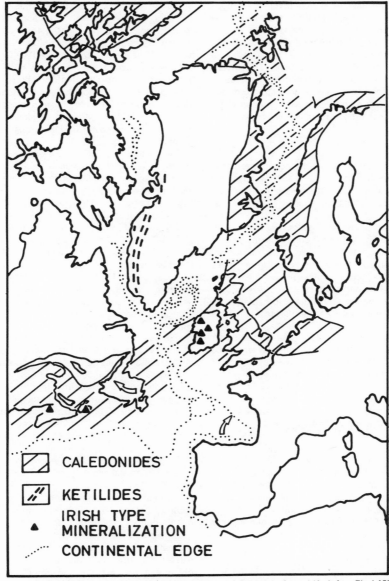

Fig. 3   Reconstruction of the North Atlantic continents before drift (after Fitch[13]) illustrating relationship of Irish-type mineralization to the continental margins and the Caledonian geosynclinal rocks. (Ketilides from Dearnley[17])

Silvermines

The Silvermines deposit[1] consists of galena, sphalerite, pyrite and barytes extending 2 miles along a complex east–west fault and associated east–northeast and north–south trending faults. Ore is also seen as disseminations in the lower dolomite of Tournaisian age. The ore is rich in silver, containing between 6 and 80 oz/ton. The fault has a throw of between 800 and 1100 ft to the north and brings Old Red Sandstones, overlying Silurian shales, in contact with the Carboniferous limestones. The largest deposit, however, is the newly discovered zinc–lead *G* orebody to the north of the fault. This is a stratiform deposit lying between a muddy reef limestone and the overlying dolomite breccia of lower Viséan age. A large strata-bound deposit of cryptocrystalline barytes occurs at the same horizon just to the east at Ballynoe mine. Gordon-Smith[4b] suggested that the *G* orebody and the barytes deposit result from sedimentary precipitation of sulphides and sulphates at different redox potentials—the barytes in oxidizing conditions and sulphides in a shallow basin with reducing conditions. The metal-bearing solutions may have emanated from the east–west fault. Colloform and sedimentary structures are common in the *G* orebody. The only evidence for volcanic activity is the footwall shale, which may be a tuff band.

Gortdrum

Bornite, tetrahedrite and chalcopyrite occur as replacements and fracture fillings in Tournaisian muddy limestones and limestone shales to the north of a fault throwing Old Red Sandstone against Carboniferous limestones.[3,4a]

This replacement deposit is associated with feldspar–quartz porphyry dykes. Four intrusions, two of which are probably volcanic necks, outcrop further west along the same fault and also comprise feldspar–quartz porphyry. The volcanic necks are the feeders for the Pallas Green volcanic series of lower Viséan age. A consideration of the spatial relationship between the ore and the intrusions implies a lower Viséan age for this deposit.

Veins containing galena, sphalerite and barytes occur at Oola Hills, half a mile north of Gortdrum. A lead–zinc deposit at Carrickittle, 5 miles to the west of Gortdrum mine, occurs in the Waulsortian Limestone and is also associated with a quartz–feldspar porphyry intrusion. This apparent zonation of a copper deposit surrounded by lead–zinc deposits may have been repeated vertically, though the evidence has been removed by erosion.

Riofinex

Galena, sphalerite, pyrite and barytes occur in Silurian banded grits and mudstones, Devonian sandstone and conglomerates and the Tournaisian silts, limestones and shales.[3,4a] The massive and disseminated mineralization is associated with an east–west fault which brings Silurian rocks overlain by Old Red Sandstone against sandstones and limestones of Lower Carboniferous age.

General features of Irish ore deposits

The low-grade high-tonnage ore deposits in Ireland consist of a low-temperature suite of sulphides; sphalerite, galena, pyrite and, less commonly, copper sulphides, often accompanied by barytes. Wallrocks are dolomitized in some cases, although dolomitization is not necessarily accompanied by mineralization. Silver of the order of 2 oz/ton of ore is present in most of the deposits, although it is appreciably higher in the vein deposits of Silvermines. Usually, the sulphides are fine-grained and often intimately mixed. Sphalerite, galena and pyrite exhibit colloform textures in some deposits.

There is no evidence to suggest one source of ore at depth for all the deposits—rather, the zoning within some of the mines, for example Tynagh and Gortdrum, points to an individual source for each area of mineralization. Igneous activity in the form of tuff bands is evident in some of the deposits and in the south the sulphides are intimately associated with feldspar–quartz porphyry intrusions. Approximately east–west trending faults which bring Old Red Sandstone or older rocks into contact with Carboniferous Limestone are a feature of each deposit. These faults are parallel to the axes of Caledonian folding. Movements along some of these faults during the Lower Carboniferous can be demonstrated.

The evidence at Tynagh, Silvermines and Gortdrum points to a period of major mineralization during late Tournaisian and early Viséan times. This coincides with the time of major igneous activity in the Carboniferous in which basic and intermediate lavas were extruded.

Lead isotope studies on these Irish ores by Moorbath[20] and Pockley[21] gave age dates concentrating in the Carboniferous, but the limited precision of these results (from ±40 to ±90 m.y.) renders them of little value in the present context.

Abbeytown, Tynagh, Silvermines and Gortdrum mines are distributed along an approximately north–south line, subsequently referred to as the 'Abbeytown–Gortdrum line'. The trend is N 7° W (Fig. 1). The Silvermines and Abbeytown deposits occur in Carboniferous rocks faulted against large inliers of older rocks. Tynagh and Gortdrum deposits are associated with small Old Red Sandstone inliers. Adjacent to this line are Old Red Sandstone and Lower Carboniferous Sandstone inliers at Mount Mary and Castlerea. The Riofinex deposit is also spatially related to an Old Red Sandstone inlier. A line drawn approximately parallel to the Abbeytown–Gortdrum line through the Riofinex deposit and adjacent to the eastern extremities of three other Old Red and Carboniferous Sandstone inliers near Longford, Moate and Ferbane, runs N 3° W and is referred to subsequently as the 'Riofinex–Ferbane line' (Fig. 1).

The major structural directions, except in the extreme south of Ireland, are Caledonian and north–south The distribution of major sulphide deposits appears to be governed by these same trends.

Smaller sulphide deposits in eastern Ireland also exhibit a north–south distribution: this is especially obvious in the Castleblayney area (Fig. 1), where epigenetic vein mineralization in Silurian rocks occurs on the possible northerly extension of the Kingscourt Fault. Veins containing sulphides in the Ballyvergin area, County Clare,[5] also have a north–south trend. According to Murphy,[22] late north–south fractures at Avoca mine, County Wicklow (a conformable copper–lead–zinc and iron sulphide deposit in Ordovician rocks), contain enough remobilized chalcopyrite to make small-scale stoping worth while.

Deposits similar to the Irish type occur in the Maritime Provinces of eastern Canada, the largest being at Walton, Nova Scotia. In this area rocks underlying the

Carboniferous are slates, argillites and quartzites of the Meguma Series, of Lower Palaeozoic age. Lower Carboniferous sandstone, shales, ferruginous limestones and conglomerates of the Horton and Cheverie Formations lie with strong unconformity on the Meguma Series. Boyle[23] noted the generally high sulphur, lead, zinc, copper and barium contents of the Horton—Cheverie group. According to Bell,[24] these rocks are equivalent to the limestones, shales and sandstones of Tournaisian age in Ireland. Fissile limestones, limestone conglomerates, and shales of the Windsor Group, which correspond to the Viséan rocks of Ireland, overlie the Horton—Cheverie Formations.

The barium—lead—zinc—silver deposit at Walton consists of fine-grained sphalerite and galena overlain by cryptocrystalline barytes. It is approximately stratiform and was formed by replacement and fracture filling of the lower Windsor limestones and the underlying Cheverie sandstones and shales.[23] The deposit is localized in a brecciated zone at the intersection of two major fault zones. The directions of these faults are east—west and northeast—southwest. The age of mineralization is unknown, but the stratigraphic and structural location of the deposit is similar to that of the Irish deposits.

The large-tonnage Irish deposits contrast strongly with the lead—zinc deposits of the English northern

Pennines.[25] The main differences between the two are shown in Table 1. The occurrence of sulphide deposits at the junction between the inner fluorite zone and outer barytes zone in the Pennines suggests that sulphide deposition only occurred after removal of a substantial proportion of the fluorine from the ore solutions. There was an estimated 20 000 000 tons of fluorite in the north Pennine ore field. The Pennine deposits are similar to those of the Erzgebirge,[26] where the source of mineralization is related to Hercynian granites and the fluorite zone is again well developed.

Fluorine is known to concentrate in the later-stage differentiates of nepheline syenites and granites and hence is involved in metasomatism and in vein mineralization related to the crystallization of the final phase of this type of magma. The absence of fluorite in the large base metal deposits in Ireland puts a magmatic hydrothermal source theory for these sulphides in some doubt.

Another possible source of the metals in the deposits is the trace concentrations in the underlying Lower Palaeozoic geosynclinal rocks. The possibility that trace quantities of metals in rocks may be dissolved by circulating thermal waters and eventually precipitated in a favourable environment for deposition is gaining in popularity. Petrascheck[27] believes that the lead, zinc and manganese deposits on each side of the Red Sea may be best explained by secondary hydrothermal derivation of metals from deeper levels.

Work on the Salton Sea brine well waters[28] is especially relevant in this respect. These brines contain high concentrations of lead, zinc, barium, iron and manganese and, according to oxygen and hydrogen isotope studies by Craig,[29] are meteoric in origin. A radiogenic tracer study[30] of lead and strontium in these deep thermal brines indicated that 80–100 per cent of the strontium and 50–100 per cent of the lead were acquired from sediments underlying the Salton Sea. The investigation also shows that the magma that formed the local rhyolites could not be the source of the lead. The large geothermal anomaly in the area suggests that a magma chamber may exist at depth.

Theory of genesis of Irish ore deposits
During Lower Carboniferous times approximately north—south fissures were formed in response to incipient continental splitting. At least two of these fissures underlie Ireland—the Abbeytown—Gortdrum and the Riofinex—Ferbane fissures. The continuity of these fissures suggests they extend from the upper mantle. The fissures intersected east—west to northeast—southwest faults formed or reactivated by early Armorican movements. Magma rose along these fractures reaching a high level in the crust at the fracture intersections and along east—west fractures in the south, where the early Armorican movements were more intense. At times some of the magma reached the surface. These volcanics and intrusives were basic and intermediate in type. Intrusions centred on the intersections acted as hot

Table 1 Comparison of Irish and north Pennine mineralization

Characteristics	Irish deposits	North Pennine deposits
Host rocks	Carboniferous sandstone, limestone shales, muddy limestones, Waulsortian reef	Carboniferous sandstone, shales and limestones
Basement	Folded geosynclinal rocks	Older granite
Age	Lower Carboniferous?	Hercynian? (Moorbath[20])
Attitude	Conformable and irregular	Veins and flats
Grain size	Fine	Coarse
Principal metal sulphides	Pb, Zn, Fe, Cu	Pb, Zn (Fe, Cu)
Principal gangue minerals	Barytes	Barytes, fluorite and ankerite
Structural control	Associated with widely separated east—west faults of large throw	Many faults of small throw
Type	Low-grade, high-tonnage	High-grade, low-tonnage
Zonation	Poor	Marked
Areal extent of mineralization from one source	Restricted	Large

spots' giving rise to a convective or partial convective system within pore waters in the Lower Palaeozoic geosynclinal rocks. Heat was maintained by the addition of juvenile water from the magma and by continued magmatic activity. Water, containing base metals dissolved from clay particles, organic matter and fine authigenic sulphide while permeating through the Lower Palaeozoic rocks, rose along the fracture intersections and to some extent along the east–west faults. The precipitation of metals from this water was dependent upon a variety of factors, including release of pressure, lowering of temperature and the reactivity of wallrocks. Thus although the intersections governed the passage of the rising fluids, the resultant ores occur as replacement or sedimentary deposits close to, but not necessarily directly above, the intersections.

Similar fractures parallel to the continental margin are present in the Maritime Provinces of Canada. Replacement deposits are found in association with these fractures in overlying Lower Carboniferous rocks. In this western counterpart of the Irish base metal province the comparatively small extent of Lower Carboniferous rocks accounts for the infrequent occurrence of sulphide deposits of this type. Where the basement consists of relatively impermeable metamorphosed rocks poor in adsorbed ions, the deposits will be small.

The coincidence in time of earth movements, and the occurrence of similar types of mineral deposits within the same stratigraphic horizon and comparable structural conditions on either side of the Atlantic, is consistent with the hypothesis that these ores were generated as a consequence of the process of continental drift.

Geochemical considerations

Experimental work, accompanied by thermodynamic calculations by Helgeson,[31] demonstrates that base metals may form water-soluble chloride complexes. Barnes and Czamanske[32] believe that metals may also enter solution as sulphide and bisulphide complexes at least up to a temperature of 250°C.

The source of the solutions giving rise to Irish mineralization is believed to be broadly similar to that of the Salton Sea area. In this case the underlying rocks are mainly Lower Palaeozoic geosynclinal rocks. To establish the concentration of metals in these rocks 63 samples of Ordovician and Silurian sediments and volcanics were collected throughout Ireland, away from known mineralized areas, and analysed with a Philips X-ray fluorescence spectrograph, PW 1540. The results of these analyses are presented in Table 2, with precision data and analyses of the standard rocks G_1 and W_1. The standard deviations from the mean values are high and the results of this reconnaissance survey demonstrate only the general order of magnitude of trace elements in these rocks. Silver was not determined, but Taylor[33] quoted a Clarke value in greywackes and shales of 0·05 ppm and a similar value is assumed for these sediments. Greywackes and silts

constitute the bulk of the Lower Palaeozoic rocks in Ireland.

Table 2 Minor-element concentrations in some Lower Palaeozoic rocks from Ireland (parts per million)

34 samples of shales and greywackes				
	Pb	Zn	Cu	Ba
Range	<17–71	8–288	<8–82	111–1270
Mean	22	109	51	697
Standard deviation	±16	±53	±19	±58
Experimental precision on ten replicate determinations				
Mean	20	105	57	720
Standard deviation	±4	±2	±4	±27
Relative deviation	20%	2·2%	7·0%	3·8%
29 samples of volcanics and intrusives				
	Pb	Zn	Cu	
Range	<15–62	5–130	<5–82	
Mean	~18	69	24	
Standard deviation	±14	±42	±22	
Experimental precision on ten replicate determinations				
Mean	33	18	5	
Standard deviation	±4	±2	±2	
Relative deviation	12·4%	9·8%	38%	
G_1 analyses	51	52	17	
G_1 recommended (Fleischer[43])	49	45	13	
W_1 analyses	~7	82	124	
W_1 recommended (Fleischer[43])	8	82	110	
Order of magnitude of concentration, presuming that the sediments make up 80% of the Lower Palaeozoic rocks				
	Pb	Zn	Cu	Ba
	20	100	45	(700)

The high thermal gradient caused by magmatic intrusion would initiate movements of pore waters towards the convective upcurrent. This movement would take advantage of cleavages in the argillites and silts and of porous and permeable sediments. The volume of Lower Palaeozoic rocks contributing metal would be elongate in the Caledonian fold direction and more narrow normal to the axis of folding. Two models consistent with these considerations are presented. Both assume uniform leaching in parallelepipeds of vertical thickness 1 mile. The top of these source volumes lie about 1000 ft below the Old Red Sandstone–Lower Palaeozoic unconformity.

The first is 10 miles long in the direction of the Caledonian axis and 6 miles wide. The vertical axis of this model corresponds to the intersections of the north–south fissures with the faults of Caledonian trend. The ends and sides of this figure are thus, respectively, 5 and 3 miles from the intersection at the shortest point. Assuming orders of magnitude for lead, zinc, copper and barium in Table 2, the following expressions may be used to calculate the percentage of each element required to form an ore deposit.

$$F = \frac{\text{wt of metal in ore deposit}}{\text{wt of source rock}}$$

Hence F is the proportion of source rock required to form the ore deposit; then the percentage of metal leached from the source volume $= \dfrac{F}{C} \times 100$, where $C = $ concentration of metal in source rocks. The proportion of the lead required to form the Tynagh ore deposit may be calculated as follows.

> Wt of lead in ore deposit = 600 000 tons (Table 3)·
> Wt of source rock = $6 \cdot 72 \times 10^{11}$ tons, calculated from a source volume 10 miles × 6 miles × 1 mile with specific gravity of $2 \cdot 73$ (Murphy[9])
> $F = 6 \times 10^5 / 6 \cdot 72 \times 10^{11} = 0 \cdot 89$ ppm
> Amount of lead in source rock = 20 ppm (Table 2)
> Percentage of lead required for ore deposit
> $= 0 \cdot 89/20 \times 100 = 4 \cdot 5\%$

Estimates of weights of various metals are presented in Table 3; they are approximations calculated from published reserves and mined ore and will therefore be underestimates of metal contributed by the upwelling solutions. Assuming the validity of this model, the formation of the Tynagh deposit involves the solution of $4 \cdot 5$ per cent by weight of the total quantity of lead in the source volume, as shown above, $1 \cdot 8$ per cent of the silver, $0 \cdot 74$ per cent of the zinc, $0 \cdot 17$ per cent of the copper and $0 \cdot 42$ per cent of the barium.

Table 3 Approximate tonnages of elements in various mineral deposits

Deposit	Pb	Zn	Cu	Ag	Ba
Abbeytown	50 000	50 000			
Tynagh	600 000	500 000	50 000	600	2 000 000
Silvermines	500 000	1 000 000	—	500	?
Gortdrum	—	—	60 000	90	—

The second model has the same general shape, but a length of 6 miles and a width of 4 miles. The formation of the Tynagh deposit for this source volume of Lower Palaeozoic rocks would require leaching of $11 \cdot 2$ per cent of the total lead, $4 \cdot 4$ per cent of the silver, $1 \cdot 86$ per cent of the zinc, $0 \cdot 42$ per cent of the copper and $1 \cdot 06$ per cent of the barium. The Silvermines deposit has approximately twice as much zinc as the Tynagh deposit and, consequently, the quantity of zinc leached increases by a similar factor. More barium is probably needed to satisfy the Ballynoe barytes deposit. The Gortdrum deposit requires the solution of $0 \cdot 20$ per cent of the copper assuming the first model and $0 \cdot 51$ per cent for the second; the absence of lead and zinc is due to zonation of sulphides in the area. The Abbeytown lead–zinc deposit is small because of the low permeability and reactivity of the underlying quartz–mica schist and gneiss.

The factors governing the solution of these metals are their availability and solubility as complexes. Some of the lead, zinc, copper and silver may be present in authigenic and diagenetic sulphides. Barnes and Czamanske,[32] working on sulphide and bisulphide complexing, have shown that zinc, copper and lead are all soluble in hot waters as bisulphide complexes. These same elements may also be associated with clays and organic matter. The ionic radii and electronegativity of lead (Pb++, $r = 1 \cdot 20$ Å and $e = 1 \cdot 55$) and silver (Ag+, $r = 1 \cdot 26$ Å and $e = 1 \cdot 42$) preclude their incorporation into octahedral or tetrahedral sites in clay mineral structures. Any lead and silver not in the sulphide phase may be adsorbed on the clays and organic matter and is thereby readily available to percolating water. The ionic radius and electronegativity of zinc are the same as those of the ferrous ion ($r = 0 \cdot 74$ Å and $e = 1 \cdot 66$); the zinc is therefore camouflaged by the ferrous ion in clay mineral structures. Barium can substitute for potassium in potash feldspar and is strongly absorbed into intersheet positions in clay minerals. Barium is sparingly soluble as a sulphide which hydrolyses to a mixture of the hydroxide and hydrosulphide.

The general statements made above go some way towards explaining the preferential solution of lead and silver compared to zinc. The zinc in the clay lattice will only be released on the breakdown of the clay mineral.

Green[45] found the contents of fluorine in sandstones and shales to be 290 and 590 ppm respectively. This fluorine substitutes for (OH) groups in phyllosilicates and is therefore not readily available for solution.

The Salton Sea brines are different from the thermal waters envisaged in relation to Irish mineralization. The temperatures at the bottom of the 5000-ft well in the Salton Sea area were reported[34] as being between 300 and 350°C. At this depth the sedimentary rocks are metamorphosed to the low green schist facies. This metamorphism may explain the high zinc to lead ratio in the brines. Analyses of reservoir brines from two wells[34] gave the significant concentrations, in ppm, as lead, 84 and 80; zinc, 790 and 500; barium, 235 and 250; silver, $0 \cdot 8$ and 2; copper, 8 and 3; and fluorine, 15 and not reported. In Ireland the temperatures of the waters are thought to have been lower and the metals were derived by leaching, although the maximum amount of leaching probably took place along major fractures and over the intrusion.

The results and considerations of the geochemistry discussed above support the theory that the metals in Irish deposits were derived from the Lower Palaeozoic basement. The actual source volume, the shape of this volume and the degree of leaching will be different in different cases. An additional factor may be the high heat flow to be expected in continental margins in times of incipient drift. This heat flow may have driven water out of sediments at deeper levels than considered here.

Boyle[23] has noted high trace-element concentrations of lead, zinc, copper and silver in the grits, arkoses, shales and sandstones of the Tournaisian Horton and

Cheverie Formations of Nova Scotia, and suggested that the mineral deposits were formed by concentration of these trace elements. In southern Ireland the Upper Old Red Sandstone is rich in copper. The explanation favoured by the writer is that these metals were deposited as sulphides from mineralizing solutions originating in the convective system. Part of the upward current could escape through porous sandstones, slowly precipitating sulphides. Interaction with ground-waters may have accelerated the process.

Comparison of North Atlantic rifting with the Red Sea Graben

Westoll[14] compared the incipient drift in the North Atlantic area in Devonian–Carboniferous times with the Red Sea area today. The Red Sea Graben is envisaged as being analogous to the Greenland–Norway rhombo-chasm, while the transcurrent faulting involving sinistral movement of 100 km along the Jordan Rift Valley[35] is similar to movement along the Great Glen Fault. The Oligocene olivine basalt lavas in this area also compare with the Carboniferous lavas of the Midland Valley of Scotland. The analogy continues with the occurrence of deposits of lead, zinc, iron and manganese minerals on either side of the Red Sea. Deposits between El Qosieir and Bir El Ranga in Egypt[36] occur mainly as lenticular bodies in metasomatized basal Miocene limey grits. They have a linear distribution adjacent and parallel to the Red Sea Graben. Gindy[37] has shown that the high uranium content of these deposits could not have been derived from Tertiary basalts. He suggested that the labile uranium was leached from the Pre-cambrian rocks and overlying sediments by uprising fluids, believing that other metal ions were remobilized in this manner and precipitated in reactive lower Miocene rocks. The syngenetic zinc and iron sulphides, associated with hot brines, discovered recently in the Red Sea,[38] may also have been derived from under-lying sediments. Here the existence of basic intrusives has been proved by Drake and Girdler.[39] According to hydrogen and oxygen isotope work by Craig,[29] the waters are of meteoric origin. The intrusives may have been responsible for initiating a convective system in these brines.

The Irish base metal deposits are envisaged as having a similar genesis to the Red Sea deposits. Their forma-tion was also connected with the formation of a continental split, but their deposition has occurred further from the rift than the Red Sea ores. The Irish deposits are related to structures removed from the original rift by approximately 270 miles. Parallel mineralized structures at this distance from the Red Sea are not known.

Any deposits close to the centre of rifting in the North Atlantic, as is the mineralization near the Red Sea (and the mineralizing fluids of the Salton Sea, another area of embryonic drift), now lie on the edge of the conti-nental shelves or, more probably, have been removed by erosion.

Economic considerations of the hypotheses

The occurrence of sulphide ore deposits in Ireland, either as epigenetic or syngenetic bodies, is not signifi-cant in respect to the origin of the mineralizing solutions. The metal-bearing waters arose along the intersection of north–south fissures with faults of Caledonian trend. Precipitation of sulphides depends on various factors, and exploration programmes should take account of all possible depositional environments. The ores can occur in sandstones as well as limestones, and syngenetic deposits may occur at some distance from the inter-section. Further syngenetic deposits may yet be found in the proximity of known epigenetic deposits.

Exploration programmes should concentrate on intersections of faults of Caledonian trend with the postulated north–south fissures: two of these fissures are the Abbeytown–Gortdrum line and the Riofinex–Ferbane line. The newly discovered mineralization at Moate[40] lies on the Riofinex–Ferbane line, and was predictable by this hypothesis. Arcuate east–west to east–northeast faults are particularly continuous in Ireland. Murphy[41] proved a length of 80 miles for a fault running northeast from near Garrykennedy (Fig. 1). Other faults of Caledonian trend may be related to Old Red Sandstone inliers. Possible intersections are marked in Fig. 1; two intersections require fuller explanation. At Garrykennedy, on the southeast shore of Lough Derg in County Tipperary, galena occurs in veinlets on an east–west fault in Silurian silts.[5] Carboniferous Limestone outcrops a mile to the north on the opposite shore of the lake and also 1 mile to the east, and is presumably present on the lake bottom. In this context it is important to note that Silurian rocks adjacent to the Silvermines and Riofinex ore deposits are also mineralized. This mineralization may represent the roots of the sulphide deposits brought to the surface by subsequent move-ment along the east–west faults. The Garrykennedy lead showings lie on the Abbeytown–Gortdrum line and it is possible that a base metal deposit lies just north of Garrykennedy on the lake bottom. It may be that a deposit here would have been partially eroded by ice or river action in Pleistocene and Recent times.

The second area is at Kilconnell, County Galway, 8 miles west of Ballinasloe. The southerly shoreline of the Galway granite is straight at N 84° E. There is no gravity contrast across this line, but the writer believes this feature is a fault with downthrow to the south. If granite is faulted against granite no density contrast would be expected. The direction of the shoreline is parallel to faults in the south of the Galway granite and to other faults in western Ireland. An arcuate fault, sub-parallel with other faults of Caledonian trend and having a downthrow to the south, would extend eastwards from the Galway granite fault and intersect the Abbeytown–Gortdrum line 8 miles west of Ballinasloe. The extrapolation of this fault direction takes it to Riofinex, Keel, where a mineralized fault with southerly throw is known. There is no outcrop of Old Red Sandstone at Kilconnell, so any deposit may be deep, but it is significant that a younger limestone outcrops to the south of this postulated fault line.

The possibility that the Kingscourt Fault is a surface manifestation of a north–south fissure is worthy of investigation. The Castleblayney lead–zinc veins may lie on the northerly extension of this fault, as was mentioned previously. The Ballyvergin sulphide deposits[42] may also relate to a north–south fissure.

Sulphide deposits may be more frequent but smaller in the south, as a result of closer spacing of east–west faults near the Armorican front.

Acknowledgement

The writer wishes to record his thanks to Dr. K. C. Dunham for introducing him to this extremely interesting ore field and to Dr. R. L. Stanton, who initiated his interest in ore deposits. He is particularly grateful to Mr. A. P. Stacey for stimulating discussions on the geophysics and structure of the North Atlantic region, and is indebted to Dr. D. M. Hirst for encouragement, advice and suggestions and to Mr. R. Phillips for a critical reading of the manuscript. Finally, he would like to express his appreciation to the National Environment Research Council for financial support in the form of an industrial fellowship.

References

1. **Rhoden H. N.** Structure and economic mineralization of the Silvermines district, County Tipperary, Eire. *Trans. Instn Min. Metall.*, **68**, Dec. 1958, 67–94.
2. **Derry D. R. Clark G. R. and Gillat N.** The Northgate base-metal deposit at Tynagh, County Galway, Ireland: a preliminary study. *Econ. Geol.*, **60**, 1965, 1218–37.
3. **OBrien M. V.** Review of mining activities in the Republic of Ireland. *Trans. Instn Min. Metall. (Sect. A: Min. industry)*, **75**, 1966, A70–84.
4. (*a*) **Snelgrove A. K.** Irish 'strata-bound' base metal deposits. *Can. Min. J.*, **87**, 1966; Nov., 47–53; Dec., 55–60.
(*b*) **Gordon-Smith J.** Personal communication to A. K. Snelgrove (cited in reference 4(*a*)).
5. **Cole G. A. J.** Memoir and map of localities of minerals of economic importance and metalliferous mines in Ireland. *Mem. geol. Surv. Ireland*, Dublin, 1922 (reprinted 1956), 155 p.
6. **OBrien M. V.** The future of non-ferrous mining in Ireland. In *The future of non-ferrous mining in Great Britain and Ireland* (London: The Institution of Mining and Metallurgy, 1959), 5–26.
7. **Pereira J.** Further reflections on ore genesis and exploration. *Min. Mag., Lond.*, **109**, Nov. 1963, 265–80.
8. **Schultz R. W.** The Northgate base-metal deposit at Tynagh, County Galway, Ireland. *Econ. Geol.*, **61**, 1966, 1443–9.
9. **Murphy T.** Measurements of gravity in Ireland: gravity survey of central Ireland. *Geophys. Mem. Dublin Inst. adv. Studies* no. 2, pt 3 1952, 34 p.

10. **Charlesworth J. K.** *Historical geology of Ireland* (Edinburgh, London: Oliver and Boyd, 1963), 565 p.
11. **Lees A.** The structure and origin of the Waulsortian (Lower Carboniferous) 'reefs' of West-Central Eire. *Phil. Trans. R. Soc.*, **247**B, 1964, 483–531.
12. **Gray F. and Stacey A. P.** Private communication.
13. **Fitch F. J.** The structural unity of the reconstructed North Atlantic continent. *Phil. Trans. R. Soc.*, **258A**, 1965, 191–3.
14. **Westoll T. S.** Geological evidence bearing upon continental drift. *Phil. Trans. R. Soc.*, **258A**, 1965, 12–26.
15. **Black R. F.** Palaeomagnetic support of the theory of rotation of the western part of the island of Newfoundland. *Nature, Lond.*, **202**, 1964, 945–8.
16. **Wilson T.** Cabot fault: an Appalachian equivalent of the San Andreas and Great Glen faults and some implications for continental displacement. *Nature, Lond.*, **195**, 1962, 135–8.
17. **Dearnley R.** Orogenic fold-belts and a hypothesis of earth evolution. *Physics Chem. Earth*, **7**, 1966, 1–114.
18. **Oswald D. H.** The Carboniferous rocks between the Ox Mountains and Donegal Bay. *Q. J. geol. Soc. Lond.*, **111**, 1955, 167–86.
19. **Schultz R. W.** Lower Carboniferous cherty ironstones at Tynagh, Ireland. *Econ. Geol.*, **61**, 1966, 311–42.
20. **Moorbath S.** Lead isotope abundance studies on mineral occurrences in the British Isles and their geological significance. *Phil. Trans. R. Soc.*, **254A**, 1961–62, 295–360.
21. **Pockley R. P. C.** Lead isotope and age studies of uranium and lead minerals from the British Isles and France. Ph.D. thesis, University of Oxford, 1961.
22. **Murphy G. J.** The Avoca enterprise: 2. The geology of the mineralised area. *Mine Quarry Engng*, **25**, Aug. 1959, 330–8.
23. **Boyle R. W.** Geology of the barite, gypsum, manganese, and lead–zinc–copper–silver deposits of the Walton–Cheverie area, Nova Scotia. *Pap. geol. Surv. Can.* 62–25, 1963, 36 p.
24. **Bell W. A.** Horton–Windsor District, Nova Scotia. *Mem. geol. Surv. Can.* 155, 1929, 268 p.
25. **Dunham K. C.** Geology of the Northern Pennine orefield. *Mem. geol. Surv. U.K.*, 1948, 357 p.
26. **Schumacher F.** Genesis des Freiberge Erzdistriktes. In *XVI Int. geol. Congr. 1933* (Washington, D.C.: The Congress, 1936), vol. 1, 399–405.
27. **Petrascheck W. E.** Typical features of metallogenic provinces. *Econ. Geol.*, **60**, 1965, 1620–34.
28. **White D. E. Anderson E. T. and Grubbs D. K.** Geothermal brine well. *Science, N.Y.*, **139**, 1963, 919–22.
29. **Craig H.** Isotopic composition and origin of the Red Sea and Salton Sea geothermal brines. *Science, N.Y.*, **154**, 1966, 1544–8.
30. **Doe B. R. Hedge C. E. and White D. E.** Preliminary investigation of the source of lead and strontium in deep geothermal brines underlying the Salton Sea geothermal area. *Econ. Geol.*, **64**, 1966, 462–83.

31. **Helgeson H. C.** *Complexing and hydrothermal ore deposition* (Oxford, etc.: Pergamon, 1964), 128 p.

32. **Barnes H. L. and Czamanske G. K.** Solubilities and transport of ore minerals. In *Geochemistry of hydrothermal ore deposits* Barnes H. L. ed. (New York: Holt, Rinehart and Winston, 1967), 334–81.

33. **Taylor S. R.** The application of trace element data to problems in petrology. *Physics Chem. Earth*, 6, 1965, 133–213.

34. **Skinner B. J.** *et al.* Sulfides associated with the Salton Sea geothermal brine. *Econ. Geol.*, 62, 1967, 316–30.

35. **Quennel A. M.** The structural and geomorphic evolution of the Dead Sea Rift. *Q. J. geol. Soc. Lond.*, 114, 1958–59, 1–24.

36. **Amin M. S.** Geological features of some mineral deposits in Egypt. *Bull. Inst. Désert*, 5, 1955, 209–39.

37. **Gindy A. R.** Radioactivity and Tertiary volcanic activity in Egypt. *Econ. Geol.*, 56, 1961, 557–68.

38. **Miller A. R.** *et al.* Hot brines and recent iron deposits in deeps of the Red Sea. *Geochim. cosmochim. Acta*, 30, 1966, 341–60.

39. **Drake C. L. and Girdler R. W.** A geophysical study of the Red Sea. *Geophys. J.*, 8, 1964, 473–95.

40. Gortdrum Mines hits zinc–lead in first Moate drill hole. *World Min.*, 21, May 1968, 72–3.

41. **Murphy T.** Some unusual low Bouguer anomalies of small extent in central Ireland and their connection with geological structure. *Geophys. Prospect.*, 10, 1962, 258–70.

42. **Hallof P. G. Schultz R. and Bell R. A.** Induced polarization and geological investigations of the Ballyvergin copper deposit, County Clare, Ireland. *Trans. Am. Inst. Min. Engrs*, 223, 1962, 312–8.

43. **Fleischer M.** Summary of new data on rock samples G-1 and W-1, 1962–1965. *Geochim. cosmochim. Acta*, 29, 1965, 1263–83.

44. *Geological map of Ireland, 3rd edn* (Dublin: Ordnance Survey, 1962).

45. **Green J.** Geochemical table for the elements for 1959. *Bull. geol. Soc. Am.*, 70, 1959, 1127–84.

12

Reprinted from *Geol. Soc. Amer. Bull.* **82**:2295–2298 (1971)

South Atlantic, Benue Trough, and Gulf of Guinea Cretaceous Triple Junction

NORMAN KENNEDY GRANT   *Department of Geology, Oberlin College, Oberlin, Ohio 44074*

[*Editor's Note:* In the original, material precedes this excerpt.]

BENUE TROUGH

The Benue trough runs northeast from the Niger delta to the Chad basin (Fig. 1). Filled with perhaps up to 5,500 m of folded Cretaceous sediments and rare volcanic rocks (Carter and others, 1963), it was interpreted as a rift valley (Cratchley and Jones, 1965) lying above a line of short-lived Cretaceous lithosphere plate separation (Burke and Whiteman, 1970). Although the folding of the infilling sediments is anomalous, there are persuasive arguments for the trough having originated by underlying Cretaceous lithosphere separation. The trough possesses, for instance, a central zone of relatively high Bouguer values flanked on either side by elongated negative anomalies, which are considered to be due to the combined effects of crustal thinning, basic to intermediate igneous rocks, and possible shallow crystalline basement (Cratchley and Jones, 1965). Further, the central zone of high Bouguer values coincides with a 560-km-long belt of Cretaceous Pb-Zn mineralization (Farrington, 1952), which was emplaced toward the end of a period of basic to intermediate igneous activity (Cratchley and Jones, 1965). It is likely that the mineralization involved geothermal heavy metal-bearing brines circulating under the influence of a deeper geothermal reservoir, such as occur at several locations along present-day lines of lithosphere separation in the Red Sea and the Salton Sea area to the north of the Gulf of California. There is lead isotope data (Jacobson and others, 1963) to show that lead in Benue galenas was derived from low U/Pb sites in the underlying or adjacent crystalline basement, and thus is analogous to the dissolved salts in the present-day geothermal brines, which deuterium and oxygen-18 measurements (Craig, 1966, 1969) show were also leached from pre-existing rocks.

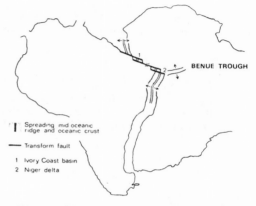

Figure 1.   Schematic separation of Africa from South America in Albian times when the Ivory Coast basin was merged to the North Atlantic by movement on the Gulf of Guinea transform faults, and when a short-lived line of spreading led to the formation of the Benue trough. From north to south the transform faults are the St. Paul's, the Romanche, and the Chain faults (*compare with* Hayes and Ewing, 1970, Fig. 5). The continental outlines are taken at the 500 fm line and are modified *from* Bullard and others (1965).

The combined evidence of marine sediments of Albian age, which extend into the upper reaches of the Benue trough, and the probable limits on igneous and mineralization activity, based on sparse radiometric data (Snelling, 1965) and geological controls (Cratchley and Jones, 1965), suggest the trough had an active spreading life from the Albian to the Santonian, a period of approximately 30 m.y. The end of the spreading movements is dated by the appearance of strong Santonian folding of the sediments in the southwestern part of the trough (Cratchley and Jones, 1965). The focus of the fold movements appears to have moved northeast, affecting the far northeast of the trough in post-Maestrichtian times (Cratchley and Jones, 1965).

[*Editor's Note:* In the original, material follows this excerpt.]

REFERENCES CITED

Bullard, E., Everett, J. E., and Gilbert Smith, E., 1965, The fit of the continents around the Atlantic: Roy. Soc. London Philos. Trans., Ser. A, v. 258, p. 41-51.

Burke, K., and Whiteman, A., 1970, The geological history of the Gulf of Guinea: Paper given at S.C.O.R. Symposium, Cambridge, March, 1970.

Craig, H., 1966, Isotopic composition and origin of the Red Sea and Salton Sea geothermal brines: Science, v. 154, p. 1544-1547.

—— 1969, Geochemistry and origin of the Red Sea brines, *in* Degens, E. T., and Ross, D. A., eds., Hot brines and recent heavy metal deposits in the Red Sea: New York, Springer-Verlag New York Inc., p. 208-242.

Cratchley, C. R., and Jones, G. P., 1965, An interpretation of the geology and gravity anomalies of the Benue valley, Nigeria: Overseas Geol. Survey, Geophys. Paper no. 1, 26 p.

Farrington, J. L., 1952, A preliminary description of the Nigerian lead-zinc field: Econ. Geology, v. 47, p. 583-608.

Hayes, D. E., and Ewing, M., 1970, North Brazilian Ridge and adjacent continental margin: Am. Assoc. Petroleum Geologists Bull., v. 54, p. 2120-2150.

Jacobson, R.R.E., Snelling, N. J., and Truswell, J. F., 1963, Age determinations in the geology of Nigeria, with special reference to the Older and Younger Granites: Overseas Geology and Mineral Resources, v. 9, p. 168-182.

[*Editor's Note:* Figure 1 in this excerpt illustrates the triple junction: three spreading ridges meet at the site of the present Niger delta. The northwestern arm is interpreted by Grant as partly of transform type, but it was more likely a normal constructive margin (cf. Burke et al., 1970).]

Part II

MINERAL DEPOSITS
AND PLATE TECTONICS

Papers applying plate tectonic concepts to metallogenesis began to appear in 1971. As many as possible of the subsequent steady flow of publications on this topic have been included here, partly because it is difficult to identify those which will ultimately prove to be of greatest significance, and partly because each one throws light on a different facet of the problem—even papers by different authors covering the same subject have a contrasted approach.

As far as possible, the papers are presented in the order in which they were written in order to illustrate the development of ideas from consideration of general principles to application of newer plate tectonic concepts. A growing number of papers deals with the application of plate tectonics to specific metallogenic provinces, a trend which is likely to continue. Later sections look at the influence of plate tectonic concepts on ideas about the origin of ore metals and the relationship between plate tectonics and mineral deposits of mainly sedimentary origin. Papers expressing doubts about the relevance of plate tectonic processes to development of mineral deposits are included in appropriate sections.

Editor's Comments
on Papers 13 Through 25

23 **HUTCHINSON**

Excerpts from *Volcanogenic Sulfide Deposits and Their Metallogenic significance*

24 **MITCHELL and BELL**

Island-Arc Evolution and Related Mineral Deposits

25 **SOLOMON**

Massive Sulphides and Plate Tectonics

METALLOGENESIS AND PLATE TECTONICS
—GENERAL PRINCIPLES

In terms of priority, the first mention of this topic to appear in print seems to have been the abstract of a paper presented at an association meeting of mining engineers in March, 1971, which is reproduced here as Paper 13, chiefly for interest.

The first papers to apply strictly plate tectonic concepts to metallogenesis, however, were by Guild and by Pereira and Dixon. They neatly complement each other and provide a useful link with Part I. Guild (Paper 14) summarizes several of the metallogenic principles and problems touched on in the previous section and suggests how they can be accommodated in a plate tectonic framework. The brief note by Pereira and Dixon (Paper 15) provides illustrations summarizing the kind of metallogenic processes that might be expected at three different types of plate margin.

In another early review paper Petrascheck (1972) commented that correlation of mineral provinces by itself does little more than provide additional evidence for continental reconstructions (I made much the same point in an earlier section). He also gave a general discussion of plate tectonics applied to metallogenesis, re-emphasizing his belief (Paper 5) that the majority of ore-forming metals are domiciled in the crust and not the mantle. A shorter version of this paper was subsequently published in English (Petrascheck, 1973b).

Another review was published at about the same time by Burk (1972), but he went no further than to remark in a general way that an understanding of plate tectonic processes would be important in the search for mineral resources.

The first detailed review of metallogenesis and plate tectonics was Guild's contribution to the Montreal I.G.C., reproduced as Paper 16.

Attention is focused on Tertiary mineralization in relation to presently or recently active plate margins—especially consuming margins. Guild also suggests that stresses at consuming margins may reactivate lineament and fracture systems ('ancient flaws in the over-riding plate'), which would control the emplacement of mineral deposits. He does not, however, commit himself on the origin of the ore metals.

The two papers that follow are both concerned with Mesozoic-Tertiary porphyry copper mineralization around the Pacific, but with complementary emphasis: Sillitoe (Paper 17) concentrates on the western American cordilleran system; whereas Mitchell and Garson (Paper 18) cover the western Pacific and also deal with tin mineralization of the Malaysian peninsula.

Sillitoe presents a well-documented case for concluding that copper (and molybdenum) derive from subducted oceanic crust, and shows how metal ore deposits above consuming plate margins probably have a three-stage origin: primary dispersion in basalts at oceanic ridges, followed by hydrothermal concentration at oceanic ridges, and then re-distribution into new concentrations when the 'conveyor belt' system reaches the subduction zone. Mitchell and Garson give less space to the problem of metal source, though they do suggest that Malaysian tin was remobilised from continental crust by volatiles from deeper levels, especially fluorine (compare D. Taylor, 1974). The authors all agree that the sulphur required to precipitate the copper is of mantle derivation (cf. Petrascheck, Paper 5 and Petrascheck, 1972, 1973a). Sillitoe's contention that pre-existing lineament and fracture systems exert little control over the location of mineral deposits represents something of a minority view.

At about the same time as these papers were published, Hodder and Hollister (1972) briefly summarized the features of porphyry copper deposits and concluded that their tectonic setting is typically tensional. After going on the describe in greater detail the porphyry coppers of British Columbia, the authors have this to say about their genesis and origin (page 26):

> (1) The westward drift of North America began perhaps as early as mid-Palaeozoic and resulted in interaction of the Canadian Shield with a Pacific oceanic plate. This convergence led to upthrusting of the Shield's leading edge by the Pacific ocean plate. . . . Northwest-trending folds and faults resulted, along with tension fractures. . . . at the bend of the downthrust plate . . . accompanied by extrusion of calc-alkaline submarine flows the Late Palaeozoic to Early Mesozoic age along the front of the downthrust plate . . . and emplacement of late-stage plutons with copper- and molybdenum-bearing porphyritic units in tensional fractures of northwest trend. An example is the group of Triassic porphyry copper deposits in granodiorites and syenites trending northwest from Copper Mountain.

(2) Plate convergence controlled the structural development of the Cordilleran fold belt until Late Cretaceous and Early Tertiary drifting carried the continent over the East Pacific Rise . . . This gave a regional tensional framework with thin crust, block faulting and direct access to the upper mantle. Lavas and porphyrytic copper-bearing intrusions from differentiated mantle material rosee along northerly trends and in the transverse structure . . . which trends easterly. The volcanic rocks are of the basalt-andesite-rhyolite association, and . . . the porphyrytic plutonic rocks are typically quartz monzonites.

Particularly interesting here is the emphasis on tensional tectonics at destructive plate margins, irrespective of whether a continental margin has over-ridden an oceanic ridge or not. It is important to recognize that the converging motions of oceanic or oceanic and continental plates do *not* necessarily involve compressive stresses as is so often assumed. Only in the continental collision setting can compressional tectonics be typically expected.

Mineral provinces change in character across the grain of orogenic belts, just as magmatic and tectonic features do. Sillitoe was the first to correlate this pattern of transverse metallogenic zoning with depth to the underlying Benioff zone (Paper 19). As supporting evidence for his subduction model Sillitoe used Noble's (1970) detailed delineation of metal provinces in western North America—Noble himself, however, as we shall see later, does not accept the model! Sawkins' review of sulphide ore deposits in relation to plate tectonics again devotes most space to the convergent plate margin setting, because this is where most sulphide deposits occur (Paper 20). He also discusses metallogenic provinces and plate tectonics from a genetic rather than a descriptive standpoint, and presents some thoughts on the origin of ore metals. This paper appears to be the first to carry the explicit suggestion that plate tectonic processes might have been responsible for pre-Mesozoic ore deposits.

In his earlier paper (Paper 17) Sillitoe showed how the metals in porphyry copper deposits formed above destructive plate margins could well have originated at oceanic ridges by the hydrothermal processes discussed earlier, to be later remobilized by partial fusion of oceanic crust during subduction.

Ophiolite complexes emplaced as tectonic slices in many orogenic belts are generally regarded as fragments of oceanic crust which have escaped subduction. Many such complexes contain important sulphide deposits and it follows that these sulphides could also be the products of hydrothermal ore-forming processes at oceanic ridges preserved by overthrusting (obduction) instead of subduction at destructive margins.

This theme is developed by Sillitoe in the first of the following four papers (Paper 21). In the second and third extracts, both Sillitoe (Paper 22) and Hutchinson (Paper 23) discuss massive sulphide deposits associated with volcanism at both ocean ridge and island arc/continental margin sites, and expand upon some of the points made earlier by Sawkins (Paper 20). Hutchinson's paper (considerably abridged for this volume) is important in that it demonstrates how the character of volcanogenic sulphide deposits has changed with time, as the processes forming and modifying the Earth's crust have become progressively more complex (compare also Hutchinson and Hodder, 1972). Mineral deposits formed at successive stages of island arc development are catalogued in some detail by Mitchell and Bell (Paper 24), whose diagrams become increasingly complicated and require careful study to follow. They confine their attention to modern arcs, particularly in the western Pacific, and go into more detail than any of the previous works dealing with volcanogenic sulphides, distinguishing no less than seven stages of island-arc evolution, and covering deposits formed by surface processes of weathering and erosion, as well as those controlled mainly by internal processes. Among the more interesting features of this paper, is that the crustal setting is shown to be entirely oceanic (basaltic) throughout. The authors envisage no contribution from continental crust, either for development of granites or for concentrations of such typically 'continental' metals as Sn and W; in other words, they believe all these materials to be of upper mantle origin.

Perhaps because of their emphasis on modern arc development, Mitchell and Bell appear to disagree with Hutchinson in their conclusion that there has been relatively little change with time in ore-forming processes of island arcs, and by implication little change in the plate tectonic processes responsible for island arc development.[1]

One of the assumptions implicit in all these papers is that ocean floor sediments are carried down into the mantle along subduction zones, where they participate in the partial melting and remobilization that leads to magmatism and mineralization.

A superficial approach to the question of whether the metal content of such sediments could contribute significantly to ore deposits above a Benioff zone was attempted by Wright and McCurry (1973a,b; compare Szadeczky-Kardoss, 1974); but the question is relatively unimportant, for the previous papers have shown that the basaltic ocean crust contains more than enough metals to supply ore-forming fluids at destructive plate margins (see also D. Taylor, 1974). The sediments need not be considered in this context, but there is a greater chance that they *will* be subducted if tensional rather than compressional tectonics characterize consuming plate margins (cf. comments following quotation from Hodder and Hollister (1972).

The final entry for this section presents a dissenting view by Solomon (Paper 25), who suggests that massive sulphide deposits occur in

---

[1] See also discussions of Mitchell and Garson (Paper 18) and Sillitoe (Paper 21) in Inst. Mining and Metallurgy Trans., B, **82**(795):B40–B45 (1973), and B, **82**(798): B76–B77 (1973).

such a wide range of geological environments that they cannot satisfactorily be accounted for in plate tectonic terms. His "geothermal model," however, requires a heat source, and his crustal cross-section shows massive sulphide deposits in several settings. Plate tectonic processes related to either constructive or destructive plate margins could have provided the energy source for all of them.

Two useful reviews which summarize much of the material in earlier papers, are provided by Guild (1974) and Tarling (1973). The first elaborates on Table 1 in Paper 14, the second is a more general classifications of plate environments and related metallogenesis.

Thonis and Burns (1975) also provided a useful review of plate tectonics and mineral deposits, but it was applied specifically to manganese ores in a variety of settings.

13

Reprinted from *Mining Eng.* **22**(12):69 (1970)

MANTLE CELLS AND MINERALIZATION

Wilfred Walker

Consulting Geologist, Willowdale, Ontario, Canada

Much is known of the geography of the Alpine orogeny because we are now at the end of it and all the geosynclines are in their terminal, cordilleran form. The principle of sea floor spreading, continental drift, and mantle cells has gained general acceptance, but the number and position of the cells, even during the Alpine orogeny, is controversial. The oceanographers have shown us the uprises and I suggest that downflows are marked by island arcs in the oceans and geosynclines (now cordillera) in the continents, for a total of about a dozen cells. Major mineral deposits associated with the Alpine downflows are porphyry copper and molybdenum, nickel, chrome, lead-zinc-silver, gold, mercury, tungsten, and tin. A more precise environment is known for some of these metals, for example, nickel, chrome, and mercury in flysch flanking mantle downturns. Ten major orogenies have taken place from 3500 m.y. to the present. Their reconstruction and metallogenesis is the subject of a continuing study.

14

Reprinted from Mining Eng. 23(1):69-72 (1971)

Metallogeny: A Key to Exploration

Philip W. Guild

Approaching exhaustion of areas where traditional prospecting methods can pay off and sharply rising costs require increasing sophistication in planning exploration. Most outcrops, not only of ore and gossan but also of obvious zones of alteration, have been examined; many of the anomalies readily detectable by geophysical means have been drilled; geochemical surveys now cover extensive tracts. How should the explorationist proceed to find the ores that will keep his company in business and supply the world with the energy and raw materials it needs? Two courses are open to him: (1) saturation coverage by geology, geophysics, geochemistry, and drilling in the expectation that something will turn up or (2) critical analysis of the factors responsible for the concentration of valuable minerals in the earth's crust. The successful explorationist applies both methods to the limit of his resources.

The second course, analysis of geologic factors, also comprises two principal lines of research: (1) ever more detailed study of the chemistry (including isotopic analysis) of the ore and gangue minerals of known deposits in an effort to understand how they were formed and (2) increasingly detailed mapping of their environment to determine the reasons for their localization. Again, both are very necessary.

There is, however, a third type of analysis that is undoubtedly used, though not necessarily consciously, by every person concerned with exploration. It is, quite simply, analogy. Where do similar conditions exist? If the individual's experience, knowledge of both the field and the literature, and ability to sort through and select pertinent facts from a multitude of data are sufficient, he may succeed in pinpointing targets or at least in getting into the ballpark. A

major stumbling block is, of course, the sheer mass of information that must be mastered as the targets become more elusive.

Global Mapping Is Under Way

Such analysis on a regional, national, continental, or worldwide basis has come to be called metallogeny, the study of the genesis of ore deposits. (Metal in this sense derives from a Greek word meaning mine; thus metallogeny treats of both metallic and nonmetallic minerals. The fossil fuels are commonly excluded.) As befits a subject of such widespread interest, a well-coordinated international project has been underway for a decade to prepare metallogenic maps on a cooperative basis. The results are beginning to appear; within a few years most of the world will be covered.

This effort has been stimulated by the Commission for the Geologic Map of the World of the International Geological Congress–International Union of Geological Sciences, now sponsored by UNESCO. The national geological surveys or their equivalents contribute the basic working materials in accordance with mutually agreed upon standards. Many countries have published or will shortly publish national maps, and continent-wide maps are now in varying stages of completion.

A distinct pattern has evolved in compilation of metallogenic maps. The first requirements are for geologic and tectonic maps on the one hand and mine or mineral spot maps on the other. Subsequently these must be modified by (1) suitable simplification of the geologic-tectonic framework with emphasis on those features considered most important for ore-deposit occurrences, and (2) elaboration of the

Table I. Proposed Relationship of Some Ore-Deposit Types to Plate Tectonics

Deposits Formed	Type, Possible Examples
At plate margins	Orientation of deposits, districts, and "provinces" tend to parallel margin
Accreting	Red Sea muds. Ancient analogs?
	Podiform Cr, (probably carried across ocean basin before incorporation in orogen)
Transform	Podiform Cr, Guatemala (?)
Consuming	Chiefly of continent/ocean, also of island arc/ocean types; deposits formed at varying distances on side opposite oceanic, descending plate
	Podiform Cr, Cuba, California, Oregon, Alaska
	FeS-Cu-Zn(Pb) stratabound massive sulfides, Cuba, California, Alaska
	Kuroko ores, Japan
	Mn of volcanogen type associated with marine sediments, Cuba, California, Olympic Peninsula
	Magnetite-chalcopyrite skarn ores, Puerto Rico, Cuba, Mexico, California, British Columbia, Alaska
	Au, Mother Lode, SE Alaska
	Bonanza Au-Ag deposits
	Cu(Mo) porphyries, Puerto Rico, Panama, Mexico, SW United States, British Columbia
	Ag-Pb-Zn, Mexico, western United States and Canada
	W, Sn, Hg, Sb
Within plates	Deposits tend to be equidimensional, distribution of districts and "provinces" less oriented
In oceanic parts	Mn(Cu, Ni, Co) nodules
	Mn-Fe sediments in small ocean basins with abundant volcanic contributions?
	Evaporites in newly opened or small ocean basins
At continental margins of Atlantic (trailing) type	Black sands, Ti, Zr, magnetite, etc.
	Phosphorite on shelf
In continental parts	Mississippi Valley-type Pb-Zn-Ba-F deposits
	Mesabi and Clinton types of iron formation
	Evaporites; Michigan Basin, Permian Basin; salt, potash, gypsum, sulfur
	Red bed Cu; Kupferschiefer and Katangan Cu
	U, U-V deposits, Colorado Plateau and elsewhere
	Ti in anorthosite
	Carbonatite-associated deposits of Nb, V, P, RE
	Diamonds in kimberlite
	Stratiform Cr, Stillwater; Cr, Fe-Ti-V, Pt, Bushveld
	Kiruna-type Fe, SE Missouri

deposit symbols so that like or similar ores can be readily distinguished from those having little or no genetic relationship, even though the metals and minerals they contain may be the same. (For example, sedimentary iron deposits must be differentiated from hydrothermal replacement deposits, and these in turn from magmatic, lateritic, or black sand detrital deposits.) The results of this phase are ordinarily called metallogenic maps, though of course they are merely compendia of rather large masses of data in visual form and do not in themselves either explain the reasons for localization of deposits or point directly to favorable prospecting areas. In effect, such maps are jumping-off points for true metallogenic studies, which may lead to what the Russians call "prognostic maps." They should be as factual as possible, with a minimum of interpretation.

No effort has been made to impose a standard symbology, largely because at the start no one could be sure what would be most effective. As a result, several basic legends, each with modifications, have evolved, but a knowledge adequate to understand basic features can be quickly acquired.

The North American map legend provides for showing the following information about a deposit or district: (1) the principal metals or minerals it contains, (2) whether it is large, medium, or small, (3)

its geologic environment with respect to (a) the enclosing sedimentary and (or) volcanic rocks and (b) associated intrusive rocks, if any, (4) its geologic class, (5) the age of mineralization and (6) the general class (oxide, sulfide, etc.) of the ore minerals. The geologic base distinguishes stratified rocks on the basis of: (1) their age, (2) their nature, especially if they have volcanic components, and (3) their general structural and metamorphic condition, *i.e.*, undeformed, folded, moderately or highly metamorphosed. Intrusive rocks are divided on the basis of their petrologic nature (alkaline, granitic, mafic, or ultramafic) and the age of intrusion relative to principal orogenies (early, main period, or late). Background (geologic-tectonic) information is in pale colors; deposit data in bright colors.

Because continent-wide maps must be compiled at small scales (80 miles to the inch or 1:5 million is the most common one), enormous compression and selection of data are required. Compilers are therefore turning to computers to store all the deposit data. Eventually it should be possible to prepare special maps by selecting the desired data and presenting them graphically at any appropriate scale. Compatability of the data bases would have obvious advantages for both theoretical and practical studies, and close contact between groups in the different countries will hopefully ensure at least a minimum ability to exchange information in magnetic form.

What practical assistance can the explorationist expect? I suspect that the greatest value of the maps will be to stimulate speculation, while at the same time providing enough data to permit simultaneous checking of theory with fact.

Ore Deposits Still Need Explaining

The uneven distribution of ores in the crust is well known; the reasons for it are not. Among the fundamental problems are: Do the concentrations in certain areas (as, for example, of copper in the southwestern United States) result from original inhomogeneities or because favorable conditions for ore genesis prevailed? Were the metals derived directly from the mantle or have multistage processes reworked crustal material to form ores? To what extent do magmas constitute the immediate sources of metals? Why are some ores (for example, nickel sulfides, stratiform chromite, and titanium in anorthosite) largely restricted to older Precambrian rocks, whereas others (such as porphyry copper and molybdenum, antimony, and mercury) are, with few exceptions, of Mesozoic or Tertiary age? Answers to these and other questions are not immediately forthcoming, of course, but documentation of the facts may help to provide them.

Within North America, two major distributional patterns are evident: (1) more or less along the depositional and structural fabric in orogenic belts, and (2) transverse to the general grain of the geology. In (1), mineralization is roughly of the same geologic age as the associated rocks; it may range from contemporaneous (syngenetic) to perhaps the close of the orogenic period following deposition of the host rocks. In (2), mineralization may be (and commonly is) much later than its host rocks; ores of Laramide and mid-Tertiary ages in Precambrian and Paleozoic rocks of the West are examples. In (1), deposits of more or less similar type and metal content occur along the geologic grain; in (2), very dissimilar ores may be in close juxtaposition, as along the Colorado Mineral Belt with the Ag-Pb-Zn-Au-Mn ores of Leadville and Gilman, the molybdenum of Climax and Urad, the tungsten of the Nederland area, gold-silver in many places, uranium, and fluorite. Two or more distinct periods of mineralization are common.

Deposits of the first category are commonly attributed to the development of geosynclines. The second category comprises the lineament-related deposits; there have been numerous attempts to correlate these with intersections, both with other lineaments and with axes of geosynclines. A zonation of metals inward from continental margins has been noted, and the general correspondence of deposit types and contained metals to igneous rock types, and of these igneous rocks to the continental framework, has long been known. The proliferation in isotopic age determinations does not seem to substantiate the eastward progression across the Cordillera of times of mineralization; indeed, increasing data suggest that the region has had an extremely complex history of repeated orogeny, metamorphism, magmatism, and mineralization.

Several recent lines of research and a rapidly developing theory of earth history give promise of resolving some of the scientific problems of ore genesis, and conceivably of providing real help in planning exploration for concealed deposits.

The use of radioactive isotopes to determine ages is well known; in addition, studies of the stable isotopes suggest that: (1) sulfur of some deposits probably comes from deep (mantle or at least well-homogenized) sources, whereas other deposits contain isotopically distinct sulfur derived by fractionation of sedimentary sulfate through biogenic or other processes; (2) strontium isotope ratios (initial Sr^{87}/Sr^{86}) of many ore-associated intrusive rocks show derivation from subcrustal sources and may indicate that the metals were also of mantle derivation; (3) variations in the isotopic compositions of galena and rock lead can reveal not only the sources of ores but also for some the geologic history of basement rocks from which they were derived; (4) ratios of the light stable isotopes (deuterium-hydrogen and O^{18}/O^{16}) in ore fluid samples recovered from liquid inclusions give clues to the nature of the water (magmatic, connate, meteoric) and hence to the probable source of the metals.

Evidence Builds for Plate Tectonics

The reality of continental drift is gaining wide acceptance through converging lines of evidence assembled by geologists, geophysicists, and oceanographers. (See, for example, "Mountain belts and the new global tectonics" by John F. Dewey and John M. Bird in the *Journal of Geophysical Research, vol. 75, no. 14, May 10, 1970.*) In the newest and most logical version of the theory, the crust of the earth (lithosphere) is believed to consist of a relatively small number of plates that are growing at ocean ridges by the addition of new material from the mantle, moving across the globe, colliding, and being consumed by descent of one plate into the mantle and remelting. The plates are as much as 150 km thick, the continents ride "piggyback" on them, and most plates comprise both continental and oceanic crust. Plate edges that are neither accreting nor being consumed slide past one another on so-called transform faults.

The concept provides a unified theory into which more and more data are being integrated rapidly. If it is valid, the origin and localization of ore deposits must also fit. A very preliminary look, preliminary both because the factual metallogenic maps are incomplete and the theory itself is still evolving, suggests to me that they do, and that many of the deposits of the world occur in places reasonably well defined in terms of plate tectonics. Most of the remaining discussion here will be devoted to sketching the broad outlines of this thesis.

Collisions of plates may involve edges of continents, oceanic crust, or volcanic island arcs. If an oceanic plate is involved, it normally descends below the continent or arc at about 45° to 50°; seismic data suggest that it can extend as deep as 700 km along a Benioff (subduction) zone before losing its identity completely, but partial fusion at much shallower depths yields basaltic, then calc-alkaline, and finally alkalic, magmas that rise to produce extrusive and intrusive rocks. Sediments and blocks of oceanic crust are scraped off in the trench that forms along the juncture of the plates; these are thrust into the sedimentary wedge at the edge of the continent or arc; the underridden block rises and sheds new clastic sediments into the adjacent troughs; and deformation and heating produce the folding, faulting, and metamorphism of the developing orogenic belt.

Dewey and Bird distinguish four principal types of collisions: continent/ocean crust (Cordilleran), island arc/ocean crust (*e.g.,* Japan), island arc/continent and continent/continent (*e.g.,* the Himalayas). The present plates developed from Triassic time onward; most of the orogen-associated endogenic mineral deposits of Mesozoic and Tertiary ages occur on the sites of Cordilleran-type (continent/ocean) collisions; some deposits are also present where island arcs collided with oceanic crust (Japan, Philippines, Antilles, southern Central America, etc.). The continent/continent collisions, in which neither plate can descend far because of its buoyancy and the force is dissipated by severe crushing, do not produce igneous rocks and are notably deficient in hydrothermal ore deposits.

Divergence in trends of the margins of continents, arcs, and oceanic plates can produce appreciable time lags along an orogen; similar events do not have to be strictly contemporaneous. As ocean floor is consumed, an initial continent/ocean collision can become a continent/island arc confrontation; the arc can be welded to the continent, a new continent/ocean collision ensue, and so forth. An axis of spreading eventually dies and a new one arises somewhere else. The possibilities are numerous, and very many details remain to be worked out.

Table I summarizes some possible relationships of ore deposits to the postulated plates. These are obviously tentative but may serve as openers to stimulate thought and discussion on the gross aspects of the environment of ore genesis. Critical appraisal of all the factors involved could provide answers to the fundamental problems raised earlier and others as well; conversely, it may either support or help refute the theory itself.

Geologic Puzzles Point Up
Work That Needs Doing

Brief consideration of a few kinds of deposits will illustrate points that need further study:

The stratabound massive sulfide (pyrite-pyrrhotite-chalcopyrite-sphalerite-galena) ores occur along orogenic belts, commonly with mafic to felsic volcanic rocks, in areas that may have little or no older continental crust. The lead ordinarily gives good model ages and is of single-stage derivation from a homogeneous source. The metals probably came from fairly deep, either directly from the mantle or from remelted oceanic floor and subcrustal material of a descending plate. In contrast, the sulfur in some deposits is moderately heavy, suggesting derivation from supracrustal rocks.

Porphyry copper deposits in Puerto Rico, Panama, and Bougainville developed in young andesitic island arcs, where there apparently was little opportunity for accumulation of the metals in a sedimentary pile. They probably represent end-stage products of differentiating calc-alkaline magmas generated by partial fusion of a descending plate (in which, however, some preliminary concentration might have occurred). Strontium-isotope ratios indicate that the intrusive rocks associated with porphyry copper and molybdenum deposits of the western United States were derived from deep sources rather than from the remelting of supracrustal rocks and suggest that the metals are also juvenile. The sulfur of the ore minerals is homogeneous and of magmatic type.

Marginal Formation Forces
Act in Various Ways

Not all the metals of deposits in and near consuming plate margins need be derived from mantle sources, however. Thermal energy generated by the collision and transmitted by metamorphic or magmatic processes, or by deep circulation of connate or meteoric water, can extract and concentrate elements previously dispersed in crustal rocks or contained in low-grade source beds. Deposits of tungsten and tin, for example, may have originated through differentiation of palingenic magmas that incorporated metals accumulated as resistates on trailing continental margins during an older cycle.

The best evidence for derivation of metals from crustal sources is provided by the lead of Mississippi Valley-type deposits. Many of these deposits have anomalous multistage leads that show by their isotopic compositions that they evolved in a crustal environment over long periods of time. The sulfur is also anomalous. Interestingly, some districts (e.g., Pine Point) have single-stage lead and heavy sulfur, suggesting the former is of deep and the latter of

shallow origin. Most of the world's major lead deposits occur in areas of old continental crust, although some of these have been involved in major orogenies.

Lead deposits are notably absent from the Antilles, in contrast to the rather numerous copper deposits there. This generalization holds also for the Pacific margin of North America, where the only lead of significance is in massive sulfide deposits.

The tendency for deposits of a given type to be distributed parallel to an orogen decreases away from the hypothetical axis of collision. Lineaments at rather large angles to it seem to exert increasing control; there is evidence that these are ancient flaws in the continent (e.g., Colorado Mineral Belt, Lewis and Clark line). It is debatable whether deposits in the broad zone from the Sierra Nevada to the Front Range should be categorized as consuming margin or continental plate deposits; very likely they are intermediate, and indeed isotope data suggest that the lead is partly deep (new?) and partly reworked. The continent seems to have overridden the East Pacific Rise during Tertiary time. This has undoubtedly introduced complications not explicable by the relatively simple model currently advanced; it may also be responsible for the unusually wide belt of mineralization in the western states, which contrasts markedly with all other areas of the world.

Prospects in a Concept

Ocean-floor spreading and resultant collisions have been spasmodic. The locations of accreting and consuming plate margins have changed with time. Details even of the more recent events are still very tentative (even controversial), but there is reason to believe that the principal metallogenic epochs of the American Cordillera—Nevadan, Laramide, and mid-Tertiary—can be correlated with times of active plate movement.

If the distribution of deposits related to recent or present-day plates is only beginning to be understandable, what of those associated with mountain chains incorporated into the continents during previous orogenic cycles? The locations and attitudes of ancient subduction zones, and hence of favorable prospecting areas, may be revealed by polarity of magmas (tholeitic–calc-alkaline–alkalic) generated in them, by type of metamorphism (especially blueschist vs. greenschist), by direction of sediment transport, or by other characteristics. Many of these criteria may be obscured by metamorphic events, erased by erosion, or hidden by younger rocks. As the ore deposits themselves are evidence of the conditions under which they formed, they constitute an important part of the record and may well provide critical evidence concerning the ancient history of the planet.

A final word. Even if the plate-tectonics concept turns out to be completely invalid and has to be discarded, the comparative study of deposits in their total geologic environment will still be essential in the search for analogous situations where orebodies may also be present. Documentation of the facts in metallogenic maps will surely be one of the fundamental steps in future exploration. ▼Ξ

15

Copyright © 1971 by Springer-Verlag

Reprinted from *Mineral. Deposita* **6**:404–405 (1971)

Mineralisation and Plate Tectonics

J. Pereira and C. J. Dixon

London, England

That these two words link an important theme which may be fundamental to many metalliferous deposits, is now becoming evident to many geologists who work in this field. This note is only intended to link some past ideas of the authors with what may emerge in the future as a key model of metallogeny.

In the early 1960's J. P. was deeply concerned by the apparent connection between certain types of mineral deposit, volcanic sequences, and shorelines. It was an empirical observation the significance of which is now becoming evident (Pereira 1963).

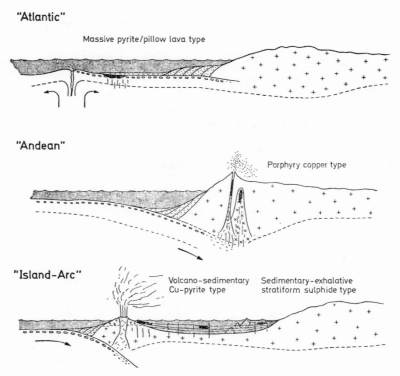

Fig. 1.   Mineral deposits and plate-edge environments

In a later series of papers (Pereira and Dixon 1964) and (Pereira and Dixon 1969), while casting some doubt on continental drift we put forward the theme that Earth is a planet on which evolution was not merely biological, but has evolved in every geological sense one thinks of, though we confined our discussions to mineral deposits.

We can now accept continental drift and many of the ideas set forth by Petraschek (1968) for instance. The theory of continental drift (ss) only covered some 5% of the Earth's history and it is now important to seek the detailed evidence suggested by the more general theory of Plate Tectonics. This we are busy doing, and we hope that other geologists are also at work on the same topic.

Mitchell and Reading (1969) have presented a classification of geosynclines based on different types of plate edge environments, and similar ideas have been used by a number of authors in geotectonic interpretations that have important metallogenic consequences (see for example Dewey (1969) on the Caledonian Appalachina belt).

Fig. 1 shows a development of Mitchell and Reading's interpretation showing the way in which some major types of metalliferous deposit may be related to observable plate edge environments. The validity of these ideas must rest to a large extent on the correct identification of fossil plate edges, and we sincerely hope that research will reveal examples. Conversely we may take models such as we suggest in Fig. 1, to use the distribution of different types of mineral deposit as an aid to the discovery of fossil plate edges.

The kind of evidence required takes time to accumulate so we would like to offer encouragement to other workers who may be able to offer new ideas in this field, and we hope to follow this note by a more extensive publication.

References

Pereira, J.: Reflections on ore genesis and exploration. Mining Mag. (London) **108**, 9—22 (1963).

— Further reflections on ore genesis and exploration. Mining Mag. (London) **109**, 265—280 (1963).

— Dixon, C. J.: Evolutionary trends in ore deposition. Trans. I. M. M. **74**, 505—527 (1964).

— — A statistical investigation of mineral occurrences in Western Europe. Proceedings of 15th Inter-University Geological Conference. Published University of Leicester 1969, 259—269.

Petrascheck, W. E.: Kontinentalverschiebung und Erzprovinzen. Mineral. Deposita, (Berlin) **3**, 57—65 (1968).

Mitchell, A. H., Reading, H. G.: Continental margins, Geosynclines and ocean floor spreading. J. Geol. **77**, 629—645 (1969).

Dewey, J.: Evolution of the Appalachian/Caledonian Orogen. Nature **222**, 1046 (1969).

Received June 1971

Jocelyn Pereira, Esqu.
20 St. Leonards Terrace, London, S. W. 3, England

C. J. Dixon
Dept. of Mining Geology, Imperial College of Science, Prince Consort-Road, London S. W. 7, England

16

Reprinted from *24th Intern. Geol. Congr. Proc.* 4:17–24 (1972)

Metallogeny and the New Global Tectonics*

PHILIP W. GUILD,
U.S.A.

ABSTRACT

Most post-Eocene metallogenic provinces of endogenic type are located on or near present-day plate boundaries as defined by geophysical and oceanographic evidence; positions of older provinces may help define plates that can no longer be observed directly. The Andean, Japanese and other Western Pacific and Tethyan provinces are near consuming plate margins, of which continent-ocean and island arc - ocean margins are most favorable for ore. The culmination of a continent-continent collision is notably unfavorable.

The apparent absence of significant deposits along some extensive arcs (e.g., the Aleutians) may reflect their minimal exposure (cover of water and recent volcanic rocks), relative lack of prospecting or some more fundamental factor.

The Red Sea brine deposits and probably others (e.g. Boleo, Mexico) formed along rifts, perhaps at intersections with transform faults. Simple transform margins generally lack deposits.

The broad metallogenic province extending through northern Mexico and western United States to latitude 41°N may constitute an exception or at least a modification to the rule of proximity. Continental overriding of the East Pacific Rise may explain some very late activity, but does not seem to account for the major mineralization unless one or more subduction zones persisted under the continent through most of Miocene time. A pronounced tendency toward lineament control along ancient basement flaws suggests for this area a transition toward the metallogeny of reactivated platforms.

METALLOGENY, the study of the genesis of ore deposits in their total geologic environment, must of necessity take into account the rapidly evolving reinterpretation of earth history embraced in the new global tectonics. The general relationship of ores to specific magmatic series (Buddington, 1933), especially the rocks of the calc-alkaline suite (Fonteilles, 1967), and of the latter to orogenic belts (McBirney, 1969), is well known. Sumner (1970) has pointed out the potential application of regional geophysical studies and new tectonic theory to exploration. Sillitoe (1970) suggested that porphyry copper deposits derive from materials regenerated in subduction zones and emplaced at high levels in the crust with stocks of calc-alkaline type. Guilbert (1971) believes that the Boleo copper deposits, Baja California, formed along a transform fault contemporaneously with or shortly before the opening of the Gulf of California. H. L. James, in his Presidential Address to the Society of Economic Geologists (Milwaukee, November, 1970), developed the theme of the relationship of mineralization to plate tectonics

---

*Publication authorized by the Director, U.S. Geological Survey.

and used the distribution of ore deposits of the eastern United States to speculate on the position and attitude of the Paleozoic subduction zone(s) during the Appalachian orogenies.

Guild (1971) proposed that many types of deposits form in environments that can be defined in terms of the postulated lithosphere plates: these include those formed at or near plate boundaries (accreting, transform and consuming) and within plates (in oceanic parts, at trailing continental margins and within continental parts). Those formed at consuming margins, chiefly of continent/ocean but also of island arc/ocean type (Dewey and Bird, 1970) seem to constitute the most varied, numerous and widespread endogenic deposits of post-Paleozoic age.

Details of pre-present-day plates are to a large extent speculative and even subject to circular reasoning. By contrast, the outlines, directions of relative movement and nature of impinging boundaries of the modern plates can be documented by geophysical, geological and oceanographic data. This paper attempts to compare the positions of the principal post-Eocene endogenic metallogenic provinces of the world and the major lithospheric plates with the purpose of (1) defining potentially favorable areas for exploration and (2) evaluating the premise that the distribution of mineral deposits can aid in unraveling the older geologic history of the globe.

The general positions of the principal plates, nature of the boundaries between them and relative motions are indicated in Figure 1, as are the post-Eocene mineralized areas. The author has borrowed from a number of sources, including McKelvey and Wang (1969) for the basic map, Dewey and Bird (1970), Isacks, Oliver and Sykes (1968), and Dietz and Holden (1970). No pretense is made for accuracy; indeed, certain details are still being worked out by specialists in this field, among whom the author cannot be counted. With respect to the mineral-deposit data, there may also be errors and omissions; information is frequently sketchy or subject to modification. However, with the foregoing caveats in mind, the general thesis of proximity of ore to plate boundaries holds; nearly all the provinces are in the overriding plate and elongated approximately parallel to its margin.

The simplest, most clearcut example of apparent relationship of mineralization to subduction of an oceanic plate is provided by the Andean province of South America. Porphyry copper deposits and related tourmaline-bearing breccia pipes extend in a narrow belt for about 4000 kilometers from Ecuador to central Chile. Most deposits are early Tertiary (hence not shown on Fig. 1), but K/Ar dating of intrusive rocks and alteration zones around associated deposits in the Copiapó district, north-central Chile, reveal a progressive decrease in ages from Lower Jurassic near the coast to upper Eocene 130 km inland (Clark and others, 1970). Sillitoe (1970) suggested the possibility that younger, even presently forming porphyry copper deposits may exist to the east under volcanic cover. The known post-Eocene deposits, largely tin-silver, tungsten, lead-zinc, etc. veins (Turneaure, 1971), are restricted to a belt about half as long, centered on Bolivia with extensions into Peru and Argentina, that is parallel to and for the most part slightly east of the older deposits. The Andean metallogenic province faithfully mirrors the coast and trench, even to seemingly dying out both to the north and south where the trench terminates. Although the limits of significant mineralization have been ascribed to depth of erosion [the deposits being either not yet exposed in the north (Peterson, 1970) or already removed in the south (Ruiz Fuller, 1965)], a better explanation may be that subduction has been absent or less active north of the Carnegie Ridge and at the southern end of the continent.

The best known example of mineralization along an island arc/ocean boundary is in Japan, which has a history of volcanic-related mineralization occurring in five principal epochs from the later Paleozoic to Quaternary (Tatsumi and others, 1970), with exceptionally varied and active deposition in the Miocene. In contrast to the Andes, the mineral belts overlap and cross; another noteworthy difference is the predominance of stratabound cupriferous pyrite deposits and apparent lack of porphyry copper deposits. There is a strong correlation between the mid-Tertiary provinces and magmatism of calc-alkaline affinities (Tatsumi and others, 1970) postulated to result from partial fusion of the descending lithosphere plate at depths of perhaps 120 to 200 km (Dewey and Bird, 1970). Comparison of seismic data (e.g. Oxburgh and Turcotte, 1970, Fig. 1) and maps of mineral distribution (Japan Geological Survey, 1957, 1960) reveal that the Quaternary, and to a considerable degree the Miocene, metallogenic provinces are above the segments of the Benioff zones that lie at depths of between 100 and 200 km. The detailed structure of the slabs is considerably more complex (Carr and others, 1971); future studies may show even greater correlation between ores and positions of subduced plates.

Post-Eocene mineralization (some only discovered very recently) is known at a number of places around the other segments of the Western Pacific Plate (Fig. 1), and most of it is in areas above the 100- to 200-km contours on the Benioff zones. This relationship to the "andesite line" and to the "deep fractures" has been recognized and commented on by Russian metallogenists (Itsikson and others, 1960). They did not, of course, have the advantage of more recent plate-tectonic concepts to aid in interpreting these facts.

Mercury (and to a lesser extent antimony) deposits constitute exceptions; they may be considerably outside the limits defined above. Moisseyev (1971) pointed out that mercury is not associated with a particular volcanic rock type and may depend on volcanism only as a source of heat to mobilize the metal from sedimentary sources. Much of the mineralization of Sakhalin and the adjacent mainland and of Kamchatka consists of mercury and minor antimony; the divergence in trend of this belt also suggests, however, a rapid shift in location and perhaps attitude of subduction in this area since earlier Neogene time.

Mineralization in Taiwan and the eastern islands of the Philippine Archipelago fits the pattern well. Relatively little important mineralization of Neogene age seems to be known in Indonesia, but the seismic data strongly suggest that much remains to be found. In particular, the 100- and 200-km contours on the Benioff zone extend throughout Sumatra and Java; many companies are exploring there now, and prospects for discoveries seem good if the cover of recent volcanic rocks can be penetrated successfully.

The OK Tedi porphyry copper deposit in New Guinea, perhaps less than 5 million years old (Paul Eimon, written communication, 1970), is within the Australian continental plate, which has only recently begun to override the Pacific plate (Dewey and Bird, 1970). The remaining contact zones of the Western Pacific are only sporadically exposed above sea level; nevertheless, porphyry copper deposits have been found in two areas, Bougainville and Fiji. The Panguna deposit on Bougainville lies over an exceptionally steeply dipping subduction zone; this probably accounts for the presence of land above the critical 100- 200-km depth, which in other parts of the arcs is covered by water. An exception is North Island of New Zealand where only sparse base- and precious-metal indications are known today; this may be a promising area for exploration.

Parts of the Tethyan zone between Eurasia and the various fragments of Gondwana have extensive metallogenic provinces of post-Eocene age; others have little or no endogenic mineralization. Figure 1 barely hints at the complexities of

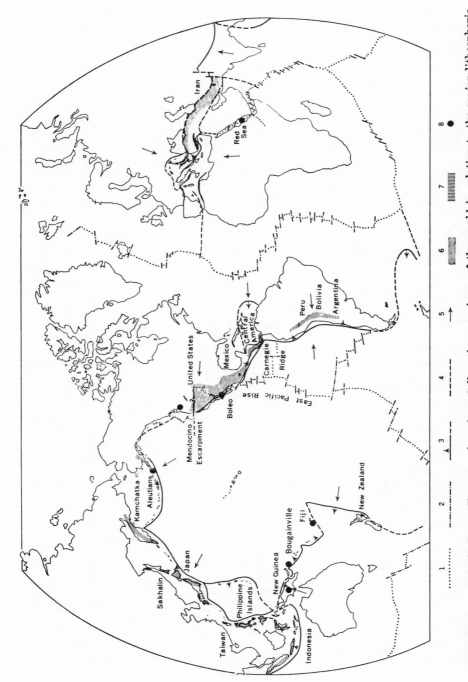

FIGURE 1 — Principal post-Eocene endogenic metallogenic provinces of the world in relation to the major lithospheric plates. Key to ornament: (1) accreting plate margin; (2) transform plate margin; (3) consuming plate margin with dip direction of downgoing plate; (4) margin of uncertain nature and (or) location; (5) relative plate motion; (6) area of mineralization of post-Eocene age; (7) minor or suspected post-Eocene mineralization; (8) major or noteworthy isolated ore deposit of post-Eocene age.

this zone, along which there has been repeated opening and closing, fragmentation into microcontinents, formation of island arcs and suturing as continental masses have driven together (Dewey and Bird, 1970). In general, the Himalayas, the Zagros in southern Iran and the Alps, all loci of severe continent/continent collision in relatively recent times, are lacking in significant known ore deposits of Tertiary age. By contrast, an extensive belt (or belts) of ore deposits extends from Baluchistan in western Pakistan to southeastern Europe. Neither the geologic nor seismic data are clearcut as in the western border of the Pacific plate or along the Andes, but numerous deposits in Iran, which has very recently been recognized as a porphyry copper province of major importance (Bazin and Hübner, 1969), are associated with Oligocene and Neogene andesitic extrusive rocks and plutonic or subvolcanic acid to intermediate intrusive rocks. It seems likely that mineralization can be related to generation of calc-alkaline magmas where oceanic crust, formed by accretion in the Tethyan seaway, was available for subduction. In the final stages of collision between continents, neither plate can descend, and neither magma generation nor endogenic mineralization occurs.

The picture in western North America differs in several respects from those sketched out above. A fairly straightforward pattern of subduction, magmatism and post-Eocene mineralization exists in Central America (Dengo and others, 1969) and southern Mexico adjoining the Middle America trench. Northward the mineralized zone increases steadily in width to near the Mexico - United States border; the eastern border of the zone turns sharply west (along the "Texas lineament"?) then north, and the zone splits around the Colorado Plateau. It terminates rather abruptly on the east near the Colorado-Wyoming border, and except for minor areas in Idaho, Washington and on Vancouver Island does not extend north of the Snake River and Columbia River volcanic fields. (Again, mercury and antimony are exception.) This pattern contrasts strongly with the Mesozoic and Laramide belts, which extend throughout western North America. In British Columbia, for example, mineralization of copper, molybdenum, tungsten, lead, zinc, gold and silver was extensive and repeated until late Eocene time (Sutherland Brown and others, 1971).

Many people have drawn attention to the tendency for deposits to be aligned in belts transverse to the dominant regional depositional and structural pattern, and perhaps to be localized at intersections of two or more lineaments or of lineaments with geosynclinal axes. This holds for Mesozoic as well as Tertiary deposits; repeated mineralization took place along the northeast-trending Colorado mineral belt, which lies along a zone of Precambrian shearing (Tweto and Sims, 1963). Five principal metallogenic epochs are recognized in north-central Nevada and southern Idaho, the last two being Oligocene, 40 m.y. to 30 m.y., and Miocene-Pliocene, 16 m.y. to 10 m.y. ago (Roberts and others, 1971). Several transverse mineral belts have been identified.

The western edge of the continent is now decoupled so that the present boundary with the Pacific plate is essentially transform along much of its extent, and little or no subduction is taking place. There is only shallow seismicity. The contrast with South America — parallel belts, little or no transverse lineament control (Sillitoe, 1970), and present-day trench and deep-focus earthquakes — and with apparent strict relationship of mineralization to relatively simple island-arc conditions in most other areas around the Pacific indicates complexities in western North America that are not present elsewhere. All but the westernmost edge has been continental crust since earlier Precambrian time and has been affected by numerous orogenies. The author suggests that this segment of the continent combined metallogenic features of both consuming plate margins and reactivated platforms until about middle Tertiary time. East-dipping subduction far under the edge of the continental plate has been suggested for the period from Jurassic to Miocene, complicated by a west-dipping zone from mid-Cretaceous on (Moores, 1970). Such a pattern might account for the unusual width of the metallogenic province. Ancient flaws in the overriding plate tended to localize intrusion and mineralization in transverse belts.

Overriding of the East Pacific Rise has been called on to explain many features of the geology of the western United States. However, although its postulated present position is within the belt of post-Eocene mineralization and it is apparently offset westward near the northern limit of the belt (Fig. 1), the relationship, if any, between the rise and the ore deposits is far from clear. Possibly an increase in geothermal gradient and extension of the crust led to emplacement of the more easily mobilized elements (mercury and antimony) or even to other elements from residual or newly generated pockets of magma. For example, the bonanza gold-silver ores may be products of remobilization of upper crustal material. Some, at least, of the most recent mineralization seems to lie along essentially north-south trends (fluorite in New Mexico and Colorado, mercury in Oregon and northern Nevada) or near major faults paralleling the continental border (mercury in California and perhaps in British Columbia).

The Pacific plate is underthrusting the Alaska Peninsula and the Aleutian island chain. To date, little mineralization is known; much of the area is very rugged and isolated, and covered by very recent volcanic deposits or by water. However, a few base- and precious-metal deposits have been found; at least one, on Unga Island near the tip of the Peninsula, has been worked; and the area deserves thorough search for additional deposits.

By comparison with mineralization that seems to be related to consuming plate margins, deposits demonstrably located along accreting margins are rare. The best known and least equivocal, of course, are the recently discovered metal-rich sediments associated with the hot brines in the Red Sea Deeps (Degens and Ross, 1969). Zinc and copper sulfides, with major quantities of goethite, some manganese oxides, anhydrite and minor amounts of lead, silver and several other metals occur in stratified accumulations along the medial axis of rifting. Some of the material is of ore grade. The deposits are believed to be forming today from solutions being discharged on the sea floor. As the accreting plate margins are generally in the ocean depths (Fig. 1), other deposits of this type may remain to be discovered. However, several lines of evidence, among them the ratios of the saline constituents and isotopic compositions of the overlying brines, indicate very strongly that the brines and their metals are in fact derived from continental sediments flanking the pools (White, 1968), and hence that such concentrations are unlikely to occur along most of the world-girdling extent of these "mid-ocean" rises. Contrary evidence as to origin may be provided by Bostrom and Peterson (1966), who found that iron, manganese, copper, chromium, nickel and lead are enriched in samples of sediments taken from the crest of the East Pacific Rise west of the Chile-Peru trench.

Copper and nearby manganese deposits on the east coast of the peninsula of Baja California may represent links between the oceanic concentrations and terrestrial deposits. Stratiform orebodies occur in clayey tuff beds of the Boleo Formation of early Pliocene age. This formation, which consists of thin nonmarine basal conglomerate, a thin marine limestone, gypsum and overlying interbedded tuff and tuffaceous conglomerate of latitic to andesitic composition, constitutes the initial deposits in the newly forming Gulf of California (Wilson, 1955). The ores have been called syn- and epigenetic and have been compared with other controversial bedded ores such as the Kupferschiefer. Wilson (1955) favored an epigenetic origin from ascending solutions, but ones related to the same igneous activity that produced the volcanic materials. As Guilbert (1971) pointed out, whether the ores are of volcanic-exhalative or replacement origin, they formed at about the time of initial separation of Baja California from the mainland. He emphasized that the deposits occur near a transform fault defined by the submarine contours and earthquake data and that the reconstruction of the pre-rift geography could lead to discovery of an offset segment. It seems entirely possible that the mineralization occurred at the intersection of the East Pacific Rise itself and an incipient transform fault when separation was in its initial stage, and that these deposits constitute a valid example of mineralization along an accreting plate margin. If this is true, some of the late Tertiary mineralization of the western United States may be cognate in the sense that it originated over a rifting rather than subducting zone. Obviously, further study is needed to determine whether this mineralization is related to the calc-alkaline suite and, if so, whether magmas of this composition can be generated in a nonorogenic environment.

In conclusion, evidence is strong for a close spatial relationship between present plate boundaries and endogenic ore deposits of many of the common metals, and the presumption that this relationship also existed in earlier times appears valid. Furthermore, although many details need confirmation, a genetic relationship also appears entirely probable for the deposits with magmatic affinities. This does not necessarily imply derivation of the metals from the descending slab with its layer of sediments; the possibility of multiple sources or even of nonmagmatic sources is not precluded, the role of the magmas being confirmed to providing energy (heat) to mobilize the ore elements and concentrate them. This subject has been widely discussed and argued; it goes far beyond the scope of this paper.

It is tempting to look back in time. The relationship of Laramide to later belts is obvious in most places, as mentioned above for South America, and to the author it seems very probable that as additional data on all aspects of earth history are analyzed in the light of the new theories, the place of mineral deposits in the total picture will be increasingly clear. The role of the economic geologist should become that of a contributor to as well as beneficiary of the more theoretical aspects of global tectonics, and the explorationist should gain immensely.

REFERENCES

Bazin, D., and Hübner, H., 1969. Copper deposits in Iran. Iran Geol. Surv., Rep. No. 13, 232 p.
Bostrom, K., and Peterson, M. N. A., 1966. Precipitates from hydrothermal exhalations of the East Pacific Rise. Econ. Geol. 61, p. 1258-1265.
Buddington, A. F., 1933. Correlation of kinds of igneous rocks with kinds of mineralization. *In* Ore deposits of the Western States. Am. Inst. Min. Metall. Eng., New York, p. 350-385.
Carr, M. J., Stoiber, R. E., and Drake, C. L., 1971. A model of upper mantle structure below Japan (Abstr.). Am. Geophys. Union, Trans., 52, p. 279.
Clark, A. H., Farrar, E., Haynes, S. J., Quirt, S., Conn, H., and Zentilli, M., 1970. K-Ar chronology of granite emplacement and associated mineralization, Copiapó mining district, Atacama, Chile. Geol. Soc. Am., Abstr. with Programs, 2, p. 521.
Degens, E. T., and Ross, D. A., 1969. Hot brines and recent heavy metal deposits in the Red Sea. Springer-Verlag, Inc., New York, 600 p.
Dengo, G., Levy, E., Bohnenberger, O., and Caballeros, R., 1969. Metallogenic map of Central America. Inst. Centroam. Invest. Technol. Indus., Guatemala.
Dewey, J. F., and Bird, J. M., 1970. Mountain belts and the new global tectonics. J. Geophys. Res., 75, p. 2625-2647.
Dietz, R. S., and Holden, J. C., 1970. Reconstruction of Pangaea: Breakup and dispersion of continents, Permian to present. J. Geophys. Res., 75, p. 4939-4956.

Fonteilles, M., 1967. Appréciation de l'intérêt métallogénique du volcanisme de Madagascar à partir de ses caractères pétrologiques. France, Bur. Res. Geol. Min., Bull. No. 1, p. 121-154.

Guilbert, J. M., 1971. Known interactions of tectonics and ore deposits in the context of new global tectonics. Am. Inst. Min. Metall. Pet. Eng., Soc. Min. Eng. Preprint 71-S-91, 19 p.

Guild, P. W., 1971. Metallogeny: a key to exploration. Min. Eng. 23, p. 69-72.

Isacks, B., Oliver, J., and Sykes, L. R., 1968. Seismology and the new global tectonics. J. Geophys. Res., 73, p. 5855-5899.

Itsikson, M. I., Kormilitsyn, V. S., Krasnyi, L. I., and Matveyenko, V. T., 1960. Osnovnye cherty metallogenii severo-zapadnoi chasti Tikhookeanskogo rudnogo poyasa (The main metallogenic features of the northwestern part of the Pacific Ocean ore belt). Geol. Rud. Mestorozh. No. 1, p. 16-44. Also in English in Econ. Geol. U.S.S.R., 1, Nos. 1-2, p. 14-43.

Japan Geological Survey, 1957. Mineral province of Japan: I. Mineralization of Quaternary Period. Map, 1:2,000,000.

————, 1960. ————: II. Mineralization of Neogene Period. Map, 1:2,000,000.

McBirney, A. R., 1969. Andesitic and rhyolitic volcanism of orogenic belts. In Hart, P. J. (Editor) The Earth's crust and upper mantle. Am. Geophys. Union, Geophys. Mon. 13, p. 501-507.

McKelvey, V. E., and Wang, F. F. H., 1969. World subsea mineral resources. U.S. Geol. Surv., Misc. Geol. Inv. Map I-632.

Moisseyev, A. N., 1971. A non-magmatic source for mercury ore deposits? Econ. Geol., 66, p. 591-601.

Moores, E., 1970. Utramafics and orogeny, with models of the U.S. Cordillera and the Tethys. Nature, 228, p. 837-842.

Oxburgh, E. R., and Turcotte, D. L., 1970. Thermal structure of island arcs. Geol. Soc. Am., Bull., 81, p. 1665-1688.

Petersen, U., 1970. Metallogenic provinces in South America. Geol. Rundschau, 59, p. 834-897.

Roberts, R. J., Radtke, A. S., and Coats, R. R., 1971. Gold-bearing deposits in north-central Nevada and southwestern Idaho. Econ. Geol. 66, p. 14-33.

Ruiz Fuller, C., 1965. Geología y yacimientos metalíferos de Chile. Inst. Invest. Geol. Chile, 305 p.

Sillitoe, R. H., 1970. South American porphyry copper deposits and the new global tectonics (Abstr.). Primer congreso latino-americano de geología, Lima, p. 254-256.

Sumner, J. S., 1970. Mining geophysics, past, present and future. Geol. Soc. Am. Abstr. with Programs, 2, p. 696-697.

Sutherland Brown, A., Cathro, R. J., Panteleyev, A., and Ney, C. S., 1971. Metallogeny of the Canadian Cordillera. Can. Inst. Min. Metall. Trans., 74, p. 121-145.

Tatsumi, T., Sekine, Y., and Kanehira, K., 1970. Mineral deposits of volcanic affinity in Japan: Metallogeny. In Tatsumi, T. (Editor) Volcanism and ore genesis. Univ. Tokyo Press, Tokyo, p. 3-47.

Turneaure, F. S., 1971. The Bolivian tin-silver province. Econ. Geol., 66, p. 215-225.

Tweto, O., and Sims, P. K., 1963. Precambrian ancestry of the Colorado mineral belt. Geol. Soc. Am., Bull., 74, p. 991-1014.

White, D. E., 1968. Environments of generation of some base-metal ore deposits. Econ. Geol., 63, p. 301-335.

Wilson, I. F., 1955. Geology and mineral deposits of the Boleo copper district, Baja California, Mexico. U.S. Geol. Survey, Prof. Pap. 273, 134 p.

17

A Plate Tectonic Model for the Origin of Porphyry Copper Deposits

Richard H. Sillitoe

Abstract

The theory of lithosphere plate tectonics, embodying the concepts of sea-floor spreading, transform faulting, and underthrusting at continental margins and island arcs, is employed as a basis for an actualistic, though speculative, model for the origin and space-time distribution of porphyry copper and porphyry molybdenum deposits.

Porphyry ore deposits, occurring in the western Americas, southwest Pacific and Alpide orogenic belts, are thought to constitute a normal facet of calc-alkaline magmatism. Chemical and isotopic data cited are consistent with the generation of the components of calc-alkaline igneous rocks and porphyry ore deposits by partial melting of oceanic crustal rocks on underlying subduction zones at the elongate compressive junctures between lithospheric plates.

It is proposed that the metals contained in porphyry ore deposits were derived from the mantle at divergent plate junctures, the ocean rises, as associates of basic magmatism, and transported laterally to subduction zones as components of basaltic-gabbroic oceanic crust and small amounts of suprajacent pelagic sediments; evidence supporting the presence of significant amounts of metals in the oceanic crust is listed.

It is suggested that the temporal and spatial distribution of porphyry ore deposits is dependent on two principal factors, namely the erosion level of an intrusive-volcanic chain, and the time and location of magma generation, and the availability of metals, on an underlying subduction zone. The erosion factor is believed to offer an explanation for the paucity of porphyry ore deposits in pre-Mesozoic orogenic belts, and for the relative abundance of exposed porphyry deposits of Upper Cretaceous-Paleogene age in post-Paleozoic orogens. Provinces with a high concentration of porphyry copper deposits, such as southern Peru-northern Chile and the southwest United States, may be interpreted as regions beneath which anomalously copper-rich oceanic crust was subducted at the time of porphyry copper emplacement; one possible explanation for the episodic formation of volumes of copper-rich oceanic crust is the presence of a heterogeneous distribution of metals in the low velocity zone of the upper mantle. Porphyry ore deposits seem to have formed during a series of relatively short, discrete pulses, perhaps correlable with changes in the relative rates and directions of motion of lithospheric plates. In some regions, such as Chile, porphyry ore deposits are arranged in parallel, linear belts, which may be explicable in terms of shifting loci of magma and included metal generation on a subduction zone, and which seem to be largely independent of control by tectonic lineament intersections. The time intervals during which the formation of porphyry deposits took place are shown to be broadly coincident with periods of lithosphere plate convergence, and porphyry deposits may still be forming above currently active subduction zones.

A number of potential regions for the discovery of porphyry ore deposits are suggested, and the importance to exploration of analyzing orogenic belts in terms of plate tectonics is emphasized.

Introduction

THE sea-floor spreading hypothesis of Dietz (1961) and Hess (1962) has recently been further developed by Isacks, Oliver and Sykes (1968), Le Pichon (1968), McKenzie and Parker (1967) and Morgan (1968), resulting in the theory of lithosphere plate tectonics (the new global tectonics), the only existing global tectonic model compatible with much recently accumulated geological and geophysical data. The model considers the Earth's surface to be divided into six large and several smaller lithospheric plates (Fig. 2), which accrete at ocean rises by uprise of subcrustal basic magmas, slide past one another along transform (horizontal shear) faults, and are destroyed at trench systems by descent into the asthenosphere down inclined subduction (Benioff) zones (Fig. 3).

The concepts of the new global tectonics have rendered the stabilist geosynclinal theory of orogeny outmoded; it is now realized that a generalized model for geosynclinal development, commencing with the accumulation of a thick volcanic and sedimentary pile, and followed by plutonism and deformation and finally by epeirogenic uplift and volcanism (e.g., Beloussov, 1962) does not satisfactorily explain the evolution of most orogens (Coney, 1970). In terms of the theory of plate tectonics, geosynclines may be equated with oceans and continental margins, and oceans and island arcs. Mountain systems are generated as a consequence of the underthrusting of oceanic lithosphere beneath an adjacent plate at continental margins (e.g., the cordilleran system of North and South America), or at island arcs (e.g., Japan). In some instances, eventual continent-continent collision (e.g., the Himalayas) or continent-island arc collision (e.g., New Guinea) are involved in orogenic development (Dewey and Bird, 1970; Dewey and Horsfield, 1970). It should be stressed, however, that each orogenic belt displays an essentially unique sequential history, even though certain sequences of events are more common than others.

Currently accepted concepts of the metallogenesis of post-Precambrian orogenic belts are based on stabilist geotectonic doctrine (Bilibin, 1968; McCartney and Potter, 1962; McCartney, 1964), and support the association of distinct types of mineralization with each stage of geosynclinal development. Porphyry copper deposits, for instance, are considered to typify the post-orogenic, late tectonic stage (McCartney and Potter, 1962). In view of the recent advances in geotectonic theory, it would therefore seem opportune to reexamine the metallogenesis of orogenic belts in terms of the new global tectonics. This paper outlines a plate tectonic model to account for the genesis and distribution in both space and time of one important class of mineralization, the porphyry copper and porphyry molybdenum deposits. Some aspects of the model were presented elsewhere as an abstract (Sillitoe, 1970). It should perhaps be stressed that the proposed model is of a speculative nature, and does not purport to embody rigorous proofs of its validity.

Geological and Genetic Characteristics of Porphyry Copper Deposits

Over one half of the world's copper production is currently derived from porphyry copper deposits, large tonnage (commonly exceeding 500 m tons), low grade, roughly equidimensional deposits of disseminated and stockwork-veinlet, pyrite-chalcopyrite mineralization, carrying at least trace amounts of molybdenum, gold and silver. They are spatially and genetically related to passively-emplaced hypabyssal felsic stocks, commonly porphyries. The geological characteristics of the porphyry molybdenum de-

posits, from which molybdenum is the principal metal recovered, are similar to those of the porphyry coppers. Hypogene ore grade in the porphyry coppers rarely exceeds 1% Cu, and is commonly below 0.5% Cu.

The host intrusions of porphyry copper deposits, and their various types of country rocks, may both be ore-bearing, and are characterized by widespread, zonally-arranged hydrothermal alteration, commonly of potassic, phyllic, argillic and propylitic types (Meyer and Hemley, 1967; Lowell and Guilbert, 1970), and by hydrothermal brecciation.

The close association of intrusion and mineralization in porphyry copper and molybdenum deposits is emphasized by K-Ar dating which has demonstrated the two processes to be temporally indistinguishable in several instances (e.g., Livingston, Mauger and Damon, 1968; Moore, Lanphere and Obradovich, 1968; Laughlin, Rehrig and Mauger, 1969). In at least some deposits, part of the mineralization may in fact be syngenetic with respect to the associated intrusive rock (e.g., Ely, Nevada; Fournier, 1967). The intimate temporal association of intrusion and mineralization lends support to the orthomagmatic model of porphyry copper genesis (Nielsen, 1968; Lowell and Guilbert, 1970). According to this model, a felsic magma, becoming water-saturated as it intrudes towards the surface zone, undergoes crystallization of its outer parts, which are subsequently brecciated by the release of accumulated fluids, which also produce the alteration and mineralization. Meteoric waters are involved in the formation of the outer zones of hydrothermal alteration and mineralization.

The Origin of Calc-Alkaline Igneous Rocks and Porphyry Copper Deposits

World Distribution of Porphyry Ore Deposits

Figure 1 shows the location of the majority of exploited porphyry copper and molybdenum deposits, and of many important prospects, which in Figure 2 are related to Mesozoic-Cenozoic orogenic belts and currently active lithospheric plate boundaries. It can be appreciated that the majority of the world's porphyry deposits are located in the circum-Pacific orogenic belts and in the central portion of the Alpide orogenic belt. The western Americas belt, containing most of the known porphyry copper deposits, extends from western Argentina and central and northern Chile, through Peru, Ecuador, Panama, Mexico, the western United States (Arizona, New Mexico, Nevada, Utah, Colorado, Idaho, Washington and Montana), to British Columbia, the Yukon and Alaska. Marked concentrations of deposits occur in Sonora-Arizona-New Mexico and in British

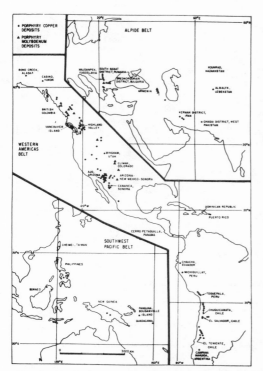

Fig. 1. Porphyry copper and molybdenum deposits and prospects in the western Americas, southwest Pacific and Alpide belts.

Columbia. Deposits in the Dominican Republic and Puerto Rico may be considered as an offshoot of the western Americas belt. Two other belts, for which published information on the porphyry deposits is scant, are located in the Taiwan, Philippines, Borneo, West Irian, Papua-New Guinea and Solomon Islands region (the southwest Pacific belt), and in the South Banat district of Romania, Yugoslavia, central Bulgaria, Armenia, Iran and West Pakistan (the Alpide belt). The only well-authenticated occurrences of porphyry deposits outside of these post-Paleozoic orogenic belts are those in Uzbekstan and Kazakhstan, USSR.

Relationships between Calc-Alkaline Igneous Rocks and Porphyry Copper Deposits

The post-Paleozoic history of the continental margins and island arcs where porphyry copper deposits are located was characterized by widespread calc-alkaline volcanism which gave rise to basalts, andesites, dacites, rhyolites and, in some parts, felsic

ignimbrites. These volcanic rocks are commonly observed to be intruded or underlain by extensive batholiths and smaller intrusions of a similar composition. Hamilton and Myers (1967) and Hamilton (1969a, b) have convincingly demonstrated the consanguinity of the calc-alkaline volcanic suite and the spatially related felsic plutonic rocks, the latter interpreted as the roots of major eruptive chains.

The location of porphyry copper orebodies in the cupolas of plutons of intermediate composition was emphasized many years ago (Emmons, 1927), and subsequent work tends to confirm the high-level, subvolcanic nature of their environment of formation. Some porphyry copper deposits may have been emplaced at very shallow depths, perhaps at less than 1,500 m (Fournier, 1968). This is emphasized by the occurrence at Bingham, Utah of a porphyry-copper stock and nearby penecontemporaneous, comagmatic volcanic rocks (Moore, Lanphere and Obradovich, 1968).

It is suggested that the accumulation of copper and molybdenum in high-level felsic stocks was a normal part of calc-alkaline magmatism in post-Paleozoic orogenic belts.

The Origin of Calc-Alkaline Igneous Rocks

Many workers (e.g., Benioff, 1954; Coats, 1962; Dickinson and Hatherton, 1967; Dickinson, 1968; Ringwood, 1969) agree that the magmas which have given rise to calc-alkaline volcanic rocks—and by analogy their plutonic equivalents (Hamilton, 1969a) —were generated on subduction zones which underlay the eruptive chains. The magmas were probably generated by partial fusion consequent upon frictional heating of subducted, water-saturated oceanic lithosphere which originally was generated at ocean rises, and transported laterally into ocean trenches (Fig. 3). Restricted volumes of fused ocean-floor sediments (layer 1) and the lowest melting fractions of layers 2 and 3 of the oceanic crust seem to be the most likely source materials for calc-alkaline magmas (Oxburgh and Turcotte, 1970). Several chemical and isotopic parameters determined for calc-alkaline volcanic and intrusive rocks are now cited in support of an origin by partial melting on a subduction zone.

Recent volcanic rocks in the circum-Pacific belt show a systematic increase in their potash to silica ratios landwards from the trench (Dickinson and Hatherton, 1967; Dickinson, 1968); these ratios correlate with the depth from the site of eruption to the Benioff zone dipping beneath the island arc or continental margin, suggesting an origin on the Benioff zone, and the absence of widespread crustal contamination. A comparable landward increase in potash has also been demonstrated for post-Paleozoic volcanic and intrusive rocks in parts of western

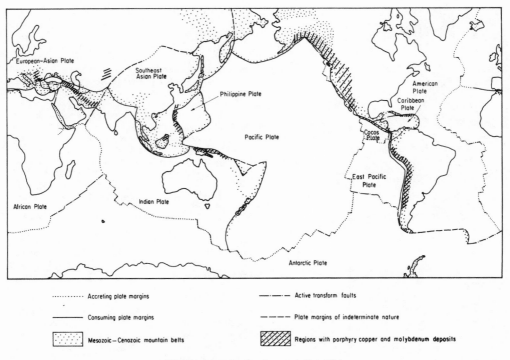

(Plate boundaries taken from Dewey and Bird, 1970)

FIG. 2. The western Americas, southwest Pacific and Alpide porphyry belts in relation to Mesozoic-Cenozoic orogenic belts and accreting and consuming plate boundaries

North America (Moore, 1959, 1962; Moore, Grantz and Blake, 1963; Bateman and Dodge, 1970).

Initial Sr^{87}/Sr^{86} ratios (0.703–0.706) obtained for andesitic volcanics (Ewart and Stipp, 1968; Pushkar, 1968; Peterman, Carmichael and Smith, 1970), and those (0.705–0.709) obtained for felsic rocks from the British Columbia, Sierra Nevada and Boulder batholiths (Fairbairn, Hurley and Pinson, 1964; Hurley et al., 1965; Doe et al., 1968) are incompatible with an origin by partial melting or wholesale assimilation of continental crust, but would seem to be in accord with a derivation by partial melting of oceanic crust on a subduction zone. The trace element content of andesites is also consistent with a Benioff-zone origin (Taylor, 1969; Taylor et al., 1969).

The Origin of Porphyry Copper and Molybdenum Deposits

In view of the close temporal and spatial relationship between the genesis of porphyry ore deposits and calc-alkaline magmatism, summarized in the preceeding sections, the components of porphyry-copper stocks, including the contained metals, are likewise postulated to possess an origin by partial melting of oceanic crust on a subduction zone. Initial strontium isotope ratios in the range 0.706–0.708 obtained for several porphyry-copper stocks in the southwest United States and for the stock associated with the Questa, New Mexico porphyry molybdenum deposit (Moorbath, Hurley and Fairbairn, 1967; Laughlin, Rehrig and Mauger, 1969) support this contention. A deep, homogenized, probably mantle-source for sulfide sulfur in porphyry copper and molybdenum deposits in the southwest United States is suggested by δS^{34} values close to the meteoric standard (Field, 1966; Jensen, 1967; Laughlin, Rehrig and Mauger, 1969).

In summation, therefore, porphyry copper and molybdenum deposits are considered to be confined to orogenic belts characterized by calc-alkaline magmatism, and resulting from plates of oceanic

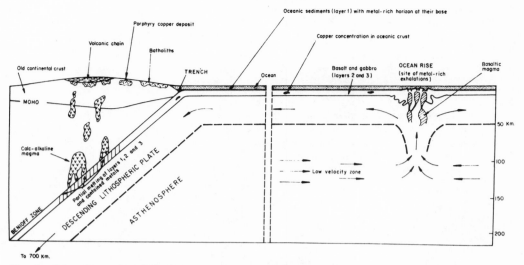

FIG. 3. Schematic representation of the genesis of porphyry copper deposits in the context of plate tectonics.

lithosphere underthrusting adjacent lithospheric plates, in some cases with ensuing continental collision—the compressive type of elongate contact between two lithospheric plates.

The Source of Metals in Porphyry Ore Deposits

The source of metals in post-magmatic sulfide ore deposits has long been a topic of contention. For the porphyry copper deposits in particular, a continental crustal provenance of copper by extraction from andesitic volcanics (Ney, 1966), geosynclinal sediments (United Nations, 1970), or shale horizons (Jensen, 1971), during igneous intrusion, has been inferred. However, considering the wide spectrum of host rock types of porphyry copper deposits, namely calc-alkaline volcanics, intrusive and metamorphic rocks, and marine sediments, partially calcareous, and taking into account the probable absence of certain of these source rock types from beneath some porphyry copper provinces, the provision of the metals by a specific rock type in the continental crust seems improbable. Furthermore, the only source of metals available to all porphyry copper deposits, which takes into consideration the thin crust and almost total absence of sialic crust beneath some porphyry copper provinces (e.g., the Solomon Islands; Coleman, 1966), would seem to be the upper mantle.

An important aspect of the model outlined above for the origin of porphyry ore deposits, is the implication that a large percentage of the metals were extracted from oceanic crust during partial melting as it sank into the mantle on a subduction zone. Collection of metals by saline fluids expelled from oceanic crust during subduction may also be operative. Initially, the metals were largely derived from the mantle at the ocean rise system and carried to th margins of ocean basins as components of layers 1, 2 and 3 of the oceanic crust (Fig. 3). The wedge of mantle above a Benioff zone may also act as a source for basaltic magmas (Oxburgh and Turcotte, 1970), and minor quantities of copper and molybdenum. The metals released during partial melting ascended as components of calc-alkaline magmas (Fig. 3), and were ultimately concentrated in chloride-rich fluid phases associated with the roof-zones of certain intrusions. The fluid phase was released upwards during consolidation of the magma to give rise to the typical upright cylinders of porphyry copper and molybdenum mineralization.

A corollary of this premise is that regions possessing high concentrations of porphyry copper deposits, such as the southwest United States and southern Peru–northern Chile, and considered as copper-rich metallogenetic provinces (e.g., Turneaure, 1955), are not regions of the Earth where the subjacent continental crust or upper mantle are enriched in copper, but regions beneath which anomalously copper-rich oceanic crust, including pelagic sediments, has been subducted. Variation in the amount of copper consumed in a zone of subduction, and therefore potentially available for mineralization, may depend on the rate of sea-floor spreading and hence the volume of oceanic crust entering the subduction zone, or the intensity of volcanism and metal production at the ocean rise, or, more fundamentally, may reflect an inhomogeneous distribution of this metal in the upper mantle, perhaps in the low velocity zone (Fig. 3); chemical heterogeneity of the mantle has recently been demonstrated (Peterman and Hedge, 1971).

Evidence Bearing on the Derivation of Metals from Oceanic Crust

The following points are thought to provide evidence favoring the oceanic crust as a source for the copper and molybdenum contents of porphyry ore deposits:

1. Recent workers (Cann, 1968; Oxburgh and Turcotte, 1968; Thayer, 1969; Christensen, 1970) have proposed that the third layer of the oceanic crust has a gabbroic composition, perhaps with dolerite dike complexes in its lower part (Dewey and Bird, 1970). It is overlain by basalts of layer 2. Basic igneous rocks have average copper and molybdenum contents of the order of 100 ppm and 1.5 ppm, respectively, five times that of granitic rocks in the case of the former element (Turekian and Wedepohl, 1961; Vinogradov, 1962). Oceanic basalts (layer 2) in the Atlantic and Pacific Oceans average 77 ppm Cu (Engel, Engel and Havens, 1965).

2. Sulfide phases containing up to approximately 10 percent copper are present in vesicles and as globules in pillow basalts from layer 2 on the ocean rises (Moore and Calk, 1971).

3. Sulfide grains of a similar nature to those occurring in ocean-floor basalts (but with up to 25 percent copper) have been discovered in recent Hawaiian basalt flows (Desborough, Anderson and Wright, 1969), and in basaltic oozes which flowed into a drill hole in the crust of Alae lava lake, Hawaii (Skinner and Peck, 1969). The Hawaiian islands are mantle-fed volcanoes in the center of the Pacific lithospheric plate (Menard, 1969).

4. Large alpine-type mafic-ultramafic (ophiolite) complexes such as those in Cyprus, Turkey, Papua and elsewhere are thought to represent fragments of surficial rinds of ancient ocean-floor plates (Dietz, 1963; Thayer, 1969; Bird and Dewey, 1970). Support for such a belief stems from records of similar rock types on the mid-Atlantic and Indian Ocean ridges (Bonatti, 1968; Engel and Fisher, 1969). The Mesozoic Troodos ophiolite complex of Cyprus is thought to correspond to layers 2 and 3, formed beneath an ocean rise in the Tethys Ocean (Gass, 1968). The well-known cupriferous pyrite deposits of Cyprus are interbedded in a succession of basaltic

pillow lavas, which, according to Gass (1968), would represent a part of layer 2 of the oceanic crust. The close genetic relation between the basaltic volcanics and the copper deposits (Vokes, 1966; Hutchinson and Searle, 1970) is evidence that concentrations of copper are present in the oceanic lithosphere.[1]

5. Manganese nodules on the ocean floors have copper contents as high as 2.5%. Analyses of nodules from the Pacific Ocean showed from 0.03 to 1.6% Cu (United Nations, 1970a).

6. Normal pelagic clays (layer 1) possess metal contents in excess of most sedimentary rocks, and average abundances of 323 ppm Cu and 18 ppm Mo have been recorded for Pacific clays (Cronan, 1969).

7. Base metals, including copper and molybdenum, of probable mantle derivation are concentrated on the East Pacific Rise (Boström and Peterson, 1966, 1969), and on the rest of the ocean rise system (Boström et al., 1969). Layer 1 sediments on the flanks of the East Pacific Rise (and presumably away from the rise beneath a blanket of normal ocean-floor sediments) have average copper and molybdenum contents of 990 ppm and 100 ppm, respectively.

8. The floor of the Red Sea, also a locus for the generation of new oceanic crust (e.g., Vine, 1966), possesses a series of metal-rich brine pools and sediments (Miller et al., 1966; Degens and Ross, 1969), possibly also charged directly from the mantle.

9. The early Pliocene Boléo bedded copper deposit in Baja California (Wilson and Rocha, 1955) may possibly represent another, albeit somewhat older, example of copper which has risen directly from the mantle, in this case related to the northwestward rotation of Baja California away from the rest of the North American continent, along segments of the East Pacific Rise mutually offset by transform faulting, a process initiated in the late Miocene to middle Pliocene (Moore and Buffington, 1968; Larsen, Menard and Smith, 1968). The high manganese- and iron oxide contents of the Boléo ores (Wilson and Rocha, 1955) provide a further similarity to the metal concentrations of the East Pacific Rise and the Red Sea.

It is also conceivable that the metals (6 ppm Cu; White, 1968) in the brines of the Salton Sea geothermal system, lying just north of the Gulf of California, and possibly underlain by the East Pacific Rise, have a comparable mantle source, although their derivation by low-grade metamorphism of clastic sediments has been proposed (Skinner et al., 1967).

[1] This suggests the possibility that other massive pyritic sulfide deposits associated with basaltic pillow lavas may have formed on the crests and flanks of ocean rise systems.

TABLE 1. Ages of Porphyry Copper and Molybdenum Deposits

Name of Porphyry Deposits or Regions	Age[2] of Porphyry Deposits	Source of Data
British Columbia		
Most porphyry copper and molybdenum deposits	Upper Triassic-Middle Jurassic Upper Jurassic Late Lower Cretaceous Upper Paleocene-Upper Eocene	White, Harakal and Carter (1968); Brown (1969)
Vancouver Island	Lower Eocene-Lower Oligocene	Carson (1969)
Western United States and Sonora		
Most porphyry copper deposits	Upper Cretaceous-Paleocene	Creasey and Kistler (1962);
Bisbee, Arizona	Middle Jurassic	McDowell and Kulp (1967);
Ely and Yerington, Nevada	Lower Cretaceous	Moorbath, Hurley and Fairbairn
Bingham, Utah	Lower Oligocene	(1967); Livingston, Mauger and
Front Range porphyry molybdenum deposits	Upper Oligocene-Lower Miocene	Damon (1968); Moore, Lanphere and Obradovich (1968); Tweto (1968); Wallace et al. (1968); Laughlin, Rehrig and Mauger (1969)
Panama		
Botija	Lower Oligocene	Ferenčić (1970)
Ecuador		
Chaucha	Upper Miocene	Müller-Kahle and Damon (1970)
Peru		
Southern Peru	Paleocene	Laughlin, Damon and Watson
Michiquillay	? Lower Miocene	(1968); Stewart and Snelling (in prep.)
Argentina	Tertiary (S)	United Nations (1970)
Farellón Negro, Catamarca	Upper Miocene-Pliocene (S)	Llambias (1970)
Chile	Upper Cretaceous Paleocene Upper Eocene-Oligocene Upper Miocene-Pliocene	Sillitoe, Quirt, Clark, Farrar and Neumann (in prep.)
Bougainville Island		
Panguna	? Pliocene (S)	Macnamara (1968)
Taiwan		
Chemei	Miocene or later	Po and Lee (1970)
Philippines	Tertiary (S)	Bryner (1969)
Atlas	Upper Paleocene	
West Pakistan		
Chagai district	? post-Oligocene (S)	Schmidt (1968)
Iran		
Kerman region	Upper Oligocene-Miocene (S)	Bazin and Hübner (1969)
Armenia	Upper Eocene Lower Oligocene Lower Miocene	Bagdasaryan, Gukasyan and Kara- myan (1969)

[2] Time scale according to Harland, Smith and Wilcock (1964).
(S) Stratigraphic estimate.

The Distribution of Ages of Porphyry Ore Deposits

Published ages for the main groups of porphyry copper and molybdenum deposits are summarized in Table 1. It is apparent that deposits were emplaced at intervals throughout the Mesozoic and Cenozoic, with a particularly large number of deposits of late Cretaceous-Paleogene age.

It is suggested that two principal factors control the space-time distribution of porphyry ore deposits. These are: 1. the level of exposure—largely dependent on erosion rate—of a plutonic-volcanic chain; and 2. the time and location of magma generation, and the quantity of metals incorporated in magmas, on a subduction zone. It is, of course, recognized that additional variables, such as the quantity of chloride-rich fluid present during the final stages of consolidation of an intrusion, are of considerable importance.

Evidence has been advanced to show that mechanisms implicit in the new global tectonics were operative in pre-Mesozoic times (Bird and Dewey, 1970). Nevertheless, with the exception of Uzbekstan and Kazakhstan, porphyry ore deposits have not yet been definitely described from the older orogenic belts. This apparent absence is tentatively attributed to the effects of Mesozoic-Cenozoic erosion, which

111

has been sufficient to remove the upper parts of batholiths, the loci of porphyry deposits. However, it is predicted that porphyry deposits will be encountered in parts of certain pre-Mesozoic orogens where erosion has been less severe.[3] The apparent predominance of late Cretaceous-Paleogene ages of porphyry deposits (Table 1) might also be dependent on the erosion factor, whereby many early Mesozoic deposits have been eroded away, and *some* post-Paleogene deposits have yet to be exhumed from beneath their volcanic cover. Support for this proposal is derived from conditions in Chile north of latitude 30°S, where the erosion level becomes progressively deeper from the recent volcanic chain on the Andean crest westwards to the Jurassic intrusions on the Pacific littoral (Sillitoe and Sawkins, 1971). The majority of the exposed porphyry copper deposits are Paleogene in age; Jurassic deposits, if they ever existed, have been lost by erosion, and deposits yet to be exposed may exist in the recent volcanic chain. However, the second factor, discussed below, may also be an important contributary, or even the dominant, cause of the relative abundance of Paleogene deposits in northern Chile.

If the importance of the erosion factor has been correctly evaluated, then porphyry deposits in regions with a high erosion rate, such as the southwest Pacific belt characterized by a tropical climate, could be expected to yield a predominance of particularly young ages; more radiometric dating is needed in order to test this proposal.

Explanations of many features of the distribution of porphyry copper and molybdenum deposits can be attempted in terms of the second factor, the lateral and secular pattern of magma generation, and the availability of copper and molybdenum, on subduction zones. In northern Chile, discrete post-Paleozoic intrusive episodes are manifested by a series of north-south-trending belts of batholiths and stocks. The ages of these belts decrease from Lower Jurassic near the coast to late Tertiary in the Andean Cordillera (Ruiz et al., 1965; Farrar et al., 1970). The ages of porphyry copper deposits in this region possess an analogous space-time distribution, although Jurassic deposits are as yet unknown. Thus the possibility arises that each discrete pulse of magma generation had the potential to give rise to porphyry ore deposits; the extended time interval and episodicity of porphyry deposit genesis, particularly in western North America, as reflected by the ages in Table 1, support this conclusion. The data

[3] Pre-Mesozoic porphyry copper-type deposits have been reported from northwest of St. John, New Brunswick (Ruitenberg, Shafiquallah and Tupper, 1970), and from east-central Queensland (Cornelius, 1969), but no particulars of the occurrences were given.

presently available suggest that the periods of porphyry copper formation in Chile were separated by quiescent intervals with durations of about 15–25 m.y. Similar pulse-like igneous intrusion has also been reported from parts of western North America, pulses lasting approximately 10–15 m.y. and being separated by 30 m.y. intervals (Damon and Mauger, 1966; Evernden and Kistler, 1970; Gabrielse and Reesor, 1964). Such magmatic periodicity may be correlable with changes in the thermal regime on subduction zones induced by variations in the relative spreading rate or motion pattern of the plates. In this context, changes in the relative motion of plates every 10–20 m.y. in the northeast Pacific, deduced by Francheteau, Sclater and Menard (1970), might be significant.

The eastward migration of the foci of intrusion and porphyry copper emplacement evident in northern Chile, which perhaps reflects a parallel trend in the position of magma generation on the underlying subduction zone, whether or not caused by a change in its position or inclination relative to the continental margin, is less well defined in the western United States. In the latter region, Gilluly (1963) recognized an overall decrease in the age of Mesozoic-Cenozoic intrusion landwards from the continental margin, but subsequent programs of radiometric dating have shown many exceptions to this generalization. A broadly comparable pattern of eastward younging is apparent from the porphyry deposits. The belt of mid-Tertiary porphyry molybdenum deposits in the Front Range lies east of the main cluster of late Cretaceous-Paleocene porphyry copper deposits in Sonora-Arizona-New Mexico (Fig. 1), and the Lower Cretaceous porphyry deposits at Ely and Yerington, Nevada are located in the western part of this porphyry province. The great concentration of late Cretaceous-Paleocene porphyry copper deposits in the southwest United States is visualized as being due to the subduction of areas of exceptionally copper-rich oceanic crust. This contention is supported by the occurrence in the same province of unmineralized (with the exception of Bingham, Utah) mid-Tertiary stocks, which I consider to have been intruded at a time when lesser amounts of copper were available on the subjacent subduction zone. Theories invoking the extraction of copper from the continental crust or upper mantle fail to account for the concentration within a limited time period of most of these porphyry copper deposits; furthermore, post-Paleozoic stocks of all ages in the province would be expected to be similarly endowed with porphyry copper deposits. Continuing the same line of argument, large amounts of molybdenum, and only minor copper, are thought to have been available on a subduction zone vertically beneath the Front Range in mid-Tertiary time. If the locus of magma generation on a subduction zone does not migrate systematically with time, then no clear pattern of porphyry copper ages is to be expected, a situation which may explain the apparently random distribution of ages in British Columbia.

Porphyry Copper Deposits/Plate Tectonics Interrelationships

Details of the world distribution of the evolving system of ocean rises and trenches which existed during Mesozoic and early Cenozoic times are not yet well known. Nevertheless, evidence derived from both continents and ocean basins has enabled the establishment of some aspects of plate tectonics at this time. The distribution of porphyry deposits in selected regions will now be related in general terms to the plate tectonic model:

Western South America

Evidence derived from the interpretation of magnetic anomaly patterns in the Pacific Ocean basin (Heirtzler et al., 1968; Morgan, Vogt and Falls, 1969) and the paleontological study of sediment cores in the South Atlantic Ocean basin (Maxwell et al., 1970) has demonstrated the convergence of the American and East Pacific plates along the western margin of South America (Fig. 2) since at least the late Cretaceous. This interval of active underthrusting embraces the period of formation of porphyry copper deposits in Chile, Argentina, Peru and Ecuador. In western South America, porphyry copper deposits may well be forming still beneath the active volcanic chains, for underthrusting is continuing (e.g., Plafker and Savage, 1970).

Central America and the Caribbean

The Panama trench is sediment-filled, and underthrusting is inactive (Le Pichon, 1968). The trench was abandoned in the Miocene when the pattern of sea-floor spreading in the North Pacific changed (Vine, 1966; Le Pichon, 1968), but was part of a continuous trench system bordering the west of the American continent intermittently during the Mesozoic and early and middle Cenozoic, during which time the Cerro Petaquilla and Botija porphyry copper deposits of Panama were emplaced.

In Puerto Rico, porphyry copper formation seems to have been associated with a phase of igneous intrusion of Eocene age (Mattson, 1965). Although present-day eastward movement of the American plate nearly parallels the Puerto Rico trench (Chase and Bunce, 1969; Molnar and Sykes, 1969), reconstructions of plate motions in this region indicate that underthrusting normal to the trench prevailed in mid-Cretaceous-early Tertiary times (Freeland and Dietz, 1971).

Western North America

It is now generally accepted (Vine, 1966; Yeats, 1968; Hamilton, 1969b; Page, 1970) that thrusting of the East Pacific ocean floor beneath the American plate in western North America took place at times during the Mesozoic and early and middle Tertiary. Underthrusting terminated in the western United States in the Miocene (Vine, 1966; Atwater, 1970) by the overrunning of the East Pacific Rise by the trench system. Porphyry copper and molybdenum deposits in the western United States range in age from middle Jurassic to Lower Miocene (Table 1), in excellent agreement with the time of plate convergence. In view of the absence of an active subduction zone off this coast, it is concluded that porphyry copper formation is not currently active in North America, north of the tip of Baja California, except in the Alaska Peninsula where a trench system is still active, and possibly landward of the Juan de Fuca plate. It is noteworthy that calc-alkaline volcanism, typical of convergent plate junctures, ceased in the western United States south of Oregon in the Miocene (Christiansen and Lipman, 1970; Lipman, Prostka and Christiansen, 1970).

The character of Mesozoic-middle Cenozoic plate interactions off British Columbia cannot be elucidated from studies of magnetic anomaly patterns (Atwater, 1970), but it is noteworthy that calc-alkaline volcanism in British Columbia terminated in the Eocene (Souther, 1970), at about the same time as the decline in the formation of porphyry copper deposits—excluding those on Vancouver Island.

The small Juan de Fuca plate, sandwiched between North America and the northeast Pacific (Fig. 2), has descended, and may still be descending, along a trench system paralleling the coast of Oregon, Washington and Vancouver Island (Morgan, 1968; Tobin and Sykes, 1968) giving rise to calc-alkaline magmatism in the Cascades. The Lower Eocene-Lower Oligocene porphyry deposits on Vancouver Island (Carson, 1969) do not fit well into the overall space-time distribution pattern of porphyry deposits in British Columbia, and might be ascribed to earlier activity in the vicinity of this localized compressive system, as might apparently young porphyry coppers in Washington. Extrapolation of plate motions back into the early Cenozoic (Atwater, 1970) has shown that subduction of the Juan de Fuca plate was preceded by more rapid underthrusting of the Farallon plate, at a trench which did not extend further north than Vancouver Island. However, her model approximately predicts the commencement of subduction in the Upper Eocene, mid-way through the interval of formation of the porphyry deposits.

Lineament Intersections and Porphyry Copper Deposits in Western America

The locations of several southwest North American porphyry copper deposits have been attributed to major orogen- and fault-zone intersections (Billingsley and Locke, 1941; Mayo, 1958; Schmitt, 1966). More specifically, the locations of several porphyry copper deposits (e.g., Ajo, Pima-Mission and Silver Bell) have been considered to have been influenced by elements of the west-northwest- trending Texas lineament, particularly by its intersection with the Wasatch-Jerome orogen (Mayo, 1958; Schmitt, 1966; Guilbert and Sumner, 1968; Wertz, 1970). Schmitt (1966) and Guilbert and Sumner (1968) have interpreted the Texas lineament as a continental manifestation of now-extinct transform faults in the North Pacific basin. Although several porphyry copper deposits outside of the southwest United States (e.g., Chuquicamata, Chile; Taylor, 1935) lie adjacent to important faults, none have been described as being located by major structural intersections. It is suggested that the control of porphyry copper emplacement by extinct transform faults and major structural intersections is not universally applicable, and is subordinate to a fundamental dependence on elongate zones of plate convergence. In Chile, for example, the linear, longitudinal array of porphyry copper deposits (Fig. 1) provides strong support for a subduction-zone origin, and no indication of control by structural intersections is evident (Sillitoe, unpublished). In the southwest United States, the less regular, disperse pattern of porphyry copper deposits may be explained in terms of partial fusion and consequent magma and metal generation over a greater downdip extension of the underlying subduction zone; this situation might be expected if the subduction zone were flat-dipping and imbricate as invoked by Lipman, Prostka and Christiansen (1971). It is not denied, however, that lineaments may have influenced locally the uprise of magma and included metals.

The Southwest Pacific Belt

In view of the young ages (Table 1) suggested for the porphyry copper deposits in Bougainville and Taiwan, it seems probable that their formation is linked to Benioff zones occupying positions closely similar to those currently active (Fig. 2). If the porphyry copper deposits in West Irian and Papua-New Guinea prove to be post-Miocene in age, then they would seem to be related to the southward underthrusting of the Pacific plate (Fig. 2). On the other hand, if the deposits were formed in pre-

Miocene times, they would be related to a northward dipping subduction zone, which became extinct during the Miocene, by the collision of its overlying island arc (Bismarck arc) with the Australian continent (Dewey and Bird, 1970).

The Alpide Belt

The Alpide belt, in terms of the new global tectonics, is one of the least known and most complex of the compressive plate boundaries. On a global scale, the compressive forces in the Alpide belt have been attributed to relative movements between the African and Eurasian plates related to sea-floor spreading in the Central and North Atlantic Oceans (Hsü, 1971; Smith, 1971). Lithosphere was consumed along the northern and northeastern edges of the Arabian plate at the Zagros thrust zone in Iran and West Pakistan and its westerly continuation in Turkey (Dewey and Bird, 1970). Porphyry copper deposits in Iran and West Pakistan, north of the Zagros zone, were emplaced while subduction was active. The porphyry deposits in Romania, Yugoslavia and Bulgaria appear to be related to a Mesozoic-Tertiary subduction zone which, according to Dewey and Bird (1970; Fig. 14), is marked by ophiolite complexes, and extended westwards from the southern shore of the Black Sea. It might be conjectured that all the porphyry ore deposits in the Alpide belt were generated during phases of subduction related to the closure of the western Tethyan-Indian Ocean.

In the case of orogenic belts in which the collision of continents with island arcs or with other continents has contributed to their development, as in the Alpine-Mediterranean system, calc-alkaline igneous rocks and associated ore deposits may have been concealed by overthrust slices or by flysch deposits during or after collision.

Concluding Remarks

In terms of the plate tectonic model outlined above for the genesis of porphyry ore deposits, several suggestions for exploration may be made. A consideration of the distribution of Mesozoic-Cenozoic subduction zones (Fig. 2) indicates several areas as potential porphyry provinces, in addition to the orogenic belts of western America and their southward continuation into the Antarctic Peninsula. Probably the most obvious of these regions are Japan [4] and New Zealand where porphyry deposits have not yet been discovered despite extensive exploration. A recent compilation of ages of magmatism in island arcs (Mitchell and Bell, 1970) shows that Upper Cretaceous-Eocene or mid-Tertiary periods of volcanism accompanied by intrusion are represented, in addition to in island arcs where porphyry copper deposits are already known, in the Aleutians, Izu-Bonin, Sumatra-Java, Banda, North and South Celebes, New Hebrides and Fiji; these areas are considered important for porphyry copper exploration. The Lesser Antilles island arc, Kamchatka and Burma and Thailand are also considered to be promising targets.

In the Alpide zone, Turkey,[4] Greece and Afghanistan seem to be likely areas for porphyry copper discovery. A detailed analysis of the Alpine-Mediterranean orogen in terms of the new global tectonics would provide a cogent tool in the search for porphyry copper deposits. For example, if a subduction zone existed during Mesozoic and early Tertiary times at the contact of the African and European-Asian plates, as depicted by Dewey and Bird (1970; Fig. 14), then porphyry copper deposits might be expected in the north of Morocco and Algeria. If the Indus suture marks the site of the subduction zone active during the closure of the eastern part of the Tethyan-Indian Ocean (Mitchell and Reading, 1969), then porphyry deposits might be present on its northern side.

It is hoped that this model for the origin of porphyry copper and molybdenum deposits, though liable to modification in the light of further information relating to porphyry deposits and plate tectonics, will stimulate confirmatory research, and attempts to apply the new global tectonics to other classes of ore deposit. A number of the stages in the proposed model would seem to be amenable to testing by further work. Potentially important research might include age-determination and Sr^{87}/Sr^{86} studies of porphyry deposits, and isotopic and chemical studies of oceanic crustal rocks. These should be accompanied by investigations to elucidate the magmatic and metal-concentrating processes operative at the ocean rise system, and the nature of lithosphere sinking, and partial melting and metal extraction, in subduction zones.

Acknowledgments

I should like to thank Dr. A. H. Clark, Dr. J. F. Dewey, Dr. W. R. Dickinson, and particularly Dr. J. W. Stewart for their useful comments on an early draft of the paper.

Instituto de Investigaciones Geológicas,
    Casilla 10465, Santiago, Chile,
Present Address:
    Department of Mining Geology,
    Royal School of Mines,
    Imperial College,
        London, S. W. 7, England,
        April 20; December 1, 1971

[4] Since this paper was submitted, porphyry copper prospects have been reported from Okinawa and Turkey.

REFERENCES

Atwater, T., 1970, Implications of plate tectonics for the Cenozoic tectonic evolution of western North America: Geol. Soc. Amer. Bull., v. 81, p. 3513–3536.

Bagdasaryan, G. P., Gukasyan, R. K., and Karamyan, K. A., 1969, Absolute dating of Armenian ore formations: Int. Geol. Review, v. 11, p. 1166–1172.

Bateman, P. C., and Dodge, F. C. W., 1970, Variations of major chemical constituents across the Central Sierra Nevada batholith: Geol. Soc. Amer. Bull., v. 81, p. 409–420.

Bazin, D., and Hübner, H., 1969, La région cuprifère à gisements porphyriques de Kerman (Iran): Mineral. Deposita, v. 4, p. 200–212.

Beloussov, V. V., 1962, Basic Problems in Geotectonics: McGraw-Hill, New York, 809 p.

Benioff, H., 1954, Orogenesis and deep crustal structure—additional evidence from seismology: Geol. Soc. Amer. Bull., v. 65, p. 385–400.

Bilibin, Y. A., 1968, Metallogenic provinces and metallogenic epochs: Geol. Bull., Dept. Geol., Queens College, Queens College Press, New York, 35 p.

Billingsley, P., and Locke, A., 1941, Structure of ore districts in the continental framework: Amer. Inst. Min. Metall. Engrs. Trans., v. 144, p. 9 64.

Bird, J. M., and Dewey, J. F., 1970, Lithosphere plate-continental margin tectonics and the evolution of the Appalachian orogen: Geol. Soc. Amer. Bull., v. 81, p. 1031–1060.

Bonatti, E., 1968, Ultramafic rocks from the Mid-Atlantic Ridge: Nature, v. 219, p. 363–364.

Boström, K., and Peterson, M. N. A., 1966, Precipitates from hydrothermal exhalations on the East Pacific Rise: Econ. Geol., v. 61, p. 1258–1265.

——, 1969, The origin of aluminium-poor ferromanganoan sediments in areas of high heat flow on the East Pacific Rise: Marine Geol., v. 7, p. 427–447.

——, Joensuu, O., and Fisher, D. E., 1969, Aluminium-poor ferromanganoan sediments on active oceanic ridges: Jour. Geophys, Res., v. 74, p. 3261–3270.

Brown, A. S., 1969, Mineralization in British Columbia and the copper and molybdenum deposits: Canad. Inst. Min. Trans., v. 72, p. 1–15.

Bryner, L., 1969, Ore deposits of the Philippines—an introduction to their geology: Econ Geol., v. 64, p. 644–666.

Cann, J. R., 1968, Geological processes at mid-ocean ridge crests: Geophys. Jour., v. 15, p. 331–341.

Carson, D. J. T., 1969, Tertiary mineral deposits of Vancouver Island: Canad. Inst. Min. Trans., v. 72, p. 116–125.

Chase, R. L., and Bunce, E. T., 1969, Underthrusting of the eastern margin of the Antilles by the floor of the western North Atlantic ocean, and origin of the Barbados Ridge: Jour. Geophys. Res., v. 74, p. 1413–1420.

Christensen, N. I., 1970, Composition and evolution of the oceanic crust: Marine Geol., v. 8, p. 139–154.

Christiansen, R. L., and Lipman, P. W., 1970, Cenozoic volcanism and tectonism in the western United States and adjacent parts of the spreading ocean floor. Part 11: Late Cenozoic: Abstracts, Geol. Soc. Amer. Cordilleran Sec. 66th Ann. Mtg., v. 2, p. 81–82.

Coats, R. R., 1962, Magma types and crustal structure in the Aleutian arc: p. 92–109, in *The Crust of the Pacific Basin*, Amer. Geophys. Union, Geophys. Monog. 6.

Coleman, P. J., 1966, The Solomon Islands as an island arc: Nature, v. 211, p. 1249–1251.

Coney, P. J., 1970, The geotectonic cycle and the New Global Tectonics: Geol. Soc. Amer. Bull., v. 81, p. 739–748.

Cornelius, K. D., 1969, The Mount Morgan mine, Queensland—A massive gold-copper pyritic replacement deposit: Econ Geol., v. 64, p. 885–902.

Creasey, S. C., and Kistler, R. W., 1962, Age of some copper-bearing porphyries and other igneous rocks in southeastern Arizona: U. S. Geol. Survey Prof. Paper 450-D, p. 1–5.

Cronan, D. S., 1969, Average abundances of Mn, Fe, Ni, Co, Cu, Pb, Mo, V, Cr, Ti and P in Pacific pelagic clays: Geochim. Cosmochim. Acta, v. 33, p. 1562–1565.

Damon, P. E., and Mauger, R. L., 1966, Epeirogeny-orogeny viewed from the Basin and Range province: Soc. Mining Engrs. Trans., v. 235, p. 99–112.

Degens, E. T., and Ross, D. A., eds., 1969, Hot Brines and Recent Heavy Metal Deposits in the Red Sea. A Geochemical and Geophysical Account: Springer-Verlag, Berlin, Heidelberg and New York, 600 p.

Desborough, G. A., Anderson, A. T., and Wright, T. L., 1969, Mineralogy of sulfides from certain Hawaiian basalts: ECON. GEOL., v. 63, p. 636–644.

Dewey, J. F., and Bird, J. M., 1970, Mountain belts and the New Global Tectonics: Jour. Geophys. Res., v. 75, p. 2625–2647.

Dewey, J. F., and Horsfield, B., 1970, Plate tectonics, orogeny and continental growth: Nature, v. 225, p. 521–525.

Dickinson, W. R., 1968, Circum-Pacific andesite types: Jour. Geophys. Res., v. 73, p. 2261–2269.

——, and Hatherton, T., 1967, Andesitic volcanism and seismicity around the Pacific: Science, v. 157, p. 801–803.

Dietz, R. S., 1961, Continent and ocean basin evolution by spreading of the sea floor: Nature, v. 190, p. 854–857.

——, 1963, Alpine serpentinites as oceanic rind fragments: Geol. Soc. Amer. Bull., v. 74, p. 947–952.

Doe, B. R., Tilling, R. I., Hedge, C. E., and Klepper, M. R., 1968, Lead and strontium isotope studies of the Boulder batholith, southwestern Montana: ECON. GEOL., v. 63, p. 884–906.

Emmons, W. H., 1927, Relations of the disseminated copper ores in porphyry to igneous intrusives: Amer. Inst. Min. Metall. Engrs. Trans., v. 75, p. 797–815.

Engel, A. E. J., Engel, C. G., and Havens, R .G., 1965, Chemical characteristics of oceanic basalts and the upper mantle: Geol. Soc. Amer. Bull., v. 76, p. 719–734.

Engel, C. G., and Fisher, R. L., 1969, Lherzolite, anorthosite, gabbro and basalt dredged from the Mid-Indian Ocean Ridge: Science, v. 166, p. 1136–1141.

Evernden, J. F., and Kistler, R. W., 1970, Chronology of emplacement of Mesozoic batholithic complexes in California and western Nevada: U. S. Geol. Survey Prof. Paper 623, 42 p.

Ewart, A., and Stipp, J. J., 1968, Petrogenesis of the volcanic rocks of the Central North Island, New Zealand, as indicated by a study of the Sr^{87}/Sr^{86} ratios and Sr, Rb, K, U and Th abundances: Geochim. Cosmochim. Acta, v. 32, p. 699–736.

Fairbairn, H. W., Hurley, P. M., and Pinson, W. H., 1964, Initial Sr^{87}/Sr^{86} and possible sources of granitic rocks in southern British Columbia: Jour. Geophys. Res., v. 69, p. 4889–4893.

Farrar, E., Clark, A. H., Haynes, S. J., Quirt, G. S., Conn, H., and Zentilli, M., 1970, K-Ar evidence for the post-Paleozoic migration of granitic intrusion foci in the Andes of northern Chile: Earth Planet. Sci. Letters, v. 10, p. 60–66.

Ferenčić, A., 1970, Porphyry copper mineralization in Panama: Mineral. Deposita, v. 5, p. 383–389.

Field, C. W., 1966, Sulfur isotope abundance data, Bingham District, Utah: ECON. GEOL., v. 61, p. 850–871.

Fournier, R. O., 1967, The porphyry copper deposit exposed in the Liberty open-pit mine near Ely, Nevada. Part 1. Syngenetic formation: ECON GEOL., v. 62, p. 57–81.

——, 1968, Depth of intrusion and conditions of hydrothermal alteration in porphyry copper deposits: Abstracts, Geol. Soc. Amer. Ann. Mtg., Mexico City, p. 101.

Francheteau, J., Sclater, J. G., and Menard, H. W., 1970, Pattern of relative motion from fracture zone and spreading rate data in the north-eastern Pacific: Nature, v. 226, p. 746–748.

Freeland, G. L., and Dietz, R. S., 1971, Plate tectonic evolution of Caribbean—Gulf of Mexico region: Nature, v. 232, p. 20–23.

Gabrielse, H., and Reesor, J. E., 1964, Geochronology of plutonic rocks in two areas of the Canadian Cordillera:

p. 96–128, *in* Geochronology in Canada, Osborne, F. F., ed., Royal Soc. Canada Spec. Publ. 8.

Gass, I. G., 1968, Is the Troodos massif of Cyprus a fragment of Mesozoic ocean floor: Nature, v. 220, p. 39–42.

Gilluly, J., 1963, The tectonic evolution of the western United States: Quart. Jour. Geol. Soc. London, v. 119, p. 133–174.

Guilbert, J. M., and Sumner, J. S., 1968, Distribution of porphyry copper deposits in the light of recent tectonic advances: Southern Arizona Guidebook III, Ariz. Geol. Soc., p. 97–112.

Hamilton, W., 1969a, The volcanic central Andes—a modern model for the Cretaceous batholiths and tectonics of western North America: Proc. Andesite Conf., Oregon Dept. Geol. Mineral Resources Bull. 65, p. 175–184.

——, 1969b, Mesozoic California and the underflow of Pacific mantle: Geol. Soc. Amer. Bull., v. 80, p. 2409–2430.

——, and Myers, W. B., 1967, The nature of batholiths: U. S. Geol. Survey Prof. Paper 554-C, p. 1–30.

Harland, W. B., Smith, A. G., and Wilcock, B., eds., 1964, The Phanerozoic time-scale: Quart. Jour. Geol. Soc., London, v. 120S, 458 p.

Heirtzler, J. R., Dickson, G. O., Herron, E. M., Pitman, III, W. C., and Le Pichon, X., 1968, Marine magnetic anomalies, geomagnetic field reversals, and motions of the ocean floors and continents: Jour. Geophys. Res., v. 73, p. 2119–2136.

Hess, H. H., 1962, History of ocean basins: p. 599-620, *in* Petrologic Studies, Engel, A. E. J., et al., eds., Geol. Soc. Amer., New York.

Hsü, K. J., 1971, Origin of the Alps and western Mediterranean: Nature, v. 233, p. 44–48.

Hurley, P. M., Bateman, P. C., Fairbairn, H. W., and Pinson, W. H., Jr., 1965, Investigation of initial Sr^{87}/Sr^{86} ratios in the Sierra Nevada plutonic province: Geol. Soc. Amer. Bull., v. 76, p. 165–174.

Hutchinson, R. W., and Searle, D. L., 1970, The mineralogy, geochemistry and origin of Cyprus sulphide deposits: IMA-IAGOD Meetings '70, Japan, Collected Abstracts, p. 89.

Isacks, B., Oliver, J., and Sykes, L. R., 1968, Seismology and the new global tectonics: Jour. Geophys. Res., v. 73, p. 5855–5900.

Jensen, M. L., 1967, Sulfur isotopes and mineral genesis: p. 143–165, *in* Geochemistry of hydrothermal ore deposits, Barnes, H. L., ed., Holt, Rinehart and Winston, New York.

——, 1971, Provenance of Cordilleran intrusives and associated metals: ECON. GEOL., v. 66, p. 34–42.

Larsen, R. L., Menard, H. W., and Smith, S. M., 1968, Gulf of California: a result of ocean-floor spreading and transform faulting: Science, v. 161, p. 781–784.

Laughlin, A. W., Damon, P. E., and Watson, B. N., 1968, Potassium-argon dates from Toquepala and Michiquillay, Peru: ECON. GEOL., v. 63, p. 166–168.

——, Rehrig, W. A., and Mauger, R. L., 1969, K-Ar chronology and sulfur and strontium isotope ratios at the Questa Mine, New Mexico: ECON. GEOL., v. 64, p. 903–909.

Le Pichon, X., 1968, Sea-floor spreading and continental drift: Jour. Geophys. Res., v. 73, p. 3661–3697.

Lipman, P. W., Prostka, H. J., and Christiansen, R. L., 1970, Cenozoic volcanism and tectonism in the western United States and adjacent parts of the spreading ocean floor. Part 1: Early and middle Cenozoic: Abstracts, Geol. Soc. Amer. Cordilleran Sec. Ann. Mtg., v. 2, p. 112–113.

——, 1971, Evolving subduction zones in the western United States, as interpreted from igneous rocks: Science, v. 174, p. 821–825.

Livingston, D. E., Mauger, R. L., and Damon, P. E., 1968, Geochronology of the emplacement, enrichment, and preservation of Arizona porphyry copper deposits: ECON. GEOL., v. 63, p. 30–36.

Llambias, E. J., 1970, Geología de los yacimientos mineros de Agua de Dionisio, Provincia de Catamarca, República

Argentina: Rev. Asoc. Argentina Min. Pet. Sed., v. 1, p. 2–32.

Lowell, J. D., and Guilbert, J. M., 1970, Lateral and vertical alteration-mineralization zoning in porphyry ore deposits: Econ. Geol., v. 65, p. 373–408.

Macnamara, P. M., 1968, Rock types and mineralization at Panguna porphyry copper prospect, Upper Kaverong Valley, Bougainville Island: Austral. Inst. Min. Metall. Proc., no. 228, p. 71–79.

Mattson, P. H., 1965, Geological characteristics of Puerto Rico: in Continental margins and island arcs, Poole, W. H., ed., Geol. Survey Canada Paper 66–15, p. 124–138.

Maxwell, A. E., Von Herzen, R. P., Hsü, K. J., Andrews, J. E., Saito, T., Percival, S. F., Jr., Milow, E. D., and Boyce, R. E, 1970, Deep sea drilling in the South Atlantic: Science, v. 168, p. 1047–1059.

Mayo, E. B., 1958, Lineament tectonics and some ore districts of the southwest: Min. Engng., v. 10, p. 1169–1175.

McCartney, W. D., 1964, Metallogeny of post-Precambrian geosynclines: p. 19–23, in *Some guides to mineral exploration*, Neale, E. R. W., ed., British Commonwealth Geol. Liaison Office Spec. Publ. S.4.

——, and Potter, R. R., 1962, Mineralization as related to structural deformation, igneous activity and sedimentation in folded geosynclines: Canad. Min. Jour., v. 83, p. 83–87.

McDowell, F. W., and Kulp, J. L., 1967, Age of intrusion and ore deposition in the Robinson mining district of Nevada: Econ. Geol., v. 62, p. 905–909.

McKenzie, D. P., and Parker, R. L., 1967, The North Pacific: an example of tectonics on a sphere: Nature, v. 216, p. 1276–1280.

Menard, H. W., 1969, Growth of drifting volcanoes: Jour. Geophys. Res., v. 74, p. 4827–4837.

Meyer, C., and Hemley, J. J., 1967, Wall rock alteration: p. 166–235, in Geochemistry of Hydrothermal Ore Deposits, Barnes, H. L., ed., Holt, Rinehart and Winston, New York.

Miller, A. R., Densmore, C. D., Degens, E. T., Hathaway, J. C., Manheim, F. T., McFarlin, P. F., Rocklington, R., and Jokela, A., 1966, Hot brines and recent iron deposits in deeps of the Red Sea: Geochim. Cosmochim. Acta, v. 30, p. 341–359.

Mitchell, A. H. G., and Bell, J. D., Volcanic episodes in island arcs: Proc. Geol. Soc. London, no. 1662, p. 9–12.

——, and Reading, H. G., 1969, Continental margins, geosynclines, and ocean floor spreading: Jour. Geol., v. 77, p. 629–646.

Molnar, P., and Sykes, L. R., 1969, Tectonics of the Caribbean and Middle America regions from focal mechanisms and seismicity: Geol. Soc. Amer. Bull., v. 80, p. 1639–1684.

Moorbath, S., Hurley, P. M., and Fairbairn, H. W., 1967, Evidence for the origin and age of some mineralized Laramide intrusives in the Southwestern United States from strontium isotope and rubidium-strontium measurements: Econ. Geol., v. 62, p. 228–236.

Moore, D. G., and Buffington, E. C., 1968, Transform faulting and growth of the Gulf of California since the late Pliocene: Science, v. 161, p. 1238–1241.

Moore, J. G., 1959, The quartz diorite boundary line in the western United States: Jour. Geol., v. 67, p. 198–210.

——, 1962, K/Na ratio of Cenozoic igneous rocks of the western United States: Geochim. Cosmochim. Acta, v. 26, p. 101–130.

——, and Calk, L., 1971, Sulfide spherules in vesicles of dredged pillow basalt: Amer. Min., v. 56, p. 476–488.

——, Grantz, A., and Blake, M. C., Jr., 1963, The quartz diorite line in northwestern North America: U. S. Geol. Survey Prof. Paper 450-E, p. E89–E93.

Moore, W. J., Lanphere, M. A., and Obradovich, J. D., 1968, Chronology of intrusion, volcanism, and ore deposition at Bingham, Utah: Econ. Geol., v. 63, p. 612–621

Morgan, W. J., 1968, Rises, trenches, great faults, and crustal blocks: Jour. Geophys. Res., v. 73, p. 1959–1982.

——, Vogt, P. R., and Falls, D. F., 1969, Magnetic anomalies and sea floor spreading on the Chile Rise: Nature, v. 222, p. 137–142.

Müller-Kahle, E., and Damon, P. E., 1970, K-Ar age of a biotite granodiorite associated with primary Cu-Mo mineralization at Chaucha, Ecuador: p. 46–48, in Correlation and chronology of ore deposits and volcanic rocks, U. S. Atomic Energy Comm. Ann. Prog. Rept. no. COO-689-130, Univ. Arizona, Tucson.

Ney, C. S., 1966, Distribution and genesis of copper deposits in British Columbia: Canad. Inst. Min. Spec. v. 8, p. 295–304.

Nielsen, R. L., 1968, Hypogene texture and mineral zoning in a copper-bearing granodiorite porphyry stock, Santa Rita, New Mexico: Econ. Geol., v. 63, p. 37–50.

Oxburgh, E. R., and Turcotte, D. L., 1968, Mid-ocean ridges and geotherm distribution during mantle convection: Jour. Geophys. Res., v. 73, p. 2643–2661.

——, and ——, 1970, Thermal structure of island arcs: Geol. Soc. Amer. Bull., v. 81, p. 1665–1688.

Page, B. M., 1970, Sur-Nacimiento fault zone of California: continental margin tectonics: Geol. Soc. Amer. Bull., v. 81, p. 667–690.

Peterman, Z. E., Carmichael, I. S. E., and Smith, A. L., 1970, Sr^{87}/Sr^{86} ratios of Quaternary lavas of the Cascade Range, northern California: Geol. Soc. Amer. Bull., v. 81, p. 311–318.

——, and Hedge, C. E., 1971, Related strontium isotopic and chemical variations in oceanic basalts: Geol. Soc. Amer. Bull., v. 82, p. 493–500.

Plafker, G., and Savage, J. C., 1970, Mechanism of the Chilean earthquakes of May 21 and 22, 1960: Geol. Soc. Amer. Bull., v. 81, p. 1001–1030.

Po, M. H., and Lee, V. C., 1970, Copper deposits in eastern Taiwan, China: IMA-IAGOD Meetings '70, Japan, Collected Abstracts, p. 70.

Pushkar, P., 1968, Strontium isotope ratios in volcanic rocks of three island arc areas: Jour. Geophys. Res., v. 73, p. 2701–2714.

Ringwood, A. E., 1969, Composition and evolution of the upper mantle: p. 1–17, in The earth's crust and upper mantle, Amer. Geophys. Union, Geophys. Monog. 13.

Ruitenberg, A. A., Shafiquallah, M., and Tupper, W. M., 1970, Late tectonic mineralized structures in southwestern New Brunswick: Geol. Assoc. Canada-Mineralog. Assoc. Canada Ann. Meeting, Programme and Abstracts., p. 46–47.

Ruiz, F. C., Aguirre, L., Corvalán, J., Klohn, C., Klohn, E., and Levi, B., 1965, Geología y yacimientos metalíferos de Chile: Inst. Invest. Geológicas, Santiago, 385 p.

Schmidt, R. G., 1968, Exploration possibilities in the western Chagai district, West Pakistan: Econ. Geol., v. 63, p. 51–60.

Schmitt, H. A., 1966, The porphyry copper deposits in their regional setting: p. 17–33, in Geology of the Porphyry Copper Deposits, Southwestern North America, Titley, S. R., and Hicks, C. L., eds., Univ. Arizona Press, Tucson.

Sillitoe, R. H., 1970, South American porphyry copper deposits and the new global tectonics: Resumenes Primer Cong. Latinoamericano Geol., Lima, Peru, p. 254–256.

——, and Sawkins, F. J., 1971, Geologic, mineralogic and fluid inclusion studies relating to the origin of copper-bearing tourmaline breccia pipes, Chile: Econ. Geol., v. 66, p. 1028–1041.

Skinner, B. J., and Peck, D. L., 1969, An immiscible sulfide melt from Hawaii: Econ. Geol. Monog. 4, p. 310–322.

Skinner, B. J., White, D. E., Rose, H. J., and Mays, R. E., 1967, Sulfides associated with the Salton Sea geothermal brine: Econ. Geol., v. 62, p. 316–330.

Smith, A. G., 1971, Alpine deformation and the oceanic areas of the Tethys, Mediterranean, and Atlantic: Geol. Soc. Amer. Bull., v. 82, p. 2039–2070.

Souther, J. G., 1970, Volcanism and its relationship to recent crustal movements in the Canadian Cordillera: Canad. J. Earth Sci., v. 7, p. 553–568.

Taylor, Jr., A. V., 1935, Ore deposits at Chuquicamata, Chile: p. 473–484, in Copper resources of the world, 16th Int. geol. Congr., Washington, v 2.

Taylor, S. R., 1969, Trace element chemistry of andesites and associated calc-alkaline rocks: Proc. Andesite Conference, Oregon Dept. Geol. Mineral Resources Bull. 65, p. 43–63.

——, Capp, A. C., Graham, A. L., and Blake, D. H., 1969, Trace element abundances in andesites. II. Saipan, Bougainville and Fiji: Contr. Mineral. Petrol., v. 23, p. 1–26.

Thayer, T. P., 1969, Peridotite-gabbro complexes as keys to petrology of mid-oceanic ridges: Geol. Soc. Amer. Bull., v. 80, p. 1515–1522.

Tobin, D. G., and Sykes, L. R., 1968, Seismicity and tectonics of the northeast Pacific Ocean: Jour. Geophys. Res., v. 73, p. 3821–3845.

Turekian, K. K., and Wedepohl, K. H., 1961, Distribution of the elements in some major units of the Earth's crust: Geol. Soc. Amer. Bull., v. 72, p. 175–192.

Turneaure, F. S., 1955, Metallogenetic provinces and epochs: Econ. Geol., 50th Anniv. Vol., Pt. 1, p. 38–98.

Tweto, O., 1968, Geologic setting and interrelationships of mineral deposits in the Mountain Province of Colorado and south-central Wyoming: p. 551–588, *in* Ore Deposits in the United States 1933/1967, Ridge, J. D., ed., Amer. Inst. Min. Metall. Petrol. Engrs., New York.

United Nations, 1970, Investigación sobre mineral de cobre porfídico en las provincias de Mendoza, Neuquén y San Juan, Argentina: United Nations, New York, 356 p.

United Nations, 1970a, Mineral resources of the sea: United Nations, New York, 49 p.

Vinogradov, A. P., 1962, Average contents of chemical elements in the principal types of igneous rocks of the earth's crust: Geochemistry, v. 7, p. 641–664.

Vine, F. J., 1966, Spreading of the ocean floor: New evidence: Science, v. 154, p. 1405–1415.

Vokes, F. M., 1966, Remarks on the origin of the Cyprus pyritic ores: Canad. Min. Metall. Bull., v. 59, p. 388–391.

Wallace, S. R., Muncaster, N. K., Jonson, D. C., Mackenzie, W. B., Bookstrom, A. A., and Surface, V. E., 1968, Multiple intrusion and mineralization at Climax, Colorado: p. 605–640, *in* Ore deposits in the United States 1933/1967, Ridge, J. D., ed., Amer. Inst. Min. Metall. Petrol. Engnrs., New York.

Wertz, J. B., 1970, The Texas lineament and its economic significance in southeast Arizona: Econ. Geol., v. 65, p. 166–181.

White, D. E., 1968, Environments of generation of some base-metal ore deposits: Econ. Geol., v. 63, p. 301–335.

White, W. H., Harakal, J. E., and Carter, N. C., 1968, Potassium-argon ages of some ore deposits in British Columbia: Canad. Min. Metall. Bull., v. 61, p. 1326–1334.

Wilson, I. F., and Rocha, V. S., 1955, Geology and mineral deposits of the Boléo copper district, Baja California, Mexico: U. S. Geol. Survey Prof. Paper 273, 134 p.

Yeats, S., 1968, Southern California structure, sea-floor spreading, and history of the Pacific Basin: Geol. Soc. Amer. Bull., v. 79, p. 1693–1702.

18

Reprinted from *Trans. Instn Min. Metall.* **81**:B10–B25 (1972)

Relationship of porphyry copper and circum-Pacific tin deposits to palaeo-Benioff zones

A. H. G. Mitchell M.A., D.Phil.
Department of Geology and Mineralogy, Oxford University
M. S. Garson B.Sc., Ph.D., M.I.M.M.
Geochemical Division, Institute of Geological Sciences, London

553.21 : 553.43 : 553.45

Synopsis

Porphyry copper deposits are emplaced in igneous belts located either on continental margins or in island arcs. These belts are related to partial melting of wet oceanic crust descending along Benioff zones at depths of 150–250 km. Deposition of the copper occurs where metal-carrying solutions rising from the descending crust meet meteoric brines. Porphyry copper deposits, and conformable massive sulphide deposits emplaced in marine environments in island arcs, suffer orogeny during continental collisions. Massive sulphides survive the resulting orogeny and are preserved in greenstone belts, but higher-level porphyry coppers are mostly destroyed by erosion.

Circum-Pacific tin–tungsten–fluorite deposits occur in granitic belts bordered either by younger igneous belts or by marginal basins and island arcs. They are emplaced during opening of marginal basins above migrating Benioff zones. Mineralization results from upward-streaming volatiles, including fluorine, which originate in dry descending crust or upper mantle along the Benioff zone at depths in excess of 300 km. The metals, probably transported initially as halides and later as fluor–hydroxyl complexes are deposited in contact with meteoric brines during changing conditions of pH, pressure and temperature. The volatile stream could also explain some epigenetic non-orogenic base-metal deposits lying on the continental side of orogenic belts.

According to the concept of plate tectonics, the earth's surface consists of a number of rigid plates of lithosphere, each plate comprising part of the upper mantle and overlying oceanic and continental crust. The concept, based initially on geophysical data, explains many of the earth's major topographical features, seismic zones and volcanic provinces by the ascent, lateral movement and descent of lithosphere.

Consideration of the geological processes occurring above the seismically defined Benioff zones, and of the inferred results of continued descent of lithosphere along them, has resulted in interpretations of several types of major lithological and structural units found in orogenic belts and mountain ranges. Thus, new explanations have recently been given for the origin or distribution of ophiolite suites, paired metamorphic belts, andesitic volcanic successions and batholiths.

The plate tectonics concept has also been used to explain the location and origin of some mineral deposits by relating them to zones of rifting and subsequent separation of continents. Examples are the deposition of evaporites at continental rift zones,[1,2] the origin of metal-rich brines near spreading centres[3] and emplacement of some base-metal sulphide deposits in major fault zones developed during break-up of continents.[4] Reassemblies of continents in their pre-Mesozoic drift positions have been used to indicate possible mineralization belts[5] and to predict extensions of ore provinces from one continent to another.[6]

Manuscript first received by the Institution of Mining and Metallurgy on 25 August, 1971 ; revised manuscript received on 1 December, 1971. Paper published in February, 1972.

Possible relationships between mineral deposits and Benioff zones have so far received little attention in the literature. The authors suggest that porphyry copper deposits and tin provinces, and also the distribution of massive sulphide deposits, are related to Benioff zones.

Location and description of porphyry copper deposits

The term porphyry copper or copper–molybdenum deposit is used in the sense of Lowell and Guilbert[7] for a zoned deposit of disseminated copper and/or molybdenum sulphides with associated hydrothermally altered host rocks. The relatively homogeneous and roughly equidimensional deposit is generally on the scale of several thousands of feet and is associated with a complex, passively emplaced stock of intermediate composition including porphyry units.

Major porphyry-type copper deposits lie, mostly at a high elevation, in either mountain belts or island arcs (Fig. 1). In mountain belts most of the deposits are of either Mesozoic or early Tertiary age, and in island arcs they are all Tertiary.

In mountain belts the best-known deposits are in western North America and the Andes. They all occur partly within porphyry intrusions which are associated with stocks of intermediate to acidic composition.[7,8,9] Some copper-bearing intrusions, such as those at Butte, Montana, form part of a batholith, and all occur within belts of volcanic and plutonic rocks.

Emplacement of porphyry copper ores and their intrusive host rocks were approximately contemporaneous in the Basin and Range Province,[10] and, recently, Moore and Lanphere[11] showed that porphyry copper mineralization in the Bingham District, Utah, took place less than 2 m.y. after emplacement of the composite Bingham stock. In contrast to this close association between major igneous episodes and the emplacement of porphyry copper deposits, the pre-ore host rocks in North and South America vary in lithology, composition and age, sometimes within a single deposit. The depth of emplacement of most deposits probably did not exceed a few kilometres.[7]

Most porphyry orebodies were formed as hollow cylinders or inverted cup-shaped shells[9] with gradational boundaries, up to 10 000 ft in length and rather less in diameter. The ore shell contains disseminated chalcopyrite and pyrite with minor molybdenite and bornite ; average copper grade is around 0·80 per cent. This shell is surrounded by a zone rich in pyrite and low in chalcopyrite.[7] The sulphides mostly occur as disseminated grains or veinlets, the proportion of disseminated mineralization increasing towards the core. The ore shell core and surrounding mineralized zones are characteristically highly fractured.

Hydrothermal alteration minerals are associated with

all porphyry copper deposits, and in most cases they occur in four main concentric zones: an innermost potassic zone, which generally coincides with the low-grade core within the ore shell; a phyllic zone, approximately coincident with the pyrite-rich zone; a poorly

sites and basalts, but may include both tholeiitic and more alkalic types. The succession is sometimes regionally metamorphosed in the zeolite facies or more rarely in the greenschist facies, and block-faulted.

Plutons, with marginal porphyry facies which may

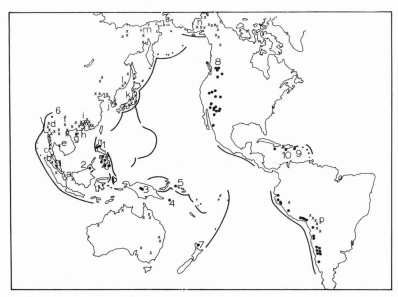

Fig. 1 Distribution of porphyry copper deposits [(1) Philippines, (2) Mamut, (3) New Guinea, (4) Misima Island, (5) Bougainville, (6) Monywa, (7) Coppermine Island, (8) British Columbia, (9) Jamaica, (10) Puerto Rico] and tin–tungsten deposits around the Pacific [(*a*) Billiton, (*b*) Kinta Valley, (*c*) Phuket, (*d*) Mawchi, (*e*) Chantaburi, (*f*) Kochiu, (*g*) Hainan Island, (*h*) Hong Kong, (*i*) Nanling Belt, (*j*) Korea, (*k*) Honshu Island, (*l*) Lifudzin–Khrustal'nyy area, (*m*) Anadyr Mountains, (*n*) Seward Peninsula, (*p*) Bolivia]. Arc systems are indicated by heavy lines

developed argillic zone; and a wide outermost propylitic zone.

Within modern island arcs, which lie within the oceans, major porphyry copper deposits are known in the Philippines, where there are at least 8 separate orebodies,[12] on Bougainville Island at the western end of the Solomon Island Chain,[13] in Jamaica and Puerto Rico, and possibly on Guadalcanal in the Solomon Islands,[14] and on Misima Island in Papua.

Island arc deposits occur partly within intrusions of intermediate composition, like those in the mountain belts of North and South America, but differ from these in that the intruded successions are broadly similar to one another. In modern island arcs these consist of a thick, predominantly volcanic, stratigraphic unit of late Cretaceous to early Tertiary age, comprising eruptive, pyroclastic and epiclastic rocks, including *aa* lavas, pillow lavas, tuffs, peperites, volcanogenic turbidites and lahar deposits; carbonate reef rocks and siltstones containing volcanic detritus may also be present. The volcanic rocks are predominantly calc-alkaline ande-

contain porphyry copper deposits, occur as high-level intrusions emplaced passively within the volcanic pile, and consist largely of diorite or dacite and andesite porphyry. Emplacement of the intrusive host rocks on Bougainville Island[13] and at all but one of the Philippine deposits[12] may have post-dated accumulation of the intruded successions by at least 10 m.y.

Island arc porphyry copper deposits are known in much less detail than deposits in western North America. On Bougainville Island, however, alteration facies are broadly comparable with the zoned facies described by Lowell and Guilbert,[7] and, as in many of the North American deposits, the orebody occurs in a zone of biotitization into which K, SiO_2 and Cu were introduced. In the Philippines, hydrothermal alteration facies are associated with most of the deposits. Fracturing is characteristic of the host rocks in island arcs.

Porphyry copper deposits, calc-alkaline igneous belts and palaeo-Benioff zones

Porphyry copper deposits in the Andes and Central

120

America, and in the Philippines and Solomon Island arcs, lie in seismically active areas, beneath which earthquake foci are concentrated along inclined Benioff zones. Recent work indicates that the submarine

coinciding with present zones (Fig. 2; see Table 1). The dioritic and granodioritic plutons in island arcs were probably formed from the same magma as the volcanic rocks, but at a deeper structural level.

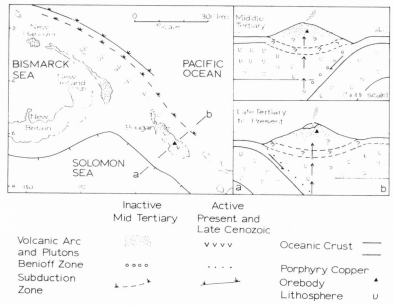

	Inactive Mid Tertiary	Active Present and Late Cenozoic	
Volcanic Arc and Plutons		v v v v	Oceanic Crust ———
Benioff Zone	o o o o	. . . .	Porphyry Copper
Subduction Zone	⊥ – – ⊥	⊥——⊥	Orebody ▲ Lithosphere U

Fig. 2 Bougainville Island and northwest Melanesia: relationship of porphyry copper deposit and igneous rocks of middle Tertiary age to inferred palaeo-Benioff zone; relationship of late Cenozoic volcanic rocks to present Benioff zones

trenches which border the Andes and island arcs are subduction zones where lithosphere bends and then descends with its upper surface lying along or near the Benioff zone. Late Cenozoic calc-alkaline volcanism in the Andes and in island arcs, concentrated in belts about 200 km from the subduction zone, is considered to be related to this descent of lithosphere.[15]

The lengths of present Benioff zones and estimated rates of descent of lithosphere along them indicate that some are at least 10 m.y. old,[16] but there is little evidence that they have occupied their present position with respect to the igneous belts significantly longer than this. Therefore igneous rocks associated with most porphyry copper deposits, which are older than late Cenozoic, are not necessarily related to present Benioff zones. The following evidence suggests, however, that the rocks were emplaced above palaeo-Benioff zones.

In island arcs the thick succession of volcanic and sedimentary rocks with associated intrusions and porphyry copper deposits are of late Mesozoic or Tertiary age. These successions resemble those accumulating around active volcanoes in island arcs and are interpreted as the deposits of now inactive volcanic arcs.[17] Thus these old volcanic arc deposits were related to palaeo-Benioff zones, not necessarily

In the Central Andes similarities between the scale, distribution and composition of Cenozoic calc-alkaline volcanic provinces and belts of late Mesozoic batholiths suggest that the batholiths were developed beneath related but now eroded calc-alkaline volcanic rocks.[18] This relationship, although not undisputed,[20] suggests that intrusive bodies, and associated porphyry copper deposits, were emplaced during igneous episodes above palaeo-Benioff zones.

Geological setting of porphyry copper deposits

If porphyry-type copper deposits in the Andes and in modern island arcs were emplaced within belts of igneous rock above palaeo-Benioff zones, they should also occur in and around similar igneous belts in ancient island arcs and in inactive and ancient mountain belts of Andean type. These igneous belts now occur in several different geological settings (Table 1).

Inactive Andean-type mountain belts

Parts of western North America have been interpreted as a Mesozoic Andean-type mountain belt, modified by Cenozoic extensional rifting and faulting.[19] With the possible exception of some deposits in the southwestern U.S.A., porphyry-type copper deposits within

Table 1 Igneous belts related to present and palaeo-Benioff zones

	Andean-type mountain belt			Island arc		
Structural setting	Modern (continental margin)		Ancient (within continents)	Modern (within ocean)		Ancient (on margin of or within continent)
Volcanism	Active	Inactive		Active	Inactive	
Benioff zone	Present (and late Cenozoic)	Palaeo (pre-late Cenozoic)	Palaeo (pre-late Cenozoic)	Present (and late Cenozoic)	Palaeo (pre-late Cenozoic)	Palaeo (pre-late Cenozoic)
Characteristic rock types	Late Cenozoic acid and intermediate volcanics	Pre-late Cenozoic volcanics and batholiths	As Inactive, but may be regionally metamorphosed and folded	Active volcanoes and late Cenozoic volcanics	Pre-late Cenozoic block-faulted volcanics and plutons	As Inactive, but commonly regionally metamorphosed and folded
Examples Igneous belts	Andes of Peru and Chile	Batholiths of western North America[1]	Early Tertiary northeast of Zagros Mountains[1]	Tonga	Green-tuff belts, Japan	Tertiary, northwest Borneo[1]
	Barisan Range, Sumatra	Peruvian batholith and early Cenozoic volcanics[2]	Granitic plutons of Hindu Kush and Karakorum	New Hebrides	Western belt, New Hebrides	Vancouver Island[2]
		New Guinea Highlands[3]		Aleutians	New Ireland–Bougainville[1]	Early Palaeozoic, Central Volcanic Belt, Newfoundland
		Late Mesozoic–early Cenozoic, Sumatra		Lesser Antilles	'Metavolcanic' belt, Philippines[2]	Borrowdale Volcanic Series, Lake District
Associated porphyry copper deposits	None exposed	[1]Yenington [2]Michiquillay [3]Fubilan	[1]Sar Chesmeh	None exposed	[1]Panguna [2]Marcopper	[1]Mamut [2]Port Hardy

igneous intrusions associated with batholiths are probably related to the development of this Andean-type belt, and, hence, to an underlying Benioff zone.

In New Guinea the Sneeuw Gebergte and Central Highlands contain several recently discovered porphyry copper deposits, for example, those at Fubilan and in the Star Mountains,[21] associated with post-Palaeozoic plutons of acidic to intermediate composition. The arcuate distribution of the plutons suggests that they were intruded above a Benioff zone which dipped southwards beneath the northern margin of the Australia–New Guinea continent. In the Tertiary, descent of lithosphere beneath New Guinea ceased, and the Andean-type mountain belt became inactive. The present poorly defined Benioff zone dips northwards beneath the Schouten Islands–New Britain active volcanic arc.

In Sumatra the belt of active volcanoes and late Cenozoic volcanic rocks, presumably related to the present Benioff zone, partly overlies late Mesozoic and Miocene volcanic and plutonic rocks of acidic and intermediate composition.[22] Analogy with the porphyry copper deposits and associated rocks in the Andes suggests that similar deposits should be present in some of the plutons in Sumatra, and also in and around the mid-Tertiary plutons in western Java.

Inactive island arcs within the oceans

Examples of porphyry copper deposits in inactive island arc successions are those of Bougainville and the Philippines. In the Philippines the distribution of the late Cenozoic and active volcanoes differs from that of the Tertiary porphyry copper deposits, indicating that the copper-bearing igneous rocks are unrelated to the present Benioff zones. On Bougainville Island late Cenozoic volcanic rocks, possibly related to the present Benioff zone, are separated from mid-Tertiary porphyry copper host rocks by a major unconformity. Structural and stratigraphic similarities between the islands of Bougainville, New Ireland and New Hanover, and their distribution in an arcuate belt, suggest that the mid-Tertiary rocks were erupted above a palaeo-Benioff zone which dipped southwestwards beneath the islands (Fig. 2).

Other islands which were formed partly in old volcanic arcs include Malekula and Espiritu Santo in the New Hebrides, and Jamaica and Puerto Rico in the Greater Antilles Islands. The present distribution of these old arc deposits, mostly in single islands rather than in arcuate or linear island belts, has resulted from rotation, faulting and rifting, accompanying the development of younger Benioff zones and movements on transform faults.

Inactive island arcs fronting infilled marginal basins

Many island arcs are separated from a continent on their concave side by a marginal basin. An example of a sediment-filled marginal basin is the central lowland area of Burma, in which a very thick Tertiary succession[23] joins western Burma to the Shan States in the east. Cenozoic volcanic rocks, containing the little known Monywa copper deposit, are present west of the Irrawaddy Valley, and are probably related to the eastward-dipping Benioff zone[74] which approaches the surface immediately west of the arcuate Indo–Burman Ranges.

Ancient island arcs welded to a continent

A continent or micro-continent located on a lithospheric plate descending beneath an island arc will eventually approach the subduction zone bordering the arc. The continent may underthrust the edge of the island arc plate, resulting in thrusting or obduction of a segment of oceanic crust and upper mantle on to the continental margin, and in welding of the island arc to the edge of the continent.[24]

A possible example of an island arc complex underthrust by a larger landmass is provided by the Cretaceous and Tertiary flysch-type sediments and volcanic rocks extending from west Sarawak to Sabah in Borneo,[25] and including the porphyry copper deposit at Mamut. Deformed ophiolites and cherts lying within this belt and at its southern margin are probably remnants of late Cretaceous ocean floor lost along a northward-dipping Benioff zone, prior to collision of the island arc complex with the older micro-continent of Kalimantan to the south.

In British Columbia, Permian and Mesozoic volcanic and sedimentary rocks, cut by intrusions containing porphyry copper orebodies, resemble successions in modern island arcs;[26] these old arcs were presumably related to palaeo-Benioff zones and welded to the North American continent during the late Mesozoic and early Tertiary.[27] In southern Central America the geological setting of the late Tertiary porphyry copper deposits[28] suggests that they were emplaced in an island arc. The porphyry copper body on Coppermine Island, New Zealand, lies in an island arc succession now attached to the New Zealand 'micro-continent'.

Andean-type mountain belts and island arcs within Himalayan-type orogenic belts

Loss of oceanic lithosphere beneath an Andean-type mountain belt on a continental margin may eventually result in collision between the Andean-type margin and a continent located on the descending lithospheric plate.

The most striking example of this is the Himalayan arc, interpreted in broad terms as the result of a late Cenozoic collision between India and Asia. The Mesozoic and early Cenozoic plutonic rocks in the Hindu Kush and Karakorum were probably related to northward descent of oceanic lithosphere[29] beneath an Andean-type mountain belt on the southern margin of Asia.

The Uralides have been interpreted as a mountain belt resulting from collision of the Russian and Siberian Platforms in the Permian or Triassic.[75] Volcanic and granodioritic rocks, related to pre-collision Benioff zones, may have contained porphyry copper ores— for example, the large disseminated copper deposits in Uzbekistan and Khazakstan.[30]

In Iran the large porphyry copper orebodies around Sar Chesmeh occur in an elongate belt of predominantly andesitic and granodioritic volcanic and intrusive rocks, mostly of early Tertiary age.[31] The belt lies parallel to and about 150 km north of the Zagros Crush Zone[32]—a line of abrupt stratigraphic discontinuity which probably overlies a Tertiary subduction zone. Descent of oceanic lithosphere along a related northeastward-dipping Benioff zone resulted in emplacement of the early Tertiary Andean-type igneous belt and associated porphyry copper deposits to the north, and in the northeastward movement of the Arabian Platform and its eventual collision with the Lut Block of Iran.

Conformable sulphide orebodies, porphyry copper deposits and erosion levels

Strata-bound massive sulphide orebodies associated with volcanic rocks are known from orogenic belts of early Precambrian to Palaeozoic age. Similar deposits in modern island arcs have received less attention in the European literature, although they occur in the Philippines, Japan and western Melanesia.

All the conformable sulphide deposits in modern island arcs occur within thick stratigraphic successions of late Mesozoic or Tertiary age, consisting largely of pillow lavas, volcanic breccias and tuffs, together with fine-grained sediments. The lithologies and facies indicate rapid accumulation of lavas and mass-flow deposits in a marine environment below wave base. The volcanic rocks vary from keratophyre and basalt in the successions around the Philippines deposits[33] to dacites around the Kuroko-type deposits in the Kosaka District, Japan.[34]

Similar successions form fault-blocks in many modern island arcs, and can be interpreted as the deposits of inactive volcanic arcs. They thus resemble the host rocks of island arc porphyry copper deposits.

Many of the conformable massive sulphide orebodies in orogenic belts of Palaeozoic age or older, such as those in the Scandinavian Caledonides and in New South Wales, occur in geological successions resembling those accumulating around reef-fringed volcanic islands in modern active island arcs.[35] The conformable orebodies associated with these successions in orogenic belts probably also accumulated within volcanic arc successions in ancient island arcs, and became folded, metamorphosed and welded to the continents as a result of migration of continents and continental collisions.

If conformable sulphide orebodies within orogenic belts were emplaced initially in ancient volcanic arcs, the presence of porphyry copper deposits might be expected around the margins of dioritic plutons within the same stratigraphic successions. These deposits have not been found, however—notably because of uplift and erosion prior to collision of the island arc with a continent. This would have removed the characteristically high-level porphyry copper deposits, but not the conformable sulphide ores, formed contemporaneously at a lower level in marine and generally deep-water successions. Schematic relationships between a porphyry copper deposit and conformable sulphide deposits in an island arc are shown in Fig. 3.

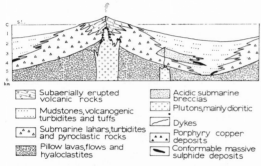

Subaerially erupted volcanic rocks
Mudstones, volcanogenic turbidites and tuffs
Submarine lahars, turbidites and pyroclastic rocks
Pillow lavas, flows and hyaloclastites
Acidic submarine breccias
Plutons, mainly dioritic
Dykes
Porphyry copper deposits
Conformable massive sulphide deposits

Fig. 3 Diagrammatic cross-section through island arc volcanic centre showing relative positions of porphyry copper, Kuroko-type and other massive conformable sulphide orebodies

One of the few Precambrian copper–molybdenum deposits of probable porphyry type, located at Ryan Lake, Ontario,[36] is associated with volcanic rocks resembling those of a modern island arc. In the Scottish Caledonides traces of copper mineralization in fractured granitic bodies within the Dalradian succession could be interpreted as the roots of eroded early Palaeozoic porphyry deposits.

The acidic composition of the intrusions suggests, however, that the deposits were of Andean type rather than island arc type.[37]

The geological setting of some mercury deposits[38] suggests that they were emplaced above Benioff zones in Andean-type mountain belts and island arcs. Similarly, gold deposits associated with the Mesozoic batholiths of western North America, Alaska and New Guinea may have developed above palaeo-Benioff

zones within and around intrusions at structural levels deeper than those of porphyry copper deposits.

Tin–tungsten–fluorite provinces and associated granitic belts

Distribution of tin, tungsten and fluorite

The abundance of these deposits in granitic belts near continental margins in the western Pacific and eastern Indian Ocean contrasts with their scarcity in the porphyry copper-bearing cordilleras of North and South America. Four main Pacific tin provinces are recognized.

Southeast Asian province

The tin and tungsten deposits of southeast Asia are separated from the Far East deposits to the north by a large-scale fracture zone trending northwest along the line of the Red River in North Vietnam.

Within this province the deposits are closely associated with sub-parallel arcuate belts of granitic rocks (Fig. 4). The westerly and most highly mineralized belt, producing more than half the non-Communist world's tin, extends from Billiton Island through the western part of West Malaysia. Prior to sinistral movement along the Khlong Marui Fault,[39] this belt continued through peninsular Thailand into Burma.

An eastern belt, with less pronounced tin mineralization, possibly starts in the Karimata Islands, west of Borneo, and passes through the east of West Malaysia, disappearing northwards beneath the sediment-filled Thailand Basin. This belt probably reappears near the Cambodian border in southeast Thailand;[40] its offset continuation may lie in the Hua Hin area,[39] extending northwards through Chiang Mai and northwestern Laos to beyond Nam Pha Tene mine.[41] Small deposits of tin occur in the Tasal area of western Cambodia[42] and in a small arcuate granitic belt south of Hanoi in North Vietnam.[43]

Fluorite deposits in southeast Asia are associated with the granitic intrusions in the tin belts. In Thailand, the world's fourth largest producer of fluorite, deposits occur near the Thai–Burma border from Chiang Mai to west of Hua Hin.[44] Fluorite is also a frequent gangue mineral in the tin–tungsten deposits of Thailand, West Malaysia, Indonesia and Burma, and is a major constituent of pipe-like deposits in the Kinta Valley of West Malaysia.[45]

Far East Province

In China there are two main belts of tin–tungsten mineralization. A northwesterly belt starts in the Kochiu area, north of the Red River Fault; Kochiu supplies 70–80 per cent of Chinese tin production.[41] The tin–tungsten–molybdenum deposits in North Vietnam may form the southern margin of this belt. The Kochiu tin belt follows the Nanling structural zone into the Kiangsi tungsten province,[46] which contributes more than 25 per cent of world tungsten output.[47]

A second tin–tungsten belt extends from Hainan

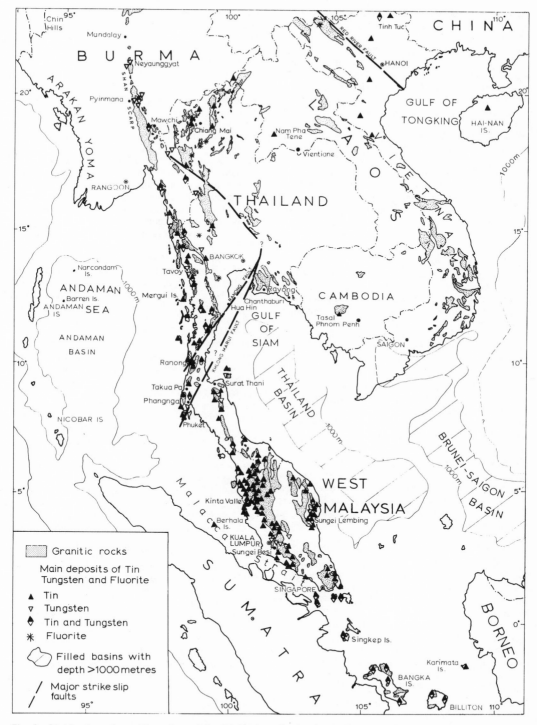

Fig. 4 Distribution of granitic rocks and Sn–W–F mineralization in southeast Asia

Island along the southeast China coast to beyond Hong Kong.

Little is known of tin deposits in northeast China, but the northwestern belt may continue through the rich zone of tungsten mineralization in Korea, which includes numerous large fluorite deposits.[49] The Korean zone trends northeast towards the Lifuzdin–Khrustal'nyy tin belt in eastern Siberia;[41] other northeast-trending tin–tungsten belts occur farther north in Siberia. An important stanniferous area in the Anadyr Mountain belt may continue into the zone of tin–tungsten–fluorite mineralization in the Seward Peninsula of Alaska.[50]

Widespread fairly minor deposits of tin and tungsten associated with granitic and rhyolitic rocks on Honshu Island in Japan[51] could be interpreted as the lateral termination of the mainland coastal China tin belt. Large fluorite deposits occur in southwest Japan and in central Honshu near Gifu City.[52]

Bolivian province

Nearly 20 per cent of the non-Communist world's tin production comes from the belt of granites and acidic volcanic rocks which extends from southeastern Peru through the Cordillera Real and Eastern Cordillera of western Bolivia[53] to the extreme north of Argentina. The richest tin–tungsten mineralization occurs near Oruro in the centre of the belt.

Australian province

The tin–tungsten deposits of the Australian continent and Tasmania are older than most of those elsewhere around the Pacific. They are related to igneous events which occurred when Australia was part of Gondwanaland, and are not considered here.

Age and nature of country rock

The pre-granite sedimentary host rocks in southeast Asia and China are mainly late Palaeozoic in age, although Mesozoic and early Palaeozoic or older rocks also occur. Most abundant are Permian limestone and fine-grained clastic terrigenous sediments forming thick successions of folded and faulted, highly indurated rocks of variable sedimentary facies. Turbidite facies are common in parts of the Thai–Malay Peninsula and Burma,[54] and schists predominate in the Indonesian islands of Singkep[55] and in the Kochiu area of China.[41] East of Pyinmana in Burma, metamorphic rocks and Mesozoic mudstones are the host rocks.[56] On Honshu Island, Japan, tin–tungsten deposits are found in host rocks ranging from early Palaeozoic metamorphosed sediments to Tertiary marine sediments and volcanic rocks. In Bolivia the mineralized rocks include Palaeozoic sandstone, shales, graywackes and Tertiary tuffs.[53]

Depth of emplacement of tin-bearing igneous rocks

Evidence for a fairly shallow depth of emplacement of the tin-bearing granites is provided by sharp contacts, narrow aureoles, roof pendants and xenoliths of country rock. In some pegmatitic areas tin mineralization may have been emplaced at slightly greater depths, whereas tin–tungsten lodes were developed at higher levels.

In Japan tin–tungsten deposits occur in deep-seated granitic intrusions and near-surface volcanic rocks.[51] In central Bolivia the ore was probably deposited in volcanic rocks within a few thousand feet of the surface, and, in the north, tin mineralization was associated with deeper granitic batholiths.[53]

Age of granites and mineralization in southeast Asia

Radiometric datings indicate that the tin-bearing granitic belts in southeast Asia have undergone more than one episode of intrusion and heating.

In the eastern granitic belt of West Malaysia most whole-rock Rb : Sr determinations yield either Upper Carboniferous or Triassic ages; late Mesozoic to early Tertiary heating events are indicated by K : Ar dates on micas. Acidic volcanic rocks of Triassic age in the south of the belt are probably related to the emplacement of granitic rocks of similar age. The inferred northward continuation of this belt at Chantaburi and Rayong has yielded whole-rock Rb : Sr ages of early Permian and late Jurassic, and K : Ar ages of late Jurassic and late Cretaceous.

In northwestern Laos, near the Thai–Burma border, tin-bearing granitic rocks lie to the east of a belt of volcanic rocks[57] which extends southwestwards into Thailand, where it is of Permo–Triassic age.[58] The north-trending granitic belt lying northeast of Chiang Mai and also the belt to the west contain plutons of both middle Palaeozoic and Lower Triassic age.

In the western or Main Range belt of West Malaysia Rb : Sr determinations have yielded a few late Carboniferous ages, and numerous Triassic ages; the latter fall in two groups separated by about 30 m.y. K : Ar dates on micas from this belt have yielded ages of 205–50 m.y., with a concentration of dates in the early Jurassic (180 m.y.), late Jurassic (135 m.y.) and late Cretaceous (70 m.y.). Granites on Billiton Island to the southwest can be provisionally dated as Triassic on Rb : Sr and U : Pb determinations.[59,60] North of the Khlong Marui Fault the western belt has yielded whole-rock Rb : Sr ages of early Cretaceous, whereas younger K : Ar ages suggest an early Tertiary heating event, which was widespread in southeast Asia.[61] At the northern extremity of the belt, in Burma, granitic rocks intrude an early Cretaceous sedimentary succession.[62]

Few data are available on the age of the mineralization. In the Kinta Valley K : Ar determinations on micas from greisen yielded late Triassic ages broadly similar to those of the granitic rocks, although the micas could be up to 20 m.y. younger than the granites.[63] Elsewhere, field relations merely indicate that the mineralization is no older than the associated granitic rocks.

Mineralization in tin–tungsten–fluorite provinces

The three main tin minerals are cassiterite, stannite and malayaite; tungsten occurs principally as wolframite and scheelite.

In southeast Asia the primary tin deposits have been recently classified as (1) magmatic disseminations including stanniferous pegmatites and aplites, (2) pyrometasomatic deposits and (3) hydrothermal deposits.[64]

Hosking[64] considered that large-scale magmatic disseminations of cassiterite are rare, and that deposits in granites such as Sungei Besi in Malaysia[65] and Haad Som Pan in Thailand[66] are forms of epigenetic mineralization beneath a metasedimentary roof. Similar origins could account for cassiterite-rich granites in the Phuket area of Thailand, although cassiterite in lepidolite pegmatites near Phangnga[67] is probably syngenetic.

Pyrometasomatic skarn-like deposits containing malayaite and calc-silicate minerals are characteristic of some granite–marble contacts, notably at Buket Besi in Malaysia and Klappa Kampit in Billiton.

Hydrothermal deposits are the main primary tin deposits of southeast Asia, comprising stockworks of cassiterite–wolframite-bearing quartz veins and pegmatites. Other indications of hydrothermal activity include tourmalinization and greisenization of apical portions of veins and of granite cusps, and development of pipe-like deposits in the Kinta Valley. Deposits at Sungei Lembing (Pahang), resembling Cornish tin lodes, may also be of hydrothermal origin with mineralization occurring during repeated opening of fault systems.[64] Some of these lodes are replacement veins carrying cassiterite, magnetite and numerous sulphides including chalcopyrite, arsenopyrite, galena and sphalerite in a gangue of quartz, carbonates, fluorite and tourmaline. Lodes with similar sulphide minerals are present in several parts of West Malaysia, in Thailand near the Malaysian border, in North Vietnam,[43] and at Mawchi in Burma, where wolframite and scheelite are present.[68]

North of Mawchi the lodes become more scattered, and wolframite supplants cassiterite as the main ore mineral. At Neyaunggyat in the extreme north of this belt cassiterite is rare and wolframite occurs in quartz lodes with copper sulphides and beryl.[69]

Many deposits contain more than one generation of tin minerals—for example, niobium/tantalum-rich early cassiterite mantled by acicular 'normal' cassiterite, pyrometasomatic malayaite replaced or bordered by cassiterite and corroded cassiterite surrounded by stannite.[64]

Hydrothermal deposits similar to those of southeast Asia are found throughout the western circum-Pacific region, tungsten minerals and associated sulphide minerals predominating in the eastern parts of the inner tin–tungsten belt of China, and in Korea; cassiterite is more important in Siberia and Alaska.

In Japan different types of tin–tungsten mineraliza-tion occurred during three separate episodes of granitic intrusion.[51] Associated with Cretaceous batholiths, the main ore minerals in vein and pyrometasomatic deposits are cassiterite, wolframite, scheelite and stannite; greisenization and tourmalinization are present. In Palaeogene granitic complexes hydro-thermal tin–tungsten ores are associated with molyb-denite and copper ores. Late Tertiary mineral deposits, associated with porphyries, andesites and rhyolites, comprise vein deposits with early copper, lead, zinc and bismuth sulphides together with later cassiterite, stannite, wolframite and scheelite. At Ashio these vein deposits are found in a funnel-shaped mass of rhyolite and tuff.

In northern Bolivia tin–tungsten deposits of the Cordillera Real occur in a belt up to 10 km wide bordering granitic batholiths. In unaltered sedimentary rocks beyond the tin–tungsten zone a zone of lead–zinc veins locally carries silver, and an outer zone contains quartz–stibnite veins. This outwards zonation is remarkably similar to the upwards zoning observed in tin deposits in Cornwall.[70] Farther south in Bolivia only local zoning occurs at some intrusions, tin–tungsten ore generally occupying the central zones. There is an apparent telescoping of tin–tungsten and porphyry-type sulphide deposits.

Fluorite deposits in the circum-Pacific regions are of two types. The first type consists of large veins and replacement deposits of considerable economic impor-ance containing mainly fluorite and quartz, with local calcite, pyrite, stibnite and chalcopyrite. Tin or tungsten minerals are generally absent, although the fluorite occurs near the main mineralized granitic belts. Typical deposits are found in western Thailand,[44] at Hsiang-Shan in China,[48] in a wide zone in Korea,[49] and near Gifu City in Japan.[52]

The second type is more directly associated with tin–tungsten mineralization, where fluorite is present with other gangue minerals; locally, it is a major ore constituent as in the Kinta Valley of West Malaysia.[45] At Kramat Pulai the ore consisted almost entirely of fluorite and 3–15 per cent scheelite with a mere 1 per cent of gangue minerals. Beatrice mine contained in depth cassiterite and copper sulphides in a gangue including fluorite and fluoborite. Near Tavoy in Burma, where fluorite is abundant, one lode at Hermyingale mine consists mainly of topaz and fluorite.[68] At Colquiri in Bolivia, fluorite and siderite are more abundant than quartz in lodes.[53]

In the Seward Peninsula of Alaska, features of both main types of fluorite mineralization are found. A strong zonation is evident, from tin deposits in greisen through transitional veins of sulphide minerals with fluorite and chrysoberyl, to fluorite–beryllium deposits and thence to barren quartz–fluorite veins.[50]

Tin-bearing granitic belts and marginal basins

In the circum-Pacific region, tin-bearing granites are all bordered either by marginal basins and island arcs or by younger granitic belts (Fig. 1). Evidence from deep-sea drilling suggests that some marginal basins are younger than the ocean floor on the convex side of the island arcs.[76] This indicates that the basins have probably been developed by the creation of new oceanic crust above the underlying Benioff zone, with related outward migration of the island arcs and bordering submarine trench.[71,72] Thus, for example, the Andaman Sea,[73] the Japan Sea and the Okhotsk Sea may all have been developed during the Cenozoic.

Development of marginal basins in the Mesozoic, followed by their later consumption along younger Benioff zones, could account for some of the sub-parallel granitic belts of eastern Asia. A modern example of a closing marginal basin is the north Philippines Basin, which is contracting as its floor is lost along an eastward-dipping Benioff zone approaching the surface west of the northern Philippines (Fig. 1). Complete closure of the Basin would result in igneous rocks of the Philippine volcanic arc lying adjacent to the igneous belt of coastal China. An ancient example of a closed marginal basin is the relatively granite-free central belt of West Malaysia. Here a marginal basin may have been developed in the late Palaeozoic or Mesozoic above an eastward-dipping Benioff zone which approached the surface to the west of the western granitic belt. The floor of this basin, of which remnants now form the discontinuous belt of ultramafic rocks in the central belt, was eventually largely lost beneath a later Benioff zone dipping eastwards beneath the eastern belt of granites.

The central lowlands of Burma may have been formed in a similar way to the Andaman Basin, with westward movement of the arcuate Indo–Burman Ranges, together with the underlying Benioff zone, away from the Shan Scarp; similarly, the Malacca Straits may have opened during westward movement of Sumatra away from West Malaysia. Development and consumption of marginal basins can probably also explain the Mesozoic granitic belts in northwest Thailand.

The mechanism by which new crust is emplaced beneath expanding marginal basins is uncertain, but it is generally considered that basaltic dykes are intruded above convective[71,72] or mechanically driven[113] circulating cells in the low-viscosity layer of the upper mantle. Heat flow above marginal basins is high—indicating that either magma or volatiles or both are transferred upwards from the mantle.

Prior to the opening of a marginal basin, the subduction zone bordering the basin probably lay adjacent to the continent, and in many cases a mountain belt of Andean type developed with volcanism and emplacement of batholiths. Circum-Pacific tin deposits all occur in igneous belts of this type which were subsequently bordered by marginal basins and island arcs—indicating a genetic relationship between basin development and mineralization.

Geochemistry of porphyry copper, massive sulphide and tin deposits

Porphyry copper and massive sulphide deposits

Magmatic sources for porphyry metals are suggested by the regular pattern of ore, gangue and hydrothermal alteration minerals around the individual centre, following similar sequences in different mineralized areas.[77,78] Although these features suggest a single sequence of physico-chemical processes involving outward movement of fluids from a centre, a common source for magma and metals is not proven. Other possible sources are metalliferous shales intruded by the magma,[79] or metalliferous connate brines entering the magma chamber in a circulatory system, and depositing metals on mixing with shallow sulphatic connate water.[80]

The presence of porphyry copper deposits in both Andean-type belts and island arcs within marine basins suggests that the metal originates in the intrusive magma rather than in the intruded crust. Copper is available in andesites and tholeiites, which contain up to 150 ppm Cu;[81] an orebody of 1 000 000 tons of metal is equivalent to a reduction of only 3 per cent of that in a 100-km³ stock.[77]

Sufficient sulphur also is available in igneous rocks (100–400 ppm) to form the various sulphide ores.[82] Sulphur accumulations at depth could result from early separation of immiscible sulphide and silicate liquids. Surface sulphur deposits occur in volcanic vents such as Mt Io-san, in Japan, where more than 100 000 tons of sulphur and large volumes of sulphur-bearing gases were emitted in 8 months.[51]

A primordial upper mantle, or lower crustal, origin for the sulphur is indicated by the near zero per mil values exhibited by δ^{34} analyses of sulphides in Cordilleran intrusives.[79] This suggests that connate sulphatic waters are not necessary to explain the ore genesis. Dispersion of hydrogen sulphide from the magma could produce contact replacement and vein deposits on meeting metalliferous brines.[83] Some of the strata-bound massive sulphide deposits may have originated in this way.

Fluid inclusion studies indicate that the ore fluids which transport base metals are dominantly chloride waters with subordinate sulphate.[84,85] Compared with common deep brines, the heavy metal content and K : Na ratios are much higher in the fluid inclusions.[86]

Hydrogen and oxygen isotope ratios in North American porphyry copper deposits indicated that hydrothermal magmatic solutions caused the K-feldspar–biotite alteration zone, but that a significant meteoric water component, rich in Na–Ca–Cl, was present during sericitization and argillization. This indicated that the ore shell lies between the central potassic–alteration core and the outer sericite–pyrite shell because this boundary is probably the magmatic–meteoric solutions interface.[87] The authors consider that the meteoric saline fluids (which may have contributed some metal to the outer zones[79]) formed part of a circulatory system, with downward and inward movement of increasingly saline fluids towards the late potassic sulphur and metal-bearing phase of the intrusion.

Tin–tungsten deposits

Tin and tungsten contents of less than 3 ppm in rocks ranging from average ultramafite to average granite,[82] and similar low values in sedimentary rocks, have led to hypotheses of pre-mineralization concentrations in the lower crust[88] or mantle.[89] Turneaure,[90] however, demonstrated similarities between some supposed metallogenetic provinces in different parts of the world, suggesting '. . . a certain uniformity in metal distribution in the crust and in processes of ore concentration'.

In granitic provinces tin and tungsten values tend to increase within the typical sequence quartz–diorite, granodiorite, hornblende–biotite–granite, biotite–granite and muscovite-rich granite. For example, in this sequence in the Shakhtaminski–Soktuiski area of Russia, values increase from $2 \cdot 5$ to $7 \cdot 4$ ppm concomitantly with increases in Li, Rb, Be and F.[91] Similarly, in the Phuket area of peninsular Thailand, rocks ranging from adamellite to muscovite–granite contain from 3 to 25 ppm Sn, some biotite–granites carrying up to 80 ppm Sn.[92] Tauson *et al.*[93] demonstrated regular increases of F, Sn, Li, Mo, Be and Rb in biotites from a similar sequence; at Phuket hornblendes, biotites and muscovites show corresponding increases in Sn from 50 to 400 ppm and in Nb from 30 to 360 ppm.[94] Maximum tin values (500–8000 ppm) occur in micaceous greisens, the last phase of the tin-bearing granitic activity.

Barsukov[95] showed that 80 per cent of the tin in tin-rich granites lies in the biotite structure into which tin and lithium are introduced simultaneously, tin replacing iron and lithium replacing magnesium. In the later muscovite, growing ions of tin, tungsten and niobium or their complexes are trapped between sheets and in crystal defect sites. These ions, however, in contrast to those in biotite, are easily leached out to form ore concentrations.

Volatiles are important transporters of ore metals, and Tauson[96] suggested that degassing of volatile-rich granitic melts would enrich the apical parts of the intrusions in Sn, W, Li, Be and Rb. Abundance of fluorine in the tin–granite environment is indicated by the presence of extensive quartz–fluorite deposits, fluorite in gangues, and fluorine-bearing minerals such as lepidolite, zinnwaldite, topaz, tourmaline and apatite. Increase of tin with fluorine in minerals would be explained if the metal is removed from the magma as volatile tin tetrafluoride and/or tetrachloride,[97,98] and precipitated later as cassiterite; some volcanic gases carry traces of ore metals including tin, and high tin concentrations can be obtained by passing gases through silicate melts.[99] Alternatively, it has been proposed that tin is transported in alkaline solutions as sodium or potassium hydroxy stannate $(Na,K)_2$ $Sn(OH, F)_6$ or alkali thiostannate[98] and that cassiterite is deposited during changing temperature and pH conditions when the solutions rise from the magma body.[100] Hydrogen fluoride, liberated on deposition of cassiterite, would react with feldspar to produce topaz and muscovite, and remove Na, K and some Ca (greisenization).

Fluid inclusions in quartz, cassiterite and early fluorite in Bolivian tin deposits consist of complex NaCl-rich brines with a little CO_2;[101] estimated temperatures declined from about 530°C in the earliest vein stages to below 70°C, and salinities decreased from more than 46 per cent to a few per cent. Kelly and Turneaure[101] suggested that these brines were the ore-forming fluids of magmatic hydrothermal origin. Analyses of fluid inclusions in quartz from cassiterite deposits show, however, that the hydrothermal solutions contained ample fluorine to transport tin as a fluoro–hydroxy complex.

Similarities with the genesis of porphyry copper deposits suggest that the inclusion brines are of meteoric origin. Significantly, in the granitic pluton the zone of tin–tungsten mineralization, with accompanying muscovitization and greisenization, corresponds to the sulphide deposition zone in porphyries at the boundary between the potassic–alteration core and the outer sericite–pyrite shell. Tin and tungsten were probably carried initially as high-temperature tetrafluorides; at high levels, on meeting hypersaline meteoric brines, the transporting medium changed to aqueous alkaline solutions carrying alkali fluoro–hydroxy stannates or similar complexes. Changes in temperature, pressure and pH resulted in deposition of ore minerals and the formation of fluorine-bearing minerals.

Relationship of mineralization to Benioff zones

The association of copper, and of tin, tungsten and fluorite, with igneous belts emplaced above palaeo-Benioff zones indicates that the mineralization is related to physico-chemical changes occurring during subduction of a lithospheric slab.

Porphyry copper deposits

During descent at rates of 5 cm/year, or more, the slab remains cooler internally than the surrounding mantle to depths exceeding 500 km.[102] Toksöz, Minear and Julian[103] showed that for a descending slab, 80 km thick, melting occurs along the slab's upper surface at 180 km, but extends about 100 km in each direction because basalt liquidus temperatures are lowered in the presence of water.[104] They concluded that the first magma would contain a significant fraction of crustal rocks, whereas at greater depths the high-temperature zone would broaden to affect the mantle more than the slab.

Volcanic centres in active island arcs mostly lie 200–300 km from trenches, and most porphyry copper deposits were emplaced at similar distances from ancient trenches. This corresponds in the computed models to temperatures up to 1600°C at depths of 250–300 km.[103] Wyllie[105] showed that this melting can occur only if water, derived probably from oceanic sediments, is present in trace amounts up to 0·1 per cent. He suggested that water ascending into the mantle above the slab would facilitate diapiric uprise of andesitic magma well above the Benioff zone. Addition of volatiles such as P_2O_5, HF and CO_2 would lower melting points,[106,107] and assist in transporting ore metals to the upper crust, where they would be deposited on contact with meteoric brines. Possible sources of volatiles are the melted basaltic layers 2 and 3 and the upper sedimentary layer 1 of the oceanic crust; the latter could include either local concentrations of phosphatic nodules, or bottom muds which contain up to 1100 g/ton fluorine.[108] Copper could be derived from the descending tholeiitic basalts, which contain more than 100 ppm Cu, or from sulphides or native Cu, observed in present oceanic sediments.[109]

Tin, tungsten and fluorite deposits

The relationship between tin-bearing granitic belts and

bordering marginal basins and island arcs can be explained by considering the evolution of the Andaman–Nicobar arc, the Andaman Sea, and the tin belt in peninsular Thailand (Fig. 5).

Temperature fields in the descending lithospheric slab in Fig. 5(*a*) are from the model by Toksöz, Minear and Julian[103] for a spreading rate of 8 cm/year. Predominantly calc-alkaline volcanic centres, possibly including porphyry copper deposits, overlie batholiths; the igneous rocks were derived either from the water-bearing zone of melting at about 1000–1600°C or from partial melting of the overlying basal continental crust. Within the upper parts of the batholith the normal igneous series would be tonalite–granodiorite–granite, any fluorine present in late residuals assisting in some concentration of tin in hornblende and biotite. The Gulf of Thailand is omitted from Fig. 5, and at this stage in its development the continental margin, with an active igneous belt, is of Andean type.

Continued descent of the slab (Fig. 5(*b*)) results in tectonic emplacement of sediments, mostly continent-derived, on the landward side of the trench, forming an extending belt of melanges and glaucophane schists.[110] This forces the trench outwards, resulting

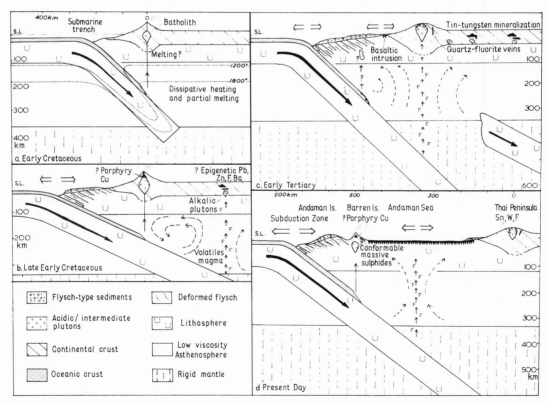

Fig. 5 Cretaceous–present evolution of Andaman Sea and peninsular Thailand, showing relationship of known Sn–W–F mineralization, and inferred porphyry copper and base-metal deposits, to Benioff zones

in a decrease in slope of the slab, so that the igneous activity remains in approximately the same position.

At depths of around 350 km melting above 1600°C occurs in a dry environment; here dry peridotite is near its fusion point.[111] Fluor-apatite and chlor-apatite, which, respectively, melt at about 1650°C[112] and 1530°C, could be concentrated either in the upper levels of gabbroic bodies in the descending crust, or in ultramafic mantle rocks. These minerals could react with partially melted silicate rocks, resulting in liberation of fluorine and perhaps chlorine.

A flow pattern in the asthenosphere, generated by the down-going slab,[113] results in upward transfer of hot volatiles and possibly magma. The near-surface expression of the rising volatiles could explain the quartz–fluorite bodies on the continent side of the igneous belt. Decrease in slope of the descending slab results in increased friction with the overlying lithosphere, and eventually movement ceases and a new high-angle Benioff zone develops (Fig. 5(c)).

The upward stream of volatiles now partly underlies the continental margin, where the drag-induced flow pattern results in low horizontal stress at the surface. Basaltic magmas generated in this zone are emplaced as intrusive bodies and lavas, forming new oceanic-type crust and resulting in opening of the marginal basin.

At this stage the upward volatile stream also underlies the still hot calc-alkaline or acidic igneous belt. Remelting of late alkali-rich fractions within the batholith and upward movement of volatile fluorides of alkalis, silicon, tin, tungsten and associated metals take place; finally, alkali hydroxy stannates or thiostannates are formed by reaction with hydrous phases near the apical margins of the magma body. Changing conditions of pH and falling temperatures result in deposition of tin or tungsten minerals, and hydrofluoric acid reacts with available lime to form fluorite and with other minerals to form lithian muscovite, lepidolite and topaz. Transport by alkali thiostannates to high levels could result in telescoped or zoned deposits of base-metal sulphides and tin–tungsten mineralization.

Continued opening of the marginal basin (Fig. 5(d)) is accompanied by development of a volcanic arc, represented by Barren and Narcondam Islands. Within this arc, porphyry copper deposits and conformable massive sulphides could be emplaced.

Uplift of the continental margin, during the stages shown in Fig. 5(c) or (d) leads to exposure of the tin–tungsten–fluorite deposits.

Epigenetic non-orogenic base-metal deposits
Deposition of tin, tungsten and fluorite around granites underlain by hot rising volatiles suggests that other metal ores might accumulate elsewhere in the continental crust above a deep Benioff zone.

In the eastern Andes late Cenozoic alkalic plutons and volcanic rocks, some of which are mineralized, are underlain by, and probably related to, the present Benioff zone at depths exceeding 250 km.[115] In the Mawson lead province of Burma, east of the tin and tungsten-bearing granitic belt, lead–barite veins are widespread in late Mesozoic limestones.[69] These ores were probably emplaced above the westward-migrating rising hot volatile stream which resulted in tin and wolfram mineralization in the granitic rocks. Their positions, and those of inferred underlying alkalic plutons, are shown schematically in Fig. 5(b and c).

Similarly, many epigenetic base-metal deposits lying within 200–700 km of an orogenic belt—for example, some Mississippi Valley type lead–zinc deposits possibly underlain by alkalic plutons[114]—were perhaps emplaced above Benioff zones which approached the surface in the orogenic belts.

Acknowledgement
Thanks are due to Dr. H. G. Reading and Dr. J. D. Bell for discussions and comments on the manuscript. Mr. M. J. Russell, Mr. P. J. Moore and Dr. T. Deans also provided helpful suggestions, and Mr. M. D. Forrest assisted with the diagrams. The paper is published by permission of Rio Tinto–Zinc Corporation, Ltd., and Dr. K. C. Dunham, Director of the Institute of Geological Sciences, London.

References
1. **Pautot G. Auzende J. M. and Le Pichon X.** Continuous deep sea salt layer along North Atlantic margins related to early phase of rifting. *Nature, Lond.,* **227,** 1970, 351–4.
2. **Hutchinson R. W. and Engels G. G.** Tectonic significance of regional geology and evaporite lithofacies in northeastern Ethiopia. *Phil. Trans. R. Soc.,* **267A,** 1970, 313–29.
3. **Tooms J. S.** Review of knowledge of metalliferous brines and related deposits. *Trans. Instn Min. Metall. (Sect. B : Appl. earth sci.),* **79,** 1970, B116–26.
4. **Russell M. J.** Structural controls of base metal mineralization in Ireland in relation to continental drift. *Trans. Instn Min. Metall. (Sect. B : Appl. earth sci.),* **77,** 1968, B117–28.
5. **Schuiling R. D.** Tin belts around the Atlantic ocean: some aspects of the geochemistry of tin. In *A technical conference on tin* (London: International Tin Council, 1967), vol. 2, 531–47.
6. **Crawford A. R.** Continental drift and un-continental thinking. *Econ. Geol.,* **65,** 1970, 11–6.
7. **Lowell J. D. and Guilbert J. M.** Lateral and vertical alteration—mineralization zoning in porphyry ore deposits. *Econ. Geol.,* **65,** 1970, 373–408.
8. **Stringham B.** Igneous rock types and host rocks associated with porphyry copper deposits. In *Geology of the porphyry copper deposits southwestern North America* **Titley S. R. and Hicks C. L. eds.** (Tucson: University of Arizona Press, 1966), 35–40.
9. **James A. H.** Hypothetical diagrams of several porphyry copper deposits. *Econ. Geol.,* **66,** 1971, 43–7.

10. **Damon P. E. and Mauger R. L.** Epeirogeny–orogeny viewed from the Basin and Range province. *Trans. Am. Inst. Min. Engrs*, **235**, 1966, 99–112.

11. **Moore W. J. and Lanphere M. A.** The age of porphyry-type copper mineralization in the Bingham mining district, Utah—a refined estimate. *Econ. Geol.*, **66**, 1971, 331–4.

12. **Bryner L.** Notes on the geology of the porphyry copper deposits of the Philippines. *Mineral Engng Mag.*, **19**, 1968, 12–23.

13. **Macnamara P. M.** Rock types and mineralization at Panguna porphyry copper prospect, Upper Kaverong Valley, Bougainville Island. *Proc. Australas. Inst. Min. Metall.* no. 228, 1968, 71–9.

14. Copper and bauxite prospects in British Solomon Islands. *Aust. Min.*, **60**, Nov. 1968, 58–9.

15. **Coats R. R.** Magma type and crustal structure in the Aleutian arc. In *The crust of the Pacific basin. Am. Geophys. Union Geophys. Monograph* 6, 1962, 92–109.

16. **Isacks B. Oliver J. and Sykes L. R.** Seismology and the new global tectonics. *J. geophys. Res.*, **73**, 1968, 5855–99.

17. **Mitchell A. H. [G.] and Reading H. G.** Evolution of island arcs. *J. Geol.*, **79**, 1971, 253–84.

18. **Hamilton W.** The volcanic central Andes—a modern model for the Cretaceous batholiths and tectonics of western North America. *Bull. Oregon Dep. geol. miner. Ind.*, **65**, 1969, 175–84.

19. **Hamilton W.** Mesozoic California and the underflow of Pacific mantle. *Bull. geol. Soc. Am.*, **80**, 1969, 2409–29.

20. **Kistler R. W. Evernden J. F. and Shaw H. R.** Sierra Nevada Plutonic cycles: part 1, origin of composite granitic batholiths. *Bull. geol. Soc. Am.*, **82**, 1971, 853–68.

21. What's going on in world mining . . . Oceania. *World Min.*, **23**, 1970, July, 66–7; Aug. p. 79.

22. **van Bemmelen R. W.** *The geology of Indonesia, volume 1A* (The Hague: Government Printing Office, 1949), 756 p.

23. **Evans P. and Sansom C. A.** The geology of British oilfields 3. The oilfields of Burma. *Geol. Mag.*, **78**, 1941, 321–50.

24. **Dewey J. F. and Bird J. M.** Origin and emplacement of the ophiolite suite: Appalachian ophiolites in Newfoundland. *J. geophys. Res.*, **76**, 1971, 3179–206.

25. **Haile N. S.** Geosynclinal theory and the organizational pattern of the North-west Borneo Geosyncline. *Q. J. geol. Soc. Lond.*, **124**, 1969, 171–94.

26. **Dickinson W. R.** Petrogenetic significance of geosynclinal andesitic volcanism along the Pacific margin of North America. *Bull. geol. Soc. Am.*, **73**, 1962, 1241–56.

27. **Souther J. G.** Volcanism and its relationship to recent crustal movements in the Canadian Cordillera. *Can. J. Earth Sci.*, **7**, 1970, 553–68.

28. **Ferenčić A.** Metallogenic provinces and epochs in southern Central America. *Mineralium Deposita*, **6**, 1971, 77–88.

29. **Dickinson W. R.** Plate tectonic models for orogeny at continental margins. *Nature, Lond.*, **232**, 1971, 41–2.

30. **Tamrazyan G. P.** Siberian continental drift. *Tectonophysics*, **11**, 1971, 433–60.

31. **Bazin D. and Hübner H.** Copper deposits in Iran. *Rep. geol. Surv. Iran* no. 13, 1969, 232 p.

32. **Stöcklin J.** Structural history and tectonics of Iran: a review. *Bull. Am. Ass. Petrol. Geol.*, **52**, 1968, 1229–58.

33. **Bryner L.** Ore deposits of the Philippines—an introduction to their geology. *Econ. Geol.*, **64**, 1969, 644–66.

34. **Horikoshi E.** Volcanic activity related to the formation of the Kuroko-type deposits in the Kosaka District, Japan. *Mineralium Deposita*, **4**, 1969, 321–45.

35. **Stanton R. L.** Lower Palaeozoic mineralization near Bathurst, New South Wales. *Econ. Geol.*, **50**, 1955, 681–714.

36. **Hutchinson R. W. Ridler R. H. and Suffel G. G.** Metallogenic relationships in the Abitibi belt, Canada: a model for Archean metallogeny. *CIM Bull.*, **64**, April 1971, 48–57.

37. **Russell M. J.** Personal communication, 1971.

38. **White D. E. Muffler L. J. P. and Truesdell A. H.** Vapor-dominated hydrothermal systems compared with hot water systems. *Econ. Geol.*, **66**, 1971, 75–97.

39. **Garson M. S. and Mitchell A. H. G.** Transform faulting in the Thai Peninsula. *Nature, Lond.*, **228**, 1970, 45–7.

40. **Hughes I. G. and Bateson J. H.** Reconnaissance geological and mineral survey of the Chanthaburi area of South-East Thailand. *Rep. I.G.S. overseas spec. Surv.* no. 7, 1967, 33 p.

41. **Sainsbury C. L.** Tin resources of the world. *Bull. U.S. Geol. surv.* 1301, 1969, 55 p.

42. **Lasserre M.** *et al.* Géologie, chimie et géochronologie du granite de Tasal (Cambodge occidental). *Bull. Bur. Rech. géol. Min.* 2nd Ser. Sect. IV, no. 1, 1970, 5–13.

43. **Losert J.** Nerostné suroviny Vietnamské deomokratické republiky. *Vestn. Ustred. ustavu geol.*, **41**, no. 1 1966, 59–66; *Ref. Zh., Geol.*, 1966, 6D 28.

44. **Gardner L. S. and Smith R. M.** Fluorspar deposits of Thailand. *Rep. Invest. Dep. Mines Res. Thailand* no. 10, 1965, 42 p.

45. **Ingham F. T. and Bradford E. F.** The geology and mineral resources of the Kinta Valley, Perak. *Distr. Mem. geol. Surv. Fed. Malaya* 9, 1960, 347 p.

46. **Hsu K. C. and Ting I.** Geology and tungsten deposits of southern Kiangsi. *Mem. geol. Surv. China*, Ser. A. no. 17, 1943, 360 p.

47. **Wang. K. P.** The Far East: China. In *Mining annual review* (London: Mining Journal, 1971), 396–401.

48. **Igarashi Z.** Fluorite deposits in the vicinity of Hsiang-Shan Che-Kiang Province. In *Geology and mineral resources of the Far East, volume 1* **Ogura T.** ed. (Tokyo: University of Tokyo Press, 1967), 471–7.

49. **Gallagher G.** *Mineral resources of Korea, volume VIA, non-metallics and miscellaneous metals* (Mining

Branch Indust. Mining Div. USO M/Korea, 1963), 121 p.

50. **Sainsbury C. L.** Geology and ore deposits of the central York Mountains, western Seward Peninsula, Alaska. *Bull. U.S. geol. Surv.* 1287, 1969, 101 p.

51. **Sekine Y.** Metallogenetic epochs and provinces of Japan. In *Geology and mineral resources of Japan, 2nd edn* **Saito M.** *et al.* eds (Kawasaki-shi: Geological Survey of Japan, 1960), 141–68.

52. **Ueno M.** On the fluorite deposits in the northern part of Gifu Prefecture. *Bull. geol. Surv. Japan,* **14**, no. 5 1963, 61–72.

53. **Turneaure F. S.** The Bolivian tin–silver province. *Econ. Geol.,* **66**, 1971, 215–25.

54. **Mitchell A. H. G. Young B. and Jantaranipa W.** The Phuket Group, Peninsular Thailand: a Palaeozoic ? geosynclinal deposit. *Geol. Mag.,* **107**, 1970, 411–28.

55. **van Overeem A. J. A.** Offshore tin exploration in Indonesia. *Trans. Instn Min. Metall. (Sect. A: Min. industry),* **79**, 1970, A81–5.

56. **Bateson J. H. Mitchell A. H. G. and Clarke D. A.** Geological and geochemical reconnaissance of the Seikphudaeung–Padatgyaung area of central Burma. *Rep. Inst. geol. Sci.,* 1971, unpublished.

57. **Page B. G. N. and Workman D. R.** Geological and geochemical investigations in the Mekong Valley, between Vientiane and Sayabouri and at Ban Houei Sai. *Rep. Inst. geol. sci.* no. 9, 1968, unpublished.

58. **Baum F.** *et al.* On the geology of northern Thailand. *Beihefte geol. Jb,* 102, 1970, 24 p.

59. **Edwards G. and McLaughlin W. A.** Age of granites from the tin province of Indonesia. *Nature, Lond,* **206**, 1965, 814–6.

60. **Snelling N. J. Bignell J. D. and Harding R. R.** Ages of Malayan granites. *Geologie Mijnb.,* **47**, 1968, 358–9.

61. **Burton C. K. and Bignell J. D.** Cretaceous–Tertiary events in Southeast Asia. *Bull. geol. Soc. Am.,* **80**, 1969, 681–8.

62. **Garson M. S. Amos B. J. and Mitchell A. H. G.** Geology of the Neyaungga–Yengan area, Southern Shan States, Burma. *Overseas Mem. Inst. geol. Sci.,* 2, in press.

63. **Snelling N. J.** Personal communication, 1971.

64. **Hosking K. F. G.** The primary tin deposits of south-east Asia. *Minerals Sci. Engng,* **2**, no. 4, Oct. 1970, 24–50.

65. **Jones M. P.** The tin industry. *Geologie Mijnb.,* **48**, 1969, 451–65.

66. **Aranyakanon P.** The cassiterite deposit of Haad Som Pan, Ranong Province, Thailand. *Rep. Invest. R. Dep. Mines, Thailand,* no. 4, 1961, 182 p.

67. **Garson M. S. Bradshaw N. and Rattawong S.** Lepidolite pegmatites in the Phangnga area of Peninsular Thailand. In *Second technical conference on tin* (London: International Tin Council, 1969), 327–39.

68. **Brown J. C. and Heron A. M.** The distribution of ores of tungsten and tin in Burma. *Rec. geol. Surv. India,* **50**, 1919, 101–21.

69. **Garson M. S.** *et al.* Economic geology and geochemistry of the Neyaungga–Yengan area, Southern Shan States, Burma. *Rep. overseas Div., Inst. geol. Sci.,* 22, 1971, unpublished.

70. **Dewey H.** The mineral zones of Cornwall. *Proc. geol. Ass.,* **36**, 1925, 107–35.

71. **Karig D. E.** Origin and development of marginal basins in the western Pacific. *J. geophys. Res.,* **76**, 1971, 2542–61.

72. **Matsuda T. and Uyeda S.** On the Pacific-type orogeny and its model—extension of the paired belts concept and possible origin of marginal seas. *Tectonophysics,* **11**, 1971, 5–27.

73. **Rodolfo K. S.** Bathymetry and marine geology of the Andaman Basin, and tectonic implications for Southeast Asia. *Bull. geol. Soc. Am.,* **80**, 1969, 1203–30.

74. **Fitch T. J.** Earthquake mechanisms in the Himalayan, Burmese and Andaman regions and continental tectonics in central Asia. *J. geophys. Res.,* **75**, 1970, 2699–709.

75. **Hamilton W.** The Uralides and the motion of the Russian and Siberian Platforms. *Bull. geol. Soc. Am.,* **81**, 1970, 2553–76.

76. Deep sea drilling project: Leg 6. *Geotimes,* **14**, Oct. 1969, 13–6.

77. **Krauskopf K. B.** Source rocks for metal-bearing fluids. In *Geochemistry of hydrothermal ore deposits* **Barnes H. L.** ed. (New York: Holt, Rinehart and Winston, 1967), 1–33.

78. **Barnes H. L.** Mechanisms of mineral zoning. *Econ. Geol.,* **57**, 1962, 30–7.

79. **Jensen M. L.** Provenance of Cordilleran intrusives and associated metals. *Econ. Geol.,* **66**, 1971, 34–42.

80. **Solomon M. Rafter T. A. and Dunham K. C.** Sulphur and oxygen isotope studies in the northern Pennines in relation to ore genesis. *Trans. Instn Min. Metall. (Sect. B: Appl. earth sci.),* **80**, 1971, B259–75.

81. **Taylor S. R.** Geochemistry of andesites. In *Origin and distribution of the elements* **Ahrens L. H.** ed. (Oxford, etc.: Pergamon, 1968), 559–83.

82. **Vinogradov A. P.** Average contents of chemical elements in the principal types of igneous rocks of the earth's crust. *Geokhimiya,* no. 7 1962, 555–71; *Geochemistry, Ann Arbor,* no. 7 1962, 641–64.

83. **White D. E.** Environments of generation of some base-metal ore deposits. *Econ. Geol.,* **63**, 1968, 301–35.

84. **Hall W. E. and Friedman I.** Composition of fluid inclusions, Cave-in-Rock fluorite district, Illinois, and Upper Mississippi Valley zinc-lead district. *Econ. Geol.,* **58**, 1963, 886–911.

85. **Roedder E. Ingram B. and Hall W. E.** Studies of fluid inclusion III: Extraction and quantitative analysis of inclusions in the milligram range. *Econ. Geol.,* **58**, 1963, 353–74.

86. **Dunham K. C.** Mineralization by deep formation waters: a review. *Trans. Instn Min. Metall. (Sect. B: Appl. earth sci.),* **79**, 1970, B127–36.

87. **Sheppard S. M. F. Nielsen R. L. and Taylor H. P.**

Jr. Hydrogen and oxygen isotope ratios in minerals from porphyry copper deposits. *Econ. Geol.*, **66**, 1971, 515–42.

88. **Schuiling R. D.** Tin belts on the continents around the Atlantic Ocean. *Econ. Geol.*, **62**, 1967, 540–50.

89. **Noble J. A.** Metal provinces of the Western United States. *Bull. geol. Soc. Am.*, **81**, 1970, 1607–24.

90. **Turneaure F. S.** Metallogenetic provinces and epochs. *Econ. Geol. 50th Anniversary volume*, 1955, 38–98.

91. **Kuzmin M. I.** The function of the rare element distribution in granitoids and their parameters. In *Origin and distribution of the elements* **Ahrens L. H. ed.** (Oxford, etc.: Pergamon, 1968), 641–8.

92. **Garson M. S.** *et al.* Geology of the tin belt around Phuket, Phangnga and Takua Pa in peninsular Thailand. *Mem. overseas Div. Inst. geol. Sci.* 1, 1972, in press.

93. **Tauson L. V.** *et al.* Distribution of rare-earth elements (RE), yttrium, beryllium and tin in alkaline granitoids and their metasomatites. In *Origin and distribution of the elements* **Ahrens L. H. ed.** (Oxford, etc.: Pergamon, 1968), 663–77.

94. **Smith S.** Geochemistry of granites in the Phuket Region, Thailand and a model for tin-bearing granites. M.Sc. thesis, University of Leeds, 1969.

95. **Barsukov V. L.** The geochemistry of tin. *Geokhimiya*, no. 1 1957, 41–52; *Geochemistry, Ann Arbor*, no. 1 1957, 41–52.

96. **Tauson L. V.** Distribution regularities of trace elements in granitoid intrusions of the batholith and hypabyssal types. In *Origin and distribution of the elements* **Ahrens L. H. ed.** (Oxford, etc.: Pergamon, 1968), 629–39.

97. **Vogt J. H. L.** Magmas and igneous ore deposits. *Econ. Geol.*, **21**, 1926, 207–33, 309–32, 469–97.

98. **Smith F. G.** Transport and deposition of the non-sulphide vein minerals. II. Cassiterite. *Econ. Geol.*, **42**, 1947, 251–64.

99. **Ovchinnikov L. N.** On the role of gases in post-magmatic ore formation. In *Symposium: problems of post-magmatic ore deposition* **Kutina J. ed.** (Prague: Geological Survey of Czechoslovakia, 1963), vol. 1, 492–6.

100. **Barsukov V. L. and Kuril'chikova G. Y.** On the forms in which tin is transported in hydrothermal solutions. *Geokhimiya*, 1966, 943–8; *Geochem. int.*, **3**, 1966, 759–64.

101. **Kelly W. C. and Turneaure F. S.** Mineralogy, paragenesis and geothermometry of the tin and tungsten deposits of the Eastern Andes, Bolivia. *Econ. Geol.*, **65**, 1970, 609–80.

102. **Hasebe K. Fujii N. and Uyeda S.** Thermal processes under island arcs. *Tectonophysics*, **10**, 1970, 335–55.

103. **Toksöz M. N. Minear J. W. and Julian B. R.** Temperature field and geophysical effects of a down-going slab. *J. geophys. Res.*, **76**, 1971, 1113–38.

104. **Yoder H. S. Jr. and Tilley C. E.** Origin of basalt magmas: an experimental study of natural and synthetic rock systems. *J. Petrol.*, **3**, 1962, 342–532.

105. **Wyllie P. J.** Role of water in magma generation and initiation of diapiric uprise in the mantle. *J. geophys. Res.*, **76**, 1971, 1328–38.

106. **Wyllie P. J. and Tuttle O. F.** Experimental investigation of silicate systems containing two volatile components. Part II. The effects of NH_3 and HF, in addition to H_2O on the melting temperatures of albite and granite. *Am. J. Sci.*, **259**, 1961, 128–43.

107. **Wyllie P. J. and Tuttle O. F.** Experimental investigation of silicate systems containing two volatile components. Part III. The effects of SO_3, P_2O_5, HCl, and Li_2O, in addition to H_2O, on the melting temperatures of albite and granite. *Am. J. Sci.*, **262**, 1964, 930–9.

108. **Rankama K. and Sahama Th. G.** *Geochemistry* (Chicago: University of Chicago Press, 1950), 912 p.

109. Deep sea drilling project: Leg II. *Geotimes*, **15**, Sept. 1970, 14–6.

110. **Oxburgh E. R. and Turcotte D. L.** Origin of paired metamorphic belts and crustal dilation in island arc regions. *J. geophys. Res.*, **76**, 1971, 1315–27.

111. **Green D. H. and Ringwood A. E.** The stability fields of aluminous pyroxene peridotite and garnet peridotite and their relevance in upper mantle structure. *Earth planet. Sci. Lett.*, **3**, 1967, 151–60.

112. **Biggar G. M.** Phase relationships in the join $Ca(OH)_2$–$CaCO_3$–$Ca_3(PO_4)_2$–H_2O at 1000 bars. *Mineralog. Mag.*, **37**, 1969, 75–82.

113. **Sleep N. and Toksöz M. N.** Evolution of marginal basins. *Nature, Lond.*, **233**, 1971, 548–50.

114. **Heyl A. V.** Some aspects of genesis of zinc–lead–barite–fluorite deposits in the Mississippi Valley, U.S.A. *Trans. Instn Min. Metall. (Sect. B: Appl. earth sci.)*, **78**, 1969, B148–60.

115. **Stewart J. W.** Neogene peralkaline igneous activity in eastern Peru. *Bull. geol. Soc. Am.*, **82**, 1971, 2307–12.

19

Reprinted from *Geol. Soc. Amer. Bull.* 83:813–817 (1972)

Relation of Metal Provinces in Western America to Subduction of Oceanic Lithosphere

RICHARD H. SILLITOE   *Instituto de Investigaciones Geológicas, Agustinas 785, Casilla 10465, Santiago, Chile*

ABSTRACT

In the orogenic belts of western North and South America, metal provinces are aligned approximately parallel to the continental margins, and, despite irregularities, a general pattern of provinces comprises the following sequence from west to east: Fe; Cu (with some Au and Mo); Pb, Zn, and Ag; and in some regions Sn or Mo. The genesis of these metal provinces is attributed to the release of metals or associations of metals from basaltic oceanic crust and pelagic sediments during partial melting at progressively deeper levels on subduction zones which dipped eastward beneath the continent; the metals subsequently ascended as components of calc-alkaline magma. Initially, the metals were released from the mantle at the East Pacific Rise, transported to the margins of the Pacific basin and thrust beneath the continental margins by the process of sea-floor spreading. This model surmounts the problem of envisaging the existence in the crust or upper mantle of long, narrow zones characterized by a concentration of an individual metal or association of metals.

In the context of this model, possible explanations may be advanced for several features of the distribution in both space and time of metal provinces in western America, including: the occurrence of multiple metallogenic epochs within a given metal province; the difference in age of the dominant metallogenic epoch from one region to another; and the concentration or scarcity of metal deposits in certain regions.

INTRODUCTION

The origin of metal provinces and the source of their contained metals are controversial topics, and even recently, satisfactory solutions to the problems were still apparently lacking (Krauskopf, 1967). Attention was focused on these fundamental questions by Noble (1970), who accurately delimited the metal provinces of the western United States. He considered metal provinces to reflect primitive heterogeneities of metal distribution in the underlying upper mantle, and to have been little influenced by continental crustal structures or processes. Whereas the proposal that the continental crust played only a minor role in the formation of metal deposits of magmatic affiliation in western America is accepted by the writer, the notion that the mantle distribution of metals is directly reflected in the surface configuration of metal provinces merits reconsideration in the light of the recently formulated theory of lithosphere plate tectonics (Isacks and others 1968; Le Pichon, 1968; Morgan, 1968). This theory implies that active orogens, such as the Andean Cordillera, which lie parallel to compressive plate junctures, are not underlain by an immobile column of mantle material, but by a mantle through which a cold, inclined slab of oceanic lithosphere is constantly sinking (Fig. 1).

The classical theory of geosynclinal development has been reinterpreted in terms of plate tectonics (Dewey and Bird, 1970). In view of the close interrelation between orogenic development and ore deposition (for example, Bilibin, 1968), it should prove instructive to apply the concepts of plate tectonics to interpretations of metallogenesis. The model presented here attempts to explain the origin and distribution of metal provinces in western North and South America in terms of the theory of plate tectonics.

METAL PROVINCES IN WESTERN AMERICA

In the western United States, Noble (1970) recognized an over-all change in the metal-content of ore deposits eastward from the

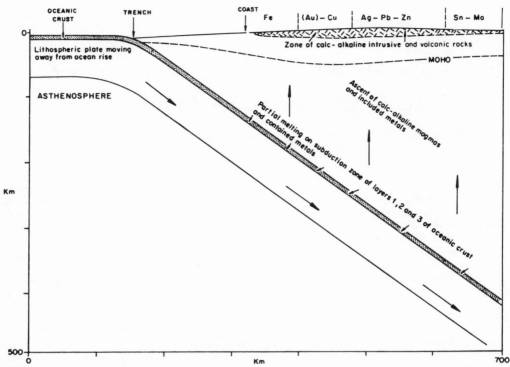

Figure 1. A diagrammatic representation of the generalized sequence of metal provinces in western America in the context of a plate tectonics–subduction model.

continental margin, in the sequence Hg, Cu, Au, Ag, W, Pb, and Mo. Although this pattern of metal provinces cannot be exactly duplicated in other parts of western North and South America, similar transverse changes in the metal contents of ore deposits are nonetheless evident.

In British Columbia, Brown (1969) demonstrated the sequence Fe, Cu, Mo, Zn, and Pb from west to east. In Peru, Bellido and others (1969) described a discontinuous Fe province along the Pacific littoral, followed eastward by a Cu province with some Au, a polymetallic province dominated by Pb, Zn, Ag, and Cu, and still farther east a less important province in which Au, Pb, Cu, and Sn are the principal metals of economic interest. The Fe and Cu provinces can be recognized farther south in Chile (Ruiz and Ericksen, 1962; Ruiz and others, 1965), and the Pb, Zn, Ag, and Cu province extends southward into the Altiplano

of western Bolivia (Ahlfeld, 1967) and western Argentina (Stoll, 1964, 1965). The porphyry copper deposits in the Cu province of Chile and Peru possess important quantities of Mo, and, in Chile, Au deposits are located in the western part of the Cu province. The easternmost polymetallic province in Peru, in which Sn is a component metal, continues southward into the Bolivian Sn-W province (Ahlfeld, 1967; Stoll, 1965). In Ecuador and Colombia, the smaller number of known ore deposits renders metal provinces less easy to define, but it can be appreciated that a western Cu province (with some Mo and Au) is flanked landward by a Pb-Zn province, which in Ecuador possesses important quantities of Ag (Goossens, 1969; Singewald, 1950); these two provinces are apparently northward extensions of the Cu and Pb, Zn, Ag, and Cu provinces of Peru.

In Mexico, Noble (1970) recognized a change eastward in the dominant metal in ore deposits, from Cu to Ag to Pb, and Gabelman and Krusiewski (1968) depicted a Pacific coastal Fe province, followed to the east by a Au-Cu province and still farther east by a province containing Cu, Pb, Zn, and Ag.

Differences in the sequence of metal provinces eastward from different parts of the Pacific continental margin of America are apparent, but the similarities are considered sufficiently striking to indicate a general pattern from Fe, to Cu with some Au and Mo, to Pb, Zn, and Ag, and perhaps finally to Sn or Mo.

ORIGIN OF METAL PROVINCES

It is proposed that post-Paleozoic metal provinces in western North and South America are related to subduction zones which were active beneath the western American continental margin at times during the Mesozoic and early and middle Cenozoic (for example, Atwater, 1970; Hamilton, 1969), and that are still active beneath Central America, the Andean Cordillera, the Cascade Range of Oregon and Washington, and the Alaskan Peninsula.

It is further proposed that much of the metals contained in post-Paleozoic magmatogene ore deposits in western America were derived from the mantle at the East Pacific Rise and its predecessors, and associated with basic magmatism. From the ocean rise, the metals were carried toward the margins of the Pacific Ocean basin as components of basaltic-gabbroic oceanic crust and overlying pelagic sediments, and thrust beneath the continents along inclined Benioff zones. Metals were released from the underthrust oceanic crust and sediments during partial melting, and incorporated in ascending bodies of calc-alkaline magma. The metals attained high crustal levels as components of the magmas, finally to be concentrated in fluid phases associated with the roof-zones of intrusive masses and also with comagmatic extrusive rocks (Fig. 1).

Evidence for derivation of metals from the mantle at ocean rises is provided by the occurrence of anomalously high concentrations of metals in pelagic sediments on the crests and flanks of the East Pacific Rise and other ocean rises (Boström and Peterson, 1969; Boström and others, 1969). Brines and sediments rich in metals are also found along an active divergent plate margin in the Red Sea (Degens and Ross, 1969). The cupriferous pyrite deposits of Cyprus correspond to metal concentrations in layer 2 of the oceanic crust, in the context of Gass' (1968) interpretation of the Troodos ophiolite complex as part of the ocean rise system in the Tethys Ocean.

Noble (1970) explained the close spatial association of metal deposits and igneous rocks in the western United States by proposing that metals rose from the mantle along the same conduits as those previously utilized by bodies of magma. However, mineralogical and age-determination studies of porphyry copper, porphyry molybdenum, and magmatic-hydrothermal Pb-Zn deposits have shown that mineralization and alteration possess a close temporal, as well as spatial, relation to igneous rocks (Fournier, 1967; Livingston and others, 1968; Laughlin and others, 1969; Ohmoto and others, 1966). Therefore, the metals contained in these types of ore deposits were more likely emplaced as integral parts of calc-alkaline magmas, a conclusion supported by experimental studies of Burnham (1967).

In view of the apparent close genetic tie between igneous rocks and spatially related ore deposits, the landward change in the metals characterizing western American metal provinces is considered to be analogous to the systematic increase in the potash-to-silica ratios of andesitic volcanic rocks landward from the circum-Pacific continental margin; the increase is apparently unrelated to crustal composition and thickness, and dependent on the depth of the underlying Benioff zone (Dickinson, 1968). Also, comparable variations in potash content have been documented for post-Paleozoic calc-alkaline volcanic and intrusive rocks in western North America (Moore, 1959, 1962; Bateman and Dodge, 1970). These transverse changes in the compositions of calc-alkaline igneous rocks and metal deposits are visualized as being dependent on processes of partial melting on an underlying subduction zone, as proposed for compositional changes in andesitic volcanic rocks by Dickinson (1968). Recent workers (for example, Oxburgh and Turcotte, 1970) have envisaged the attainment of melting temperatures by frictional heating due to slippage on a subduction zone. Such volumes of poorly consolidated sediments as escaped being scraped off and added to the continental margin would melt at the lowest temperatures, followed at greater depths and higher temperatures by the lowest melting fractions of the basaltic-gabbroic oceanic crust (Oxburgh and Turcotte,

1970). It is here proposed that products of partial fusion at shallow depths on a subduction zone may be enriched in Fe and Cu (with some Au and Mo), giving way at deeper levels to a predominance of Pb, Zn, and Ag, and possibly at the deepest levels to Sn or Mo (Fig. 1).

The relative narrowness and notable north-south orientation of the long axes of the metal provinces would seem to support their dependence on processes of partial melting, since the thermal regime on a subduction zone might be expected to possess similar longitudinal continuity and to undergo relatively abrupt transverse changes.

On the other hand, it is difficult to visualize the existence, parallel to the continental margin, of a series of long, narrow zones in the upper mantle, each enriched in an individual metal or metal association, as demanded by existing theories for the mantle-derivation of metals. A similar problem is encountered by theories in support of a crustal origin for the metals in western American metal provinces.

The concept of the dependence of metal provinces on partial melting on zones of subduction is further supported by the existence of a similar, though more complicated, series of metal provinces in the northern Appalachians (Gabelman, 1968), where orogenic evolution included the operation of subduction zones (Bird and Dewey, 1970).

DISCUSSION

Several problematic features of the spatial and temporal distribution of metal provinces are considered to be explicable in terms of the above model:

In view of the relative permanency of compressive plate junctures, ore deposits assignable to more than one metallogenic epoch might be expected in a single metal province. Good examples of this phenomenon would seem to be the Sn province of Bolivia, where Sn deposition has taken place in late Triassic, Miocene, Pliocene and possibly Pleistocene times (Turneaure, 1971), and the Cu province of Chile, in which copper deposits range in age from Jurassic to Pliocene.

Regions with particularly high concentrations of ore deposits, such as southern British Columbia, the southwest United States and southern Peru-northern Chile—in the case of Cu deposits—might be thought of as zones beneath which higher than normal quantities of metals were subducted, due to a rapid rate of

sea-floor spreading, or to an above-average rate of volcanism and metal-production on the corresponding segment of ocean rise, or, more fundamentally, to an inhomogeneous distribution of metals in the upper mantle beneath the ocean rise (Sillitoe, in prep.).

In British Columbia, Brown (1969) described two broad east-trending zones characterized by a paucity of ore deposits; these zones cut across the longitudinal metal provinces and tectonic units. It might be suggested that such zones face lengths of ocean rise along which metal production has been consistently low. Similarly, the restriction of important Sn mineralization in western South America to the Bolivian Sn province may reflect a concentration of Sn in the upper mantle beneath the East Pacific Rise over the latitudes spanned by Bolivia.

A decrease in age of the most productive period of mineralization southward from British Columbia through the western United States to Mexico, noted by Noble (1970), may reflect a similar, though of course earlier, migration of the main episode of metal production at the ocean rise.

This brief consideration of the possibility of relating metal provinces to activity on underlying zones of subduction indicates that basic research in economic geology should be directed toward the world ocean rise system and the oceanic crust, in an attempt to assess the viability of some of the suggestions advanced above.

ACKNOWLEDGMENTS

I am grateful to Professor Konrad B. Krauskopf and Dr. James W. Stewart for reading the manuscript.

REFERENCES CITED

Ahlfeld, F., 1967, Metallogenetic epochs and provinces of Bolivia: Mineralium Deposita, v. 2, p. 291–311.

Atwater, T., 1970, Implications of plate tectonics for the Cenozoic tectonic evolution of western North America: Geol. Soc. America Bull., v. 81, p. 3513–3536.

Bateman, P. C., and Dodge, F.C.W., 1970, Variations of major chemical constituents across the Central Sierra Nevada batholith: Geol. Soc. America Bull., v. 81, p. 409–420.

Bellido, B. E., de Montreuil, D. L., and Girard, P. D., 1969, Aspectos generales de la metalogenía del Perú: Lima, Peru, Servicio Geol. Min., 96 p.

Bilibin, Y. A., 1968, Metallogenic provinces and

metallogenic epochs: New York, Queens College Press, Geol. Bull., Dept. Geol., 35 p.

Bird, J. M., and Dewey, J. F., 1970, Lithosphere plate–continental margin tectonics and the evolution of the Appalachian orogen: Geol. Soc. America Bull., v. 81, p. 1031–1060.

Boström, K., and Peterson, M.N.A., 1969, The origin of aluminium-poor ferromanganoan sediments in areas of high heat flow on the East Pacific Rise: Marine Geol., v. 7, p. 427–447.

Boström, K., Peterson, M.N.A., Joensuu, O., and Fisher, D. E., 1969, Aluminium-poor ferromanganoan sediments on active oceanic ridges: Jour. Geophys. Research, v. 74, p. 3261–3270.

Brown, A. S., 1969, Mineralization in British Columbia and the copper and molybdenum deposits: Canadian Inst. Mining and Metallurgy Trans., v. 72, p. 1–15.

Burnham, C. W., 1967, Hydrothermal fluids at the magmatic stage, in Geochemistry of hydrothermal ore deposits: New York, Holt, Rinehart & Winston, Inc., p. 34–76.

Degens, E. T., and Ross, D. A., eds., 1969, Hot brines and recent heavy metal deposits in the Red Sea: A geochemical and geophysical account: New York, Springer-Verlag New York, Inc., 600 p.

Dewey, J. F., and Bird, J. M., 1970, Mountain belts and the New Global Tectonics: Jour. Geophys. Research, v. 75, p. 2625–2647.

Dickinson, W. R., 1968, Circum-Pacific andesite types: Jour. Geophys. Research, v. 73, p. 2261–2269.

Fournier, R. O., 1967, The porphyry copper deposit exposed in the Liberty open-pit mine near Ely, Nevada, Part I: Syngenetic formation: Econ. Geology, v. 62, p. 57–81.

Gabelman, J. W., 1968, Metallotectonic zoning in the North American Appalachian region: Internat. Geol. Cong., 23rd, Prague, v. 7, p. 17–33.

Gabelman, J. W., and Krusiewski, S. V., 1968, Regional metallotectonic zoning in Mexico: Soc. Mining Engineers Trans., v. 241, p. 113–128.

Gass, I. G., 1968, Is the Troodos massif of Cyprus a fragment of Mesozoic ocean floor?: Nature, v. 220, p. 39–42.

Goossens, P. J., 1969, Mineral index map, Republic of Ecuador: Quito, Ecuador, Servicio Nac. Geol. Min.

Hamilton, W., 1969, Mesozoic California and the underflow of Pacific mantle: Geol. Soc. America Bull., v. 80, p. 2409–2430.

Isacks, B., Oliver, J., and Sykes, L. R., 1968, Seismology and the New Global Tectonics: Jour. Geophys. Research, v. 73, p. 5855–5900.

Krauskopf, K. B., 1967, Source rocks for metal-bearing fluids, in Geochemistry of hydro-

thermal ore deposits: New York, Holt, Rinehart & Winston, Inc., p. 1–33.

Laughlin, A. W., Rehrig, W. A., and Mauger, R. L., 1969, K-Ar chronology and sulfur and strontium isotope ratios at the Questa Mine, New Mexico: Econ. Geology, v. 64, p. 903–909.

Le Pichon, X., 1968, Sea-floor spreading and continental drift: Jour. Geophys. Research, v. 73, p. 3661–3697.

Livingston, D. E., Mauger, R. L., and Damon, P. E., 1968, Geochronology of the emplacement, enrichment and preservation of Arizona porphyry copper deposits: Econ. Geology, v. 63, p. 30–36.

Moore, J. G., 1959, The quartz diorite boundary line in the western United States: Jour. Geology, v. 67, p. 198–210.

—— 1962, K/Na ratio of Cenozoic igneous rocks of the western United States: Geochim. et Cosmochim. Acta, v. 26, p. 101–130.

Morgan, W. J., 1968, Rises, trenches, great faults, and crustal blocks: Jour. Geophys. Research, v. 73, p. 1959–1982.

Noble, J. A., 1970, Metal provinces of the western United States: Geol. Soc. America Bull., v. 81, p. 1607–1624.

Ohmoto, H., Hart, S. R., and Holland, H. D., 1966, Studies in the Providencia area, Mexico, II, K-Ar and Rb-Sr ages of intrusive rocks and hydrothermal minerals: Econ. Geology, v. 61, p. 1205–1213.

Oxburgh, E. R., and Turcotte, D. L., 1970, Thermal structure of island arcs: Geol. Soc. America Bull., v. 81, p. 1665–1688.

Ruiz, F. C., and Ericksen, G. E., 1962, Metallogenetic provinces of Chile, S.A.: Econ. Geology, v. 57, p. 91–106.

Ruiz, F. C., Aguirre, L., Corvalán, J., Klohn, C., Klohn, E., and Levi, B., 1965, Geología y yacimientos metalíferos de Chile: Santiago, Chile, Inst. Invest. Geológicas, 385 p.

Singewald, Q. D., 1950, Mineral resources of Colombia: U.S. Geol. Survey Bull. 964–B, 204 p.

Stoll, W. C., 1964, Metallogenetic belts, centers, and epochs in Argentina and Chile: Econ. Geology, v. 59, p. 126–135.

—— 1965, Metallogenic provinces of South America: Mining Mag., v. 112, p. 22–23, 90–99.

Turneaure, F. S., 1971, The Bolivian tin-silver province: Econ. Geology, v. 66, p. 215–225.

Manuscript Received by the Society July 19, 1971

Revised Manuscript Received September 14, 1971

Present Address: Department of Mining Geology, Royal School of Mines, Imperial College, Prince Consort Road, London, S.W. 7, England

20

Copyright © 1972 by The University of Chicago Press

Reprinted from *J. Geol.* **80**(4):377–397 (1972)

SULFIDE ORE DEPOSITS IN RELATION TO PLATE TECTONICS[1]

FREDERICK J. SAWKINS

Department of Geology and Geophysics, University of Minnesota,
Minneapolis, Minnesota 55455

ABSTRACT

The dynamics of lithospheric plate motions, as defined by plate tectonic theory, appear to exert a fundamental control on many geologic processes, including sulfide ore deposition. An analysis of basic types of sulfide ore deposits, in relation to various plate tectonic regimes reveals a systematic pattern. The calc-alkaline magmatism apparently generated via the subduction process at convergent plate boundaries gives rise, under favorable circumstances, to Kuroko-type (conformable massive sulfide) deposits in submarine volcanic environments, and Cordilleran-type (postmagmatic) ore deposits in epizonal plutonic environments. Porphyry copper deposits, an important subclass of Cordilleran-type deposits, exhibit a remarkable spatial relationship to present or former convergent plate boundary regimes. Certain copper-nickel deposits occurring in ultramafic extrusive or penecontemporaneous intrusive rocks also appear to be related to plate convergence. Sulfide ore deposits generated at divergent plate boundaries (spreading-center systems) are apparently rare, but this may be a function of their submarine habitat. No examples of sulfide ore deposits formed at transform plate boundaries are known. Stratiform copper deposits and most Mississippi Valley–type lead-zinc deposits are generated in continental intraplate environments. Magmatic deposits in layered mafic complexes also occur in intraplate environments, where spreading-center activity was initiated but failed to develop further. Many gold deposits in volcanic arc sequences of greenstone type can be related to a two-stage concentration process operative at inferred convergent plate boundaries. Plate tectonic theory may provide insights into the generation of many metallogenic provinces, and detailed studies of these provinces and their constituent ore deposits will probably provide insights into the processes operative at some plate boundaries. Finally, the plate tectonic approach to the generation and distribution of sulfide ore deposits has much to offer the exploration geologist.

INTRODUCTION

It is becoming increasingly clear that few if any of the subdisciplines of earth science will remain unaffected by the power and breadth of plate tectonic theory. The geology of ore deposits is no exception.

This attempt to explore the space-time systematics of sulfide ore deposits with respect to different plate tectonic regimes necessarily involves speculation, especially with respect to ore deposits of pre-Mesozoic age. Despite this, an internally consistent model emerges that provides insights to the space-time relationships among various basic types of sulfide ore deposits, illuminates genetic problems, and lays the groundwork for a meaningful classification of sulfide deposits.

Previous attempts to understand the space-time distribution of ore deposits have focused on the interrelationship of ore deposits to regional tectonics and geosynclinal cycles (McCartney and Potter 1962; McCartney 1964; Bilbin 1968; Sawkins and Petersen 1969). In particular, McCartney and Potter's analysis of the metallogeny of the Canadian Appalachians amply demonstrated the general utility of this approach. Recently, however, a major reassessment of

[1] Manuscript received November 26, 1971; revised January 31, 1972.

[JOURNAL OF GEOLOGY, 1972, Vol. 80, p. 377–397]

regional tectonics, and in particular the geotectonic cycle, has resulted from the integration of plate tectonic and orogenic concepts (Coney 1970; Dewey and Bird 1970). It is thus clear that any current attempt to understand the interrelationship of sulfide ore deposits to their regional tectonic setting must take cognizance of plate tectonic theory.

If plate interactions and regimes do in fact control the formation of ore deposits, then an understanding of these controls should provide a powerful tool to aid the exploration geologist. Furthermore, the geochemical and isotopic data obtained from ore deposits related to specific plate regimes may well provide insights into some aspects of the plate tectonic mechanism itself.

This paper is restricted to sulfide ore deposits, rather than all concentrations of elements that have economic value, because of the undue complexities involved in a comprehensive scheme, and because economic deposits of sulfide ores can be considered geochemically distinctive as a group.

PLATE TECTONICS IN RELATION TO MAGMATIC AND OROGENIC PROCESSES

The recent literature indicates that the broad tenets of plate tectonic theory (Dickinson 1970a) are now accepted by a large majority of earth scientists, and do not require restatement here. Use of these concepts has allowed the integration of a large mass of diverse data, and defined a number of problems that require solution. For example, the relationship of much magmatism to plate motions and interactions is empirically demonstrable (Gilluly 1971), but the precise details of the magma generation processes in each case remain to be worked out. Much work also remains to be done before Mesozoic and earlier plate boundaries and interactions can be defined with any real degree of confidence.

Magmatism, sedimentation, and orogenesis related to subduction zones.—Convergence of two lithospheric plates leads to plate consumption along subduction zones. In nearly all cases, the plate undergoing subduction is capped by oceanic crust and is descending beneath plates capped either by oceanic crust (e.g., island arcs of western Pacific) or by continental crust (e.g., Andean arc-trench system). Convergent plate boundaries involving two plates capped by continental crust do occur (e.g., Himalayan and Alpine belts), but in both cases continental collision was apparently preceded by subduction of lithosphere of oceanic type.

In the last year or so, increasing emphasis has been placed on the relation of subduction zones to volcano-plutonic orogenic magmatism and mountain building (Hamilton 1969a, 1969b; Dewey and Bird 1970; Dickinson 1970b; James 1971). Dickinson (1970a, 1970b), using data available from the circum-Pacific orogenic belt, is able to demonstrate the fundamental coherence of the volcano-plutonic magmatism, sedimentation, and orogenesis that characterize subduction-related mobile belts. Of particular importance with respect to sulfide ore deposits is the contention that circum-Pacific andesitic and granitic suites are essentially comagmatic (see also Hamilton and Myers 1967) and were generated by processes closely related to the subduction of slabs of lithosphere at convergent plate boundaries (see figs. 1A, 1B).

Supporting evidence for this model for circum-Pacific magmatism is provided by the spatial relationships of Benioff zones to recent volcanism (Isacks et al. 1968; Hatherton and Dickinson 1969), the relationship of the potash content of andesitic suites to depth of the inclined seismic zone beneath (Hatherton and Dickinson 1969), and the strontium isotopic composition of circum-Pacific magmatic rocks (Hurley et al. 1965; Pushkar 1968). Furthermore, both the geometry of elongate batholithic belts and their contact relations suggest their emplacement in the deeper parts of complex volcano-plutonic arcs.

Dickinson (1970b) also demonstrates that the greywacke-arkose sedimentary suites of orogenic belts are first-cycle derivatives of volcano-plutonic complexes and record in their successions the progressive

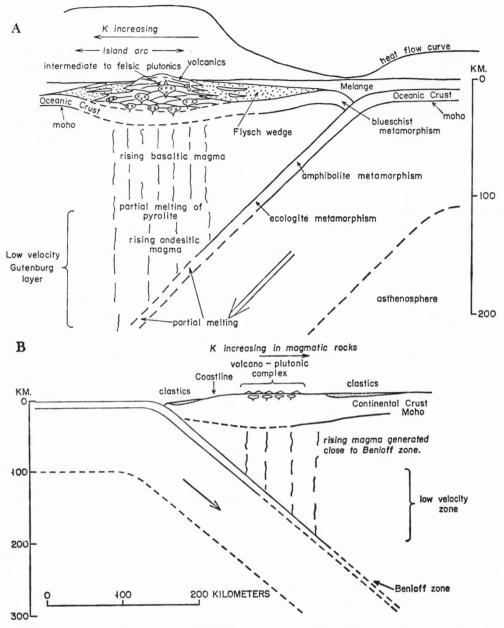

FIG. 1.—*A*, diagrammatic representation of inferred structure and processes taking place in a subduction-related oceanic island-arc system; no vertical exaggeration; note that distance from trench to volcano-plutonic complex in island arc is in excess of 100 km; modified from Dewey (1971, fig. 2, p. 232). *B*, diagrammatic representation of inferred structure and processes taking place at a subduction-related continental margin (e.g., western South America); modified from Hamilton (1969*b*, fig. 5, p. 182).

142

unroofing of such complexes. Orogeny and metamorphism generally accompany the magmatism and sedimentation associated with lithosphere consumption (subduction), and when such consumption occurs in areas where only oceanic crust is initially present, new crust with continental characteristics tends to be created. Such appears to be the case in the Fiji-Tonga-Kermadec areas of the Pacific (see Dewey and Horsefield 1970; Karig 1970). Furthermore, the earliest continental crust now preserved in the central parts of Precambrian shields around the world was presumably created by essentially similar processes associated with plate tectonics. This assertion is supported by the following data:

1. Petrochemical similarity of volcanic rocks preserved in Precambrian greenstone belts to modern island-arc and continental-margin volcanic suites (see Wilson et al. 1965; Engel 1968; Goodwin 1968; Baragar and Goodwin 1969).

2. Petrochemical similarity of Precambrian calc-alkaline granitic rocks spatially associated with greenstone belts, to dioritic-granodioritic intrusives found in modern island-arc and subduction-related–continental-margin environments (Anhaeusser et al. 1968; Goodwin 1968; Dickinson 1970b).

3. Close age relationships of volcanic and plutonic rocks in both Precambrian and modern circum-Pacific mobile zones (Hamilton and Myers 1967; Hart and Davis 1969; Dickinson 1970b; Hanson and Goldich 1970; Peterman and Goldich 1970).

4. The broad equivalence in the areal extent of Precambrian age provinces (e.g., Superior age province of Canada) and areas of Tertiary subduction-related–volcano-plutonic activity in the southwestern Pacific (see fig. 2). The point I wish to make here is that the relatively large size of some Precambrian age provinces ceases to be a puzzle when compared with modern areas in the southwestern Pacific where radiometric clocks over large areas are being, or have been, set within relatively short spans of geologic time (∼100 m.y.).

5. The size relationships between individual greenstone belts and modern volcanic arc systems. Even the largest greenstone belts are smaller than many modern volcanic arc systems, and this, taken in conjunction with some of the points made above, leads to the suggestion that

greenstone belts are preserved remnants of Precambrian volcanic arc systems (see Goodwin and Riddler 1971).

Recent isotope age dating of zircons from volcanics in the greenstone belts of the Superior age province, Canadian Shield (Krogh and Davis 1971), has demonstrated a progressive decrease in age of these belts from north to south. This relationship can be interpreted in terms of a southward migration of subduction zone sites in this area during the early Archean (S. R. Hart, personal communication).

Anhaeusser et al. (1969) and Dewey and Horsefield (1970) point out that later Precambrian orogenic events tend to produce more elongate belts of high-grade metamorphic rocks (e.g., Grenville belt and Mozambique belt) marginal to the earlier formed cratonic nuclei. These appear to represent the root zones of orogenic belts formed by subduction and possible continental collision at a continental margin (see Dewey and Bird 1970; Dewey and Horsefield 1970). Clearly, more work is needed to fully validate these concepts, but the bulk of presently available evidence strongly indicates that plate tectonic mechanisms hold the key to Precambrian as well as Phanerozoic geology.

Magmatism associated with other types of plate boundaries.— The magmatism associated with spreading-center ridges at mid-oceanic divergent plate boundaries is predominantly of the tholeiitic type (Kay et al. 1970; McBirney 1971). Spreading-center systems marked by high heat flow, basaltic magmatism, and crustal extension also occur behind many of the western Pacific island-arc complexes (Karig 1970; Matsuda and Uyeda 1971). Our knowledge of the petrologic makeup of the oceanic crust has been enormously aided by the recognition that a number of ultramafic complexes (e.g., Troodos Complex, Cyprus) (Moores and Vine 1971) are overthrust or obducted slices of oceanic crust and underlying mantle (Coleman 1971; Dewey and Bird 1971). From the details of such ultramafic complexes, it appears that the magmatism oc-

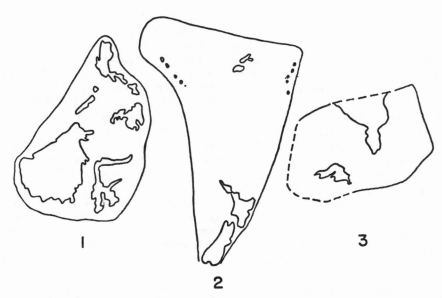

FIG. 2.—Equal area projection of (1) Borneo, Celebes, Philippines area; (2) New Zealand, New Hebrides, Fiji, Tonga area; and (3) Superior age province of Canadian Shield. Note relatively smaller size of Superior age province in relation to area 1, which is mainly underlain by continental crust, and area 2, mainly underlain at present by oceanic crust.

curring along spreading centers produces primitive tholeiitic pillow basalts underlain by dike complexes, and, at a deeper level, by gabbros and coarse-grained ultramafic rocks. Felsic igneous rocks generated along spreading-center systems are less common, but not unknown (Aumento 1969; Gilluly 1971).

McBirney (1971, fig. 6) suggests that some slow spreading-center systems may essentially lack volcanism with crustal extension mainly related to the formation of dike complexes. The presence or absence of volcanism in spreading-center regimes is, however, presumably dictated by the relationship of spreading rate to magma supply.

Magmatic activity along transform plate boundaries is generally rare, but Gilluly (1971, p. 2388) has pointed out that the north-south transform boundaries of the Indian plate (Arakan-Yoma and Quetta lines) exhibit Tertiary and even recent volcanic activity.

Intraplate magmatism.—The interiors of lithospheric plates, whether covered by oceanic or continental crust, are now generally quiescent areas, in terms of magmatism and orogenesis. However, three important types of magmatic rocks are produced in intraplate environments:

1. Flood basalts such as the Deccan Traps, India; the Siberian Traps; the Paraná Basalts, South America; and the Karroo Dolerites and Stormberg Lavas, South Africa. The Keweenawan basalts and related intrusives are interpreted as a Precambrian example of similar intraplate basaltic magmatism. The large tholeiitic, mafic, layered complexes such as the Bushveld Igneous Complex, the Stillwater Complex, the Duluth Complex, and the Sudbury Irruptive are considered to be the intrusive equivalents of continental flood basalts.

2. Oceanic basalts. Most oceanic basalts are generated at spreading centers (divergent plate boundaries), but many oceanic island and seamount chains represent examples of oceanic intraplate magmatism (McBirney 1971, p. 538–543).

3. Alkalic complexes, carbonatites, and kimberlites. These magmatic types are

widely distributed in stable intraplate continental areas and, although of relatively limited volumetric importance, have provided considerable information relating to the composition of the subcontinental mantle.

The causal relationship of these rocks to plate dynamics, if any, is as yet unclear. However, the generation and upward movement through sialic crust of the large volumes of mafic magma represented by continental flood basalts and mafic layered complexes suggest regional tension, perhaps related to incipient spreading centers that for some reason did not develop further.

SULFIDE ORE DEPOSITS IN RELATION TO PLATE BOUNDARY REGIMES

In the preceding sections, the geologic activity, and in particular the magmatism, associated with various types of plate boundaries and with the interiors of lithospheric plates has been reviewed briefly. The central question now becomes Does the space-time distribution of sulfide ore deposits exhibit a meaningful causal relationship to specific geologic environments generated at plate boundaries and within plates?

Kuroko deposits: examples of ore deposits related to convergent plate boundary volcanism.—The Japanese arc exhibits all the characteristics of a convergent plate boundary: (1) presence of a trench on the convex (ocean-facing) side of the arc; (2) active Benioff zone dipping inward below the arc from the trench; and (3) recent, predominantly andesitic, volcanism along the arc. Virtually all interpretations of the geotectonics of the Japanese arc invoke the presence of active subduction in this area (e.g., Matsuda and Uyeda 1971). Furthermore, the Cenozoic and even Mesozoic geology of Japan indicates that plate convergence (i.e., subduction) must have been going on in the general area for tens, if not hundreds, of millions of years (Miyashiro 1967; Matsuda and Uyeda 1971).

The western portion of Hokkaido and much of northern Honshu are underlain by extensive Miocene stratified volcanic rocks (Jenks 1966), the so called "green-tuff" region of northern Japan. The Japanese Kuroko ore bodies typically occur in these Miocene submarine andesitic, dacitic, and rhyolitic sequences. According to Jenks (1966, p. 469): "The ore bodies are mostly in volcanic rocks of Middle Miocene age, in the vicinity of centers of volcanism where flows and pyroclastic rocks are associated with intrusives, domes, breccia zones, and pre- and post-ore structural disturbances."

Detailed studies of the volcanic rocks in the Kosaka district enabled Horikoshi (1969) to identify a number of eruptive cycles in the volcanic sequences; he concludes that formation of Kuroko-type ores is related to the waning stage of any single eruptive cycle.

In general terms, each Kuroko ore body is semiconcordant to concordant with respect to its enclosing rocks and is composed of three ore types: siliceous, yellow, and black (see fig. 3). In detail, these three ore types do not always lie directly above one another (see Jenks 1966, figs. 5, 6; Horikoshi 1969, fig. 16), but the sequence diagrammed is well established. Additional noteworthy features of Kuroko deposits are spatially associated masses of gypsum or barite, the strong alteration associated with the siliceous ore, and the spatial relationship to rhyolite or dacite domes in many instances.

Recent stable isotope studies of Kuroko ores and fluid inclusions in minerals from these deposits (Ohmoto, Kajiwara, and Date 1970) strongly suggest that the formation of Kuroko ores is closely related to the interaction of metal-rich solutions of magmatic origin with seawater. Although some uncertainty still exists whether these ores formed at the seawater interface, or somewhat below it in seawater-saturated tuffs, the close genetic association of these ores with their surrounding volcanic host rocks appears to be incontrovertible. Thus, a direct connection between Kuroko-type sulfide deposits and plate tectonics can be

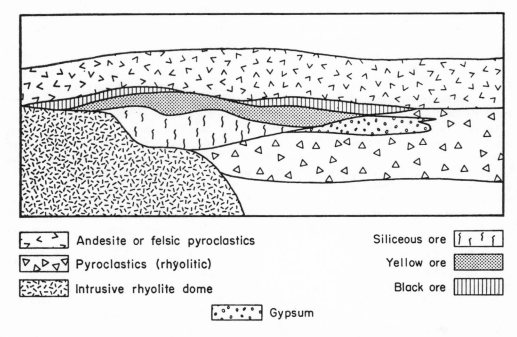

Andesite or felsic pyroclastics   Siliceous ore

Pyroclastics (rhyolitic)   Yellow ore

Intrusive rhyolite dome   Black ore

Gypsum

FIG. 3.—Generalized section of a typical Kuroko ore deposit; modified from Jenks (1966, figs. 5, 6)

postulated via subduction-related magmatism.

Extension of Kuroko-type ores to older deposits.—A large number of conformable massive sulfide deposits occur in andesitic volcanic and volcaniclastic sequences considered to be analogous to those found in circum-Pacific, subduction-related volcanic arcs (table 1).

Although many of these deposits cannot be considered to be exact analogs of Kuroko-type deposits, they exhibit significant similarities not only in lithological environment but also in form, composition, isotopic systematics, and zoning patterns. Furthermore, most modern interpretations of their genesis invoke some facet of the volcanic-exhalative theory (Oftedahl 1958). Also, allowance must be made for the fact that many of the older massive sulfide districts listed occur in rocks that have undergone metamorphism (Kalliokoski 1965). The conformable massive sulfide deposits of the Superior province in Canada are at the opposite end of the geologic time scale to the Kuroko ores

and yet exhibit startling similarities to them (cf. fig. 4 with fig. 3; note similarity).

Although the former presence of subduction zones can only be inferred in relation to the andesitic volcanism and associated sulfide deposits in older terranes, no evidence counter to the suggestion that all the deposits in table 1 were in fact related to earlier convergent plate boundaries has been found. If these inferences are correct, the detailed study of areal petrologic and petrochemical variations in young, subduction-related volcanic arcs (e.g., western Pacific) may well provide important keys for exploration for massive sulfide deposits in older calc-alkaline volcanic terranes and their metamorphosed equivalents.

Some important conformable, massive sulfide deposits occur in lithological sequences where the volcanic component is minor or even absent, for example, Rammelsberg (Kraume et al. 1955), Ducktown (Magee 1967), Sullivan (Carswell 1961), and Mount Isa (Stanton 1962). In the case of the Broken Hill NSW lead-zinc ores,

TABLE 1

SOME MASSIVE SULFIDE DEPOSITS IN VOLCANIC ENVIRONMENTS (ANDERSON 1969)

Area	Host Rocks	Age	Reference
Philippines	Andesitic volcanics	Late Mesozoic to early Cenozoic	Bryner 1969
Northern California (Shasta County)	Andesites	Mesozoic	Albers and Robertson 1961
Caucasus	Andesitic volcanics	Mesozoic	Skripchenko 1967
Newfoundland	Andesitic volcanics	Paleozoic	Swanson and Brown 1962
New Brunswick	Greenstones-felsic volcanics	Paleozoic	McAllister 1960
Norway	Greenstones, amphibolites	Paleozoic	Vokes 1968
Urals	Basalt-andesite	Paleozoic	Yarosh 1965; Zavaritsky 1948
Spain	Slate-dacite porphyry	Paleozoic	Williams 1962
Czechoslovakia	Basalt-quartz keratophyre volcanics	Paleozoic	Pouba 1971
Arizona	Metavolcanics-slates	Proterozoic	Gilmour and Still 1968
Ontario	Andesitic-rhyolitic volcanics	Archean	Wilson 1967 Goodwin 1965 Hutchinson 1965
Saudi Arabia	Andesite—rhyolite volcanics	Precambrian	Earhart 1971

NOTE.—If Kuroko deposits were included in this list, all time periods from Archean to Cenozoic would be represented.

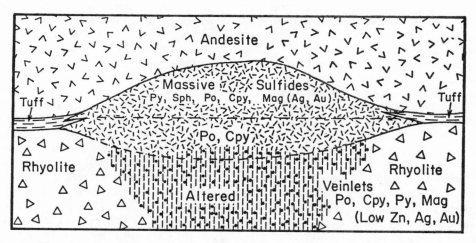

FIG. 4.—Diagrammatic section of typical massive sulfide deposit in Noranda area, Quebec; modified after Roscoe (1965, fig. 2). Note that dimensions of massive sulfide deposits can vary considerably but the large majority have the form of elongate lenses.

metamorphism has completely obliterated the original host-rock lithologies. If these deposits are in fact related to exhalative processes, they may mark the sites of former behind-arc, nonvolcanic spreading-center systems (see above).

Cordilleran ore deposits of Peru, Bolivia, *and Chile: examples of sulfide ores related to plutonic arc magmatism.*—The Andean mountain belt from northern Peru to central Chile is particularly well endowed with metal sulfide ore deposits (Ahlfeld and Schneider-Scherbina 1964; Petersen 1965, 1970; Ruiz 1965; Kelly and Turneaure

1970). With the exception of the Corocoro copper deposits of Bolivia (Ljunggren and Meyer 1964) and a number of relatively small conformable copper deposits in Chile (Ruiz et al. 1970), virtually all of the sulfide ore deposits in this region fall into the class generally termed "postmagmatic," or "magmatic hydrothermal." These terms are not considered satisfactory because Kuroko-type ores and older massive sulfide deposits can also be viewed as related to postmagmatic or magmatic hydrothermal processes. Inasmuch as the epigenetic deposits under consideration in this section have their most typical and prolific development in the Cordilleran regions of both South and North America, I propose that the term "Cordilleran deposits" be applied to them.

The main features of Andean Cordilleran deposits can be itemized as follows: (1) close association in time and space with the emplacement of calc-alkaline felsic intrusives; (2) ores are epigenetic and transport and deposition of ore constituents is effected by hydrothermal solutions; (3) ore occurs either as disseminations or open space fillings in silicate host rocks, or as replacements in carbonate host rocks; (4) ore deposition occurs at relatively shallow depths in the upper crust (Kelly and Turneaure 1970; Sawkins and Rye 1970; Sillitoe and Sawkins 1971); (5) strong tendency toward zonal distribution of metals in deposits; and (6) sulfides exhibit a narrow range of sulfur isotopic values close to meteoritic standard. Two other important features of Cordilleran deposits in the Andes are their Tertiary ages (Gilleti and Day 1968; Quirt et al. 1971) and their occurrence at elevations of from 2,000 to 4,000 m above sea level. Erosion of the Andes down to current sea level would clearly eliminate this entire suite of Cordilleran ore deposits. This, as will be discussed below, provides important insights into the time-space distribution of Cordilleran deposits.

All the available data on the geology, geochemistry, and isotopic composition of Andean Cordilleran deposits (see Petersen 1965; Kelly and Turneaure 1970; Sawkins and Rye 1970; Sillitoe and Sawkins 1971) indicate a direct genetic relationship between the ore deposits and felsic intrusive rocks. Stable isotope data, where available (Sawkins and Rye 1970; Sheppard et al. 1971), suggest that mixing of primary magmatic hydrothermal fluids of sub-upper-crustal origin with waters of meteoric origin was restricted to the later or final stages of sulfide ore deposition.

As outlined in preceding sections, the generation of the felsic calc-alkaline intrusive rocks of the east Pacific borderlands is considered to be directly related to subduction of oceanic lithosphere plates (see also Gilluly 1971). Thus the occurrence of Cordilleran sulfide ore deposits in the Andean belt is viewed as an end result of plate tectonic mechanisms. That these processes are presumably still going on is indicated by the presence of the Peru-Chile trench, an active Benioff zone below the Andes; the Pliocene ages of some ore-associated intrusives in Peru (Gilleti and Day 1968) and Chile (Quirt et al. 1971); and the presence of contemporary volcanism along parts of the chain.

Space-time distribution of Cordilleran-type sulfide ore deposits.—In addition to their widespread occurrence in the Andes, Cordilleran deposits form the major sulfide ore type in Central America, the Caribbean arc, and the western United States (Jerome and Cook 1967; Ridge 1968), and are important in the Cordillera of western Canada (Gunning 1966). The great width of the Cordillera of the western United States is problematic with respect to this general model. However, the Cenozoic sulfide ore deposits of the Colorado Mineral Belt (Tweto 1967), New Mexico (Carpenter 1967), and even the Black Hills (D. M. Rye, personal communication) appear to be typical Cordilleran deposits in most aspects, despite their great distance from the margin of the North American plate. A solution to this problem may be related to an extremely shallow dip of the preexisting Benioff zone under the western United States. An alternate explanation suggested by Lipman et al. (1971)

following their detailed study of Na/K ratios in volcanics of early and middle Cenozoic age in the western United States is that subduction occurred along two subparallel imbricate zones.

In western circum-Pacific areas, such as Japan and the Philippines (Bryner 1969), this type of sulfide ore deposit is well represented, if not dominant. Furthermore, not only are all the above-mentioned areas sites of present or inferred past subduction, but the felsic calc-alkaline intrusive rocks, with which circum-Pacific Cordilleran deposits are so closely associated, appear to be typical products of subduction-related magmatism. The same broad relationships appear to hold for the tin-tungsten deposits of Indonesia and Southeast Asia (Katili 1967).

The Alpine-Himalayan mountain belt is considered to have been created by subduction of oceanic lithosphere and subsequent continental collision (Dewey and Horsefield 1970). In the Alps and the Himalayas proper, subduction-related–calc-alkaline magmatism is of minor extent, but in the intermediate area from the eastern Mediterranean to western Pakistan an important belt of Cordilleran deposits occurs (Superceanu 1970). It is noteworthy that all the world's known porphyry copper deposits, *sensu stricto*, an important subclass of Cordilleran deposits, occur either along this belt, in the circum-Pacific region or in the Antillean arc—all areas of inferred Cenozoic or Mesozoic subduction-related magmatism. The impressive correlation between Cordilleran deposits of later Mesozoic and Cenozoic age and inferred lithospheric subduction suggests that older intrusive-related sulfide ore deposits (e.g., tin-tungsten deposits of eastern Australia, Hercynian deposits of southwest England, Portugal, and Germany) are essentially of Cordilleran type, although the reconstructions relating their associated intrusive rocks to past subduction-zone magmatism have yet to be made.

The marked relative scarcity of Cordilleran deposits in pre-Mesozoic–orogenic-magmatic belts is believed to be directly related to erosion levels. Cordilleran deposits are typically emplaced at shallow levels in rising or uplifted island arcs or mountain belts, and except in rare instances they will be removed by erosion. Examples of Cordilleran deposits in Precambrian shield terranes are indeed rare, but where known, they occur in supracrustal rocks such as greenstone belts invaded by epizonal intrusives (e.g., Ferguson 1966, p. 110). The molybdenite deposits marginal to the Preissac-Lacorne batholith in Quebec (Vokes 1963, p. 105–27) appear to represent important examples of early Precambrian Cordilleran deposits.

Magmatic sulfide deposits related to subduction-zone mafic magmatism.—One puzzling, but well-documented, phenomenon is the occurrence of mafic to ultramafic volcanic rocks within the typical basalt-andesite-dacite-rhyolite volcanic sequences of greenstone belts (Anhaeusser et al. 1968; Naldrett 1970). Petrological considerations exclude the possibility that such magmatic rock types are generated by partial melting of the uppermost portions of a descending lithospheric plate, but their relatively widespread occurrence in greenstone belts indicates some relationship to the dynamics of subduction. Important copper-nickel sulfide ores are associated, in some cases, with the basal portions of ultramafic flows and penecontemporaneous intrusions of this type (Haapala 1969; Kilburn et al. 1969). In particular, the Marbridge nickel deposit in the Superior province of Canada (Clark 1965) and the Kambalda and related deposits in the western Australian Shield (Woodall and Travis 1969) occur within ultramafic rocks intercalated with typical greenstone volcanic sequences. The extrusive or near-surface emplacement of the ultramafic host rocks for these deposits is demonstrable (see Naldrett 1970), but their origin is problematic. The genesis of these ultramafic magmas may relate to episodes of rifting within island-arc systems during their growth, as has been demonstrated by Karig (1970) in the Tonga-Kermadec area of the southwestern Pacific.

The copper-nickel sulfide ores present along the important Thompson Lake–Moak Lake belt of Manitoba are related to ultramafic rocks occurring along a major zone of shearing at the boundary between the Superior and Churchill age provinces of the Canadian Shield. This boundary has been interpreted as an ancient island arc (Zurbrigg 1963), but it may well mark the site where interaction of two lithospheric plates occurred. A geologically younger example of plate interaction and ultramafic emplacement is provided by the convergence of the Russian and Siberian platforms to form the Urals during Hercynian times (Hamilton 1970). It is noteworthy that the Uralian belt also contains important magmatic sulfide ore deposits.

It is at least possible that some of the magmatic copper-nickel deposits present in suture zones may have been generated together with their enclosing rocks at spreading-center systems, and may be merely related to plate convergence in terms of their final emplacement. The present form of the Thompson ore body, however, located mainly as a concordant zone in strongly folded biotite schists (see Boldt 1967, p. 52–53), is puzzling, and a comprehensive genetic model for this important copper-nickel deposit awaits the completion of further studies.

Despite the genetic uncertainties relating to this general class of magmatic copper-nickel deposits, the exploration potential provided by ultramafic bodies both within greenstone belts and along ancient suture zones is being increasingly realized.

Sulfide ore deposits related to spreading-center systems.—Recent work in the oceans has provided intriguing insights into the ore-deposit potential of processes occurring at spreading-center ridges. Bostrom and Peterson (1966) have shown that the metal contents of sediments across the East Pacific rise demonstrate a positive correlation with heat-flow values. From this they conclude that iron, manganese, and lesser amounts of copper, lead, and zinc are being added to the oceans by volcanic-exhalative

processes along the crest of the East Pacific rise. High concentrations of iron, manganese, and nonferrous metals also occur in hot brine pools in the median depression of the Red Sea (Degens and Ross 1969). Isotopic studies of these brines and the underlying metal-rich sediments (Cooper and Richards 1969; Craig 1969) have not provided clear indications of a primary magmatic source for the constituents of the brine pools. Nevertheless, an origin for the metals involving volcanic exhalation along a slowly spreading system is not ruled out, and provides a logical explanation for the source of the heat and metals in these brine pools. This conclusion is supported by the most recent work in this area (Ross et al. 1971).

The copper- and zinc-bearing pyritic ore bodies of Cyprus occur in the pillowed basalts of the uppermost part of the Troodos Complex. As indicated earlier, this complex is considered to be a typical section of oceanic crust, and as such it must have been generated at a spreading-center system. The massive sulfide deposits associated with it are clearly genetically related to the volcanics in which they occur (Hutchinson 1965; Govett and Pantazis 1971), and thus apparently provide us with an example of ore deposition at a divergent plate boundary. The extent to which sulfide deposits of Cyprus type are formed at spreading centers is a matter for conjecture in view of the submarine habit of most oceanic crustal material.

SULFIDE ORE DEPOSITS FORMED IN INTRAPLATE REGIMES

Magmatic sulfide deposits.—Important copper-nickel sulfide ore deposits are related to certain mafic intrusions emplaced in intraplate environments. Two North American examples of such deposits are those associated with the Sudbury Irruptive, Ontario (Souch et al. 1969) and the Stillwater Complex, Montana. At some future date, the copper-nickel sulfide mineralization at the base of the Duluth Complex, Minnesota may be classified as ore-grade material. The large igneous bodies that contain these ores generally have a lopolithic shape and, as stated earlier, appear to represent the intrusive equivalents of continental flood basalts. The important platinum deposits of the Merensky Reef in the Bushveld Igneous Complex, South Africa (Cousins 1969), also provide an example of magmatic sulfide deposits formed in an intraplate environment.

Stratiform copper deposits.—A significant percentage of the world's copper is produced from stratiform deposits quite distinct from the conformable deposits associated with subduction-zone volcanism. Typical examples are the Kupferschiefer (Deans 1948), Zambian copper deposits (Mendelsohn 1961), White Pine (White 1968), and the copper deposits of Kazakhstan (Popov 1962). Such deposits appear to lack magmatic affinities, and occur in shales and siltstones either immediately overlying or intercalated with arkosic sandstones and conglomerates. Such sedimentary facies occur most typically on the inner continental side of subduction-related orogenic belts, but can be generated anywhere within the continental parts of plates where basin formation and complementary uplift occurs. The source of the copper in these deposits is problematic, but the wide lateral extent of such ores, their impressive stratigraphic control, and the absence of wall-rock alteration indicate that the copper was emplaced at the time of deposition of the host rocks or a relatively short time thereafter.

Mississippi Valley–type deposits.—Mississippi Valley–type deposits typically occur in carbonate rocks overlying Precambrian shields in intraplate environments. Examples include the midcontinent deposits of the United States (Heyl 1967), the Pine Point deposits of Canada (Campbell 1967), and the Pennine deposits of England (Dunham 1964). Deposits of this type are characterized by their emplacement in carbonate rocks, lack of clear association to igneous rocks, the extensive areas over which they occur, and the low silver content of their lead-zinc ores.

Fluid-inclusion studies on these ores (Hall and Friedman 1963; Sawkins 1966; Roedder 1967) have indicated that ore deposition occurred at low temperatures from highly saline brines derived, at least in large part, from sedimentary environments. A number of investigators have suggested a genetic link between intraplate alkaline magmatic rocks, and the lead and zinc (and fluorine and barium) in some of these deposits (e.g., Galkiewicz 1967; Heyl 1967; Sawkins 1968). The high values of lead and zinc observed in inclusion fluids trapped in minerals from Mississippi Valley deposits reported by Pinckney and Haffty (1970) are orders of magnitude greater than those observed in saline connate waters (White et al. 1963). These data do suggest that the metals in *some* Mississippi Valley–type deposits may be of deep-seated derivation. In certain instances (e.g., Pine Point, Canada), mixing of formation waters of different provenance appears to account satisfactorily for the deposition of these lead-zinc ores (Beales and Jackson 1966; Beales and Onasick 1970).

Whereas the intraplate status of most Mississippi Valley–type deposits is readily demonstrable, the zinc deposits of the Appalachians (Heyl 1967) and the Alpine deposits of Europe (Maucher and Schneider 1967) appear to correlate more closely with orogenic events at convergent plate boundaries. In general, good genetic models for both Alpine and Mississippi Valley–type deposits still remain to be hammered out.

DISCUSSION

Major sulfide-deposit types can thus be broadly divided into those associated with plate-margin environments and those formed in intraplate environments. Convergent plate boundaries are areas of prime importance for sulfide ore generation, for Cordilleran-type and Kuroko-type (massive sulfide) deposits account for a major proportion of all the metal sulfide ore mined. Although these two types of deposit are relatively distinctive, they tend to be intimately intermixed in island-arc environments, since both types of deposit appear to be generated by subduction-related felsic magmatism.

A number of sulfide deposit types have not been dealt with thus far, and some of these merit discussion. Gold deposits, excluding modern placers, are generally accompanied by sulfide minerals. Most Tertiary gold vein deposits are of Cordilleran type and are spatially associated with subduction-related intrusive rocks. However, an impressive number of older gold vein and lode deposits occur in greenstones, greenschists, and volcanigenic slates typical of inferred subduction-related volcanic sequences that have been subsequently folded and metamorphosed. The large majority of gold deposits in Alaska, the Canadian Shield, the Piedmont region of the southeastern United States, California, Venezuela, Brazil, West Africa, Rhodesia, southern India, western Australia, and southeastern Australia (see Emmons 1937) fall into this category. The vast majority of these deposits occur either as lodes or lenses conformable with their enclosing rocks, or in structural sites such as shear zones, tensional fissures, or the crests and troughs of folds. Felsic intrusive rocks are present in many of the areas cited (see Emmons 1937), but no convincing data have indicated a genetic relation between the large majority of the gold deposits and these intrusives (Viljoen et al. 1969), although some of the gold deposits of the Canadian Shield do seem to be directly related to penecontemporaneous intrusive rocks (Ferguson 1966). Furthermore, these deposits do not display the zonal features so commonly encountered in Cordilleran deposits. Genetic models involving syngenesis (Riddler 1970; Sawkins and D. M. Rye 1971) and/or lateral secretion during metamorphism (see Knight 1957) merit serious consideration with respect to the formation of many of these deposits. Stable isotope studies of the Homestake gold ores (D. M. Rye, personal communication) and the geochemical studies of the Yellowknife gold deposits, Northwest Territories, by Boyle (1961) support these

ideas. The energy and aqueous solutions needed to redistribute the gold, quartz, and accompanying sulfides, where major redistribution is indicated by the geological relationships, may well have been provided by metamorphism, felsic intrusions, or both. The important point is that the worldwide association of inferred subduction-related volcanic rocks and certain types of gold deposits is too strong to discount a genetic tie between them.

It seems clear we are dealing here with a group of deposits closely tied to a whole gamut of geological processes occurring at convergent plate boundaries: calc-alkaline volcanism, volcanic exhalation, plutonic magmatism, folding, and metamorphism.

The gold-uranium deposits of the Witwatersrand and the essentially similar occurrences of Blind River (Ontario) and Jacobina (Brazil) have a high sulfide content. The sulfides in these deposits are mostly detrital pyrite, whose preservation is presumably related to the low oxygen content of Precambrian atmospheres. Thus these deposits, at least in part due to surficial concentration mechanisms, are outside the main scope of this paper. They are, nevertheless, clearly related to sedimentary processes taking place in intraplate environments.

The important silver deposits of the Cobalt area (Ontario) are difficult to categorize. Recent work (Petruk 1968) has attempted to demonstrate that remobilization or lateral secretion of originally dispersed ore elements may provide a satisfactory genetic model for the formation of these deposits. In any event, they appear to have formed in an intraplate environment.

In the last few years, an important lead-zinc-silver province has been discovered in the Carboniferous rocks of central Ireland (Morissey et al. 1971). The deposits, which are fault-localized, appear to be epigenetic replacements of specific carbonate horizons; as such, they exhibit some similarities to Mississippi Valley–type deposits. They differ from these, however, in their relatively high silver content and indicated temperature of deposition (~250° C). Isotopic characteristics of lead in galena deposits (Greig et al. 1971) indicate a Mesozoic or even Tertiary age. The deposits formed within an essentially stable intraplate environment, but Russell (1968) has suggested that the structural control of mineralization is related to faulting allied to the breakup of North America and Europe. Clearly, more work on these deposits is needed before they can be related to geotectonics.

Another problematic group of deposits are the ore bodies at Kennecott. There, sulfide replacement deposits in carbonate rocks were apparently emplaced by low-temperature hydrothermal solutions not of typical postmagmatic character. It seems possible that the ore constituents may have been derived from the thick sequence of altered volcanics underlying the carbonate rocks in which the deposits occur. Thus, these deposits may provide a further example of a two-stage concentration process that operated at a convergent plate boundary.

The Salton Sea geothermal system contains metal-rich brines and has been the subject of a great deal of geophysical, geochemical and isotopic study (see White 1968). Gravity work by Biehler (1971) indicates that the geothermal system lies astride a short spreading-center segment connecting two offset portions of the San Andreas fault system. Stable isotope studies have indicated that the brines in this system consist predominantly of meteoric water (Craig 1966), but the origin of the metals they contain is not yet clear. What is clear is that the geotectonic setting of this system is *not* similar to that of typical Cordilleran deposits, and thus the utilization of data from the Salton Sea in formulating genetic models of Cordilleran-type sulfide deposits is hazardous.

Availability of sulfur is obviously an indispensable factor in the formation of sulfide ore deposits, and data on the isotopic ratios of the sulfur in a sulfide ore deposit permit general conclusions on its origin. In Cordilleran and magmatic deposits, the $S^{32}/$ S^{34} ratios typically exhibit a limited range, the average value of which lies close to that of the meteoritic standard. This is interpreted to indicate that the sulfur in these deposits is either of mantle origin, or has been extracted from sites sufficiently deep or extensive to permit large-scale homogenization of isotopic values.

Kuroko-type (massive sulfide) deposits (Sangster 1968), Mississippi Valley–type deposits (Jensen and Dessau 1967; Solomon et al. 1971), and stratiform copper deposits (Dechow and Jensen 1965) typically exhibit a wider spread of S^{32}/S^{34} ratios, the average value of which does not correspond to the meteoritic standard. Thus, it may be concluded that the sulfur in these types of deposits has been subjected to either inorganic or organic fractionation processes (see Jensen 1962; Ohmoto 1970) that occur in upper crustal, surficial, or marine environments.

In this paper I do not claim to have related *all* sulfide ore deposits to plate tectonic concepts, but I do claim that my approach provides insights into the space-time relationships and genesis of the great majority of important sulfide ore deposits. As the worldwide demand for metals continues to grow, plate tectonic concepts must play an increasing role in the initial planning stages of metal exploration programs.

METALLOGENIC PROVINCES AND THEIR
RELATION TO PLATE TECTONICS

If, as suggested in this paper, a meaningful relationship exists between sulfide ore-deposit types and plate tectonic environments, then some exploration of the problem of metallogenic provinces is perhaps warranted (Turneaure 1955). Metallogenic provinces (see Bolivian and Southeast Asian tin-tungsten provinces; Chilean and southwestern U.S. copper provinces; Mexican and Peruvian silver, lead, and zinc provinces; Zambian copper province) represent major geochemical anomalies of a rather specific nature and some explanation for their fundamental cause is clearly desirable.

In the intraplate environment, sulfide ore deposition is apparently not directly related to subcrustal processes, except in the case of magmatic ore deposits. The major copper-nickel deposits associated with the Sudbury Irruptive can be considered to represent a metallogenic province restricted to a single series of magmatic events. The large amounts of copper and nickel in these deposits presumably indicate an underlying area in the mantle containing either anomalous amounts of copper and nickel, or above-average amounts of sulfur, that was able to extract the copper and nickel from the silicate phases in the source area.

The copper deposits of Zambia and the Congo Republic clearly define a metallogenic province, which is also reflected in the high copper contents of basement rocks in the area (Pienaar 1961). Here, however, ore deposition appears to have been related to sedimentary processes and thus this metallogenic province must have been defined by earlier geologic events.

The orogenic belts of western North and South America contain a number of geochemically distinct groups of Cordilleran deposits that make up individual metallogenic provinces. The central thesis of this paper is that these metallogenic provinces have a direct relationship to subduction-controlled magmatism, but it is recognized that a consensus has yet to be reached among economic geologists concerning the fundamental cause of Cordilleran (post-magmatic) ore deposition. Three major factors present themselves as basic determinants for the formation of Cordilleran deposits singly or in groups making up metallogenic provinces: (1) metal content of the intruded country rocks, (2) geochemical dynamics of magma generation and crystallization, and (3) regional geochemical anomalies at the site of magma generation.

The first alternative would seem to be ruled out by empirical data relating to the wide variety of country-rock types, each with distinctive trace metal contents in which similar deposits (e.g., porphyry coppers) occur (see Titley and Hicks 1966). Certainly, the areal extent of many metallogenic provinces is such that highly varied lithologies of country rocks occur within them.

The second alternative has many advocates who point to the large amounts of trace metals contained in any magmatic body of reasonable size. Acceptance of this alternative as the root cause of Cordilleran ore deposition only aggravates the problem of the generation of metallogenic provinces. For example, why does magmatic evolution lead to groups of tin-silver deposits in some areas, silver-lead-zinc deposits in others, and copper-molybdenum deposits in still others?

The last alternative, that the trace-metal geochemistry of Cordilleran deposits associated with intermediate and felsic intrusives is largely a function of the geochemistry of the materials present at the site of magma generation, is the conclusion reached by Krauskopf (1971) in his discussion on the source of ore metals. At present, this hypothesis is not amenable to rigorous scrutiny, because of the lack of data on the regional trace-metal geochemistry of magmatic provinces in Cordilleran belts. However, when linked with plate tectonic concepts, it does provide a reasonable explanation for the general features of metallogenic provinces, as will be discussed below.

The trace-metal composition of subduction-related magmas will be determined in large part by the trace-metal geochemistry of the materials carried down along and below the Benioff zone (see fig. 5, 1). Recent work on the isotopic and trace-element chemistry of island-arc volcanic rocks (Armstrong 1971; Sinha and Hart 1971) indicates that some sedimentary material must be involved, along with oceanic volcanic crust, in the production of subduction-related magmas. Thus the possibilities for variation of the trace-metal geochemistry of the primary material involved in subduction-zone magma generation are considerable, especially in a lateral sense, along arc systems.

Another less likely possibility, that cannot be excluded, is that subduction-generated magmas pass through a volume of overlying mantle containing anomalous contents of specific trace metals (see fig. 5, 2). This

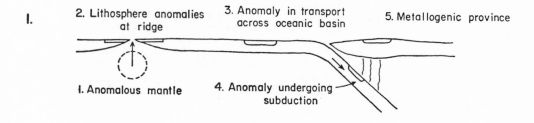

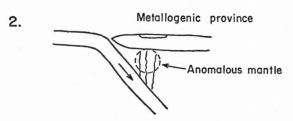

FIG. 5.—Two possible models to explain the generation of metallogenic provinces generated at convergent plate boundaries. See text for details.

situation might cause sufficient enrichment of the ascending magmas in trace metals to result in the formation of a metallogenic province in the upper crust.

Finally, the possibility of geochemical, and thus ore deposit, variation *across* subduction-related arc systems must be considered. Variation of the major- and minor-element chemistry of volcanic rocks across subduction-related magmatic arcs has been demonstrated by Hatherton and Dickinson (1969) and Kuno (1960). It seems inevitable that similar relationships should pertain for the trace-element geochemistry of these magmatic rocks, and current work apparently supports this conclusion (Rhodes and Ridley 1971; Jakes and White 1972).

With respect to Cordilleran ore deposits, the relationships are not clear-cut, which is not surprising considering the multiplicity of factors that affect the formation and preservation of such deposits. Nevertheless, in the Andes of Chile and Argentina (Stoll 1964; Ruiz 1965) and in British Columbia (Gunning 1966), a crude west-to-east pattern from iron → copper → lead-zinc deposits may be present. Clearly much more data —geologic, geochemical, and geotectonic— are needed before a realistic test of these ideas can be attempted. (NOTE ADDED IN PROOF: Sillitoe [1972] has documented across-arc variations in the Cordilleran ore deposits of North and South America more fully, and concluded that these variations are dependent on subduction-related processes.)

CONCLUSIONS

The interrelationship of the formation of many sulfide ore deposits to plate tectonic processes is demonstrable. Furthermore, different basic types of sulfide ore deposits can be related to specific plate tectonic regimes; as our understanding of past plate configurations and interactions grows, this relationship should provide a powerful tool for predicting the possible metal potential of areas that, for various reasons, have seen little exploration activity.

The distribution of various types of sulfide ore deposits with respect to geologic time is partially clarified by the approach used, which may also provide insights into the fundamental causes for the development of many metallogenic provinces.

Finally, it is clear that careful geochemical and isotopic studies of sulfide ore deposits related to convergent plate boundary magmatism may elucidate the nature of subduction-related processes.

ACKNOWLEDGMENTS.—I am greatly indebted to W. D. Normark, University of Minnesota, for introducing me to many aspects of plate tectonic theory and for many helpful suggestions during preparation of the manuscript.

Many of the ideas relating to the systematics of sulfide ore deposits were hammered out with U. Petersen, Harvard University, during preparation of a manuscript entitled "A Tectonic-genetic Classification of Sulfide Ore Deposits," and his contribution is gratefully acknowledged. Critical comments and helpful suggestions during preparation of the manuscript were also received from C. Chase, W. Dickinson, and J. Wilson.

However, I accept full responsibility for the ideas expressed in this paper. Work on sulfide ore deposits in the Andes, which aided in the formulation of these ideas, was supported by NSF grant no. GA 4355.

REFERENCES CITED

AHLFELD, FEDERICO, and SCHNEIDER-SCHERBINA, ALEJANDRO, 1964, Los yacimientos minerales y de hidrocarburos de Bolivia: Dept. Nac. Geología, La Paz, Bolivia, Bol. 5 (Especial), 388 p.

ALBERS, H. P., and ROBERTSON, J. F., 1961, Geology and ore deposits of East Shasta copper-zinc district, Shasta County, California: U.S. Geol. Survey Prof. Paper 338, 107 p.

ANDERSON, C. A., 1969, Massive sulfide deposits and volcanism: Econ. Geology, v. 64, p. 129–146.

ANHAEUSSER, C. R.; MASON, R.; VILJOEN, M. J.; and VILJOEN, R. P., 1969, Reappraisal of some aspects of Precambrian shield geology: Geol. Soc. America Bull., v. 80, p. 2175–2200.

———; ROERING, C.; VILJOEN, M. J.; and VILJOEN, R. P., 1968, The Barberton Mountain Land: a model of the elements of evolution of an Archean fold belt: Geol. Soc. South Africa, Annexure, v. 71, p. 225–254.

ARMSTRONG, R. L., 1971, Isotopic and chemical constraints on models of magma genesis in volcanic arcs: Earth and Planetary Sci. Letters, v. 12, p. 137–142.

AUMENTO, F., 1969, Diorites from the Mid-Atlantic ridge at 45° N: Science, v. 165, p. 1112–1113.

BARAGAR, W. R. A., and GOODWIN, A. M., 1969. Andesites and Archean volcanism of the Canadian Shield, in Proceedings of the andesite conference: Oregon Dept. Geology and Mineral Industries Bull. 65, p. 121–142.

BEALES, F. W., and JACKSON, S. A., 1966, Precipitation of lead-zinc ores in carbonate reservoirs as illustrated by Pine Point ore field, Canada: Inst. Mining Metallurgy Trans., v. 75, p. B278–285.

———, and ONASICK, E. P., 1970, Stratigraphic habitat of Mississippi Valley-type ore bodies: Inst. Mining Metallurgy Trans., v. 79, p. B145–154.

BIEHLER, S., 1971, Gravity models of the crustal structure of the Salton Trough (Abs.): Geol. Soc. America Cordilleran Sec., p. 81–82.

BILBIN, Y. A., 1968, Metallogenic provinces and metallogenic epochs: Dept. Geology, Queen's College, N.Y., Geol. Bull., 35 p.

BOLDT, J. R., JR., 1967, The winning of nickel: D. Van Nostrand Co., Princeton, N.J., 487 p.

BOSTRÖM, K., and PETERSON, M. N. A., 1966, Precipitates from hydrothermal exhalation on the East Pacific Rise: Econ. Geology, v. 61, p. 1258–1265.

BOYLE, R. W., 1961, The geology, geochemistry and origin of the gold deposits of the Yellowknife district: Geol. Survey Canada Mem. 310.

BRYNER, L., 1969, Ore deposits of the Philippines: an introduction to their geology: Econ. Geology, v. 64, p. 644–666.

CAMPBELL, N., 1967, Tectonics, reefs and stratiform lead-zinc deposits of the Pine Point area, Canada: Econ. Geology Mon. 3, p. 59–70.

CARPENTER, R. H., 1967, Geology and ore deposits of the Questa molybdenum mine area, Taos County, New Mexico, in Ore deposits of the United States 1933/1967 (Graton-Sales volume): New York, Am. Inst. Mining Metall. Engineers, p. 1328–1350.

CARSWELL, H. T., 1961, Origin of the Sullivan lead-zinc-silver deposits, British Columbia: Unpub. Ph.D. thesis, Queens University, Canada.

CLARK, L. A., 1965, The geology and geothermometry of the Marbridge nickel deposit, Malartic, Quebec: Econ. Geology, v. 60, p. 792–811.

COLEMAN, R. G., 1971, Plate tectonic emplacement of upper mantle peridotites along continental edges: Jour. Geophys. Research, v. 76, p. 1212–1222.

CONEY, P. J., 1970, The geotectonic cycle and the new global tectonics: Geol. Soc. America Bull., v. 81, p. 739–748.

COOPER, J. A., and RICHARDS, J. R., 1969, Lead isotope measurements on sediments from Atlantis II and Discovery deep areas, in DEGENS, E. T., and ROSS, D. A., eds., Hot brines and recent heavy metal deposits in the Red Sea: New York, Springer-Verlag, p. 499–511.

COUSINS, C. A., 1969, The Merensky Reef of the Bushveld Igneous Complex: Econ. Geology Mon. 4, p. 329–251.

CRAIG, H., 1966, Isotopic composition and origin of the Red Sea and Salton Sea geothermal brines: Science, v. 154, p. 1544–1548.

——— 1969, Geochemistry and origin of the Red Sea brines, in DEGENS, E. T., and ROSS, D. A., eds., Hot brines and recent heavy metal deposits in the Red Sea: New York, Springer-Verlag, p. 208–242.

DEANS, T., 1948, The Kupferschiefer and the associated lead-zinc mineralization in the Permian of Siberia, Germany and Poland: Internat. Geol. Cong., 18th, London, no. 7, p. 340–351.

DECHOW, E., and JENSEN, M. L., 1965, Sulfur isotopes of some central African sulfide deposits: Econ. Geology, v. 60, p. 894–941.

DEGENS, E. T., and ROSS, D. A., eds. 1969, Hot brines and recent heavy metal deposits in the Red Sea: New York, Springer-Verlag, 600 p.

DEWEY, J. F., 1971, A model for the lower Paleozoic evolution of the southern margin of the early Caledonides of Scotland and Ireland: Scottish Jour. Geology, v. 7, p. 219–240.

———, and BIRD, J. M., 1970, Mountain belts and the new global tectonics: Jour. Geophys. Research, v. 75, p. 2625–2647.

———, ——— 1971, Origin and emplacement of the ophiolite suite: Appalachian ophiolites in Newfoundland: Jour. Geophys. Research, v. 76, p. 3179–3206.

———, and HORSEFIELD, B., 1970, Plate tectonics, orogeny and continental growth: Nature, v. 225, p. 521–526.

DICKINSON, W. R., 1970a, Global tectonics: Science, v. 168, p. 1250–1259.

——— 1970b, Relations of andesites, granites, and derivative sandstones to arc-trench tectonics: Rev. Geophysics, v. 8, p. 813–860.

DUNHAM, K. C., 1964, Neptunish concepts in ore genesis: Econ. Geology, v. 59, p. 1–21.

EARHART, R. L., 1971, Geologic setting of massive sulfide deposits in the Wadi Bidah District, Kingdom of Saudi Arabia: Internat. Mineralog. Assoc.-Internat. Assoc. Genesis Ore Deposits, Tokyo-Kyoto 1970, Proc., p. 310–318.

EMMONS, W. H., 1937, Gold deposits of the world: New York, McGraw-Hill Book Co., 562 p.

ENGEL, A. E. J., 1968, The Barberton Land: clues to the differentiation of the earth: Geol. Soc. South Africa, Annexure v. 71, p. 255–270.

FERGUSON, S. A., 1966, The relationship of mineralization to stratigraphy in the Porcupine and Red Lake areas, Ontario: Geol. Assoc. Canada, Spec. Paper 3, p. 99–120.

GALKIEWICZ, T., 1967, Genesis of Silesian-Cracovian zinc-lead deposits: Econ. Geology Mon. 3, p. 156–168.

GILLETI, B. J., and DAY, H. W., 1968, Potassium-argon ages of igneous intrusive rocks of Peru: Nature, v. 220, p. 570–572.

GILLULY, J., 1971, Plate tectonics and magmatic evolution: Geol. Soc. America Bull., v. 82, p. 2383–2396.

GILMOUR, P., and STILL, A. R., 1967, The geology of the Iron King Mine, in Ore deposits of the United States 1933/1967 (Graton-Sales volume): New York, Am. Inst. Mining Metall. Engineers, p. 1238–1257.

GOODWIN, A. M., 1965, Mineralized volcanic complexes in the Porcupine-Kirkland Lake-Noranda region, Canada: Econ. Geology, v. 60, p. 955–97.

——— 1968, Evolution of the Canadian Shield: Geol. Assoc. Canada Proc., v. 19, p. 1–14.

———, and RIDLER, R. H., 1971, The Abitibi orogenic belt, in BAUER, A. J., ed., Basins and geosynclines of the Canadian Shield: Geol. Survey Canada Paper 70-40, p. 1–24.

GOVETT, G. J. S., and PANTAZIS, TH. M., 1971, Distribution of Cu, Zn, Ni and Co in the Troodos Pillow Lava Series, Cyprus: Inst. Mining Metallurgy Trans., v. 80, p. B27–46.

GREIG, J. A.; BAADSGAARD, H.; GUMMING, G. L.; FOLINSBEE, R. E.; KROUSE, H. R.; OHMOTO, H.; SASAKI, A.; and SMEJKAL, V., 1971, Lead and sulfur isotopes of the Irish base metal mines in Carboniferous carbonate host rocks: Internat. Mineralog. Assoc.-Internat. Assoc. Genesis Ore Deposits, Tokyo-Kyoto 1970, Proc., p. 84–92.

GUNNING, H. C., senior ed., 1966, Tectonic history and mineral deposits of the western Cardillera: Canadian Inst. Mining and Metallurgy, spec. v. 8, 353 p.

HAAPALA, P. S., 1969, Fennoscandian nickel deposits: Econ. Geology Mon. 4, p. 262–275.

HALL, W. E., and FRIEDMAN, I., 1963, Composition of fluid inclusions Cave-In-Rock fluorite district, Illinois, and upper Mississippi Valley zinc-lead district: Econ. Geology, v. 58, p. 886–911.

HAMILTON, W., 1969a, Mesozoic California and the underflow of Pacific mantle: Geol. Soc. America Bull., v. 80, p. 2409–2429.

——— 1969b, The volcanic central Andes: a modern model for the Cretaceous batholiths and tectonics of western North America: Oregon Dept. Geology and Mineral Industries Bull. 65, p. 175–184.

——— 1970, The Uralides and the motion of the Russian and Siberian platforms: Geol. Soc. America Bull., v. 81, p. 2553–2576.

———, and MYERS, B. W., 1967, The nature of batholiths: U. S. Geol. Survey Prof. Paper 554-C, p. 1–30.

HANSON, G. N., and GOLDICH, S. S., 1970, Early Precambrian geology of the Saganaga-Northern Light lakes area, Minnesota-Ontario (Abs.): Inst. Lake Superior Geology Ann. Mtg., 16th, Thunder Bay Ontario, p. 18.

HART, S. R., and DAVIS, G. L., 1969, Zircon U-Pb and whole rock Rb-Sr ages and early crustal development near Rainy Lake, Ontario: Geol. Soc. America Bull. v. 80, p. 595–616.

HATHERTON, T., and DICKINSON, W. R., 1969, The relationship between andesitic volcanism and seismicity in Indonesia, the Lesser Antilles, and other island arcs: Jour. Geophys. Research, v. 74, p. 5301–5309.

HEYL, A. V., JR., 1967, Some aspects of genesis of stratiform lead-zinc-fluorite-barite deposits in the United States: Econ. Geology Mon. 3, p. 20–32.

HORIKOSHI, E., 1969, Volcanic activity related to the formation of the Kuroko-type deposits in the Kosaka Dist. Japan: Mineralium Deposita, v. 4, p. 321–345.

HURLEY, P. M.; BATEMAN, P. C.; FAIRBAIRN, H. W.; and PINSON, W. H., JR., 1965, Investigation of initial Sr^{87}/Sr^{86} ratios in the Sierra Nevada plutonic province: Geol. Soc. America Bull., v. 76, p. 165–174.

HUTCHINSON, R. W., 1965, Genesis of Canadian massive sulfides reconsidered by comparison to Cyprus deposits: Canadian Inst. Mining Metallurgy Trans., v. 68, p. 286–300.

ISACKS, B.; OLIVER, J.; and SYKES, L. R., 1968, Seismology and the new global tectonics: Jour. Geophys. Research, v. 73, p. 5855–5899.

JAKES, P., and WHITE, A. J. R., 1972, Major and trace element abundances in volcanic rocks of orogenic areas: Geol. Soc. America Bull., v. 83, p. 29–40.

JAMES, D. E., 1971, Plate tectonic model for the evolution of the central Andes: Geol. Soc. America Bull., v. 82, p. 3325–3346.

JENKS, W. F., 1966, Some relations between Cenozoic volcanism and ore deposition in northern Japan: New York Acad. Sci., Trans., v. 28, p. 463–477.

JENSEN, M. L., ed., 1962, Biogeochemistry of sulfur isotopes: Natl. Sci. Found. Symposium, Yale Univ., Proc. 193 p.

———, and DESSAU, G., 1967, The bearing of sulfur isotopes on the origin of Mississippi Valley type deposits: Econ. Geology Mon. 3, p. 400–409.

JEROME, S. E., and COOK, D. R., 1967, Relation of some metal mining districts in the western United States to regional tectonic environments and igneous activity: McKay School Mines, Nevada, Bull. 69, 35 p.

KALLIOKOSKI, J., 1965, Metamorphic features in North American massive sulfide deposits: Econ. Geology, v. 60, p. 485–505.

KARIG, D. E., 1970, Ridges and basins of the Tonga-Kermadec island arc system: Jour. Geophys. Research, v. 75, p. 239–254.

KATILI, J. A., 1967, Structure and age of the Indonesian tin belt with special reference to Bangka: Tectonophysics, v. 4, p. 403–418.

KAY, R.; HUBBARD, N. J.; and GAST, P. W., 1970, Chemical characteristics and origin of oceanic

ridge volcanic rocks: Jour. Geophys. Research, v. 75, p. 1585–1613.

KELLY, WM. C., and TURNEAURE, F. S., 1970, Mineralogy, paragenesis and geothermometry of the tin and tungsten deposits of the eastern Andes, Bolivia: Econ. Geology, v. 65, p. 609–680.

KILBURN, L. C.; WILSON, H. D. B.; GRAHAM, A. R.; and RAMAL, K., 1969, Nickel sulfide ores related to ultrabasic intrusions in Canada: Econ. Geology Mon. 4, p. 276–293.

KNIGHT, C. L., 1957, Ore genesis: the source bed concept: Econ. Geology, v. 52, p. 808–817.

KRAUME, E.; DAHLGRUN, F., RANDHOR, P.; and WILKE, A., 1955, Die Erzlager des Rammelsberges bei Goslar: Monographien der deutschen Blei-Zink Erzlagerstatten 4, Hannover, 394 p.

KRAUSKOPF, K. B., 1971, The source of ore metals: Geochim. et Cosmochim. Acta, v. 35, p.643–659.

KROGH, T. E., and DAVIS, G. L., 1971, Zircon U-Pb ages of Archean metavolcanic rocks in the Canadian Shield: Geophys. Lab. Washington Ann. Rept., p. 241–242.

KUNO, H., 1960, High alumina basalt: Jour. Petrology, v. 1, p. 121–145.

LIPMAN, P. W.; PROSTKA, H. J.; and CHRISTIANSEN, R. L., 1971, Evolving subduction zones in the western United States, as interpreted from igneous rocks: Science, v. 174, p. 821–825.

LJUNGGREN, P., and MEYER, H. C., 1964, The copper mineralization of the Corocoro Basin, Bolivia: Econ. Geology, v. 59, p. 110–125.

McALLISTER, A. L., 1960, Massive sulfide deposits in New Brunswick, in Symposium on the occurrence of massive sulfide deposits in Canada: Canadian Mining and Metall. Bull., v. 53, p. 88–89.

McBIRNEY, A. R., 1971, Oceanic volcanism: a review: Rev. Geophysics Space Physics, v. 9, p. 523–555.

McCARTNEY, W. D., 1964, Metallogeny of post-Precambrian geosynclines: British Commonwealth Geol. Liaison Office, Spec. Pub. 5.4, p. 19–23.

———, and POTTER, R. R., 1962, Mineralization as related to structural deformation, igneous activity, and sedimentation in folded geosynclines: Canadian Mining Jour., v. 83, p. 83–87.

MAGEE, M., 1967, Geology and ore deposits of the Ducktown district, Tennessee, in Ore deposits of the United States, 1933/1967 (Graton-Sales volume): New York, Am. Inst. Mining Metall. Engineers, p. 207–241.

MATSUDA, T., and UYEDA, S., 1971, On the Pacific-type orogeny and its model-extension of the paired belts concept and possible origin of marginal seas: Tectonophysics, v. 11, p. 5–27.

MAUCHER, A., and SCHNEIDER, H. J., 1967, The Alpine lead-zinc ores: Econ. Geology Mon. 3, p. 71–89.

MENDELSOHN, F., ed., 1961, Geology of the Northern Rhodesian copperbelt: London, Macdonald, 523 p.

MIYASHIRO, A., 1967, Orogeny, regional metamorphism, and magmatism in the Japanese islands: Medd. Dansk Geol. Fören., v. 17, p. 390–446.

MOORES, E. M., and VINE, F. J., 1971, The Troodos Massif, Cyprus and other ophiolites as oceanic crust, evaluation and implications: Royal Soc. (London) Proc., A, v. 268, p. 443–466.

MORRISEY, C. J.; DAVIS, G. R.; and STEED, G. M., 1971, Mineralization in the lower Carboniferous of Central Ireland: Canadian Inst. Mining Metallurgy Trans., v. 80. p, B174–185.

NALDRETT, A. J., 1970, Syn-volcanic ultramafic bodies: a new class (Abs.): Geol. Soc. America Ann. Mtg., Milwaukee, p. 633.

OFTEDAHL, C., 1958, A theory of exhalative-sedimentary ores: Geol. Fören. Stockholm Förh., v. 80, p. 1–19.

OHMOTO, H., 1970, Influence of pH and P_{O_2} of hydrothermal fluids on the isotopic composition of sulfur species (Abs.): Geol. Soc. America Ann. Mtg., Milwaukee, p. 640.

——; KAJIWARA, Y.; and DATE, J., 1970, The Kuroko ores of Japan: products of sea water? (Abs.): Geol. Soc. America Ann. Mtg., Milwaukee, p. 640–641.

PETERMAN, Z. E., and GOLDICH, S. S., 1970, Early Precambrian geology of the Rainy Lake District (Abs.): Inst. Lake Superior Geology Ann. Mtg., 16th, Thunder Bay, Ontario, p. 34.

PETERSEN, U., 1965, Regional geology and major ore deposits of central Peru: Econ. Geology, v. 60, p. 407–476.

—— 1970, Metallogenic provinces in South America: Geol. Rundschau, v. 59, p. 834–897.

PETRUK, W., 1968, Mineralogy and origin of the Silverfields silver deposit in the cobalt area, Ontario: Econ. Geology, v. 63, p. 512–531.

PIENAAR, P. J., 1961, Mineralization in the basement complex, in MENDELSOHN, F., ed., The geology of the Northern Rhodesian copperbelt: London, Macdonald, p. 30–41.

PINCKNEY, D. M., and HAFFTY, J., 1970, Content of zinc and copper in some fluid inclusions from the Cave-in-Rock District, Southern Illinois: Econ. Geology, v. 65, p. 451–458.

POPOV, L., 1962, Geologic regularities in the distribution of cupriferous sandstones in central Kazakhstan: Internat. Geol. Rev., v. 1, p. 393.

POUBA, A., 1971, Relations between iron and copper-lead-zinc mineralizations in submarine volcanic ore deposits in the Jeseniky Mts., Czechoslovakia: Internat. Mineralog. Assoc.-Internat. Assoc. Genesis Ore Deposits, Tokyo-Kyoto 1970, Proc., p. 186–192.

PUSHKAR, P., 1968, Strontium isotopic ratios in volcanic rocks of three island arc areas: Jour. Geophys. Research, v. 73, p. 2701–2714.

QUIRT, S.; CLARK, A. H.; FARRAR, E.; and SILLITOE, R. H., 1971, Potassium-argon ages of porphyry copper deposits in northern and central Chile (Abs.): Geol. Soc. America Ann. Mtg., Washington, p. 676–677.

RHODES, J. M., and RIDLEY, W. I., 1971, Trace element abundances in some Andean andesites (Abs.): Geol. Soc. America Ann. Mtg., Washington, p. 681–682.

RIDGE, J. D., ed., 1968, Ore deposits of the United States, 1933/1967 (Graton-Sales volumes): New York, Am. Inst. Mining Metall. Engineers.

RIDLER, R. H., 1970, Relationship of mineralization to volcanic stratigraphy in the Kirkland-Larder lakes area, Ontario: Geol. Assoc. Canada, v. 21, p. 33–42.

ROEDDER, E., 1967, Environment of deposition of stratiform (Mississippi Valley-type) ore deposits, from studies of fluid inclusions: Econ. Geology Mon. 3, p. 349–362.

ROSCOE, S. M., 1965, Geochemical and isotopic studies, Noranda and Matagamie areas: Canadian Inst. Mining Metall. Trans., v. 68, p. 279–285.

ROSS, D. A.; MILLIMAN, J. D.; BREWER, P. G.; MANHEIM, F. T.; and HATHAWAY, J. C., 1971, The Red Sea brine area: revisited (Abs.): Geol. Soc. America Ann. Mtg., Washington, p. 689–670.

RUIZ, C., 1965, Geologia y yacimientos metalliferos de Chile: Inst. Inv. Geol., Chile, 305 p.

—— et al., 1970, Strata-bound copper deposits of Chile (Abs.): Internat. Mineralog. Assoc.–Internat. Assoc. Genesis Ore Deposits, Tokyo-Kyoto 1970, Collected Abs., p. 97.

RUSSELL, M. J., 1968, Structural controls of base metal mineralization in Ireland in relation to continental drift: Inst. Mining Metallurgy Trans. v. 77, p. B117–128.

SANGSTER, D. F., 1968, Relative sulfur isotope abundances of ancient seas and strata-bound sulfide deposits: Geol. Assoc. Canada Proc., v. 19, p. 79–91.

SAWKINS, F. J., 1966, Ore genesis in the North Pennine orefield, in the light of fluid inclusion studies: Econ. Geology, v. 61, p. 385–401.

—— 1968, The significance of Na/K and Cl/SO$_4$ ratios in fluid inclusions and subsurface waters, with respect to the genesis of Mississippi Valley-type ore deposits: Econ. Geology, v. 63, p. 935–942.

——, and PETERSEN, U., 1969, A tectonic-genetic classification of sulfide ore deposits (Abs.): Geol. Soc. America Ann. Mtg., Atlantic City, p. 197–198.

——, and RYE, D. M., 1971, On the relationship of certain Precambrian gold deposits to iron formation (Abs.): Geol. Soc. America Ann. Mtg., Washington, p. 694.

——, and RYE, R. O., 1970, The Casapalca silver-lead-zinc-copper deposit, Peru: an ore deposit formed by hydrothermal waters of deep-seated origin? (Abs.): Geol. Soc. America Ann. Mtg., Milwaukee, p. 674–675.

SHEPPARD, S. M. F.; NIELSEN, R. L.; and TAYLOR, H. P., JR., 1971, Hydrogen and oxygen isotope ratios in minerals from porphyry copper deposits: Econ. Geology, v. 66, p. 515–542.

SILLITOE, R. H., 1972, Relation of metal provinces in western America to subduction of oceanic lithosphere: Geol. Soc. America Bull., v. 83, p. 813–818.

——, and SAWKINS, F. J., 1971, Geologic, mineralogic and fluid inclusion studies of copper-bearing tourmaline breccia pipes, Chile: Internat. Mineralog. Assoc.-Internat. Assoc. Genesis Ore Deposits, Tokyo-Kyoto 1970, Proc., p. 100–105.

SINHA, A. K., and HART, S. R., 1971, Geochemical relationship of sediments and volcanic rocks from the Tonga area (Abs.): Geol. Soc. America Ann. Mtg., Washington, p. 707–708.

SKRIPCHENKO, N. S., 1967, Massive pyritic deposits related to volcanism and possible methods of emplacement: Econ. Geology, v. 62, p. 292–293.

SOLOMON, M.; RAFTER, T. A.; and DUNHAM, K. C., 1971, Sulfur and oxygen isotope studies in the northern Pennines in relation to ore genesis: Inst. Mining and Metallurgy, v. 80, p. B259–275.

SOUCH, B. E.; PODOLSKY, T.; and GEOLOGICAL STAFF, 1969, The sulfide ores of Sudbury: their particular relationship to a distinctive inclusion-bearing facies of the nickel irruptive: Econ. Geology Mon. 4, p. 252–261.

STANTON, R. L., 1962, Elemental constitution of the Black Star orebodies, Mt. Isa, Quebec, and its interpretation: Inst. Mining Metallurgy (London) Bull., v. 72, p. 69–124.

STOLL, W. C., 1964, Metallogenic belts, centers, and epochs in Argentina and Chile: Econ. Geology, v. 59, p. 126–135.

SUPERCEANU, C. I., 1970, The Eastern Mediterranean-Iranian Alpine copper-molybdenum belt: Internat. Mineralog. Assoc.-Internat. Assoc. Genesis Ore Deposits, Tokyo-Kyoto 1970, Proc., p. 393–400.

SWANSON, E. A., and BROWN, R. L., 1962, Geology of the Buchans orebodies: Canadian Inst. Mining Metallurgy Bull., v. 55, p. 618–626.

TITLEY, S. R., and HICKS, C. L., eds., 1966, Geology of the porphyry copper deposits, southwestern North America (Wilson volume): Univ. Arizona Press, Tucson, Ariz., 287 p.

TURNEAURE, F. S., 1955, Metallogenic provinces and epochs: Econ. Geology, 50th Anniversary volume, p. 38–98.

TWETO, O., 1967, Geologic setting and interrela-

tionships of mineral deposits in the mountain province of Colorado and south-central Wyoming, in Ore deposits of the United States 1933/1967 (Graton-Sales volume): New York, Am. Inst. Mining Metall. Engineers, p. 552–565.

VILJOEN, R. P.; SAAGER, R.; and VILJOEN, M. F., 1969, Metallogenesis and ore control in the Steynsdorp Goldfield, Barberton Mountain Land, South Africa: Econ. Geology, v. 64, p. 778–798.

VOKES, F. M., 1963, Molybdenum deposits of Canada: Geol. Survey Canada ·Econ. Geology Rept. 3, 332 p.

——— 1968, Regional metamorphism of the Paleozoic geosynclinal sulfide ore deposits of Norway: Inst. Mining Metallurgy (London) Bull., v. 77, pp. 353–359.

WHITE, D. E., 1968, Environments of generation of some base metal ore deposits: Econ. Geology, v. 63, p. 301–335.

———; HEM, J. D.; and WARING, G. A., 1963, Data of geochemistry Chapter F: Chemic Composition of Subsurface Waters: U.S. Geol. Survey Prof. Paper 440-F.

WILLIAMS, D., 1962, Further reflections on the origin of porphyries and ores of Rio Tinto, Spain: Inst. Mining Metallurgy (London) Bull., v. 71, p. 265–266.

WILSON, H. D. B., 1967, Volcanism and ore deposits in the Canadian Archean: Geol. Assoc. Canada Proc., v. 18, p. 11–31.

———; ANDREWS, P.; MOXHAN, R. L.; and RAMLAL, K., 1965, Archean volcanism in the Canadian Shield: Canadian Jour. Earth Sci., v. 2, p. 161–175.

WOODALL, R., and TRAVIS, G. A., 1969, The Kambalda nickel deposits, Western Australia: Commonwealth Mining Metallurgy Cong., 9th Paper 26.

YAROSH, P. Y., 1965, Metasomatism and metamorphism of pyrite ores in the Zyuzel 'skoye deposit (English Abs.): Econ. Geology, v. 60, p. 1545.

ZAVARITSKY, A. N., 1948, Metasomatism and metamorphism in the pyrite deposits of the Urals: Internat. Geol. Cong., 18th, London 1948, Proc., sec. B, pt. 3, p. 192–208.

ZURBRIGG, H. F., 1963, Thompson Mine geology: Canadian Inst. Mining Metallurgy Bull., v. 56, p. 451–460.

Formation of certain massive sulphide deposits at sites of sea-floor spreading

551.3 : 553.21 : 553.661.21

R. H. Sillitoe B.Sc., Ph.D., A.M.I.M.M.
Department of Mining Geology, Royal School of Mines, Imperial College, London

Synopsis

In the context of the theory of plate tectonics, it is suggested that many cupriferous massive sulphide deposits associated with ophiolite complexes are generated in ocean basins at sites of sea-floor spreading—the ocean rises. This interpretation supersedes the view that these deposits are formed during the initial stages of development of eugeosynclines. The massive sulphide deposits are temporally and spatially related to tholeiitic basalts, which constitute layer 2 of the oceanic crust. This layer overlies coeval dolerite dyke swarms and gabbro complexes (layer 3), which, in turn, overlie ultrabasic rocks (upper mantle). The gabbro complexes may represent the chambers from which magma to form layer 2 basalts and hydrothermal fluids to form massive sulphide deposits were expelled. As components of oceanic lithosphere, massive sulphide deposits of ophiolite affiliation are transported away from ocean rises by means of sea-floor spreading. On reaching the margins of ocean basins, the massive sulphide deposits are underthrust along subduction zones beneath continental margins and island arcs. During subduction, some of the deposits are incorporated mechanically into the continental crust as parts of slices of oceanic lithosphere.

It is suggested that massive sulphide deposits in Cyprus, southwest Japan, Newfoundland, Turkey, the Urals, the Philippines and northern Italy were formed at loci of sea-floor spreading. Two further sites for the generation of massive sulphide deposits are recognized : shallow marine environments above subduction zones at continental margins or island arcs (e.g. the Kuroko deposits of Japan and Rio Tinto, Spain), and, less certainly, sites of slow extension and basalt extrusion in the marginal ocean basins behind island arcs.

It is now widely accepted that stratiform massive sulphide deposits, consisting of pyrite and/or pyrrhotite, with various amounts of chalcopyrite, sphalerite and galena, are genetically related to submarine volcanism in so-called 'eugeosynclinal' environments.[1-4] Emplacement of the massive sulphide deposits seems to have been accomplished on or immediately beneath the sea floor during volcanism, and is not attributable to subsequent non-volcanogenic processes of intracrustal replacement. Massive sulphide deposits and their submarine volcanic host rocks have been assigned to the initial tectonic stage of geosynclinal development by Bilibin[5] and later workers.[6] They considered that the initial tectonic stage was characterized by submarine, 'eugeosynclinal' volcanism, largely basic in type, and by the emplacement of basic and ultrabasic intrusive masses.

Development of the theory of plate tectonics has, however, led to a reinterpretation of such stabilist geosynclinal concepts,[7] and to the recognition of geosynclines as continental margins or island arcs and the adjoining ocean basins. Orogenic belts form at continental margins and island arcs during the underthrusting of oceanic lithosphere (crust and upper mantle) along subduction zones and during consequent intercontinental or continent/island arc collisions.[8,9,10]

Manuscript first received by the Institution of Mining and Metallurgy on 14 March, 1972; revised manuscript received on 30 March, 1972. Paper published in August, 1972.

In the light of these recent advances in the interpretation of mountain belts, the orogenic setting of massive sulphide deposits may be reassessed.

This paper concentrates attention on massive sulphide deposits which accompany mafic volcanic sequences, although deposits associated with calc-alkaline volcanic sequences, consisting largely of andesites, dacites and rhyolites, are briefly mentioned. The mafic volcanic host rocks to massive sulphide deposits form an integral part of alpine-type mafic–ultramafic, or ophiolite, complexes in orogenic belts.

Ophiolites and plate tectonics

The notion that ophiolites are extruded in the axial regions of deep, linear 'eugeosynclinal' troughs has recently been discredited, and ophiolites are now considered as slices of oceanic crust and upper mantle tectonically emplaced in orogenic belts.[11,12,13] Linear belts of ophiolites lie along boundaries of lithospheric plates and are believed to mark ancient sites of lithospheric plate consumption.[8,14,15]

These contentions are based on the close similarity between the bulk compositions, structures and sequences of ophiolite complexes and of oceanic crust–mantle. Ophiolite complexes have been studied in Turkey, Greece, Oman, Iran, Cyprus, the Himalayas, Papua, Celebes, Philippines, the Coast Ranges of California, and elsewhere in the Tethyan and circum-Pacific belts (see, for example, references 16–18). Knowledge of the constitution of oceanic crust–mantle is derived from interpretations of geophysical data,[19] supported by the results of deep-sea drilling[20] and dredging in the vicinity of ocean rises (mid-ocean ridges).[21,22]

From the top downwards, ophiolite sequences and oceanic lithosphere consist of thin sequences of chert, argillite and pelagic limestone (layer 1 of the oceanic crust), underlain by dominantly extrusive tholeiitic basalt and some hyaloclastite and alkali basalt (layer 2), which commonly display pillow structures and conversion to spilite. These pass downwards into gabbro with subsidiary diorite and trondhjemite (layer 3). In some places the lower part of layer 2 and the upper part of layer 3 are intruded by conspicuous sheeted complexes composed of swarms of dolerite dykes. Layers 1, 2 and 3 of the oceanic crust overlie partly serpentinized dunite, harzburgite, lherzolite and garnet peridotite of the upper mantle. The effects of metamorphism are apparent in layer 2 downwards. The massive sulphide deposits that constitute the subject of this paper are associated with basalts of layer 2.

Support for the equivalence of ophiolite complexes and oceanic lithosphere is provided by their similar contents of major elements[15] and of Ti, Zr and Y.[23]

If oceanic lithosphere and ophiolite complexes can be equated, then ophiolites may be considered to be generated as oceanic crust—mantle at sites of sea-floor spreading—the ocean rises (Fig. 1). The magma that ascends at ocean rises to form the basalts of layer 2 and the underlying intrusions of dolerite and gabbro is derived by the partial fusion of diapiric masses of peridotitic asthenosphere rising vertically upwards between the trailing edges of the two outward-moving plates of lithosphere (Fig. 1).[24] Upper-mantle harzburgite and dunite may represent the coevally produced residue of this partial fusion process. The dolerite dykes represent the feeders to the suprajacent basalt flows.

The oceanic lithosphere formed at ocean rises is transported laterally as quasi-rigid plates by sea-floor spreading, and, finally, thrust beneath continental margins or island arcs along subduction zones (Fig. 2). Layer 1 sediments accumulate on layer 2 basalts from the time of lithosphere generation at ocean rises until consumption takes place at ocean trenches. During subduction, relatively small slabs of oceanic lithosphere may be tectonically emplaced behind or beneath trenches, giving rise to ophiolites, which are subsequently exposed, sometimes as components of *mélanges* (Fig. 2),[25] by isostatic uplift and erosion. Additional mechanisms for incorporating oceanic lithosphere into continental crust seem to have operated,[14] including the overthrusting (obduction) of large slabs of oceanic lithosphere on to the edges of continents or island arcs,[15] perhaps induced by the collision of a continent with a subduction zone dipping away from it.[14]

Massive sulphide deposits in ophiolite terrains
Massive sulphide deposits are known to be associated with basaltic volcanic rocks of a wide variety of ages. Those in Cyprus have undergone only minor metamorphism and deformation and are therefore used as a type example. The other deposits to be mentioned have been tectonized to varying degrees.

Cyprus
In Cyprus some 90 massive sulphide deposits and prospects are genetically related to the Troodos ophiolite complex of probable late Cretaceous age. Studies of the Troodos complex by Gass and Masson-Smith[26,27] and Moores and Vine[28] described the centrally located Plutonic Complex composed of harzburgite, dunite, pyroxenite, gabbro and trondhjemite, which grades upwards into a Sheeted Complex consisting almost entirely of dolerite dykes. This is overlain by the Lower and Upper Pillow Lavas, mainly of tholeiitic basaltic composition. The basalts are overlain by ferruginous and manganiferous sediments (umbers), grits, shales and cherts, comprising the Perapedhi Formation, which pass upwards into marls and limestones.

It is considered that harzburgite and dunite represent the depleted mantle; gabbro and the lower part of the Sheeted Complex represent layer 3; and the upper part

of the Sheeted Complex and the Lower and Upper Pillow Lavas represent layer 2.[28] The Perapedhi Formation and overlying marls and limestones may be interpreted as layer 1. Compositional differences between the pillow lavas and layer 2 of the oceanic crust are attributed to subsequent diagenetic, metamorphic and other changes.[28,29]

The massive sulphide deposits[30-33] occur as irregular, elongate lenticles, being commonest at the level of the contact between the Lower and Upper Pillow Lavas but also occurring throughout the pillow lava succession. The hanging-walls of the orebodies are characterized by pyrite-bearing chemical sediments (ochres), enriched in silica and hydrated Fe oxides. Overlying flows are unaltered and unmineralized. The sediments are underlain by massive pyrite, with some chalcopyrite, sphalerite and marcasite, with minor galena, pyrrhotite, Au and Ag. The massive parts of the orebodies pass downwards into stockworks of sulphide veinlets, in which pyrite predominates, accompanied by brecciation, silicification and propylitization. Some orebodies are cut by dolerite dykes that are feeders to higher and later flows.

Massive sulphide deposits seem to have been generated episodically during the formation of the layer 2 basalts, prior to the deposition of the Perapedhi Formation. The massive, stratiform deposits are presumed to have been formed initially as accumulations of brine-soaked sulphide mud in depressions on the sea floor. Mineralizing fluids were supplied through subjacent conduits now represented by the stockworks of epigenetic mineralization.

It is postulated that the Troodos ophiolite complex was generated at an ocean rise in the Tethyan Ocean,[27] and subsequently emplaced by collision of the African plate to the south with a subduction zone dipping northwards.[28]

Southwest Japan
Conformable, copper-bearing massive sulphide deposits are widespread in belts of late Palaeozoic ophiolites in southwest Japan.[34,35] More than 150 deposits have been worked in the Sanbagawa metamorphic zone on the Pacific side of Japan,[36] and other deposits exist in the Sangun metamorphic zone on the west side of Japan. Additional massive sulphide deposits of ophiolite affiliation are found in the similar Shimanto zone of Jurassic—Cretaceous age on the Pacific side of Japan. The deposits, in which pyrite, chalcopyrite and some sphalerite are the principal sulphides, occur in basic schists and greenstones, which were probably originally deposited as submarine volcanic rocks of dominantly basaltic composition. Volcanic members are interlayered with pelitic and limy rocks. Ferruginous and manganiferous cherts commonly overlie the orebodies and also occur as thin beds elsewhere in the volcanic successions. Ultrabasic and gabbroic intrusive rocks are present in the same areas. The copper deposits were deformed and metamorphosed with their wallrocks.

The characteristics of the ophiolite terrains of southwest Japan, including the low K contents and high K/Rb ratios of the basalts,[37] suggest that they represent oceanic crust and mantle that have been strongly deformed and metamorphosed subsequent to their incorporation within the continental lithosphere.

Philippines
In the Philippines, cupriferous massive sulphide deposits are associated with a succession of Cretaceous–Palaeogene, dominantly mafic, submarine pillow lavas, especially near the tops of lava piles, where they pass stratigraphically upwards into sedimentary rocks[38] (? layer 1). Sphalerite, barite, Au and Ag are also present in various proportions in the orebodies. The deposits of Balabac Island[39] in the southwest of the Philippines provide typical examples. Ultramafic rocks, gabbro, trondhjemite and diorite accompany the volcanics and massive sulphide deposits and stratiform manganese deposits in three north–south-trending ophiolite belts in the Philippines.[40]

U.S.S.R.
Available English-language descriptions of massive sulphide deposits in the U.S.S.R. indicate that some are related to typical ophiolite terrains, whereas others are probably related to calc-alkaline magmatism (see later). Greenstone belts of Palaeozoic age in the Urals, West Sayan and elsewhere are hosts to numerous massive sulphide deposits.[4] One hundred deposits occur on the east side of the Urals. In the southern Urals conformable lenses of massive sulphides, grading laterally into thin sheet-like orebodies, occur in spilitic host rocks.[41] The massive ores are underlain by sulphide disseminations in silicified and sericitized footwall rocks. Nearby clastic sulphide accumulations are thought to represent massive deposits fragmented by subsequent eruptive processes. Some deposits grade both outwards and upwards into silica–hematite rock.

In a recent plate tectonic reconstruction of the late Precambrian and Palaeozoic mountain belts that lie between the Russian and Siberian platforms,[42] it was suggested that the greenstone belts of the Urals represent oceanic crust. During the approach and eventual collision of the Russian and Siberian sub-continents, the greenstones and associated basic and ultrabasic intrusives were added to continental crust as the intervening ocean was progressively eliminated by the operation of subduction zones.

Newfoundland
In the Appalachians of Newfoundland there is a series of typical ophiolite complexes[14,43,44] formed in the early Ordovician, and perhaps also in earlier times, as in the Notre Dame Bay and Bay of Islands areas. At Betts Cove a typical ophiolite sequence is observable with basal dunite, harzburgite and pyroxenite, overlain by gabbro, sheeted dolerites and pillow lavas and hyaloclastites.[45] In the layer 2 pillow lava sequence,

copper-bearing massive sulphide deposits containing pyrite and chalcopyrite, and some sphalerite, Au and Ag, are common.[46] The most productive deposits include those at Tilt Cove and Betts Cove on Burlington Peninsula and that at Little Bay on Notre Dame Bay.[47] In some places original relations have been largely obscured by deformation of the orebodies and their wallrocks, which led Baird[46] to propose a replacement origin for the mineralization. At the Tilt Cove deposit both massive lenses and stockworks of pyrite–chalcopyrite are present.[46]

It has been proposed that the ophiolites represent oceanic lithosphere eliminated by plate consumption in a subduction zone on the northwest margin of a proto-Atlantic Ocean.[14,43]

Turkey
Cupriferous massive sulphide deposits are widespread in Turkey and, along with stratiform accumulations of Mn and Fe oxides, are thought to be genetically related to ophiolite magmatism.[48] Deposits in the Küre district on the Black Sea coast and at Ergani Maden in the southeast of the country are two of the best known examples.

At Ergani Maden,[49] near-horizontal massive orebodies containing Cu, significant Co, Au and Ag, and minor Zn pass downwards into lower-grade disseminated ore in silicified and chloritized rock. Dolerite and gabbro occur immediately below the mineralization, and ultrabasic rocks are present at still deeper levels. The massive ore is capped by Upper Cretaceous black-red limestone, basic tuff and spilitic agglomerate.

In the Küre district[50] several massive and disseminated pyrite–chalcopyrite–Au orebodies are located in basic pillow lavas interbedded with hyaloclastites, agglomerates, argillites and greywackes. The succession is intruded by diorite, gabbro and ultrabasic rocks.

Italy
Further examples of cupriferous massive sulphide deposits in ophiolite terrains are known from the Mesozoic of the western Alps of Italy.[51]

Suggested origin of massive sulphide deposits associated with ophiolites
In the ocean basins volcanism is most active at ocean rises along divergent plate margins, which therefore appear to be the most likely sites for the formation of massive sulphide deposits accompanying submarine basalts (Fig. 1). Volcanic activity is concentrated in the axial valleys of ocean rises. Moreover, the fact that orebodies were emplaced prior to the deposition of layer 1 (e.g. in Cyprus) precludes other parts of the ocean basins as potential loci of formation. If the Cyprus deposits are typical, then metals were probably deposited on, and in feeder channels immediately beneath, the sea floor from hydrothermal fluids genetically related to the temporally and spatially related basaltic lavas. Anoxic conditions at the layer 2–

sea water interface must have been maintained during sulphide precipitation. Either an upward increase in Eh during mineralization due to mixing of hydrothermal fluids with overlying sea water, or syn- and post-depositional oxidation of sulphides by sea water,[33] may be invoked to explain the formation of the horizons enriched in hydrated Fe oxides which commonly cap massive ore. It is tentatively suggested that hydrothermal fluids were expelled during the final stages of crystallization from magma chambers now represented by the gabbro complexes underlying dyke complexes and layer 2 basalts (Fig. 1). Cumulate textures exhibited by many

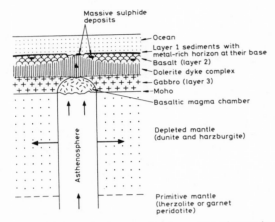

Fig. 1 Schematic diagram showing origin of oceanic crust and massive sulphide deposits at an ocean rise along a divergent plate margin. Uppermost ~ 25 km of the oceanic lithosphere are depicted

ophiolite gabbros probably developed in such shallow magma reservoirs beneath ocean rises.[16,28] In addition to a direct derivation from magma chambers as components of magmatically derived fluids, metals in massive sulphide deposits could also be released from layers 2 and 3 during metamorphism beneath the ocean floor, or during leaching by magmatogene fluids or circulating heated sea water. The estimated content of 800 ± 150 ppm sulphur in layer 2 of the oceanic crust[52] suggests that this element in massive sulphide deposits may also have a volcanogenic source.

From their site of origin at, or near to, the locus of active sea-floor spreading on the crest of an ocean rise, the massive sulphide deposits were probably transported laterally towards the margins of the ocean basins, to be overlain by progressively greater thicknesses of normal ocean-floor sediments (layer 1) (Fig. 1). It is postulated that most massive sulphide deposits and their enclosing oceanic crust are subducted at convergent plate margins, and that only a minority are incorporated into continental margins or island arcs (Fig. 2). It seems likely that some massive sulphide deposits and their host rocks may undergo deformation

and metamorphism either during transportation away from ocean rises, at the time of tectonic emplacement at continental margins, or during subsequent phases of orogenesis.

Evidence from the ocean basins

Massive sulphide deposits have not yet been found outside the continents. The existence of unusually high metal values in sediments on the floors of the oceans, however, seems to corroborate the suggestion that massive sulphide deposits are emplaced at ocean rises. Anomalously high concentrations of Mn, Fe, Cu, Zn, Pb and other elements occur in layer 1 sediments on the crest and flanks of the East Pacific Rise,[53,54,55] and elsewhere on the ocean-rise system.[56,57,58] Their origin is ascribed dominantly to precipitation from hydrothermal emanations accompanying basaltic volcanism, although small contributions from continental outwash and cosmic sources are undoubtedly present. Corliss[59] has argued that the hydrothermal emanations consist largely of heated sea water that has derived its content of metals during circulation through flows of layer 2 basalt.

Metal-rich brines and muds occur in the Red Sea in three axial deeps along a divergent plate margin.[60,61] The presence in the muds of a sulphide facies containing sphalerite and lesser chalcopyrite, pyrite and marcasite overlain by layers enriched in iron montmorillonite, goethite and Mn-bearing minerals is comparable with the situation associated with many massive sulphide deposits. Compaction of these metalliferous muds might yield deposits comparable in many ways with small massive sulphide deposits. Metalliferous brines are also found beneath the Salton Sea trough,[62] recently shown by geophysical investigations to be a sediment-filled northward extension of the Gulf of California[63]—a site of active sea-floor spreading. Some of the metals in the Red Sea and Salton Sea brines may have a volcanogenic source, although their derivation from sediments has been proposed.[62,64]

In addition to metal concentrations at active ocean rises, buried layers of metal-rich sediment of early Miocene to mid-Eocene age have been encountered by drilling in the eastern Pacific basin away from the East Pacific Rise;[65,66] these occurrences have been transported from the Rise to their present positions by sea-floor spreading.

These metal-rich sediments, in which Fe and Mn are the dominant metals, on the ocean floors may be analogous to the previously mentioned stratiform accumulations of Fe- and Mn-rich sediments in ophiolite terrains, including those at the base of the Perapedhi Formation in Cyprus.

In the ocean basins copper is also present in basalts of layer 2. Sulphide phases containing up to about 10 per cent Cu are common as vesicle fillings and globules in pillow basalts from the ocean rises.[67] Veinlets carrying quartz and Cu-bearing sulphides have been observed in basalt and dolerite dredged from the

Carlsberg Ridge in the Indian Ocean.[68] These sulphide phases are perhaps comparable with disseminated Cu-bearing sulphides which occur in the Upper Triassic submarine basalts of the Karmutsen Group on Vancouver Island, British Columbia. The major- and trace-element contents of the Karmutsen Group basalts demonstrate that they were formed as oceanic crust.[69]

Further implications

Many massive sulphide deposits are associated with volcanic rocks of calc-alkaline affinity. Volcanic rocks of this type characterize orogenic belts along continental margins or island arcs which were, and perhaps still are, underlain by subduction zones. Therefore, a second class of massive sulphide deposits, located at convergent plate junctures, may be recognized; these deposits were emplaced as a component of submarine, calc-alkaline volcanism (Fig. 2),

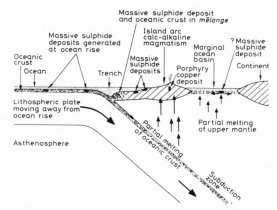

Fig. 2 Conceptual sketch illustrating interrelationships between massive sulphide deposits in the oceanic crust, in an island arc and in a marginal ocean basin, and porphyry copper deposits in an island arc. Vertical dimension is approximately 300 km and horizontal dimension is approximately 800 km

and appear generally to possess greater amounts of Pb, Zn, Ag and Ba than deposits associated with ophiolite complexes.

Typical of massive sulphide deposition at convergent plate margins are the Kuroko deposits of northeast Japan—clearly an integral part of an episode of Neogene calc-alkaline volcanism in a shallow marine environment.[70] As a somewhat older representative of the class, the belt of massive sulphide deposits extending from southern Spain to Portugal, and including Rio Tinto, that is related to a series of rhyolitic volcanic rocks of Lower Palaeozoic age,[2] can be cited. The modern analogue of the sites of massive sulphide deposition at convergent plate junctures is probably represented by the hot springs related to volcanism which debouch into shallow marine environ-

ments above active subduction zones at Vulcano, Italy[71] and Talasea Harbour and Matupi Harbour, New Britain.[72] Sediments at Vulcano and Talasea Harbour are enriched in iron sulphides and at Matupi Harbour they carry iron oxides and hydroxides and up to 1·2 per cent Mn and 1400 ppm Zn.

As was noted above, it seems likely that a large percentage of massive sulphide deposits emplaced at ocean rises are eventually subducted (Fig. 2). As was discussed elsewhere,[73,74] metals contained in these subducted deposits, and also those in layers 1 and 2, may be extracted during dehydration and partial melting along subduction zones. Alternatively, metal-bearing brines may be expelled from layers 1 and 2 on subduction zones, and during their subsequent ascent induce partial fusion of relatively dry upper mantle or lower crustal rocks. From subduction zones or overlying loci of partial fusion, the metals ascend as components of calc-alkaline magmas to the near-surface zone, where they are concentrated in hydrothermal fluids and precipitated to form ore deposits. In orogenic belts where calc-alkaline volcanism has an important submarine component, as is common in island arcs, the metals may be incorporated in massive sulphide deposits, whereas in regions where volcanism is mainly subaerial, as in the present-day Andean Cordillera, the metal-bearing fluids give rise to magmatic-hydrothermal deposits, typified by porphyry copper deposits.[74]

Discussion

In addition to the probable association of massive sulphide deposits with layer 2 of the oceanic crust, the basic and ultrabasic intrusive members of ophiolite complexes, representing layer 3 of the oceanic crust and the underlying upper mantle, also contain ore deposits. Disseminated Cu—Ni deposits in gabbros (e.g. the Urals), and alpine-type (podiform) chromite deposits and asbestos in ultrabasic rocks (e.g. Cyprus, Greece and the Philippines) may be cited as examples.

In many respects the metallogeny of Archaean greenstone belts is similar to that of ophiolite terrains. There an abundance of Cu- and Zn-bearing massive sulphide deposits is associated with Cu and Ni deposits in basic and ultrabasic intrusive rocks and with stratiform exhalative-sedimentary iron deposits (see, for example, reference 75). The occurrence of calc-alkaline andesitic and rhyolitic volcanic rocks as an integral part of most greenstone belts, and other characteristics, however, distinguishes them from ophiolite terrains. Furthermore, the evidence presently available suggests that Archaean volcanism and the coeval formation of massive sulphide deposits were the products of a unique stage in the development of the earth's crust, unrelated either to divergent or convergent plate margins of the types presently active.

A source at ocean rises cannot necessarily be ascribed to all massive sulphide deposits associated with basaltic volcanic rocks, since similar rocks may be generated in at least two other tectonic environments.

First, tholeiitic basalts, comparable in many respects with those generated at ocean rises, but differing from them in silica mode, alkali content, isotopic composition, and MgO, FeO, Ni and Cr contents, are known to occur as a facet of calc-alkaline magmatism along convergent plate margins. There they overlie subduction zones on the trench side of island arcs, and were erupted at an early stage in island arc development.[76] These island arc tholeiites could be spatially related to other members of the ophiolite assemblage, especially if the products of partial melting of upper mantle overlying subduction zones contribute to their genesis.[76] At present, the possibility that some ophiolite complexes form above subduction zones in this way, and not at ocean rises, remains to be investigated.

Secondly, submarine tholeiitic basalts and dolerites compositionally similar to those generated at ocean rises, and probably subjacent ultrabasic rocks, are produced by slow extensional processes in marginal ocean basins situated between island arcs and the facing continents (Fig. 2).[77] Extension and basalt extrusion is probably attributable to a high heat flow and consequent mobilization of mantle material above the deeper parts of subduction zones.[78] No information is currently available on the contents of metals in sediments and basalts in marginal ocean basins. Nevertheless, the possibility exists that the diffuse spreading centres in marginal ocean basins may represent a third type of environment for the generation of massive sulphide deposits (Fig. 2). With the present state of knowledge, it is difficult to ascertain whether some massive sulphide deposits in ophiolite terrains were formed at ocean rises or in marginal ocean basins. It has, however, been suggested that at least some of the ophiolite complexes in Newfoundland, with which massive sulphide deposits are associated, represent crust and upper mantle generated beneath marginal ocean basins.[14]

Now that drilling and dredging technologies are fairly well advanced, and neutron activation techniques are being developed for submarine exploration, it may be opportune to attempt to locate massive sulphide deposits on the ocean floors. The detection of such deposits on the 60 000-km long ocean-rise system should be easier than in the ocean basins, since the sediment cover there is considerably thinner and in parts absent. It must be realized, however, that massive sulphide deposits at ocean rises may be concealed by later lava flows, as in the case of the Cyprus deposits. Furthermore, the implication that massive sulphide deposits are a normal facet of ophiolite magmatism indicates that land-based exploration should pay more attention to Phanerozoic ophiolite terrains along both active and inactive convergent plate margins as potential sources of copper and lesser quantities of other metals. Such terrains might include the ophiolite slices in Papua, New Caledonia and Borneo, the Semail Complex in Oman, the Franciscan Formation in the Coast Ranges of California, and occurrences of ophiolites in the Highland Boundary region and the Southern Uplands of Scotland.

Acknowledgement
The author is very grateful to Professor David Williams for his comments on the manuscript. This work was undertaken during the tenure of a Shell postdoctoral Research Fellowship.

References
1. **Kinkel A. R. Jr.** Massive pyritic deposits related to volcanism and possible methods of emplacement. *Econ. Geol.*, **61**, 1966, 673–94.
2. **Williams D.** Volcanism and ore deposits. *Freiberg. ForschHft.*, **C210**, 1966, 93–111.
3. **Anderson C. A.** Massive sulfide deposits and volcanism. *Econ. Geol.*, **64**, 1969, 129–46.
4. **Smirnov V. I.** Pyritic deposits. Parts 1 and 2. *Int. Geol. Rev.*, **12**, 1970, 881–908, 1039–58.
5. **Bilibin Yu. A.** Metallogenic provinces and metallogenic epochs. *Geol. Bull. Queen's Coll., N.Y.* no. 1, 1968, 35 p.
6. **McCartney W. D. and Potter R. R.** Mineralization as related to structural deformation, igneous activity and sedimentation in folded geosynclines. *Can. Min. J.*, **83**, April 1962, 83–7.
7. **Coney P. J.** The geotectonic cycle and the new global tectonics. *Bull. geol. Soc. Am.*, **81**, 1970, 739–47.
8. **Dewey J. F. and Bird J. M.** Mountain belts and the new global tectonics. *J. geophys. Res.*, **75**, 1970, 2625–47.
9. **Dickinson W. R.** Plate tectonic models of geosynclines. *Earth planet. Sci. Lett.*, **10**, 1971, 165–74.
10. **Dickinson W. R.** Plate tectonics in geologic history. *Science, N.Y.*, **174**, 1971, 107–13.
11. **Dietz R. S.** Alpine serpentines as oceanic rind fragments. *Bull. geol. Soc. Am.*, **74**, 1963, 947–52.
12. **Thayer T. P.** Peridotite–gabbro complexes as keys to petrology of mid-oceanic ridges. *Bull. geol. Soc. Am.*, **80**, 1969, 1515–22.
13. **Hamilton W.** Mesozoic California and the underflow of Pacific mantle. *Bull. geol. Soc. Am.*, **80**, 1969, 2409–30.
14. **Dewey J. F. and Bird J. M.** Origin and emplacement of the ophiolite suite: Appalachian ophiolites in Newfoundland. *J. geophys. Res.*, **76**, 1971, 3179–206.
15. **Coleman R. G.** Plate tectonic emplacement of upper mantle peridotites along continental edges. *J. geophys. Res.*, **76**, 1971, 1212–22.
16. **Davies H. L.** Papuan ultramafic belt. In *Rep. 23rd Int. geol. Cong.* (Prague: Academia, 1968), pt 1, 209–20.
17. **Moores E. M.** Petrology and structure of the Vourinos ophiolitic complex of northern Greece. *Spec. Pap. geol. Soc. Am.* 118, 1969, 74 p.
18. **Bailey E. H. Blake M. C. Jr. and Jones D. L.** On-land Mesozoic oceanic crust in California Coast Ranges. *Prof. Pap. U.S. geol. Surv.* 700–C, 1970. C70–81.

19. **Shor G. G. Jr. and Raitt R. W.** Explosion seismic refraction studies of the crust and upper mantle in the Pacific and Indian Oceans. In *The earth's crust and upper mantle* **Hart P. J. ed.** (Washington D.C. : American Geophysical Union, 1969), 225–30. (*Geophys. Monog.* 13)

20. **Ewing M.** *et al.* *Initial reports of the Deep-Sea Drilling Project, volume 1* (Washington D.C. : Government Printing Office, 1969).

21. **Bonatti E.** Ultramafic rocks from the mid-Atlantic Ridge. *Nature, Lond.,* **219**, 1968, 363–4.

22. **Engel C. G. and Fisher R. L.** Lherzolite, anorthosite, gabbro and basalt dredged from the Mid-Indian Ocean Ridge. *Science, N.Y.,* **166**, 1969, 1136–41.

23. **Pearce J. A. and Cann J. R.** Ophiolite origin investigated by discriminant analysis using Ti, Zr and Y. *Earth planet. Sci. Lett.,* **12**, 1971, 339–49.

24. **Kay R. Hubbard N. J. and Gast P. W.** Chemical characteristics and origin of oceanic ridge volcanic rocks. *J. geophys. Res.,* **75**, 1970, 1585–613.

25. **Hsü K. J.** Franciscan mélanges as a model for eugeosynclinal sedimentation and underthrusting tectonics. *J. geophys. Res.,* **76**, 1971, 1162–70.

26. **Gass I. G. and Masson-Smith D.** The geology and gravity anomalies of the Troodos massif, Cyprus. *Phil. Trans. R. Soc.,* **A255**, 1963, 417–67.

27. **Gass I. G.** Is the Troodos massif of Cyprus a fragment of Mesozoic ocean floor? *Nature, Lond.,* **220**, 1968, 39–42.

28. **Moores E. M. and Vine F. J.** The Troodos Massif, Cyprus and other ophiolites as oceanic crust: evaluation and implications. *Phil. Trans. R. Soc.,* **A268**, 1971, 443–66.

29. **Govett G. J. S. and Pantazis Th. M.** Distribution of Cu, Zn, Ni and Co in the Troodos Pillow Lava Series, Cyprus. *Trans. Instn Min. Metall.* (*Sect. B : Appl. earth sci.*), **80**, 1971, B27–46.

30. **Hutchinson R. W.** Genesis of Canadian massive sulphides reconsidered by comparison to Cyprus deposits. *Trans. Can. Inst. Min. Metall.,* **68**, 1965, 286–300.

31. **Hutchinson R. W. and Searle D. L.** Stratabound pyrite deposits in Cyprus and relations to other sulphide ores. In *Proc. IMA–IAGOD meetings '70, IAGOD volume* (Tokyo : Society of Mining Geologists of Japan, 1971), 198–205.

32. **Clark L. A.** Volcanogenic ores : comparison of cupriferous pyrite deposits of Cyprus and Japanese Kuroko deposits. In *Proc. IMA–IAGOD meetings '70, IAGOD volume* (Tokyo : Society of Mining Geologists of Japan, 1971), 206–15.

33. **Constantinou G. and Govett G. J. S.** Genesis of sulphide deposits, ochre and umber of Cyprus. *Trans. Instn Min. Metall.* (*Sect. B : Appl. earth sci.*), **81**, 1972, B34–46.

34. **Tatsumi T. Sekine Y. and Kanehira K.** Mineral deposits of volcanic affinity in Japan : metallogeny. In *Volcanism and ore genesis* **Tatsumi T. ed.** (Tokyo : University of Tokyo Press, 1970), 3–47.

35. **Imai H.** Geology of the Okuki mine and other related cupriferous pyrite deposits in southwestern Japan. *Neues Jb. Min. Abh.,* **94**, 1960, 352–89.

36. **Yamaoka K.** Studies on the bedded cupriferous iron sulfide deposits occuring in the Sanbagawa metamorphic zone. *Sci. Reps Tohoku Univ.,* 3rd Ser., **8**, 1962, 1–68.

37. **Sugisaki R. Tanaka T. and Hattori H.** Rubidium and potassium contents of geosynclinal basalts in the Japanese islands. *Nature, Lond.,* **227**, 1970, 1338–9.

38. **Bryner L.** Ore deposits of the Philippines—an introduction to their geology. *Econ. Geol.,* **64**, 1969, 644–66.

39. **John T. U.** Geology and mineral deposits of east-central Balabac Island, Palawan Province, Philippines. *Econ. Geol.,* **58**, 1963, 107–30.

40. **de Guzman R. A.** Geology and remobilized aspects of the massive sulfide deposits of Port Bicobian, Ilagan, Isabela and other similar Philippine deposits. *Philipp. Geol.,* **22**, 1968, 109–32.

41. **Ivanov S. N.** Sheet-like deposits of pyrite ores of eugeosynclinal regions. In *Proc. IMA–IAGOD meetings '70, IAGOD volume* (Tokyo : Society of Mining Geologists of Japan, 1971), 193–7.

42. **Hamilton W.** The Uralides and the motion of the Russian and Siberian Platforms. *Bull. geol. Soc. Am.,* **81**, 1970, 2553–76.

43. **Church W. R. and Stevens R. K.** Early Paleozoic ophiolite complexes of the Newfoundland Appalachians as mantle–oceanic crust sequences. *J. geophys. Res.,* **76**, 1971, 1460–66.

44. **Williams H.** Mafic ultramafic complexes in western Newfoundland Appalachians and the evidence for their transportation : A review and interim report. *Proc. geol. Ass. Can.,* **24**, 1971, 9–25.

45. **Upadhyay H. D. Dewey J. F. and Neale E. R. W.** The Betts Cove ophiolite complex, Newfoundland : Appalachian oceanic crust and mantle. *Proc. geol. Ass. Can.,* **24**, 1971, 27–34.

46. **Baird D. M.** Massive sulphide deposits in Newfoundland. *Trans. Can. Inst. Min. Metall.,* **63**, 1960, 39–42.

47. **Rose E. R. Sanford B. V. and Hacquebard P. A.** Economic minerals of southeastern Canada. In *Geology and economic minerals of Canada* **Douglas R. J. W. ed.** (Ottawa : Queen's Printer, 1970) 307–64. (*Econ. Geol. Rep. geol. Surv. Can.* no. 1)

48. **Borchert H.** Die Chrom- und Kupfererzlagerstaetten des initialen ophiolitischen Magmatismus in der Türkei. *Maden Tetkik. Arama Enstit. Yayinl.* no. 102, 1958, 189 p.

49. **Gümüs A.** Genesis of some cupreous pyrite deposits of Turkey. In *Symposium on mining geology and the base metals* (Ankara : Office of U.S. Economic Coordinator for CENTO Affairs, 1964), 147–54.

50. **Bailey E. H. Barnes J. W. and Kupfer D. H.** Geology and ore deposits of the Küre district, Kastamonu Province, Turkey. In *CENTO summer training program in geological mapping techniques* (Ankara : Office of

U.S. Economic Coordinator for CENTO Affairs, 1966), 17–33.

51. **Natale P. and Zucchetti S.** Studi sui giacimenti piritoso—cupriferi stratiformi delle Alpi Occidentali. Nota 1 : compendio delle conoscenze attuali sulle piriti stratiformi. *Boll. Ass. Min. Subalp.*, 1, 1964, 49–70.

52. **Moore J. G. and Fabbi B. P.** An estimate of the juvenile sulfur content of basalt. *Contr. Mineral. Petrol.*, 33, 1971, 118–27.

53. **Boström K. and Peterson M. N. A.** Precipitates from hydrothermal exhalations on the East Pacific Rise. *Econ. Geol.*, 61, 1966, 1258–65.

54. **Boström K. and Peterson M. N. A.** The origin of aluminum-poor ferromanganoan sediments in areas of high heat flow on the East Pacific Rise. *Marine Geol.*, 7, 1969, 427–47.

55. **Dasch E. J. Dymond J. R. and Heath G. R.** Isotopic analysis of metalliferous sediment from the East Pacific Rise. *Earth planet. Sci. Lett.*, 13, 1971, 175–80.

56. **Boström K.** *et al.* Aluminium-poor ferromanganoan sediments on active oceanic ridges. *J. geophys. Res.*, 74, 1969, 3261–70.

57. **Boström K. and Fisher D. E.** Volcanogenic uranium, vanadium and iron in Indian Ocean sediments. *Earth planet. Sci. Lett.*, 11, 1971, 95–8.

58. **Horowitz A.** The distribution of Pb, Ag, Sn, Tl, and Zn in sediments on active oceanic ridges. *Marine Geol.*, 9, 1970, 241–59.

59. **Corliss J. B.** The origin of metal-bearing submarine hydrothermal solutions. *J. geophys. Res.*, 76, 1971, 8128–38.

60. **Miller A. R.** *et al.* Hot brines and recent iron deposits in deeps of the Red Sea. *Geochim. cosmochim. Acta*, 30, 1966, 341–59.

61. **Bischoff J. L.** Red Sea geothermal brine deposits : their mineralogy, chemistry, and genesis. In *Hot brines and recent heavy metal deposits in the Red Sea* **Degens E. T. and Ross D. A. eds** (New York : Springer-Verlag, 1969), 368–401.

62. **Skinner B. J.** *et al.* Sulfides associated with the Salton Sea geothermal brine. *Econ. Geol.*, 62, 1967, 316–30.

63. **Biehler S.** Gravity models of the crustal structure of the Salton trough. *Geol. Soc. Am. Ann. Mtgs Absts*, 3, no. 7, 1971, 506.

64. **Kaplan I. R. Sweeney R. E. and Nissenbaum A.** Sulfur isotope studies on Red Sea geothermal brines and sediments. In *Hot brines and recent heavy metal deposits in the Red Sea* **Degens E. T. and Ross D. A. eds** (New York : Springer-Verlag, 1969), 474–98.

65. **Cook H. E.** Iron and manganese rich sediments overlying oceanic basaltic basement, equatorial Pacific, leg 9, D.S.D.P. *Geol. Soc. Am Ann. Mtgs Absts*, 3, no. 7, 1971, 530–1.

66. **Cronan D. S.** *et al.* Iron-rich basal sediments from the eastern equatorial Pacific : Leg 16, Deep Sea Drilling Project. *Science, N.Y.*, 175, 1972, 61–3.

67. **Moore J. G. and Calk L.** Sulfide spherules in vesicles of dredged pillow basalt. *Am. Miner.*, 56, 1971, 476–88.

68. **Dmitriev L. Barsukov V. and Udintsev G.** Rift-zones of the ocean and the problem of ore-formation. In *Proc. IMA–IAGOD meetings '70, IAGOD volume* (Tokyo : Society of Mining Geologists of Japan, 1971), 65–9.

69. **Kuniyoshi S.** Chemical composition of the Karmutsen volcanics, Vancouver Island, British Columbia. *Geol. Soc. Am. Ann. Mtgs Absts*, 3, no. 7, 1971, 628–9.

70. **Horikoshi E.** Volcanic activity related to the formation of the Kuroko-type deposits in the Kosaka district, Japan. *Mineralium Deposita*, 4, 1969, 321–45.

71. **Honnorez J.** La formation actuelle d'un gisement sous-marin de sulfures fumerolliens a Vulcano (mer Tyrrhénienne). Part 1. *Mineralium Deposita*, 4, 1969, 114–31.

72. *Baas Becking geobiological laboratory annual report 1970,* (Canberra, Australia : Bureau of Mineral Resources, 1971), 17 p.

73. **Sillitoe R. H.** Relation of metal provinces in western America to subduction of oceanic lithosphere. *Bull. geol. Soc. Am.*, 83, 1972, 813–17.

74. **Sillitoe R. H.** A plate tectonic model for the origin of porphyry copper deposits. *Econ. Geol.*, 67, 1972, 184–97.

75. **Hutchinson R. W. Ridler R. H. and Suffel G. G.** Metallogenic relationships in the Abitibi belt, Canada : a model for Archean metallogeny. *CIM Bull.*, 64, April 1971, 48–57.

76. **Jakeš P. and Gill J.** Rare earth elements and the island arc tholeiitic series. *Earth planet. Sci. Lett.*, 9, 1970, 17–28.

77. **Karig D. E.** Structural history of the Mariana island arc system. *Bull. geol. Soc. Am.*, 82, 1971, 323–44.

78. **Sclater J. G.** *et al.* Crustal extension between the Tonga and Lau Ridges : petrologic and geophysical evidence. *Bull. geol. Soc. Am.*, 83, 1972, 505–18.

ENVIRONMENTS OF FORMATION OF VOLCANOGENIC MASSIVE SULFIDE DEPOSITS

Introduction

A concensus of current opinion, as expressed by Stanton (1972, Ch. 15), holds that stratiform, pyritic, massive sulfide deposits in sequences of submarine volcanic rocks originated by deposition on, or replacement immediately beneath, the sea floor, in many cases at volcanic centers where submarine fumaroles or hot springs debouched. The formation of massive sulfide deposits may be considered as an integral part of volcanic activity. Deposits of this type are a conspicuous feature of the metallogeny of Phanerozoic terranes and are also abundant in certain Precambrian shield areas.

Despite a measure of accord in favor of a syngenetic, submarine mode of genesis for volcanogenic massive sulfide deposits, widespread confusion still exists concerning the location of their sites of formation in terms of present-day submarine environments. Much of the problem seems to stem from the continued application of imprecise geosynclinal terminology to the description and classification of their formational environments (e.g., de Bretizel and Foglierini, 1971; Gilmour, 1971; Hutchinson and Hodder, 1972; Jenks, 1971). Here is is proposed that the theory of plate tectonics provides a more satisfactory framework into which the sites of formation of Phanerozoic massive sulfide deposits can be inserted.

Submarine Volcanism and Plate Tectonics

In the context of the theory of plate tectonics, four main tectonic environments in which submarine volcanism takes place may be recognized (Fig. 1). These comprise:

(1) Ocean rises (mid-ocean ridges) along divergent plate margins where volcanism is an essential part of the process of accretion of new oceanic crust at the trailing edges of outward-spreading lithospheric plates. Volcanic products are mainly tholeiitic basalts in the form of pillow lavas and minor quantities of hyaloclastites.

(2) Marginal ocean basins, such as those of the western Pacific, where volcanism accompanies rather irregular basin spreading above the deeper parts of subduction zones. Spreading leads to separation of island arcs and the facing continents, or disruption of island arcs themselves to produce interarc basins (Karig, 1970). Volcanic products consist largely of tholeiitic basalts.

(3) Continental margins and island arcs along convergent plate margins where volcanism is active above subduction zones and may largely represent the eruption of material generated by the partial fusion of the upper parts of underthrust slabs of oceanic lithosphere. The calc-alkaline suite is typical of convergent plate margins and consists of basalts, andesites, dacites, and rhyolites; pyroclastic units are particularly common. Island-arc volcanism is more commonly subaqueous than that at continental margins, where submarine activity is normally restricted to the early stages of orogenic development.

(4) Intraplate oceanic island chains in main ocean basins, where volcanism may be an expression of the ascent of deep mantle plumes over which lithospheric plates drift (Morgan, 1972). Sequences of tholeiitic and alkali basalts are typical.

Preliminary chemical data presented by Hart et al. (1972) suggest that tholeiitic basalts from oceanic islands and from the trench side of certain island arcs (Jakeš and Gill, 1970) may be distinguished from those generated at ocean rises and in marginal ocean basins.

Since the theory of plate tectonics seems to provide a firm basis for the interpretation of the geologic record in orogenic belts (Dickinson, 1972), it seems reasonable, applying the principle of uniformitarianism, to extrapolate the observed environments of submarine volcanism back to the beginning of the Paleozoic era. There is still considerable discussion, however, concerning the tectonic regime which operated in the Precambrian and it cannot yet be concluded whether or not a plate-tectonic model, at least of the present type, is applicable. Consequently, this note is restricted to a consideration of Phanerozoic terranes and Precambrian environments are omitted.

Sites of Formation of Phanerozoic Massive Sulfide Deposits

The observed geologic associations of Phanerozoic massive sulfide deposits suggest that their generation was an integral part of volcanic activity at spreading centers in main and in marginal ocean basins, and in island arcs (and perhaps locally at continental margins). The salient features of, and a few examples of deposits in, these environments

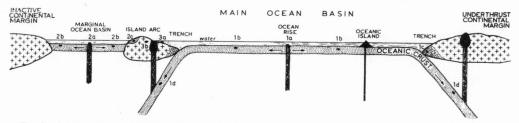

Fig. 1. A schematic model depicting the sites of submarine volcanism, the proposed environments of formation of massive sulfide deposits and the later transport and tectonic emplacement of the deposits. Massive sulfide deposits, shown as small black lenses, form at ocean rises (1), at spreading centers in marginal ocean basins (2), and at island arcs (3). Deposits generated at ocean rises (1a) and in marginal ocean basins (2a) are transported laterally (1b and 2b) during sea-floor spreading and underthrust at subduction zones where these are present. Deposits may be subducted (1d) or scraped off at the continental margin and incorporated in subduction mélanges behind and beneath the trench (1c). Deposit 2c was formed in a marginal ocean basin and emplaced as part of an obducted ophiolite slab during a subsequent arc/arc collision. Deposits formed in shallow water above subduction zones at island arcs (3a) may be buried during subsequent episodes of volcanism and related sedimentation (3b).

are now given (Fig. 1). No evidence supports the formation of massive sulfide deposits in the vicinities of oceanic islands, although this situation cannot yet be excluded as a potential site for massive sulfide formation.

Spreading-center situations

Ophiolite complexes along ancient or active convergent plate margins have been shown to represent allochthonous slabs and slivers of oceanic lithosphere that have been mechanically incorporated into continents and island arcs (Coleman, 1971; Dewey and Bird, 1971; Moores and Vine, 1971). Emplacement of fragments of oceanic lithosphere seems to have occurred behind and beneath ocean trenches during subduction and larger slices appear to have been obducted (overthrust) onto continental margins or island arcs during their collision with subduction zones dipping away from them (Dewey and Bird, 1971). The generation and subsequent obduction or subduction of oceanic lithosphere may take place in either main or marginal ocean basins (Dewey and Bird, 1971; Karig, 1972), but methods for discriminating between these two environments are still only in their infancy.

Massive sulfide deposits are found as integral components of sequences of tholeiitic basalts in ophiolite complexes at several localities on the continents. The deposits were probably formed from hydrothermal fluids related to volcanism at spreading centers and were transported to their present positions largely by sea-floor spreading (Fig. 1). Further details of this class of massive sulfide deposit and suggestions concerning genesis have been presented elsewhere (Sillitoe, 1972).

Massive sulfide deposits of ophiolite affiliation are probably best represented by the little-deformed examples in the Troodos complex of Cyprus. Evi-

dence that the Cyprus deposits were formed prior to the deposition of a series of sediments (Perapedhi Formation), which probably corresponds to the sedimentary layer 1 of the oceanic crust, supports an origin at an ocean rise, probably one active in the Tethyan ocean (Gass, 1968). Emplacement of the slab-like Troodos complex was considered by Moores and Vine (1971) to have occurred during the collision of the African plate with a north-dipping subduction zone. Most of the Cyprus deposits are characterized by massive ore, consisting largely of pyrite with lesser quantities of chalcopyrite and subordinate sphalerite, overlain by iron oxide-rich mudstones and underlain by pyritic stockworks representing the conduits up which hydrothermal fluids were supplied (Hutchinson and Searle, 1971; Constantinou and Govett, 1972). The high Cu : Zn ratio of the Cyprus ores seems to be typical of massive sulfide deposits of ophiolite affiliation, although two deposits carrying high Zn values are known in Cyprus.

In the Notre Dame Bay area of Newfoundland, massive sulfide deposits similar to those in Cyprus, but more highly deformed, occur in basaltic layer 2 of ophiolite complexes and include those at Tilt Cove, Little Bay, and Whalesback (Williams, 1963). Dewey and Bird (1971) made a strong case for the origin of the Newfoundland ophiolites in marginal ocean basins behind early Paleozoic island arcs on the northwest side of a proto-Atlantic Ocean. Church (1972) has, however, stressed that until more definitive evidence is available, both marginal ocean basins and ocean rises must be considered as possible sites for the generation of the Newfoundland ophiolites.

The Girilambone flysch in New South Wales is considered to have been deposited in a Cambrian trench on the eastern margin of the Australian-Antarctic continent and contains slivers of ophiolites generated by spreading in the paleo-Pacific Ocean (Scheibner, 1972). The basalts in these ophiolites host copper-bearing massive sulfide deposits (R. Skrzeczynski, pers. commun., 1972).

In southwest Japan, highly metamorphosed and deformed massive sulfide deposits are found in belts of late Paleozoic and Mesozoic ophiolites and sedimentary rocks (Kanehira and Tatsumi, 1970) and were perhaps formed at spreading centers. At least some of the ophiolitic rocks in these belts are considered as scraps of oceanic lithosphere obducted onto the late Paleozoic continental edge of Japan and so juxtaposed with a prism of continental rise sediments (Ernst, 1972).

Massive sulfide deposits in ophiolite complexes in the Philippines, the Urals, Turkey, and northern Italy have also been tentatively interpreted as products of volcanism at spreading centers (Sillitoe, 1972) and occurrences in Corsica (Bouladon and Picot, 1968) may be added.

Island-arc (and continental-margin) situations

Stanton (1955) was the first to recognize that certain massive sulfide deposits were formed in ancient island arcs that are now known to have been generated at convergent plate margins. As recently mentioned by several workers (Mitchell and Garson, 1972; Sawkins, 1972; Sillitoe, 1972), the Miocene Kuroko deposits in northeast Japan are good examples of massive sulfide deposits at convergent plate margins. The Kuroko deposits are relatively undeformed and are found intimately associated with calc-alkaline rocks in a currently active island arc. Typical deposits are related to dacitic or rhyolitic domes and explosive vents emplaced in a shallow marine environment and comprise, in descending sequence, zones of syngenetic massive ore enriched in barite, barite-zinc-lead-copper-silver, and copper-pyrite, and root zones characterized by silicified stockwork and disseminated copper-pyrite mineralization that represent the feeder channels for the hydrothermal fluids. Massive gypsum-anhydrite zones are associated with the lower parts of ore bodies (Matsukuma and Horikoshi, 1970).

A linear belt of subaqueous rhyolitic, and some spilitic, tuffs and lavas of Upper Paleozoic age extends from southwest Spain to southern Portugal. Many stratiform, massive sulfide deposits containing copper, lead, zinc, silver, and gold, such as those at Rio Tinto and Tharsis (Williams, 1966), form an integral part of the volcanic sequence. The belt may represent part of an ancient island arc that was incorporated into the Eurasian plate by collision or part of an Andean-type continental margin orogenic belt (Nicolas, 1972).

In eastern Australia, Cambrian (Mt. Lyell),

Ordovician-Silurian (Captains Flat), and Permian (Halls Peak) massive sulfide deposits carrying copper, lead, zinc, silver, gold, and barium occur in calc-alkaline volcanic piles that were formed above subduction zones at various stages in the development of the Tasman orogenic zone (Solomon et al., 1972).

As a final example of a district possessing massive sulfide deposits of island-arc origin, Buchans in Newfoundland may be cited. The deposits are related to a sequence of Lower Paleozoic andesites and dacites and contain sphalerite, galena, chalcopyrite, pyrite, and barite with significant silver and gold contents (Swanson and Brown, 1962). The proximity of Buchans to the Newfoundland deposits of ophiolite affiliation, mentioned above, may be ascribed to the closure of an intervening main or marginal ocean basin by subduction.

Although at this stage the exact sites of formation of massive sulfide deposits in island arcs cannot be located with any precision, it is suggested that appropriate submarine environments might include the trench sides of marginal ocean basins (including interarc basins), the slopes between island arcs and trenches (arc-trench gaps), and areas between island-arc volcanoes.

It is evident from the examples cited above that the lead, zinc, silver, and barium contents of deposits formed at convergent plate margins exceed those found in deposits formed at spreading centers. This may reflect the concentration of these elements produced by processes operating during partial melting on subduction zones.

A further possibility is that massive sulfide deposits are formed in association with island-arc tholeiites. If this is so, then such deposits may well be characterized by the metal content found in spreading-center deposits rather than by that typical of deposits generated with the calc-alkaline suite. This is because island-arc tholeiites are probably formed by the partial fusion, due to hydration, of ultrabasic rocks at shallow depths above subduction zones and not by the partial fusion of subducted oceanic crust (Nicholls and Ringwood, 1972).

Modern Analogues of Massive Sulfide Deposits

The copper- and zinc-enriched muds overlying layer 2 of the oceanic crust along a spreading center in the Red Sea have been compared with the Cyprus deposits (Hutchinson and Searle, 1971) and may be genetically related to massive sulfide deposits in ophiolite complexes. It has been suggested by Blissenbach (1972), however, that metalliferous muds of the Red Sea-type might only accumulate at ocean rises during the initial stages of ocean formation, when thick marine evaporite sequences on the nearby ocean margins can act as a source for brines to leach the metals from still-hot layer-2 basalts. Moreover, the Cu:Zn ratio of the Red Sea metal accumulations is much lower than that of most massive sulfide deposits of ophiolite affiliation and might suggest that their environment of formation is somewhat different from that at spreading centers during massive sulfide generation.

A modern analogue of Phanerozoic massive sulfide deposits of island-arc type has yet to be located, although at Matupi Harbor, New Britain, thermal springs discharge into a near-shore environment above a subduction zone and give rise to the enrichment of sediments in iron, manganese, and zinc (Ferguson and Lambert, 1972).

Transport and Deformation of Massive Sulfide Deposits

Many massive sulfide deposits have undergone extensive metamorphism and deformation, a feature which has commonly caused severe difficulties in understanding the genesis of this class of deposit. I would like to try to show here, in a very general way, that the processes of transport and deformation of Phanerozoic massive sulfide deposits can be more easily predicted, but are potentially more complex, than those recently described by Jenks (1971), and that the deformational histories of deposits formed at spreading centers are likely to be distinct from those formed at convergent plate margins. A specific treatment of the structural, textural, and mineralogic modifications induced during metamorphism (e.g., Vokes, 1969) is not attempted.

Massive sulfide deposits formed in any of the proposed geotectonic environments may undergo submarine sliding and brecciation immediately following their formation while still in a relatively unconsolidated state (e.g., Kuroko deposits of Japan; Matsukuma and Horikoshi, 1970, and upper Paleozoic deposits of southwest Spain-Portugal; Schermerhorn, 1970).

Deposits formed at spreading centers may be subjected to zeolite and greenschist facies metamorphism induced by high heat flow beneath ocean rises (Miyashiro, 1972). They may also undergo block faulting and the local development of tectonite fabrics by transform faulting (Dewey and Bird, 1971) during lateral transport away from ocean rises. These phenomena are unrelated to orogenic processes acting at the final sites of emplacement of the ore deposits. Recently Williams and Malpas (1972) have ascribed greenschist facies metamorphism in the allochthonous Bay of Islands ophiolite complex (which contains massive sulfide deposits) in Newfoundland to thermal processes operative at the

time of its formation at an early Paleozoic spreading center.

Upon their approach to convergent plate margins, deposits generated at ocean rises may be deformed during obduction or subduction along with the enclosing oceanic crust (Fig. 1), and obducted slabs, perhaps faulted and brecciated during tectonic emplacement, may continue their travels during uplift by gravity sliding as parts of olistostromes (Dewey and Bird, 1971; Church, 1972). Subducted deposits that eventually reappear at surface due to the effects of isostatic uplift or faulting and consequent erosion may have suffered blueschist facies metamorphism (e.g., some of the late Paleozoic deposits in southwest Japan) and have been extensively disrupted by incorporation in subduction mélanges.

Deposits formed at convergent plate margins may also be structurally disturbed and undergo thermal metamorphism during related, but subsequent, volcanic and intrusive events. These deposits are also susceptible to extensive deformation and metamorphism if involved in the splintering, thickening, and nappe transport of continental crust during the collision of island arcs and continents. Furthermore, older deposits, irrespective of their environments of formation, are likely to have been involved in several orogenic events, attributable to processes attendant upon later episodes of subduction and collision. The Upper Ordovician Gullbridge deposit in Newfoundland which was formed at a spreading center and subsequently underwent thermal metamorphism and deformation during Devonian plutonism (Upadhyay and Smitheringale, 1972) may be cited as an example.

Conclusions

This note presents a preliminary scheme for use in the subdivision of volcanogenic massive sulfide deposits of Phanerozoic age, based on present-day environments of submarine volcanism which are classified in terms of the theory of plate tectonics. It is proposed that Phanerozoic massive sulfide deposits were formed, and are probably still forming, with tholeiitic basaltic volcanics at spreading centers in main oceans (at ocean rises) and in marginal ocean basins, and with calc-alkaline volcanics at island arcs (and perhaps also at continental edges) along convergent plate margins.

Despite the many problems which still exist, it is contended that a scheme of this sort is potentially useful in directing exploration for massive sulfide deposits both on the continents and under the oceans. The solution to some of the existing problems and the further refinement of the scheme may be facilitated by elucidation of the sequence and nature of rock types beneath marginal ocean basins and by the development of diagnostic field and chemical criteria for the recognition of environments of formation of submarine basaltic volcanic rocks, which are hosts to massive sulfide deposits, in orogenic belts. Once the tectonic regime that operated in the Precambrian is better understood, it is hoped that a comparable scheme can be devised for pre-Phanerozoic massive sulfide deposits.

Acknowledgments

Thanks are due to Drs. C. Halls, R. Stabbins, and F. J. Vine for reading the manuscript. The note was prepared while the writer was supported by a Shell postdoctoral Research Fellowship at the Royal School of Mines, Imperial College, London.

RICHARD H. SILLITOE
DEPARTMENT OF MINING GEOLOGY
ROYAL SCHOOL OF MINES
IMPERIAL COLLEGE OF SCIENCE AND TECHNOLOGY
LONDON SW7 2BP, ENGLAND
January 30, May 2, 1973

REFERENCES

Blissenbach, E., 1972, Continental drift and metalliferous sediments, *in* Oceanology Internat. 72 Conf. Papers: Brighton, England, p. 412–416.

Bouladon, J., and Picot, P., 1968, Sur les minéralisations en cuivre des ophiolites de Corse, des Alpes françaises et de Ligurie: Bur. Recherches Geol. Min. Bull., ser. 2, no. 1, p. 23–41.

Church, W. R., 1972, Ophiolite: its definition, origin as oceanic crust, and mode of emplacement in orogenic belts, with special reference to the Appalachians: Pub. Earth Phys. Branch, Dept. Energy Mines Res., Ottawa, Canada, v. 42, no. 3, p. 71–85.

Coleman, R. G., 1971, Plate tectonic emplacement of upper mantle peridotites along continental edges: Jour. Geophys. Research, v. 76, p. 1212–1222.

Constantinou, G., and Govett, G. J. S., 1972, Genesis of sulphide deposits, ochre and umber of Cyprus: Inst. Mining Metallurgy Trans., v. 81, sec. B, p. 34–46.

de Bretizel, P., and Foglierini, F., 1971, Les gites sulfurés concordants dans l'environnement volcanique et volcano-sédimentaire: Mineralium Deposita, v. 6, p. 65–76.

Dewey, J. F., and Bird, J. M., 1971, Origin and emplacement of the ophiolite suite: Appalachian ophiolites in Newfoundland: Jour. Geophys. Research, v. 76, p. 3179–3206.

Dickinson, W. R., 1972, Evidence for plate-tectonic regimes in the rock record: Am. Jour. Sci., v. 272, p. 551–576.

Ernst, W. G., 1972, Possible Permian oceanic crust and plate junction in central Shikoku, Japan: Tectonophysics, v. 15, p. 233–239.

Ferguson, J., and Lambert, I. B., 1972, Volcanic exhalations and metal enrichments at Matupi Harbor, New Britain, T.P.N.G.: ECON. GEOL., v. 67, p. 25–37.

Gass, I. G., 1968, Is the Troodos massif of Cyprus a fragment of Mesozoic ocean floor?: Nature, v. 220, p. 39–42.

Gilmour, P., 1971, Strata-bound massive pyritic sulfide deposits—a review: ECON. GEOL., v. 66, p. 1239–1244.

Hart, S. R., Glassley, W. E., and Karig, D. E., 1972, Basalts and sea floor spreading behind the Mariana island arc: Earth Planet. Sci. Letters, v. 15, p. 12–18.

Hutchinson, R. W., and Hodder, R. W., 1972, Possible tectonic and metallogenic relationships between porphyry copper and massive sulphide deposits: Canadian Mining Metall. Bull., v. 65, no. 718, p. 34–40.

Hutchinson, R. W., and Searle, D. L., 1971, Stratabound pyrite deposits in Cyprus and relations to other sulphide ores, *in* Takeuchi, Y., ed., Proc. IMA-IAGOD Mtgs. '70,

IAGOD Vol.: Soc. Mining Geol. Japan Spec. Issue no. 3, Tokyo, p. 198–205.

Jakeš, P., and Gill, J., 1970, Rare earth elements and the island arc tholeiite series: Earth Planet. Sci. Letters, v. 9, p. 17–28.

Jenks, W. F., 1971, Tectonic tranport of massive sulfide deposits in submarine volcanic and sedimentary host rocks: ECON. GEOL., v. 66, p. 1215–1224.

Kanehira, K., and Tatsumi, T., 1970, Bedded cupriferous iron sulphide deposits in Japan, *in* Tatsumi, T., ed., Volcanism and ore genesis: Tokyo, Univ. Tokyo Press, p. 51–76.

Karig, D. E., 1970, Ridges and basins of the Tonga-Kermadec island arc system: Jour. Geophys. Research, v. 75, p. 239–255.

—— 1972, Remnant arcs: Geol. Soc. America Bull., v. 83, p. 1057–1068.

Matsukuma, T., and Horikoshi, E., 1970, Kuroko deposits in Japan, a review, *in* Tatsumi, T., ed., Volcanism and ore genesis: Tokyo, Univ. Tokyo Press, p. 153–179.

Mitchell, A. H. G., and Garson, M. S., 1972, Relationship of porphyry copper and circum-Pacific tin deposits to palaeo-Benioff zones: Inst. Mining Metallurgy Trans., v. 81, sec. B, p. 10–25.

Miyashiro, A., 1972, Pressure and temperature conditions and tectonic significance of regional and ocean-floor metamorphism, *in* Ritsema, A. R., ed., The upper mantle: Tectonophysics, v. 13, p. 141–159.

Moores, E. M., and Vine, F. J., 1971, The Troodos massif, Cyprus and other ophiolites as oceanic crust: evaluation and implications: Royal Soc. [London] Philos. Trans., v. A268, p. 443–466.

Morgan, W. J., 1972, Deep mantle convection plumes and plate motions: Amer. Assoc. Petroleum Geologists Bull., v. 56, p. 203–213.

Nicholls, I. A., and Ringwood, A. E., 1972, Production of silica-saturated tholeiitic magmas in island arcs: Earth Planet. Sci. Letters, v. 17, p. 243–246.

Nicolas, A., 1972, Was the Hercynian orogenic belt of Europe of the Andean type?: Nature, v. 236, p. 221–223.

Sawkins, F. J., 1972, Sulfide ore deposits in relation to plate tectonics: Jour. Geology, v. 80, p. 377–397.

Scheibner, E., 1972, Actualistic models in tectonic mapping: Int. Geol. Cong., 24th, Montreal, sec. 3, p. 405–422.

Schermerhorn, L. J. G., 1970, The deposition of volcanics and pyritite in the Iberian pyrite belt: Mineralium Deposita, v. 5, p. 273–279.

Sillitoe, R. H., 1972, Formation of certain massive sulphide deposits at sites of sea-floor spreading: Inst. Mining Metallurgy Trans., v. 81, sec. B., p. 141–148.

Solomon, M., Groves, D. I., and Klominsky, J., 1972, Metallogenic provinces and districts in the Tasman orogenic zone of eastern Australia: Australasian Inst. Mining Metallurgy Proc., no. 242, p. 9–24.

Stanton, R. L., 1955, Lower Paleozoic mineralization near Bathurst, New South Wales: ECON. GEOL., v. 50, p. 681–714.

—— 1972, Ore petrology: New York, McGraw-Hill, Inc., 713 p.

Swanson, E. A., and Brown, R. L., 1962, Geology of the Buchans orebodies: Canadian Inst. Mining Metallurgy Trans., v. 55, p. 288–296.

Upadhyay, H. D., and Smitheringale, W. G., 1972, Geology of the Gullbridge copper deposit, Newfoundland: Volcanogenic sulfides in cordierite-anthophyllite rocks: Canadian Jour. Earth Sci., v. 9, p. 1061–1073.

Vokes, F. M., 1969, A review of the metamorphism of sulphide deposits: Earth-Sci. Rev., v. 5, p. 99–143.

Williams, D., 1966, Volcanism and ore deposits: Freiberg. Forschungshefte, v. C210, p. 93–111.

Williams, H., 1963, Relationship between base metal mineralization and volcanic rocks in northeastern Newfoundland: Canadian Mining Jour., v. 84, Aug., p. 39–42.

—— and Malpas, J., 1972, Sheeted dikes and brecciated dike rocks within transported igneous complexes, Bay of Islands, western Newfoundland: Canadian Jour. Earth Sci., v. 9, p. 1216–1229.

23

Reprinted from *Econ. Geol.* **68**:1223-1225, 1230-1234, 1240-1246 (1973)

Volcanogenic Sulfide Deposits and Their Metallogenic Significance

R. W. HUTCHINSON

Abstract

Volcanogenic sulfide deposits may be described as stratabound, lenticular bodies of massive pyritic mineralization, containing variable amounts of chalcopyrite, sphalerite, and galena in layered volcanic rocks. Often they are found to be immediately overlain by thin-bedded siliceous and iron-rich sedimentary rocks, and they are commonly underlain by extensive zones of altered, sulfide-impregnated lava. They are believed to have formed subaqueously by volcanic-fumarolic activity which occurred periodically during volcanism.

Three distinct varieties of such deposits can be distinguished by their compositions, relative and absolute ages, and rock associations. Pyrite-sphalerite-chalcopyrite bodies are found in differentiated, mafic-to-felsic volcanic rocks; pyrite-galena-sphalerite-chalcopyrite bodies occur in more felsic, calc-alkaline volcanic rocks, and pyrite-chalcopyrite bodies occur in mafic, ophiolitic volcanic rocks.

The time-tectonic-stratigraphic interrelationships of these varieties can be related to evolutionary processes of crustal development. Deposits of the pyrite-sphalerite-chalcopyrite variety are numerous, important, and best developed in Archean greenstones, suggesting that they were generated under conditions of thin "proto-crust," possibly by degassing of as yet poorly differentiated "proto-mantle." Although they recur in younger volcanic successions, they become scarcer and smaller in later geologic time. However, their place is taken by the two other varieties which are notably rare or absent in the Archean. Thus, in Proterozoic volcanic rocks the pyrite-galena-sphalerite-chalcopyrite type appears. In Phanerozoic orogens, pyrite-chalcopyrite bodies are common in the ophiolites that typify an early stage of orogenic activity and are probably generated in oceanic ridge-rift environments during the initial stages of separation of continental crustal blocks. Both of the earlier varieties reappear in Phanerozoic belts; the pyrite-sphalerite-chalcopyrite type in early stages of subduction along continent margins, and the pyrite-galena-sphalerite-chalcopyrite type in later, more felsic, calc-alkaline volcanics that characterize somewhat later tectonism.

In addition to their obvious application in mineral exploration, these concepts may have certain scientific applicability. If, on a very broad or general basis, mineral deposits (like fauna) are products of evolutionary change, then in the absence of fossils or other correlation aids they might be used as gross-scale indicators for correlation or age determination purposes. This could be particularly applicable to highly metamorphosed Precambrian terrane throughout the world.

Introduction

A mind nimble and versatile enough to catch the resemblance of things, which is the chief point, and at the same time steady enough to fix and discern their subtle differences: . . . Francis Bacon in "The Scientific Mind."

WALDEMAR LINDGREN (1919, p. 819) considered certain massive base metal sulfide bodies to be "the most enigmatic of ore deposits." Subsequent work has substantiated Lindgren's view but still has not entirely resolved the enigma to which he referred. More recently, H. C. Gunning (1959) considered the question of the origin of these ores one of the major unresolved problems facing ore deposits geologists.

Differences among massive base metal sulfide deposits have previously been noted and discussed. Stanton (1958) was one of the earliest to present data on the different base metal contents of various deposits. Both Saager (1967) and Waltham (1968) classified the numerous deposits of the early Paleozoic Norwegian Caledonide belt into various types based mainly on their metal content, sulfide mineralogy, and texture. Gräbe (1972) attempted to explain the differences in base metal content and mineralogy by variations in surficial processes, including differences in their paleotopographic, lithofacies, and Eh environments of deposition. Gilmour (1971) related differences in the deposits to differing conditions of sedimentation or erosion, to differing petrochemistry and character of igneous activity, and to evolutionary tectonics, an approach similar to that followed here.

One purpose of this paper is to review briefly recent ideas concerning the origin of certain of these ore bodies and to summarize the broad similarities which identify them as members of a major family. A second purpose is to outline a number of important differences among these deposits that permit division of the major family into distinct types. A final purpose is to consider the significance of these differences and to interpret them in relationship to evolutionary processes of crustal tectonism and related volcanic activity.

Volcanogenic Massive Base Metal Sulfide Deposits

Latest studies of the massive, pyritic base metal sulfide deposits in volcanic rocks suggest that they are of volcanogenic origin, formed in recurrent episodes of sea-floor fumarolic activity during prolonged periods of subaqueous volcanism (Horikoshi, 1969, p. 322; Schermerhorn, 1970, p. 277–278). All members of this major volcanogenic family of ore deposits are therefore broadly related, comparable to one another, and identifiable by their common characteristics. All occur as irregularly shaped but stratabound lenses of dense pyritic sulfide containing various amounts of zinc, copper, and lead as well as important silver and gold in layered subaqueous volcanic sequences. These lenses are accompanied by two other types of mineralization. The lenses are commonly overlain by thin, siliceous iron- and manganese-rich sedimentary strata (Hutchinson and Searle, 1971, p. 199 and fig. 3; Anderson and Nash, 1972, p. 853–855) and are usually underlain by extensive zones of altered and sulfide-impregnated lava (Purdie, 1967; Suffel, 1948, p. 760). Metal zoning is prominent and the stratigraphic tops of the massive lenses are pyritic and zinc- or zinc-lead rich, whereas their stratigraphic bottoms are copper-rich and often pyrrhotitic (Price and Bancroft, 1948, p.

749–750; Scott, 1948, p. 774; Hutchinson and Searle, 1971, p. 202; Horikoshi and Sato, 1970, p. 188; Anderson and Nash, 1972, p. 855 and fig. 6). So consistent is this relationship that it apparently provides a reliable means of top determination in deformed sequences, as, for example, in the Ming deposit at the Rambler Mine in Baie Verte, Newfoundland where zinc-rich mineralization occurs on the footwall of the body (Heenan and Truman, 1972, p. 3), but where structural information suggests that the section is overturned (Kennedy, 1972). Primary breccia textures are also common, particularly at the tops of the massive bodies. These are apparently related to some form of irruptive volcanic activity, such as steam explosions (Horikoshi, 1969, p. 322) in the underlying lavas, with accompanying fragmentation of earlier deposited sulfides. This was apparently followed by redeposition involving turbidite activity, gravity-induced sedimentary slump, and downward movement on the old volcanic paleoslope (Schermerhorn, 1970, p. 276). Sharp contacts are also remarkably common and may be present at the top of the massive bodies, within the bodies between the zinc-lead- and copper-rich massive mineralization, or at the base of the massive lenses (Hutchinson and Hodder, 1970, p. 35; Hutchinson, 1965, p. 976 and 985).

These primary relationships, which characterize all members of the family, are commonly disrupted and complicated to varying degrees or even obliterated by subsequent metamorphism (Suffel et al., 1971; Vokes, 1969; p. 1130; Kalliokoski, 1965). Where metamorphism has been minimal, however, evidence for a synvolcanic or intravolcanic origin is compelling. This evidence includes soft-sediment slump and deformation structures in the overlying sedimentary strata in which the sulfides clearly behaved as heavy, primary components and also graded bedding within pyrite-rich beds (Hutchinson and Searle, 1971, p. 201). Brecciation in the tops of the massive lenses and in the underlying lavas is of "primary synvolcanic" rather than "tectonic" generation (Sinclair, 1970); primary structures, for example pillows in the nearby lavas, although fractured and brecciated, are not otherwise dislocated (Duke, 1971, p. 74). Moreover, the ores in undeformed successions, as on the north flank of the Troodos complex in Cyprus are cut by postmineralization dikes (Hutchinson and Searle, 1971, p. 201; Bear, 1960, p. 97, 99). These are chilled against ore and occasionally can be traced upward where they become feeders to fresh lavas overlying the sulfide bodies. The dikes are petrographically similar to the overlying lavas, and commonly the lavas above massive ore bodies are extremely fresh, unaltered, and barren of sulfides, as in Cyprus and at the Lake Dufault mine near Noranda,

BASE METAL TYPE	PRECIOUS METAL ASSOCIATION	ASSOCIATED VOLCANIC ROCK TYPES	TYPE OF VOLCANISM	TYPE OF SEDIMENTATION	TECTONISM	AGE	EXAMPLES
① Zn-Cu-pyrite	both Au (with high Cu) and Ag (with high Zn)	- fully differentiated suites of intermediate bulk composition(?); -tholeiitic to calc-alkaline -basalt-andesite-dacite-rhyolite, etc.	-initial deep, subaqueous mafic platform; with differentiation toward felsic volcanism, building domical centres	-chemical; cherts, iron formations -clastic; immature, first cycle, volcanogenic greywackes, volcanoclastics	- early eugeosynclinal-orogenic stage; - major subsidence	Archean / Proterozoic(?)	Timmins, Ont. Noranda, Que. / United Verde, Ariz.
Zn-Cu-pyrite	ditto	ditto	ditto	ditto	ditto; early subduction	pre-Ordovician mid-Devonian	Rambler, Nfld. W.Shasta, Calif.
② Pb-Zn-Cu-pyrite	mainly Ag	-intermediate to felsic calc-alkaline volcanic suites; -andesite-dacite-rhyolite-porphyry-crystal tuff, etc.	-felsic centres of explosive, pyroclastic and ignimbritic activity; subaqueous to subaerial	-epiclastic predominates; immature volcanogenic greywackes, manganiferous shales, graphitic shales and argillites, siltstones -chemical minor, cherts, iron formations -sulphate gangues common	- later eugeosynclinal-orogenic stage; infilling with uplift balances subsidence(?)	Proterozoic / Ordovician	Mt. Isa, Queensland Errington, Vermilion, (Sudbury Basin) / Bathurst, New Brunswick
Pb-Zn-Cu pyrite	ditto	ditto	ditto	ditto	ditto; later subduction	Triassic Tertiary	E. Shasta, Calif. Kuroko, Japan
③ Cu-pyrite	mainly Au	- poorly differentiated mafic-ultramafic(ophiolitic)suites, -tholeiitic -basaltic pillow lavas, serpentinite, etc.	-deep subaqueous, quiescent fissure eruptions	-chemical predominates, cherts, ironstones, manganstones -clastic insignificant	-early stage of continental plate rifting; tension, separation, graben	l-Ordovician u-Cretaceous Jura-Cretaceous Cret-Eocene	W.Newfoundland Cyprus Island Mountain, California Phillipines

FIG. 1. Comparative Chart: Some geological characteristics of different volcanogenic sulfide deposits.

Quebec (Hutchinson and Searle, 1971, p. 201; Purdie, 1967).

Differences Among Volcanogenic Massive Base Metal Sulfides

Despite these broad, unifying similarities of all members of the family, there are readily recognizable differences among the various deposits which serve to divide the family into three distinct types. The distinguishing geological features include: first, the base metal and precious metal assemblages present in the ores themselves; second, the petrographic character of the associated volcanic rocks and the inferred nature of the volcanism that deposited them; third, the nature and proportion of intercalated sedimentary strata within the host volcanic succession; fourth, the tectonic environment in which sulfide bodies, volcanic rocks, and sedimentary rocks alike were originally laid down; and finally, but not least, their greatest abundances through time and also their ages relative to one another within Phanerozoic orogenic belts. These criteria are tabulated in Figure 1 which also lists examples of the types.

[*Editor's Note:* Material has been omitted at this point.]

Comparison of Subvarieties and Interpretation

If, as suggested, these ores are all members of a major family distinguishable by its coeval volcanic affiliations, then differences among the three types should be relatable to differences in volcanic petrochemistry. Moreover, if the age relationships are correct or even approximately so, the variations among the ores should be explainable by evolutionary changes in volcanism, its style, and its products. These changes in volcanism, in turn, result presumably from evolutionary tectonism. Conversely, it follows that the ores themselves should provide another clue to the processes of evolutionary volcanism and tectonism that have affected the earth's crust through time and during the development of Phanerozoic orogenic belts.

Based on all the foregoing geological relationships, it is suggested that the zinc-copper type was formed under relatively thin, primitive crustal conditions. The associated volcanic rocks were derived from magmatic activity in an underlying, relatively thick, and poorly differentiated upper mantle that may have been of intermediate "average composition" (Baragar and Goodwin, 1969, p. 140). These conditions are diagrammatically represented in Figure 2. Magmas originating in this mantle during prolonged periods of Keewatin volcanism therefore underwent a complete cycle (or cycles) of differentiation. In this manner, the characteristic Archean volcanic successions were laid down, commencing with widespread basaltic lavas of the mafic platform and progressing toward felsic, domal activity in the later differentiation stages (Goodwin, 1965; 1968). Judg-

ing from the extent and repetition of these successions in Archean terrane throughout the world (Anhaeuser et al., 1969, p. 2192), these conditions must have prevailed over much of the earth's surface during Archean time, and the large and numerous zinc-copper sulfide deposits that are so characteristic of these rocks in Canada may represent the products of extensive degassing (Cloud, 1972, p. 543) of the primitive upper mantle on a scale not since repeated. Presumably it was by these processes, combined with subsequent Kenoran orogeny and its attendant granitic intrusion and gneissification, that earth generated a thick, supracrustal sialic plate by the close of Archean time.

It is suggested that in Proterozoic time eruptive volcanism breaching this thickened continental plate became rarer and more restricted, whereas epiclastic sedimentation involving degradation of the plate became more extensive and important. However, in deeper, eugeosynclinal portions of what were perhaps broad, rift-controlled basins (Card and Hutchinson, 1972, p. 69) or eugeosynclinal belts marginal to old Archean cratonic blocks (Hoffman et al., 1970), volcanism from the now differentiated, ultramafic mantle produced tholeiitic and basaltic lavas (Dimroth et al., 1970, p. 45; Frarey and Roscoe, 1970, p. 146). Locally, in shallower shelves flanking these basins and geosynclines, volcanism, presumably derived from anatectic melting of the sialic plate, occurred in centers of more felsic, explosive, and possibly even subaerial eruption (Stevenson, 1971; Williams, 1957; Sauvé, 1953). The latter conditions are diagrammatically illustrated in Figure 3. Here the resulting volcanic sequences were of more felsic, calc-alkaline

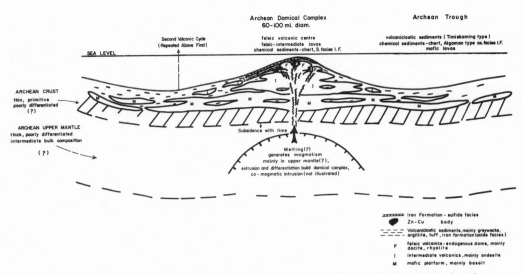

Fɪɢ. 2. Diagrammatic illustration: Zn-Cu type of volcanogenic massive sulfide deposit in Archean tectonic setting.

composition and contained a greater proportion of fine epiclastic sedimentary strata which graded laterally into increasingly coarse clastic shelf-facies rocks deposited at the margins of the subsiding basins (Dimroth et al., 1970, p. 67–68). The filling in of these basins and troughs by combined volcanism and

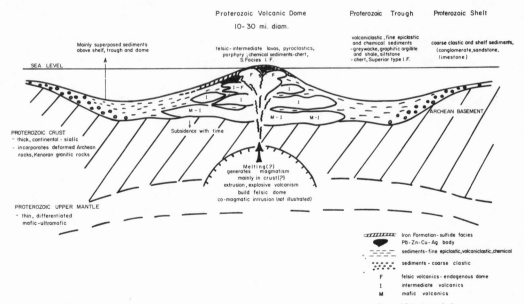

Fɪɢ. 3. Diagrammatic illustration: Pb-Zn-Cu-Ag type of volcanogenic massive sulfide deposit in Proterozoic tectonic setting.

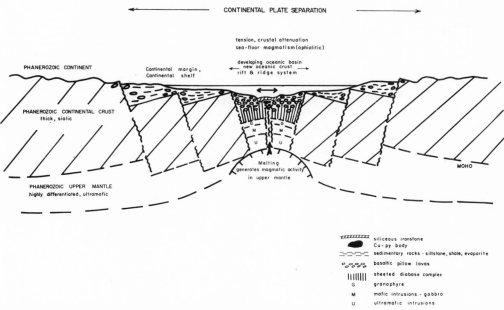

FIG. 4. Diagrammatic illustration: Cupreous pyrite type of volcanogenic massive sulfide deposit in Phanerozoic, separating oceanic rift environment.

sedimentation more or less kept pace with subsidence. These conditions were favorable for the generation of the second lead-zinc-copper-silver type of volcanogenic massive base metal sulfide deposits.

Finally, it is suggested that in still later Phanerozoic time, plate-tectonic activity in its presently known form (Isacks et al., 1968), accompanied by crustal attenuation, rifting, and separation of the rifted plates, created conditions favorable for generation of the third type of volcanogenic massive base metal sulfide deposits. The conditions are illustrated diagrammatically in Figure 4. Cupreous pyrite deposits formed under these tectonic conditions along the developing oceanic rift-ridge systems (Hutchinson and Searle, 1971, p. 198; Smitheringale, 1972, p. 586) from sea-floor fumarolic activity associated with ophiolitic magmatism. The magmatism included eruptive activity, which deposited the pillowed, spilitic basalts and intercalated radiolarian cherts and siliceous ironstones, subvolcanic feeder-dike intrusion which formed the sheeted diabase complexes, and deeper intrusion of layered gabbro and ultramafic rocks (Strong, 1972a; Upadhyay et al., 1971, p. 28, 32; Pantazis, 1967; Dewey and Bird, 1971). Serpentinites are also formed in this setting (Wyllie, 1969), and tholeiitic basalts derived from the now highly differentiated and ultramafic

mantle (Engel et al., 1965) predominate in this newly formed oceanic crust. The common association of all these oceanic crustal rock types with cupreous pyrite deposits is thereby explainable. Subsequently, complex plate interactions (Bird and Dewey, 1970, p. 1044) and underthrusting of this oceanic crust by sialic blocks (Hsu, 1970; Gass, 1967, p. 130–131) occurred along plate boundaries. Slices of oceanic crust were thereby thrust onto the continent margins where they were preserved, along with their cupreous pyrite ores, from destruction in subduction zones.

The reappearance of both earlier types in Phanerozoic orogenic belts indicates that the mechanics of plate interaction somehow recreate the earlier tectonic environments, thus permitting their local regeneration. Current investigations of plate boundary relationships (James, 1971) and of magmatism in continent margin-island arc environments (Jakes and White, 1971; Kuno, 1966) suggest that the permissive developments occur in subduction zones where continental crustal plates override descending oceanic crustal plates. Successive stages of this process are apparently involved in regenerating the two differing types. The distinction between the two in this tectonic setting may be most useful for classification and for recognition of various evolutionary stages, insofar as the broad process of subduction is presumably continuous in nature.

Nevertheless, extensive magmatism accompanies the earlier stages of subduction during initial infilling of unstable, subsiding eugeosynclinal troughs along continental margins. These early eugeosynclinal troughs may have formed above subduction zones along which oceanic crustal plates were underthrust beneath island-arc systems (Dewey and Bird, 1970, figs. 2c, d, 12a). These arcs were presumably separated by marginal ocean basins from the advancing continent margin itself, possibly like those rimming the western Pacific today (Karig, 1971), and it is interesting that zinc-rich fumarolic volcanism is active in such an environment in New Britain (Ferguson and Lambert, 1972). Whether from anatectic melting and consumption of the descending oceanic plate in the Benioff zone or from partial melting of mantle between the top of the underthrust plate and the crust, this magmatism generates extensive basaltic-andesitic volcanic successions (James, 1971, p.

3343, 3342, fig. 10). These conditions are shown diagrammatically in Figure 5. This environment resembles the primitive Archean conditions already discussed, involving thin and unstable crust, major subsidence, and extensive volcanism of intermediate overall composition, perhaps spanning the entire compositional range from mafic to felsic with progressing differentiation. These tectonic conditions favor generation of the zinc-copper type of volcanogenic massive base metal sulfide deposits, which consequently reappears with the volcanic rocks deposited during the early eugeosynclinal stages of Phanerozoic orogenic belts.

In later stages of subduction, when the continental crustal block has farther overridden the oceanic plate, melting may involve increasing amounts of sialic crustal material, either by anatexis of sialic blocks underthrust at the trench and carried downward by subduction (James, 1971, p. 3343) or by partial melting in the base of the thickened continental plate itself. These stages may have involved

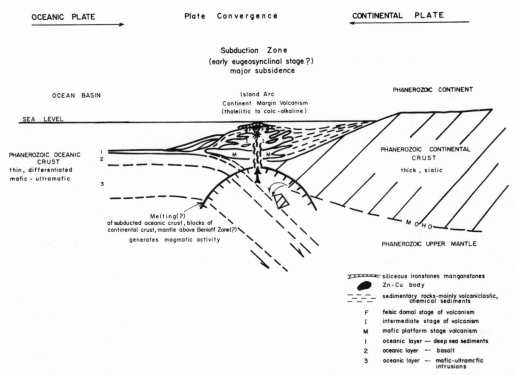

FIG. 5. Diagrammatic illustration: **Zn-Cu** type of volcanogenic massive sulfide deposit regenerated in early stage of Phanerozoic continent margin mobile belt.

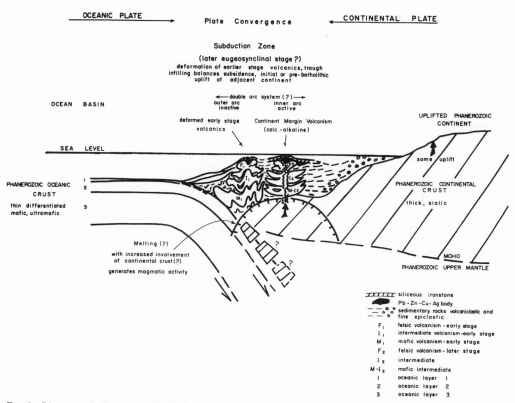

FIG. 6. Diagrammatic illustration: Pb-Zn-Cu-Ag type of volcanogenic massive sulfide deposit regenerated in later stages of Phanerozoic continent margin mobile belt.

island arc-continent collisions (Dewey and Bird, 1970, p. 2640) with underthrusting of the oceanic plate beneath major continental blocks, perhaps like the present Andean margin of South America or the Cordilleran margin of North America during Mesozoic time. These conditions are diagrammatically illustrated in Figure 6. Whatever their magmatic lineage, the resulting extrusives are more felsic and calc-alkaline and are usually displaced continentward from the earlier, more mafic volcanic sequences (James, 1971, p. 3342). These earlier sequences, with their contained copper-zinc deposits may now have undergone varying deformation due to initial uplift and batholithic intrusion that accompanied this later evolutionary stage (James, 1971, p. 3342). Increased epiclastic sedimentation, derived both from the earlier deformed volcanics and from the adjacent and intruded continental margin, deposited increasing amounts of sedimentary rocks which are intercalated

with the felsic volcanic strata. Under these conditions, infilling of the trough would keep pace, more or less, with subsidence, and the lead-zinc-copper-silver type of volcanogenic massive base metal deposit is apparently regenerated.

[*Editor's Note:* Material has been omitted at this point.]

Speculations Concerning Metal Source

If three types of massive volcanogenic base metal sulfide deposits have evolved with crustal tectonics as suggested herein, some broad speculations about the derivation of metals and their behavior in the geochemical cycle are relevant.

The abundance of copper and iron in cupreous pyrite deposits suggests that these metals in the volcanogenic family of deposits are mainly of "oceanic crust-mantle" derivation. Apparently they remain sufficiently abundant in the mantle, despite 4 b.y. of crustal evolution, to provide an adequate source of copper and iron for the cupreous pyrite ores in the ophiolitic environment.

Conversely, the virtual absence of lead in the cupreous pyrite type, compared to its abundance with zinc, but lower copper in the lead-zinc-copper-silver type, suggests the reverse situation for lead. Apparently lead is mainly of supracrustal derivation in these ores and has perhaps been recycled into them through reworking by leaching, anatectic melting, or palingenesis of sialic crust. After 4 b.y. of crustal evolution lead has apparently been impoverished in the mantle and differentiated into supracrustal rocks.

Low lead values in the Archean zinc-copper ores contrasted with high lead content in the younger Proterozoic-Phanerozoic lead-zinc-copper-silver type (Table 1) immediately suggest the possibility that in Archean time lead had not yet been generated through radioactive decay in sufficient abundance to be important in these early ores. Lead isotopic data, however, does not wholly support this speculation (Kanasewich, 1968, p. 215). Additions of new radiogenic lead from early Proterozoic to late Mesozoic time appear to have been minor and when extrapolated back to Archean time are probably insufficient to explain the major shift in metal content of the ores. Nevertheless, radiogenic lead additions must have contributed to the greater abundance of this metal in the younger deposits. It will be necessary to compare isotopic lead ratios from Archean deposits with those of Proterozoic and younger age to evaluate the extent of these additions.

If the increase in lead in younger rocks cannot be explained by radiogenic lead additions, it follows that lead was simply not concentrated and deposited under Archean tectonic conditions. The "fixing" of lead in later tectonic environments must somehow be related to evolutionary changes in lithosphere, hydrosphere, and atmosphere that accompanied tectonic evolution. The change from a reducing, carbon dioxide-rich atmosphere-hydrosphere in Archean time to oxidizing, carbon dioxide-poor conditions in Proterozoic time (Cloud, 1972, p. 543–546) apparently caused increased deposition of carbonate rocks (Cloud, 1968, p. 455) and must also have affected the solubility and stability relationships of the various lead carbonate, sulfate, and sulfide species in the oceans. The abundance of stratiform sulfates with the younger lead-zinc-copper-silver deposits also suggests an oxygenated depositional environment and contrasts with the rarity of sulfates in the Archean zinc-copper type formed under earlier, more reducing conditions. Finally, higher heat flow and surface temperatures presumably accompanied the more extensive Archean volcanism, causing lead to enter preferentially the silicate and oxide structures, rather than the very insoluble sulfide form (Rankama and Sahama, 1950, p. 733).

The abundance of zinc with copper in the Archean deposits, with lead in the lead-zinc-copper-silver type, and occasionally also with copper in the cupreous pyrite type, suggests that zinc has both mantle and supracrustal affinities. In the primitive tectonic environment of Archean time it accompanied copper and iron and was presumably of mantle derivation. During evolutionary crust-mantle differentiation zinc was mainly partitioned into supracrustal rocks, there joining lead. In Phanerozoic orogenic belts zinc was subsequently recycled with lead into the younger lead-zinc-copper-silver deposits, where the proportions of these metals, and of copper, may depend on the extent of epicrustal vs. mantle involvement at successive stages of subduction in arc-trench environments. The evolutionary partitioning of zinc between crust and mantle has apparently not been so complete as that of lead however, and enough zinc remained in the mantle to accompany occasionally iron and copper in cupreous pyrite deposits formed at spreading rift-ridge systems.

There is an alternative possibilty for the derivation of zinc in a few, small, zinc-rich deposits of cupreous pyrite type. In Cyprus, where the stratigraphic positions of numerous sulfide bodies within the 3,000 foot-thick pillow lava sequence is reasonably well-known, the zinc-rich bodies lie at deeper horizons in Basal Group or Lower Pillow Lavas (Bear, 1963, tables 10, 11; p. 61, 71, 73), whereas larger, copper-rich bodies lie at shallower levels in or at the top of Upper Pillow Lavas. Insofar as lower stratigraphic levels in ophiolitic pillow basalt successions may represent early stages of sea-floor spreading, it may be inferred from the relationships in Cyprus (Gass, 1968) that zinc-rich subaqueous fumarolic activity accompanied these earlier stages, whereas copper-rich activity accompanied later stages. Thus the formation of zinc-rich cupreous pyrite bodies may occur at an early stage of rifting when there has been only slight separation between rifted continental plates. In this setting, zinc may have been derived, not from mantle, but from the adjacent sialic plate, zinc content decreasing correspondingly in later (higher stratigraphic) bodies formed at subsequent stages of greater separation.

Some supporting evidence for this possibility is available. The metal-rich sediments in the Red Sea deeps are zinc-rich (Bischoff, 1969), forming in the initial stages of rifting in a very young ocean where there has been minimum separation between the adjacent Arabian and Nubian sialic plates (Hutchinson and Engels, 1970). These relationships led Blissenbach (1972) to suggest that deposits of this type form only during early stages of separation, when thick evaporite sequences on the nearby continent margin provide a source for metal-carrying saline brines. The stratigraphic relationships in Cyprus lend support to this suggestion. Conversely, other examples of metal-bearing sea-floor sediments are rich in iron and manganese and have copper greatly in excess of lead, although zinc was not determined (Bostrom and Peterson, 1966). These are considerably more remote from continental blocks, lying on the crest of the East Pacific Rise in a mid-oceanic rift-ridge environment. Moreover they may not be comparable to those in the Red Sea because they are probably deep-sea pelagic sediments rather than direct precipitates from heated, heavy metal-rich brines.

Iron is abundant in all three types, but greatly predominates in the cupreous pyrite deposits and is least abundant, proportionate to the other metals, in the lead-zinc-copper-silver deposits. This reflects the overall abundance of iron and its stability in both mantle and supracrustal environments. It remains abundant in the mantle after 4 b.y. of crustal evolution and from here is directly contributed into cupreous pyrite deposits. It has also, however, been extensively concentrated into supracrustal rocks through time, and can be recycled from these into deposits of the other two types.

Speculations Concerning Age

Despite the need for additional information about the absolute and relative ages of their volcanic host rocks, the indicated time-age relationships of the three types of volcanogenic massive base metal sulfide deposits are interesting. These relationships suggest that each type occurs throughout a certain time range and, moreover, that there may have been optimum times for the formation of each. These time ranges and optima reflect various stages of the earth's crustal tectonic evolution.

Deposits of the zinc-copper type are commonest, largest, and richest in Archean rocks, are of less importance in earlier Proterozoic (Aphebian) rocks, but are apparently absent in younger Proterozoic (Helikian-Hadrynian) rocks. They reappear in the earliest Paleozoic stages of Phanerozoic orogenic belt evolution, but are again apparently lacking in rocks younger than mid-Paleozoic age, perhaps representing an "extinction" of this type. Deposits of the lead-zinc-copper-silver type appear first in early Proterozoic time, may reach a maximum in early Paleozoic orogenic belts (New Brunswick), and continue into belts of Teritary age (Japan). Finally, deposits of the cupreous pyrite type appear first in earliest Paleozoic or Eocambrian orogenic belts, apparently reach a peak in the Jura-Cretaceous, and also continue in ophiolitic environments of Tertiary, and perhaps Quaternary (Red Sea) age.

These relationships are represented diagrammatically in Figure 8. They are reminiscent of faunal ranges, first appearances, greater proliferations, and perhaps extinctions. Thus they, too, suggest an evolutionary development of the three types of massive base metal sulfide deposits in which

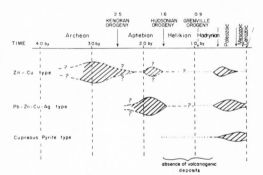

Fig. 8. Apparent time ranges and maxima of massive volcanogenic sulfide deposits.

the zinc-copper type appeared earlier, the other two following later, in response to later stages of crustal evolution. If the optimum times for generation of the three types can be confirmed by additional studies, they would be useful in area selection for mineral exploration and in evaluation of mineral potential. It would be particularly interesting to know whether there was an optimum time for generation of the important zinc-copper type *within* the long span of poorly understood and subdivided Archean time. This information would be helpful in appraising mineral possibilities in Archean terrane in Africa, Australia, and Brazil where this type of deposit is not yet known.

The similar sequence in relative ages of the three types, both through time and within Phanerozoic orogens, suggests that tectonic evolution in Phanerozoic orogenic belts along plate margins above subduction zones recapitulates, on a shortened time scale, the same broad changes involved in earth's overall crustal evolution. This situation, too, is reminiscent of paleontology where "ontogeny duplicates phylogeny." Perhaps it affords scope, within the current dynamic concepts of modern plate theory, to accommodate the older static views of classic uniformitarianism.

A particularly puzzling aspect of the time-age relationships is the apparent absence of volcanogenic massive base metal sulfide deposits of any type in younger Proterozoic (Helikian-Hadrynian) rocks, from 1.6 to about 0.7 b.y. (Fig. 8). This might be expected from the rarity of subaqueous volcanic sequences of this age. It must arise from, hence offer some indication of the tectonic processes and the stage of crustal evolution during this time (Hutchinson, 1973). It is probably significant that base metal deposits of two other important families seem commonest in the two sedimentary lithofacies that characterize Helikian-Hadrynian successions. These are the massive, but probably sedimentary, exhalative (Oftedahl, 1958) family of lead-zinc ores such as the Sullivan and Faro-Dynasty deposits of western Canada and the Broken Hill deposits of New South Wales and secondly, the sedimentary copper deposits of Northern Michigan, the African Copper Belt, and possibly Australia (Johns, 1965).

Summary and Conclusions

The volcanogenic family of massive base metal sulfide ore deposits can be divided into three different types. These are identifiable by their base and precious metal assemblages, the petrochemistry and petrogenesis of their associated volcanic rocks, the nature and relative amounts of their associated volcaniclastic, epiclastic, and chemical sedimentary rocks and their ages, both absolute and relative. All these features are coproducts of differing tectonic environments in which the rocks and ores alike were deposited.

A zinc and copper-rich type containing both silver and gold appears oldest and characteristic of Archean rocks. It reappears in earliest Paleozoic stages of young orogenic belt evolution but is apparently rare or absent in younger Phanerozoic rocks. It is considered a primitive, poorly differentiated ancestor of the other two types. A lead-zinc-copper-silver-rich type appears in early Proterozoic time, reoccurs in somewhat younger stages of Phanerozoic orogenic belt evolution than the zinc-copper type, and continues into young Tertiary rocks. A cupreous pyrite type is first recognizable in early Paleozoic ophiolitic rocks, is most important in similar Mesozoic rocks, and also occurs in Tertiary rocks. These two types are considered to be later, more specialized evolutionary derivatives of the earlier, primitive zinc-copper type. They developed in response to the more diverse tectonic processes that evolved in later geologic time. The changes in the tectonic environment that govern the type of base metal deposit formed, result from broad crustal tectonic evolution, explaining the progression in absolute ages of the different types and also their relative age progression in Phanerozoic orogenic belts.

The analogy between organic evolution and mineral deposit evolution from primitive and less diversified forms in earlier time to more varied and specialized forms in later time is striking. As in paleontology, the evolutionary changes are "unidirectional"; primitive families and types of mineral deposits reappear in younger time when primitive tectonic environments are somehow reproduced, but later, more specialized mineral deposits are lacking in very ancient rocks. When the evolutionary succession and relationships of the broad spectrum of mineral deposits is more completely understood through additional broad metallogenic investigations, then mineral deposits, like fossils, may prove usable as gross-scale age indicators and correlation tools in very old, or very highly metamorphosed terrain. Insofar as they reflect the tectonic environments in which they were originally formed, mineral deposits may also provide a useful added tool in unravelling the complex tectonic history of highly deformed orogenic belts. Metallogenic studies also have obvious practical uses in helping to define the most favorable regions, rock types, and rock ages for mineral exploration for a wide variety of important ore deposits.

Finally, it is evident that metallogenic studies of this type are interdisciplinary, bridging several fields of geoscience specialization. Igneous and sedimentary petrology are fundamental to studies of the volcanic-sedimentary rocks in which the ore bodies are found; stratigraphy and structural geology are basic to unravelling their age relationships, correlations, and tectonic history. Mineral deposits geology is necessary to provide basic data about the nature and geological setting of the ore deposits. Geochemistry is essential to proper interpretation of the magmatic and metallic evolution that is involved, and tectonophysics is probably the broad base to which all the former must be related. Economic geologists who seek to use metallogenic studies in mineral exploration or for other applied or basic purposes must become familiar with these fields. Conversely, those in specialized fields of geology who ignore mineral deposits and metallogenesis may forego another tool that could be usefully applied in their own specializations.

Acknowledgments

Funds for this study were provided by the National Research Council of Canada and facilities by the Department of Geology, University of Western Ontario. Drawings were prepared by Mrs. R. Ringsman.

Several colleagues have given advice and assistance during manuscript preparation including G. G. Suffel, R. W. Hodder, B. J. Fryer, W. R. Church, and W. S. Fyfe. Numerous students have contributed to these ideas through thesis studies, field investigations, and stimulating discussions, particularly R. H. McMillan, W. E. Johnson, G. A. Harron, and N. A. Duke.

Many other geologists have also contributed comments, information, and discussions. These include M. J. Kennedy and D. F. Strong, Memorial University, St. Johns, Newfoundland; B. M. Page, Stanford University, Palo Alto, California; R. L. Brown, ASARCO Canadian Exploration, Toronto; P. A. Kavanagh, Kerr-Addison Mines Ltd., Toronto; J. Bratt and J. Payne Jr., Anaconda-Britannia, Britannia Beach, B. C.; L. Bryner, American Exploration and Mining Co., San Francisco; W. T. Irvine, COMINCO Ltd., Vancouver; D. F. Sangster and R. H. Ridler, Geological Survey of Canada, Ottawa: A. S. Brown and E. W. Grove, British Columbia Department of Mines, Victoria, B. C.; P. A. Gilmour, Tucson, Arizona; and C. A. Anderson and A. R. Kinkel, formerly of the United States Geological Survey.

My thanks are extended to all these individuals and organizations, although not all of them will agree with the interpretations in this paper.

DEPARTMENT OF GEOLOGY
    UNIVERSITY OF WESTERN ONTARIO
    LONDON, ONTARIO, CANADA
        January 18, April 10, 1973

REFERENCES

Albers, J. P., and Robertson, J. F., 1961, Geology and ore deposits of the East Shasta copper zinc district, Shasta County, California: U. S. Geol. Survey Prof. Paper 338, 107 p.

Anderson, C. A., and Creasey, S. C., 1958, Geology and ore deposits of the Jerome area, Yavapai County, Arizona: U. S. Geol. Survey Prof. Paper 308, 185 p.

Anderson, C. A., and Nash, J. T., 1972, Geology of the massive sulfide deposits at Jerome, Arizona—a reinterpretation: ECON. GEOL., v. 67, no. 7, p. 845–863.

Anhaeuser, C. R., Mason, Robert, Viljoen, M. J., and Viljoen, R. P., 1969, A reappraisal of some aspects of shield geology: Geol. Soc. America Bull., v. 80, no. 11, p. 2175–2200.

Bailey, E. H., 1960, Franciscan formation of California as an example of eugeosynclinal deposition, [abs.] : Geol. Soc. America Bull., v. 71, no. 12, pt. 2, p. 2046–2047.

—— Barnes, J. W., and Kupfer, D. H., 1967, Geology and ore deposits of the Küre District, Kastamonu Province, Turkey, *in* Geological mapping techniques, CENTO summer training program, Küre, Turkey, 1966: Central Treaty Organization, p. 17–73.

Bailey, E. H., Blake, M. C., Jr., and Jones, D. L., 1970, On-land Mesozoic oceanic crust in California coast ranges: U. S. Geol. Survey Prof. Paper 700-C, p. C70–C81.

Baragar, W. R. A., 1968, Major-element geochemistry of the Noranda volcanic belt, Quebec-Ontario: Canadian Jour. Earth Sci., v. 5, p. 773–790.

—— and Goodwin, A. M., 1969, Andesites and Archean volcanism of the Canadian Shield, *in* Andesite Cong. (Eugene and Bend, Oregon, 1968) Proc.: Oregon Dept. Geology Mineral Industries Bull. 65, p. 121–142.

Bear, L. M., 1960, The geology and mineral resources of the Akaki-Lythrodondha area: Cyprus Geol. Survey Dept. Mem. 3, 122 p.

—— 1963, The mineral resources and mining industry of Cyprus: Cyprus Geol. Survey Dept. Bull. 1, 208 p.

Bennett, E. M., 1965, Lead-zinc-silver and copper deposits of Mount Isa, *in* Geology of Australian ore deposits, 2nd ed.: Melbourne, Australasian Inst. Mining Metallurgy, p. 233–246.

Bird, J. M., and Dewey, J. P., 1970, Lithosphere plate-continent margin tectonics and the evolution of the Appalachian orogen: Geol. Soc. America Bull., v. 81, no. 4, p. 1031–1059.

Bischoff, J. L., 1969, Red Sea geothermal brine deposits: their mineralogy, chemistry and genesis, *in* Degens, E. T., and Ross, D. A., eds., Hot brines and recent heavy metal deposits in the Red Sea: New York, Springer-Verlag, p. 368–401.

Blissenbach, E., 1972, Continental drift and metalliferous sediments: Oceanology Internat. 72 Conference Papers, Brighton, England, p. 412–416.

Bostrom, K., and Peterson, M. N. A., 1966, Precipitates from hydrothermal exhalations on the East Pacific Rise: ECON. GEOL., v. 61, no. 7, p. 1258–1265.

Bryner, Leonard, 1967, Geology of the Barlo mine and vicinity, Dasol, Pangasinan province, Luzon, Philippines: Philippines Bur. Mines Rept. Inv. 60, 55 p.

—— 1969, Ore deposits of the Philippines—an introduction to their geology: ECON. GEOL., v. 64, no. 6, p. 644–666.

Burchfiel, B. C., and Davis, G. A., 1972, Structural framework and evolution of the southern part of the Cordilleran orogen, western United States: Am. Jour. Sci., v. 272, p. 97–118.

Cagatay, A., 1968, Erzmikroskopische untersuchung des Weiss-Vorkommens bei Ergani Maden, Turkei, und genetische deutung der Kupfererzlagerstätten von Ergani Maden: Neues Jahrb. Mineralogie Abh., v. 109, p. 131–155.

Canadian Institute of Mining and Metallurgy, 1948, Structural geology of Canadian ore deposits, v. 1 (Jubilee Vol.), 948 p.

—— 1957, Structural geology of Canadian ore deposits, v. 2 (Congress Vol.), 524 p.

Canadian Mines Handbook (annual), Northern Miner Press Ltd., Toronto, Canada.

Card, K. D., and Hutchinson, R. W., 1972, The Sudbury structure; its regional geological setting, in Guy-Bray, J. V., ed., New developments in Sudbury Geology: Geol. Assoc. Canada Spec. Paper 10, p. 67–78.

Carlisle, Donald, 1963, Pillow breccias and their aquagene tuffs, Quadra Island, B. C.: Jour. Geology, v. 71, p. 48–71.

Carrière, Gilles, 1957, Suffield Mine, in Structural geology of Canadian ore deposits, v. 2 (Congress Vol.): Montreal, Canadian Inst. Mining Metallurgy, p. 466–469.

Chase, R. L., 1969, Basalt from the axial trough of the Red Sea, in Degens, E. T., and Ross, D. A., eds., Hot brines and recent heavy metal deposits in the Red Sea: New York, Springer-Verlag, p. 122–130.

Church, W. R., 1969, Metamorphic rocks of Burlington Peninsula and adjoining areas of Newfoundland, and their bearing on continental drift in North Atlantic: Am. Assoc. Petroleum Geologists Mem. 12, p. 212–233.

—— 1972, Ophiolite: its definition, origin as oceanic crust and mode of emplacement in orogenic belts with special reference to the Appalachians, in The ancient oceanic lithosphere: Canadian contribution no. 6 to the Geodynamics Project, v. 42, Pubs. of Earth Physics Branch, Dept. of Energy, Mines and Resources, Canada, p. 71–85.

Cloud, Preston, 1968, Atmospheric and hydrospheric evolution on the primitive earth: Science, v. 160, p. 729–736.

—— 1972, A working model of the primitive earth: Am. Jour. Sci., v. 272, p. 537–548.

Coats, C. J. A., Clark, L. A., Buchan, R., and Brummer, J. J., 1970, Geology of the copper-zinc deposits of Stall Lake Mines Ltd., Snow Lake area, N. Manitoba: Econ. Geol., v. 65, No. 8, 970–984.

Cornelius, K. D., 1969, The Mount Morgan mine, Queensland—a massive gold-copper pyritic replacement deposit: Econ. Geol., v. 64, no. 8, p. 885–902.

Degens, E. T., and Ross, D. A., 1969, eds., Hot brines and recent heavy metal deposits in the Red Sea: New York, Springer-Verlag, 600 p.

Dewey, J. F., and Bird, J. M., 1970, Mountain belts and the new global tectonics: Jour. Geophys. Research, v. 75, no. 14, p. 2625–2647.

—— 1971, Origin and emplacement of the ophiolite suite: Appalachian ophiolites in Newfoundland: Jour. Geophys. Research, v. 76, no. 14, p. 3179–3206.

Dimroth, E., Baragar, W. R. A., Bergeron, R., and Jackson, G. D., 1970, The filling of the Circum-Ungava geosyncline, in Basins and geosynclines of the Canadian Shield: Canada Geol. Survey Paper 70–40, p. 45–143.

Dowa Mining Company, 1970, Geology and ore deposits of the Hanaoka Mine: Company report released for IMA-IAGOD Field Excursion, Akita Area, Japan, August 1970, 49 p.

Duke, N. A., 1971, A detailed study of the relationship between massive sulphide bodies and their host rocks, York Harbour, Newfoundland: Unpub. B.Sc. thesis, University of Western Ontario, 87 p.

Engel, A. E. J., Engel, C. G., and Havens, R. G., 1965, Chemical characteristics of oceanic basalts and the upper mantle: Geol. Soc. America Bull., v. 76, no. 7, p. 719–734.

Eric, J. H., 1948, Tabulation of copper deposits of California. in Copper in California: California Div. Mines Bull. 144, p. 199–357.

—— and Cox, M. W., 1948, Zinc deposits of the American Eagle-Blue Moon area, Mariposa County, California, in Copper in California: California Div. Mines Bull. 144, p. 133–150.

Ferguson, J., and Lambert, I. B., 1972, Volcanic exhalations and metal enrichments at Matupi Harbor, New Britain, T.P.N.G.: Econ. Geol., v. 67, no. 1, p. 25–37.

Frarey, M. J., and Roscoe, S. M., 1970, The Huronian Supergroup north of Lake Huron, in Basins and geosynclines of the Canadian Shield: Canada Geol. Survey Paper 70–40, p. 143–158.

Gass, I. G., 1967, The ultramafic volcanic assemblage of the Troodos massif, Cyprus, in Wyllie, P. J., ed., Ultramafic

and related rocks: New York, John Wiley and Sons Inc., p. 121–134.

—— 1968, Is the Troodos massif of Cyprus a fragment of Mesozoic ocean floor?: Nature, v. 220, p. 39–42.

—— 1970, Tectonic and magmatic evolution of the Afro-Arabian dome, in Clifford, T. N., and Gass, I. G., eds., African magmatism and tectonics: Edinburgh, Oliver and Boyd, p. 285–300.

Geological Staff, Hudson Bay Mining and Smelting Company, Limited, 1957, Cuprus Mine, in Structural geology of Canadian ore deposits, v. 2 (Congress Vol.): Montreal, Canadian Inst. Mining Metallurgy, p. 253–258.

Geology Staff, Hudson Bay Mining and Smelting Company, Limited, and Stockwell, C. H., 1948, Flin Flon Mine, in Structural geology of Canadian ore deposits, v. 1 (Jubilee Vol.): Montreal, Canadian Inst. Mining Metallurgy, p. 295–301.

Gill, J. B., 1970, Geochemistry of Viti Levu, Fiji, and its evolution as an island arc: Contr. Mineralogy Petrology, v. 27, p. 179–203.

Gilluly, James, 1965, Volcanism, tectonism and plutonism in the western United States: Geol. Soc. America Spec. Paper 80, 69 p.

Gilmour, Paul, 1971, Strata-bound massive pyritic sulfide deposits—a review: Econ. Geol., v. 66, no. 8, p. 1239–1244.

—— and Still, A. R., 1968, The geology of the Iron King Mine, in Ridge, J. D., ed., Ore deposits in the United States 1933–1967 (Graton-Sales vol.): New York, Am. Inst. Mining Metall. Petroleum Engineers, v. 2, p. 1238–1257.

Glasson, K. R., and Paine, V. R., 1965, Lead-zinc-copper ore deposits of Lake George Mines, Captains Flat, in Geology of Australian ore deposits, 2nd ed.: Melbourne, Australasian Inst. Mining Metallurgy, p. 423–431.

Goodwin, A. M., 1965, Mineralized volcanic complexes in the Porcupine-Kirkland Lake-Noranda region, Canada: Econ. Geol., v. 60, no. 5, p. 955–971.

—— 1968, Evolution of the Canadian Shield: Geol. Assoc. Canada Proc., v. 19, p. 1–14.

—— and Ridler, R. H., 1970, The Abitibi orogenic belt, in Basins and geosynclines of the Canadian shield: Canada Geol. Survey Paper 70–40, p. 1–30.

Gräbe, Reinhold, 1972, Analyse der metallogenetischen Faktoren stratiformer sulfidischer Geosynklinallagerstätten: Zeitschr. angew. Geologie, v. 18, no. 7, p. 289–300.

Griffitts, W. R., Albers, J. P., and Omer, Öner, 1972, Massive sulfide copper deposits of the Ergani-Maden area, southeastern Turkey: Econ. Geol., v. 67, no. 6, 701–716.

Gunning, H. C., 1959, Origin of massive sulphide deposits: Canadian Mining Metall. Bull., v. 52, no. 570, p. 610–613.

Hall, Graham, Cottle, V. M., Rosenhain, P. B., McGhie, R. R., and Druett, J. G., 1965, Lead-zinc ore deposits of Read-Rosebery, in Geology of Australian ore deposits, 2nd ed.: Melbourne, Australasian Inst. Mining Metallurgy, p. 485–489.

Hamilton, Warren, 1969, Mesozoic California and the underflow of Pacific mantle: Geol. Soc. America Bull., v. 80, no. 12, p. 2409–2429.

Heenan, P. R., and Truman, M. P., 1972, The discovery and development of the Ming Zone, Consolidated Rambler Mines Ltd., Baie Verte, Newfoundland [abs.]: Canadian Mining Metall. Bull., v. 65, no. 719, p. 58 [and preprint 10 p.].

Heyl, G. R., 1948a, Foothill copper-zinc belt of the Sierra Nevada, California, in Copper in California: California Div. Mines Bull. 144, p. 11–29.

—— 1948b, Ore deposits of Copperopolis, Calaveras County, California, in Copper in California: California Div. Mines Bull. 144, p. 93–110.

—— Cox, M. W., and Eric, J. H., 1948, Penn zinc-copper mine, Calaveras Co., California, in Copper in California: California Div. Mines Bull. 144, p. 61–84.

Hoffman, P. F., Fraser, J. A., and McGlynn, J. C., 1970, The Coronation geosyncline of Aphebian age, District of Mackenzie, in Basins and geosynclines of the Canadian Shield: Canada Geol. Survey Paper 70–40, p. 200–212.

Horikoshi, Ei, 1969, Volcanic activity related to the formation of the kuroko-type deposits in the Kosaka district, Japan: Mineralium Deposita, v. 4, p. 321–345.

—— and Sato, Takeo, 1970, Volcanic activity and ore deposition in the Kosaka mine, *in* Tatsumi, Tatsuo, ed., Volcanism and ore genesis: Tokyo, Univ. Tokyo Press, p. 181–195.

Hsu, K. J., 1970, Franciscan melanges as a model for eugeosynclinal sedimentation and underthrusting tectonics: Jour. Geophys. Research v. 75, no. 5, p. 886–901.

Hutchinson, R. W., 1965, Genesis of Canadian massive sulphides reconsidered by comparison to Cyprus deposits: Canadian Mining Metall. Bull., v. 58, no. 641, p. 972–986.

—— 1973, Metallogenic relationships of massive base metal sulfide deposits in sedimentary rocks [abs.]: Econ. Geol., v. 68, no. 1, p. 138.

—— and Engels, G. G., 1970, Tectonic significance of regional geology and evaporite lithofacies in northeastern Ethiopia: Royal Soc. London Philos. Trans., A, v. 267, no. 1181, p. 313–329.

Hutchinson, R. W., and Hodder, R. W., 1970, Possible tectonic and metallogenic relationships between porphyry copper and massive sulfide deposits: Canadian Mining Metall. Bull., v. 65, no. 718, p. 34–40.

Hutchinson, R. W., and Searle, D. L., 1971, Stratabound pyrite deposits in Cyprus and relations to other sulphide ores: Soc. Mining Geologists Japan Spec. Issue 3, p. 198–205.

Isacks, Bryan, Oliver, Jack, and Sykes, L. R., 1968, Seismology and the new global tectonics: Jour. Geophys. Research, v. 73, no. 18, p. 5855–5900.

Jakes, P., and White, A. J. R., 1971, Composition of island arcs and continental growth: Earth Planet. Sci. Letters, v. 12, no. 2, p. 224–230.

—— 1972, Major and trace element abundances in volcanic rocks of orogenic areas: Geol. Soc. America Bull., v. 83, no. 1, p. 29–40.

James, D. E., 1971, Plate tectonic model for the evolution of the central Andes: Geol. Soc. America Bull., v. 82, no. 12, p. 3325–3346.

Johns, R. K., 1965, Copper and manganese ore deposits of Pernatty Lagoon–Mt. Gunson District, *in* Geology of Australian ore deposits, 2nd ed.: Melbourne, Australasian Inst. Mining Metallurgy, p. 297–300.

Johns, T. U., 1963, Geology and mineral deposits of east-central Balabac Island, Palawan Province, Philippines: Econ. Geol., v. 58, p. 107–130.

Kalliokoski, J., 1965, Metamorphic features in North American massive sulfide deposits: Econ. Geol., v. 60, no. 3, p. 485–505.

Kanasewich, E. R., 1968, The interpretation of lead isotopes and their geological significance, *in* Hamilton, E. I., and Farquhar, R. M., eds., Radiometric dating for geologists: New York, Interscience Publishers, John Wiley and Sons, p. 147–223.

Kanehira, K., and Bachinski, D., 1968, Mineralogy and textural relationships of ores from the Whalesback mine, northwestern Newfoundland: Canadian Jour. Earth Sci., v. 5, no. 6, p. 1387–1396.

Karig, D. E., 1971, Origin and development of marginal basins in the western Pacific: Jour. Geophys. Research, v. 76, no. 11, p. 2542–2561.

Kennedy, M. J., and DeGrace, J. R., 1972, Structural sequence and its relationship to sulphide mineralization in the Ordovician Lush's Bight Group of western Notre Dame Bay, Newfoundland: Canadian Mining Metall. Bull., v. 65, no. 728, p. 42–50.

Kennedy, M. J., Neale, E. R. W., and Phillips, W. E. A., 1972, Similarities in the early structural development of the northwestern margin of the Newfoundland Appalachians and Irish Caledonides: Internat. Geol. Cong., 24th, Montreal 1972, sec. 3, p. 516–531.

Kinkel, A. R., Hall, W. E., and Albers, J. P., 1956, Geology and base metal deposits of the West Shasta copper-zinc district, Shasta County, California: U. S. Geol. Survey Prof. Paper 285, 156 p.

Kuno, H., 1966, Lateral variation of basalt magma type

across continental margins and island arcs: Bull. Volcanol., v. 29, p. 195–222.

Leney, G. W., and Loeb, E. E., 1971, The geology and mining operations at Pacific Asbestos Corp., Copperopolis, California [abs.]: Mining Eng., v. 23, no. 12, p. 78.

Lindgren, Waldemar, 1919, Mineral deposits, 2nd ed.: New York, McGraw-Hill, 957 p.

Martin, W. C., 1957, Errington and Vermilion Lake mines, *in* Structural Geology of Canadian ore deposits, vol. 2 (Congress Vol.): Montreal, Canadian Inst. Mining Metallurgy, p. 363–376.

Matsukuma, Toshinori, and Horikoshi, Ei, 1970, Kuroko deposits in Japan, a review, *in* Tatsumi, Tatsuo, ed., Volcanism and ore genesis: Tokyo, Univ. Tokyo Press, p. 153–180.

McAllister, A. L., 1960, Massive sulphide deposits in New Brunswick: Canadian Mining Metall. Bull., v. 53, no. 574, p. 88–98.

Mt. Morgan Ltd. Staff, 1965, Copper-gold ore deposit of Mount Morgan, *in* Geology of Australian ore deposits, 2nd ed.: Melbourne, Australasian Inst. Mining Metallurgy, p. 364–369.

Newhouse, W. H., 1931, The geology and ore deposits of Buchans, Newfoundland: Econ. Geol., v. 26, no. 4, p. 399–414.

Oftedahl, Christoffer, 1958, A theory of exhalative-sedimentary ores: Geol. fören. Stockholm Förh., v. 80, pt. 1, no. 492, p. 1–19.

Pantazis, T. M., 1967, The geology and mineral resources of the Pharmakas-Kalavasos area: Cyprus Geol. Survey Dept. Mem. 8.

Peters, H. R., 1965, Mineral deposits of the Hall Bay area, Newfoundland: Geol. Assoc. Canada Spec. Paper 4, p. 171–179.

Price, Peter, and Bancroft, W. L., 1948, Waite-Amulet mine, *in* Structural geology of Canadian ore deposits, v. 1 (Jubilee Vol.): Montreal, Canadian Inst. Mining Metallurgy, p. 748–756.

Purdie, J. J., 1967, Lake Dufault Mines Ltd.: Canadian Inst. Mining Metallurgy Centennial Field Excursion, northwestern Quebec and northern Ontario (Hodge, H. J., chairman), p. 52–57.

Ramsay, B. A., and Swail, E. E., 1967, Manitou Barvue Mines Ltd.: Canadian Inst. Mining Metallurgy Centennial Field Excursion, northwestern Quebec and northern Ontario (Hodge, H. J., chairman), p. 19–21.

Rankama, K., and Sahama, Th. G., 1950, Geochemistry: Chicago, University of Chicago Press, 911 p.

Reid, J. A., 1907, The ore deposit of Copperopolis, Calaveras Co., California: Econ. Geol., v. 2, no. 4, p. 380–417.

Riccio, L. M., 1972, The Betts Cove ophiolite: Unpub. M.Sc. thesis, Univ. of Western Ontario, 91 p.

Ridler, R. H., 1970, Relationship of mineralization to volcanic stratigraphy in the Kirkland-Larder Lakes area: Geol. Assoc. Canada Proc., v. 21, p. 33–42.

Saager, R., 1967, Drei typen von kieslagerstätten im Mofjell-Gebiet, Nordland, und ein neuer vorschlag zur gliederung der Kaledonischen kieslager Norwegens: Norsk Geol. tidsskr., v. 47, p. 333–358.

Sangster, D. F., 1972, Isotopic studies of ore leads in the Hanson Lake-Flin Flon-Snow Lake mineral belt: Canadian Jour. Earth Sci., v. 9, no. 5, p. 500–513.

Sauvé, P., 1953, Clastic sedimentation during a period of volcanic activity, Astray Lake, Labrador: Unpub. M.Sc. thesis, Queen's Univ.

Schermerhorn, L. J. G., 1970, The deposition of volcanics and pyritite in the Iberian pyrite belt: Mineralium Deposita, v. 5, no. 3, p. 273–279.

Scott, J. S., 1948, Quemont Mine, *in* Structural geology of Canadian ore deposits, v. 1 (Jubilee Vol.): Montreal, Canadian Inst. Mining Metallurgy, p. 773–776.

Shenon, P. J., 1933, Geology and ore deposits of the Takilma-Waldo district, Oregon, including the Blue Creek district: U. S. Geol. Survey Bull. 846, p. 141–194.

Sinclair, W. D., 1970, A volcanic origin for the no. 5 zone of the Horne Mine, Noranda, Quebec: Econ. Geol., v. 66, no. 8, p. 1225–1231.

Slawson, W. F., and Russell, R. D., 1973: A multistage history for Flin Flon lead: Canadian Jour. Earth Sci., v. 10, p. 582–583.

Smith, Alexander, 1948, Tulsequah area, in Structural geology of Canadian ore deposits. v. 1 (Jubilee Vol.): Montreal, Canadian Inst. Mining Metallurgy, p. 112–121.

Smith, C. H., and Skinner, Ralph, 1958, Geology of the Bathurst-Newcastle mineral district, New Brunswick: Canadian Mining Metall. Bull., v. 51, no. 551, p. 150–155.

Smitheringale, W. G., 1972, Low-potash Lush's Bight tholeiites: Ancient oceanic crust in Newfoundland?: Canadian Jour. Earth Sci., v. 9, no. 5, p. 574–588.

Solomon, M., and Elms, R. G., 1965, Copper ore deposits of Mt. Lyell, in Geology of Australian ore deposits, 2nd ed.: Melbourne, Australasian Inst. Mining Metallurgy, p. 478–484.

Stanton, R. L., 1955, Lower Paleozoic mineralization near Bathurst, New South Wales: ECON. GEOL., v. 50, no. 7, p. 681–714.

—— 1958, Abundances of copper, zinc and lead in some sulfide deposits: Jour. Geology, v. 66, p. 484–502.

Stevens, R. K., 1970, Cambro-Ordovician flysch sedimentation and tectonics in west Newfoundland and their possible bearing on a proto-Atlantic Ocean: Geol. Assoc. Canada Spec. Paper 7, p. 165–177.

Stevenson, J. S., 1971, The Onaping ash-flow sheet, Sudbury, Ontario, in Guy-Bray, J. V., ed., New developments in Sudbury geology: Geol. Assoc. Canada Spec. Paper 10, p. 41–48.

Stinson, M. C., 1957, Geology of the Island Mountain copper mine, Trinity County, California: California Jour. Mines Geology, v. 53, no. 1, p. 9–33.

Strong, D. F., 1972a, Sheeted diabase of central Newfoundland: New evidence for Ordovician sea floor spreading: Nature, v. 235, p. 102–104.

—— 1972b, The importance of volcanic setting for base metal exploration in central Newfoundland [abs.]: Canadian Mining Metall. Bull., v. 65, no. 719, p. 45 [and preprint 9 p.].

Suffel, G. G., 1948, Waite Amulet Mine, Amulet Section, in Structural geology of Canadian ore deposits, v. 1 (Jubilee Vol.): Montreal, Canadian Inst. Mining Metallurgy, p. 757–763.

—— Hutchinson, R. W., and Ridler, R. H., 1971, Metamorphism of massive sulphides at Manitouwadge, Ontario, Canada: Soc. Mining Geologists Japan Spec. Issue 3, p. 235–240.

Swanson, E. A., and Brown, R. L., 1962, Geology of the Buchans orebodies: Canadian Mining Metall. Bull., v. 55, no. 605, p. 618–626.

Taliaferro, N. L., 1943, Manganese deposits of the Sierra Nevada, their genesis and metamorphism: California Div. Mines Bull. 125, p. 277–331.

Tatsumi, Tatsuo, Yoshihiro, Sekine, and Kanehira Keiichiro, 1970, Mineral deposits of volcanic affinity in Japan: Metallogeny, in Tatsumi Tatsuo, ed., Volcanism and ore genesis: Tokyo, Univ. Tokyo Press, p. 3–50.

Taylor, Bert, 1957, Quemont Mine, in Structural geology of Canadian ore deposits, v. 2 (Congress Vol.): Montreal, Canadian Inst. Mining Metallurgy, p. 405–413.

Upadhyay, H. D., Dewey, J. F., and Neale, E. R. W., 1971, The Betts Cove ophiolite complex, Newfoundland: Appalachian oceanic crust and mantle: Geol. Assoc. Canada Proc., v. 24, no. 1, p. 27–34.

Upadhyay, H. D., and Smitheringale, W. G., 1972, Geology of the Gullbridge copper deposit, Newfoundland: Canadian Jour. Earth Sci., v. 9, no. 9, p. 1061–1073.

Voisey, A. H., 1965, Geology and mineralization of eastern New South Wales, in Geology of Australian ore deposits, 2nd ed.: Melbourne, Australasian Inst. Mining Metallurgy, p. 402–410.

Vokes, F. M., 1966, On the possible modes of origin of the Caledonian sulphide ore deposits at Bleikvassli, Nordland, Norway: ECON. GEOL., v. 61, no. 6, p. 1130–1139.

—— 1969, A review of the metamorphism of sulphide deposits: Earth-Sci. Rev., v. 5, p. 99–143.

Waltham, A. C., 1968, Classification and genesis of some massive sulphide deposits in Norway: Inst. Mining Metallurgy, Trans., sec. B, v. 77, p. 153–161.

Wells, F. G., Hotz, P. E., and Cater, E. W. Jr., 1949, Preliminary description of the geology of the Kerby quadrangle, Oregon: Oregon Dept. Geology Mineral Industries Bull. 40, 23 p.

Williams, Howel, 1957, Glowing avalanche deposits of the Sudbury basin: Ontario Dept. Mines Ann. Rept. 1956, v. 65, part 3, p. 57–89.

Williams, H., 1967, Silurian rocks of Newfoundland: Geol. Assoc. Canada, Spec. Paper 4, p. 93–138.

—— Kennedy, M. J., and Neale, E. R. W., 1972, Economic significance of tectonic-stratigraphic subdivisions of the Newfoundland Appalachians [abs.]: Canadian Mining Metall. Bull., v. 65, no. 719, p. 37 (and preprint 16 p.).

Wilson, H. D. B., Andrews, P., Moxham, R. L., and Ramlal, K., 1965, Archean volcanism of the Canadian Shield: Candian Jour. Earth Sci., v. 2, p. 161–175.

Wilson, R. A. M., and Ingham, F. T., 1959, The geology and mineral resources of the Xeros-Troodos area: Cyprus Geol. Survey Dept. Mem. 1, 136 p.

Wyllie, P. J., 1969, The origin of ultramafic and ultrabasic rocks: Tectonophysics, v. 7, p. 437–455.

24

Reprinted from *J. Geol.* 81(4):381–405 (1973)

ISLAND-ARC EVOLUTION AND RELATED MINERAL DEPOSITS[1]

ANDREW H. MITCHELL AND J. D. BELL

Department of Geology and Mineralogy, University of Oxford,
Parks Road, Oxford OX1 3PR, England

ABSTRACT

In ensimatic arcs, initial submarine eruptions of island-arc tholeiites are succeeded by subaerial and submarine volcanism which is either calcalkaline or island-arc tholeiitic. Besshi-type massive stratiform sulfides develop in deep water on the submarine flanks of islands. Pluton emplacement beneath waning volcanoes is accompanied by mercury, porphyry copper, and gold mineralization. Renewed calc-alkaline or island-arc tholeiitic volcanism commonly follows arc reversal or splitting; Kuroko-type massive sulfides form in shallow-water clastic dacitic rocks and gold is concentrated around monzonites and in meta-andesites. Reef limestone deposition, block faulting, and uplift may be followed by formation of bauxite on karstic limestones, and of stratiform manganese deposits near the limestone base. Upper mantle and ocean crust rocks, emplaced as ophiolites in mélanges on the arc side of the trench and as obducted slices during arc-arc collision, contain Cyprus-type stratiform massive sulfides, podiform chromite, and nickel sulfides; nickeliferous laterites may develop on the upper mantle rocks. Increase in island-arc crustal thickness and emplacement of granitic plutons is accompanied by tin-tungsten-molybdenum-bismuth mineralization. Following arc-continent collision, massive sulfides, gold, tin, and ores associated with ophiolites are preserved.

INTRODUCTION

Island arcs, particularly those in the Western Pacific, have in the past few years attracted the interest of mining companies and state geological surveys as favorable prospects for metallic ore deposits. A number of publications have stressed the ore potential of these arcs, with varying emphasis on the origin of the ore bodies (e.g., Thompson and Fisher 1965; Liddy 1972; Stanton 1972).

Widespread acceptance of the plate tectonics hypothesis, together with increasing geological and geophysical data on island arcs, has led to new explanations of

volcanic, metamorphic, and tectonic processes and their relationship to arc evolution. The hypothesis requires that the inclined seismic plane, or Benioff zone, above which arcs are located, is the zone along which a descending rigid lithospheric slab consisting of ocean floor and upper mantle is consumed (Oliver and Isaacs 1967). This was implied by Hess (1962) and suggested by Coats (1962) for the Aleutian arc. A corollary of the hypothesis is that ancient island arcs occur within continents and interpretation of parts of orogenic belts within continents in terms of island arc and ocean floor successions is now quite commonly encountered. Suggestions that some ore deposits occurring within continents developed initially in island arcs (e.g., Stanton 1960, 1972) are thus now becoming widely accepted.

[1] Manuscript received January 4, 1973; revised March 28, 1973.

[JOURNAL OF GEOLOGY, 1973, Vol. 81, p. 381–405]

The aims of this paper are to consider relationships between the stages of arc evolution and the formation of types of metallic mineral deposit for which an island arc forms a particularly favorable environment. Examples of most of these deposits are found in modern island arcs lying within the oceans, but a few are known only in ancient arcs now located within continents.

MODERN ISLAND ARCS AND NATURE OF UNDERLYING CRUST

Modern island arcs are separated from continents by marine basins underlain by oceanic crust. Active island arcs (Mitchell and Reading 1971) include a belt of active volcanoes, are underlain by a seismic zone, and are bordered by a submarine trench. Inactive arcs lack these features but contain older volcanic or volcaniclastic rocks (e.g., Greater Antilles, Lau Islands).

Some modern arcs contain rocks which indicate that they lay initially adjacent to, or perhaps on, a continental margin, and moved oceanward with development of oceanic crust on the continental side of the arc. Examples of these ensialic arcs are Japan, where pebbles of mid-Pre-Cambrian rock occur in a Permian conglomerate (Sugisaki et al. 1971), and New Zealand (Fleming 1969) and New Caledonia, where continent-derived sediments occur. In most other arcs the nature of the oldest exposed rocks suggests that the arcs originated on oceanic crust, and are ensimatic.

Cenozoic igneous rocks in Japan and New Zealand resemble those in many arcs which lack evidence of continental crust. This suggests that the development of initially ensialic arcs containing fragments of continental crust is similar to that of ensimatic arcs. In this paper we consider the evolution of an ensimatic arc, but in discussing ore bodies emplaced late in its development we use examples from Japan and from other arcs which could have formed initially on, or adjacent to, continental crust.

STAGE 1: PRE-ARC GENERATION OF OCEANIC CRUST AND UPPER MANTLE

GEOLOGICAL EVENTS

The plate tectonics hypothesis requires that oceanic crust form above ascending upper mantle generated at ocean ridge spreading centers and in marginal basins (figs. 1A and 2A). Generation of marginal basin crust is subject to different interpretations (e.g., Karig 1971a; Matsuda and Uyeda 1971), but may resemble that of normal oceanic crust (Sclater et al. 1972).

Igneous rocks are emplaced either continuously or intermittently within a narrow axial zone along an ocean ridge crest. Tholeiitic magma, derived from the partial melting of mantle peridotite, forms layered gabbroic intrusions passing upward into dolerite sheet complexes. These intrude and are overlain by pillow lavas, hyaloclastites, and local thin pelagic sediments. "Burial metamorphism" of pillow basalts to greenschist and zeolite facies and of dolerites and gabbros to amphibolites (Miyashiro et al. 1971) probably takes place soon after emplacement in water-rich and water-deficient environments, respectively (Cann 1970); spilitization of basalts could result from post-cooling hydrothermal alteration (Cann 1969).

During lateral movement of this rock pile down the ridge flanks, it is locally intruded and overlain by alkali basalts and thinly mantled by cherts, pelagic mudstones, and limestones. Local volcanic and carbonate turbidites are derived from oceanic volcanic islands, either single seamounts or chains, which often consist of pedestals of tholeiitic or transitional basalt topped by alkali basalt and related fractionates.

FORMATION OF MINERAL DEPOSITS

The formation of mineral deposits within ophiolites, now interpreted as tectonically emplaced oceanic crust or upper mantle (table 1), is unrelated to arc formation, but exposure of the deposits may occur only subsequent to arc development. Deposits of this type include some of the basaltophilic metals considered typical of an early stage of geosynclinal evolution (Smirnov 1968).

Cyprus-type massive sulfides.—The origins of some massive sulfide deposits associated with submarine or "eugeosynclinal" volcanism (e.g., Anderson 1969) have recently been related to oceanic ridge environments (Pereira and Dixon 1971). Well-known examples of ore bodies in this setting are those of the Troodos ophiolite complex in Cyprus (Sillitoe 1972*a*; Dunham 1972); we term massive sulfides located within similar ophiolite associations "Cyprus-type deposits."

Cyprus-type ore bodies lie either within the tholeiitic pillow lava, metabasalt, or spilite succession—representing layer 2 of the oceanic crust—or between the pillow lavas and overlying pelagic sediments and turbidites of layer 1. The deposits in Cyprus contain massive pyrite together with chalcopyrite, sphalerite, and marcasite, and minor galena, pyrrhotite, gold, and silver (Hutchinson 1965). Strong (1972) suggested that central Newfoundland ores in oceanic tholeiites—such as Whalesback, Little Bay, and Tilt Cove—are characterized by simple mineralogy of pyrite-pyrrhotite-chalocpyrite. Other examples are those of Ergani Maden and Kure in Turkey and Island Mountain in California (Hutchinson 1973).

Evidence that Cyprus-type deposits are emplaced syngenetically during ocean ridge or marginal basin volcanism is provided by the Red Sea metal-rich brines and muds. Unless these originate on the adjacent continental margins (Davidson 1966), their presence in the Red Sea and the occurrence of concentrations of metals in oceanic sediments (Anon 1970; Cronan et al. 1972) suggest that some sulfides originate either at spreading centers or on the sea floor. Alternatively, possible early epigenetic, or subsurface, formation of Cyprus-type deposits might take place in aquagene tuffs, pillow breccias, and volcaniclastic sediments, which could form

traps susceptible to magmatic hydrothermal or metamorphic hydrothermal mineralization (Smitheringale 1972). A possible factor accounting for the formation of Cyprus-type deposits is high trace-metal discharge associated with periods of exceptional global volcanism resulting from active plume convection (Vogt 1972).

Massive sulfides in Hawaii-type volcanoes.—The probable formation of Cyprus-type ore bodies at ocean ridges suggests that stratiform massive sulfides might also develop near submarine vents in intraoceanic tholeiitic shield volcanoes. Possible environments include the Hawaiian ridge, where minor quantities of iron, copper, and nickel sulfides occur within phenocrysts in subaerially erupted tholeiites (Desborough et al. 1968). Ore minerals of Hawaii-type deposits would probably be indistinguishable from those of Cyprus-type; Smitheringale (1972) suggested that the volcanic rocks and copper sulfide deposits of the early Ordovician Lush's Bight Group in Newfoundland probably formed either at an intraoceanic volcano or at an oceanic ridge.

Podiform chromite.—In island arcs, chromite or chrome spinel occurs mainly as podiform deposits within deformed Alpine-type dunites or harzburgite bodies (Thayer 1964), some of which are overlain by gabbros, basaltic lavas, and cherts. Economic chromite deposits in arcs are known only in Cuba and the Philippines.

In Cuba, refractory chromite deposits occur in the northeast of Oriente Province, where sacklike layered bodies of massive chromite within dunite pods are surrounded by peridotite and locally cut by gabbro dikes (Park and MacDiarmid 1964). In the Philippines, refractory podiform chromite ores of metallurgical grade occur in the Zambales ultramafic complex on Luzon as lenticular layered bodies within dunite, and are surrounded by saxonite and intruded by dolerite dike swarms (Bryner 1969).

Both the Cuban and Philippine deposits

TABLE 1

MINERAL DEPOSIT SETTINGS IN ISLAND ARCS

MINERAL DEPOSIT	FORMATION			TECTONIC EMPLACEMENT		EXPOSURE			
	Active Magmatic Arc	Ocean Marginal Basin Rise	Ocean Marginal Basin Floor	Within Ophiolites in Mélange	Within Ophiolites in Obducted Slices	Magmatic Arc Active	Magmatic Arc Inactive	Within Ophiolites in Mélange	Within Ophiolites in Obducted Slices
Endogenous deposits:									
Cyprus-type massive sulfides	…	1–7	?1–7	2–7	6	…	…	2–7	7
Podiform chromites	…	1–7	?1–7	2–7	6	…	…	2–7	7
Nickel sulfides	…	?1–7	?1–7	?2–7	?6	…	…	?2–7	?7
Hawaii-type massive sulfides	…	…	1–7	2–7	6	…	…	2–7	7
Island-arc tholeiite-type massive sulfides	2	…	…	…	…	…	…	…	…
Mercury	4(5, 7)	…	…	…	…	4(5, 7)	5–7	…	…
Besshi-type massive sulfides	3(4, 5, 7)	…	…	…	…	…	5–7	…	…
Porphyry copper	4(5, 7)	…	…	…	…	…	5–7	…	…
Gold around granodiorites	4(5, 7)	…	…	…	…	…	5–7	…	…
Pyrometasomatic deposits	4(5, 7)	…	…	…	…	…	5–7	…	…
Kuroko-type massive sulfides	5, ?3, (7)	…	…	…	…	…	6, 7	…	…
Gold in andesites and around monzonites	5, ?4, (7)	…	…	…	…	…	6, 7	…	…
Tin-tungsten-wolfram-molybdenum	7	…	…	…	…	…	After 7	…	…

MINERAL DEPOSIT	Formation and Exposure
Exogenous deposits:	
Bauxite on karstic limestones	Commonly overlying inactive magmatic arc 5–7
Stratiform manganese	Commonly overlying inactive magmatic arc 5–7
Nickeliferous laterites	Overlying ultrabasic rocks, mostly obducted 7

NOTE.—1, 2, 3, etc. indicate stage of arc evolution with mineralization; (5, 7), etc. indicate stage of evolution with possible mineralization; ?3, etc. indicate stage of evolution with doubtful mineralization.

arc considered to be of early magmatic origin (Guild 1947; Bryner 1969). Their presence in rocks interpreted as upper mantle slices suggests formation at either an oceanic or marginal basin spreading center beneath contemporaneously erupted abyssal tholeiites. Recent work involving the discovery of the incongruent melting of chromian diopside (Dickey et al. 1971) indicates that chromium might be released from the silicate phases of lherzolitic mantle rock by incongruent partial fusion. Ore genesis by crystal fractionation of basic magma leading to the formation of ultrabasic cumulates is favored by Thayer (1969); this could occur either beneath an ocean ridge, or beneath an oceanic volcano as shown in figure 2B.

Nickel sulfides.—Economic deposits of nickel sulfides in island arcs are known only in the Acoje Mine in the Philippines, within the ultrabasic complex on Luzon. The ore occurs with platinum sulfides as irregular blebs in sepentinized dunite, and contains pyrrhotite, troilite, pentlandite, and variolarite (Bryner 1969). Like the chromite, this ore presumably developed near the contact of upper mantle rocks with oceanic crust. Its origin probably differs from that of nickel sulfides in Archaean and early Proterozoic shield areas which are characteristically associa-

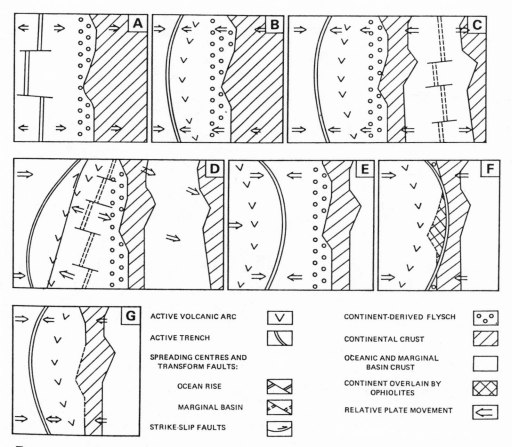

Fig. 1.—Diagrammatic plan views showing stages of arc evolution: *A–G* refers to corresponding cross section in fig. 2*A–G*.

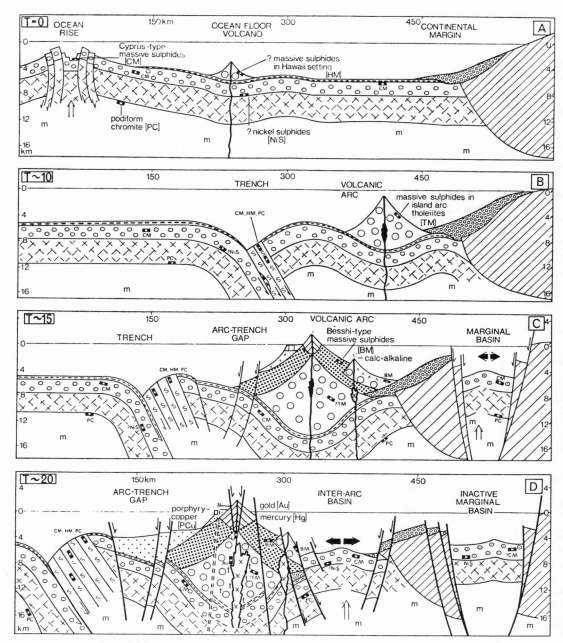

Fig. 2.—Diagrammatic cross sections through an evolving island arc; $T \sim 15$, etc. = time in m.y. since start of arc evolution: *A, Stage 1.—Pre-arc emplacement of oceanic crust and upper mantle.* Spreading ocean ridge migrates away from passive continental margin: formation of Cyprus-type and Hawaii-type massive sulfides, podiform chromite, and possibly nickel sulfides (ocean ridge after Osmaston 1971). *B, Stage 2.—Submarine volcanism and initial arc development.* Ensimatic arc develops on ocean floor near continental margin; possible massive sulfides formed in island-arc tholeiite lavas. *C, Stage 3.—Subaerial and submarine volcanism.* Volcanic arc builds up to sea level; Besshi-type massive sulfides formed on flanks of volcanic arc; rifting near continental margin and development of marginal basin with associated mineralization. *D, Stage 4.—Plutonic activity, waning volcanism, faulting and arc rifting.* Rise of granodioritic plutons in volcanic arc; caldera development with formation of porphyry copper, gold, and mercury deposits; sedimentation in arc-trench gap; development of interarc basin. (Note that porphyry copper [PCu] should be shown occurring *below* mercury [Hg] deposit.) *E, Stage 5.—Arc reversal and development of new volcanic arc.* Reversal of Benioff zone and loss of marginal basin crust; formation of Kuroko-type massive sulfides in dacites, and

194

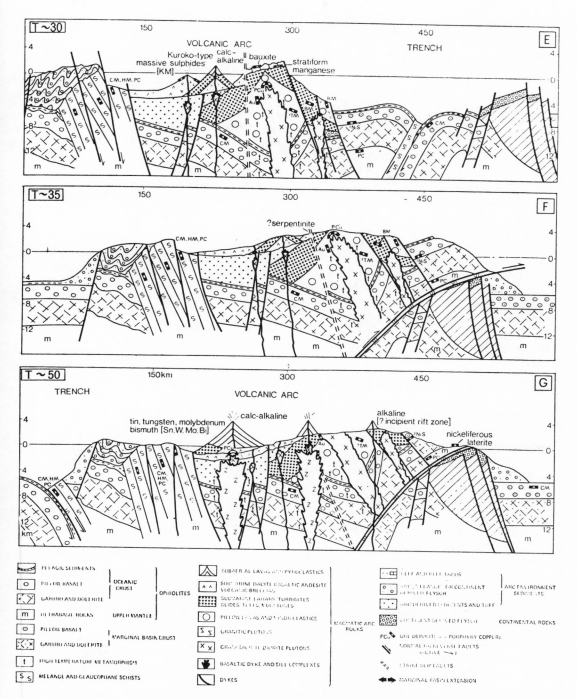

gold mineralization; elevation of trench-fill deposits; bauxite and manganese deposits form on uplifted volcanic arc. *F, Stage 6.—Collision of arcs.* Arc-arc collision follows loss of marginal basin crust; emplacement of Cyprus-type massive sulfides, podiform chromite, and nickel sulfides in obducted ophiolites. *G, Stage 7.—Development of new volcanic arc and emplacement of granites.* Erosion of obducted ocean floor to expose upper mantle rocks; uplift and erosion of mélange and old volcanic arc to expose paired metamorphic belts; development of new volcanic arc and trench; shoshonitic or alkaline volcanism in incipient rift zone prior to marginal basin development; formation of some types of ore body also formed in older volcanic arc; rise of granite plutons with associated tin-tungsten-molybdenum-bismuth mineralization. (Note that this mineralization [Sn, Wo, Mo, Bi] should be shown occurring at apices of granitic plutons.)

195

ted with highly magnesian silicate magma (Hudson 1972).

STAGE 2: SUBMARINE VOLCANISM DURING INITIAL ARC DEVELOPMENT

GEOLOGICAL EVENTS

In some arcs, the oldest exposed volcanic rocks are basaltic (Baker 1968) with the composition of island-arc tholeiites (Jakeš and White 1972). If island-arc tholeiitic volcanism can continue intermittently for several tens of millions of years (Gill 1970), only the oldest rocks of this composition will be erupted on the ocean floor. Submarine island-arc tholeiites and ocean floor basalts cannot be easily distinguished, but there is evidence of significant differences in the proportions of Ti, Zr, and Y (Pearce and Cann 1971) and in the K/Ba and Sr^{87}/Sr^{86} ratios (Hart et al. 1972).

Stratigraphic contacts between island-arc successions and oceanic crust are rarely exposed; consequently, it is uncertain whether the oldest exposed volcanic unit in an arc represents the first episode of island-arc volcanism, or is underlain by older arc rocks. A probable example of initial arc volcanism is the Water Island Formation, the oldest stratigraphic unit in the Lesser Antilles, with a "chemically primitive" or island-arc tholeiite composition (Donnelly et al. 1971). The Water Island Formation consists of spilitic and quartz keratophyre flows and minor volcaniclastic rocks, erupted in deep water (Donnelly 1964).

Eruption of basalts and basaltic andesites on the ocean floor results in a thick succession of pillow lavas (fig. 2B). Breccias, formed by gravitational collapse of some pillows, move as mass flows down the volcano flanks and accumulate as talus cones (Jones 1969). Changes in the erupted basaltic rocks as the volcano nears sea level probably resemble those described from basalts in Hawaii (Moore 1965; Moore and Fiske 1969) and from Icelandic intraglacial olivine basalts (Jones 1966). Vesicle

size increases upward, and vitric palagonitic tuff and breccia, with pillow breccia and peperites, become abundant in rocks erupted within a few hundred meters of the surface. Within the volcanic pile, anastomosing dikes and sills form an intrusive complex which may develop into a high-level reservoir.

Figures 1B and 2B show a volcanic arc bordered by a submarine trench developing above a Benioff zone near a continental margin.

FORMATION OF MINERAL DEPOSITS

Although stratiform sulfide deposits are not known from volcanic successions interpreted as early island-arc rocks, their presence in ocean crust tholeiites suggests that they could also occur in deep-water island-arc tholeiites. Similarities in composition of the basalts erupted in the two settings suggests that the ore minerals from each setting also would be similar.

STAGE 3: SUBAERIAL AND SUBMARINE VOLCANISM

GEOLOGICAL EVENTS

As a submarine volcano attains sea level, subaerial lava flows and tuffs are erupted, overlying predominantly clastic, shallow marine rocks (fig. 2C). Subaerially erupted successions are common in island arcs, both as stratovolcanoes in present active volcanic chains, and as block-faulted successions in active and inactive arcs (Mitchell and Reading 1971). However, conformable stratigraphic contacts between these and the oldest exposed submarine volcanic succession are rare. The examples below indicate that rocks erupted during the first subaerial episode in an arc are mostly either andesites or basalts.

In the Virgin Islands (Donnelly et al. 1971) the submarine Water Island Formation is overlain unconformably by the Louisenhoj Formation—a thick series of subaerially erupted porphyritic augite-andesite breccias similar to the pre-Robles succession in Puerto Rico. The Louisenhoj and pre-Robles Formations have chemically primitive characteristics, resembling those of the underlying Water Island Formation.

The New Hebrides islands of Malekula and Espiritu Santo contain very thick early Miocene successions of volcaniclastic basaltic andesites (Robinson 1969; Mitchell 1971). On Malekula, rocks of deep-water facies are abundant and include carbonized tree trunks and reef limestone detritus, suggesting accumulation on the submarine flanks of subaerially active volcanoes. Rapid erosion led to mass downslope transport of rock debris—as subaerial and submarine lahars, slide, and turbidites—into deep water (Mitchell 1970). Consequently, relatively small subaerial volcanoes become surrounded by a much larger volume of submarine volcaniclastic rocks.

The South Sandwich islands in the Scotia arc consist mostly of late Cenozoic subaerially erupted lavas forming basaltic shield volcanoes and are at an early stage of arc volcanism (Baker 1968). Submarine slopes probably resemble those of intra-oceanic basaltic volcanoes where flow-foot breccias, hyaloclastites, and tuff from phreatic eruptions move downslope into deep water (Moore and Fiske, 1969; Jones 1969).

Subaerial and submarine volcanism and sedimentation in the volcanic arc are accompanied by sedimentation in the arc-trench gap, commonly a topographic trough, located between the trench inner margin and the active volcanic front. Possible examples of these successions occur in the Cretaceous Median Zone of southwest Japan and in the Mesozoic Hokonui facies of New Zealand (Dickinson 1971), although distinction between these and interisland volcanic arc deposits is difficult.

The distribution of the recent subaerially erupted rocks suggests that the magmas originate along or above a Benioff zone. Migration and probable change in slope of a Benioff zone during volcanism may have occurred in the Sunda arc; there is ev-idence that the trench has been forced oceanward as scraped-off continent-derived turbidites (Hamilton 1972) were tectonically emplaced on the arc side of the trench (e.g., Oxburgh and Turcotte 1971), forming the Mentawai-Nicobar-Andaman Islands and the Indoburman ranges. The resulting tectonic mélange of ocean floor sediments, high-pressure metamorphic rocks, and upper mantle material forms a belt parallel to the trench and is commonly partly overlain by arc-derived sediments.

Accumulation of a thick pile of volcanic rocks, accompanied by subsidence due partly to isostatic adjustments (Moore 1971), results in deep burial and regional high-temperature metamorphism of the volcanic arc rocks and underlying oceanic crust. Possible examples of metamorphosed oceanic crust are the amphibolites of the Bermeja Complex, Puerto Rico (Donnelly et al. 1971), amphibolite rafts in serpentinite on Pentecost Island, New Hebrides (Mallick 1970), and "basement" greenschists and amphibolites on Yap Island in the Palau arc (Shiraki 1971). Metamorphosed submarine island-arc tholeiitic rocks form part of the Wainimala Group of greenschist facies in Fiji (Gill 1970).

In some arcs, the presence of rocks typical of continents suggests that the arc has migrated oceanward away from the continental margin. Examples of rifted continental fragments within active arcs occur in New Zealand (Landis and Bishop 1972) and probably in Japan. In Figures 1C and 2C, rifting near the continental margin is followed by development of a marginal basin (Karig 1972) and oceanward migration of the arc and continental fragment.

FORMATION OF MINERAL DEPOSITS

Mineral deposits formed at this and subsequent stages of arc evolution include some copper-zinc deposits considered typical of early stages of geosynclinal development (Smirnov 1968).

Besshi-type massive sulfides.—Many massive sulfide deposits occur in association with andesitic or basaltic volcanic rocks together with significant thicknesses of carbonaceous mudstones, clastic limestones, quartz-rich sediments, or clastic volcanic rocks showing evidence of deepwater accumulation. These nonophiolitic components indicate deposition adjacent to a land mass, volcanic islands, or shallow-water volcanoes, rather than on an oceanic ridge.

Metamorphism and structural complexity of ores and host rocks are common features of these deposits. Both host-rock lithology and ore mineralogy are very variable and the deposits could probably be divided into subgroups. We adopt Kato's (1937) term "Besshi-type deposits," applied to bedded cupriferous iron sulfide deposits at Shikoku on Honshu Island, Japan.

Deposits at Besshi occur in the high-temperature Sanbagawa metamorphic belt of late Mesozoic age, mostly within a succession of isoclinally folded alternations of basic, pelitic, and quartzose schists (Kanehira and Tatsumi 1970). The sedimentary environment has been interpreted as a continental slope and shelf (Ernst 1972), although mafic tuffs and subsequent metamorphism suggest a volcanic arc environment. The stratiform sulfide ores are of three types: (1) compact pyrite, chalcopyrite, sphalerite, and gangue; (2) banded sulfides and silicates; and (3) copper-rich ore, containing chalcopyrite, minor pyrite, and gangue. Pyrrhotite is abundant toward the base of the deposits.

Similar massive sulfide deposits occur in the high-temperature Hida and Abukuma metamorphic belts of Japan (Kanehira and Tatsumi 1970). These have been interpreted as either volcanic sedimentary or early epigenetic deposits; evidence of slumping and mass flow of host sediments and ore (Jenks 1971) suggests a deep-water environment.

Besshi-type deposits in other arcs probably include the Hixbar and Bagacay deposits in the Philippines (Bryner 1969). In intracontinental greenstone belts, ore

bodies in the Bathurst-Newcastle area of New Brunswick and at Captains Flat in New South Wales—considered by Stanton (1960) to be strata-bound in island arc rocks—are probably of Besshi-type. Other ore bodies in similar settings occur at Stekenjokk (Zachrisson 1971) and Menstrask (Grip 1951) in Sweden, and possibly Mt. Isa in Queensland (Hutchinson 1973). In the Iberian pyrite belt, the presence of resedimented mudstones, slates, and conglomerates interbedded with massive sulfides (Schermerhorn 1970) suggests that these are Besshi-type deposits, rather than Kuroko-type (Sillitoe 1972a) discussed below.

Besshi-type deposits include varying assemblages of ore minerals, some of which may resemble those of Kuroko type. Strong (1972) suggested that massive sulfides associated with intermediate to basic calc-alkaline volcanic rocks—for example, Betts Cove in Central Newfoundland—comprise polymetallic ores (Cu, Pb, Zn, Ag, and Au); Hutchinson (1973) and Sillitoe (1972a) considered that deposits in calc-alkaline rocks include more Pb, Zn, Ag, and Ba than those in ophiolite complexes.

Support for a syngenetic volcanic exhalative origin for deposits of Besshi-type is provided by sulfur isotope ratios in the Bathurst-Newcastle deposits (Sangster 1968, Lusk 1969). Lusk (1972) showed that the ratios of these and some similar deposits elsewhere can be explained by mixing of sulfur in rising hot igneous fluids with marine or connate sulfate below the sediment-water interface.

Deposits associated with ophiolites.— Since similarities exist between the crust and upper mantle of the oceans and of marginal basins, it is possible that mineral deposits comparable with those considered in Stage 1 may also form during generation of marginal basin crust and upper mantle in Stages 3 and 4. However, it is not yet possible to distinguish between either rocks or mineral deposits formed in these two settings.

STAGE 4: PLUTONIC ACTIVITY, WANING
VOLCANISM, FAULTING, AND ARC RIFTING

GEOLOGICAL EVENTS

Islands in many arcs contain stocks of gabbroic to granodioritic composition intruding volcanic successions (fig. 2D). Examples are the Utuada Pluton intruding Middle Cretaceous rocks in Puerto Rico (Donnelly 1964), Upper Tertiary granodiorites intruding Miocene volcanic rocks in the Aleutians (Coats 1962), Pliocene dioritic plutons intruding early Miocene andesitic rocks in Bougainville (Blake and Miezitis 1966; Macnamara 1968), pre-Middle Miocene diorites intruding lower Miocene volcaniclastic rocks on Malekula, New Hebrides (Mitchell 1966), and diorite and granodioritic plutons intruding Miocene and Pliocene rocks in Guadalcanal and New Georgia, Solomon Islands (Stanton and Bell 1969).

Stratigraphic relations rarely indicate whether the plutons were emplaced during or following eruption of the overlying volcanic rocks. Although the presence of calderas suggests that large magma chambers lie beneath volcano summits, magnetic anomalies in the New Hebrides indicate that perched magma chambers are only a few kilometers in diameter, and could consist of dike complexes (Malahoff 1970). Possibly island-arc volcanoes, like those in anorogenic areas (Thompson 1972), are underlain by small reservoirs as in Hawaii (Wright and Fiske 1971), which develop into large magma chambers only after the main volcanic activity has ceased. Magma ascent by stoping, as in the coastal batholith of Peru (Cobbing 1972), and its passive emplacement high in the volcanic pile, are probably accompanied by regional metamorphism of the subsiding volcanic and sedimentary prism around the deeper levels of the rising plutons.

In most arcs, intense volcanic activity along any one belt during the Mesozoic or Cenozoic lasted no longer than about 10 m.y. However, after an interval volcanism was commonly either renewed along the same belt, or commenced along a different belt. These variations in volcanic activity could result from stress changes accompanying arc rotation relative to the subducting plate during marginal basin opening—for example, clockwise rotation of the New Hebrides arc, during the late Cenozoic. Major strike-slip faults trending approximately parallel to the arc, as in the Philippines or Sumatra, could also result from stress changes related either to arc rotation or to changes in ocean ridge spreading direction relative to the arc. Strike-slip faults approximately parallel to complex arcs are also known in Honshu, Japan (Miyashiro 1972), Taiwan (Biq 1971), and Sulawesi (Sarasin 1901).

Strike-slip faults oblique to the arc—for example, in Guadalcanal in the Solomon Islands (Coleman 1970) and in the west of the New Hebrides (Malahoff and Woollard 1969)—are possibly continuations of oceanic transform faults along which ultrabasic and basic rocks can be emplaced (Thompson and Melson 1972). Fault slices and pods of serpentinite, commonly showing a linear distribution, such as those on Guadalcanal and the "filons" in New Caledonia (Lillie and Brothers 1969), may have been emplaced along faults of this type.

Vertical displacements along high-angle faults in volcanic arcs, common both during and following volcanism, may result either from block faulting accompanying isostatic adjustment or from vertical movements along strike-slip faults. Control of eruptive center locations by high-angle faults is indicated by the linear distribution of parasitic cones and major craters (Kear 1957; Warden 1967, 1970), by the location of major centers of eruption and calderas where two or more linear features intersect as in Hawaii (Woollard and Malahoff 1966), and by the distribution of volcanoes along en echelon fractures oblique to the arc trend as in Tonga (Bryan et al. 1972.

Figures 1D and 2D show rifting and subsequent interarc basin development

(Karig 1972) between the volcanic arc and the continental fragment. This process may explain the intraoceanic location of rafts of continental rocks or continent-derived sediments lacking a volcanic arc—for example, the Mesozoic succession on New Caledonia (Lillie and Brothers 1969).

FORMATION AND EXPOSURE OF
MINERAL DEPOSITS

Mineral deposits formed at this stage and during the preceding Stages 2 and 3 can be exposed following uplift and erosion either during or subsequent to the fault movements described above.

Porphyry copper.—Porphyry copper, or copper-molybdenum and copper-gold, deposits are emplaced in island arcs and on Andean-type continental margins in the belt of andesitic to dacitic igneous activity above a Benioff zone (Pereira and Dixon 1971; Mitchell and Garson 1972; Taylor 1972; Sawkins 1972; Sillitoe 1972b). They occur at the summit, or around the margins, of stocks or small plutons intruded beneath contemporaneously erupted volcanic rocks.

In island arcs, the intruded host rock forms part of a thick volcanic and clastic succession which mostly accumulated below sea level, although it was probably overlain by subaerial volcanoes during mineralization. Occurrence of Miocene and younger ore bodies at a high elevation— as at Mamut in Sabah and Ok Tedi in New Guinea—suggests that a thick column of intruded host rock favors mineralization, perhaps by permitting differentiation in the rising pluton. Later uplift and erosion expose the ore body in the submarine host rocks.

Whether the volcanic host rock succession is underlain by rocks erupted in an earlier stage of arc volcanism, by oceanic crust, or by continental fragments is uncertain. Porphyry copper deposits may, therefore, develop not during the first episode of subaerial volcanism in an arc, but only during a later episode (Stage 5). Alternatively, emplacement of the deposits may be independent of the stage of

arc evolution, and occur beneath any belt of andesitic or dacitic subaerial volcanoes. It has yet to be demonstrated that porphyry coppers are associated only with calc-alkaline volcanic rocks (e.g., Sillitoe 1972b), and not with island-arc tholeiitic volcanic rocks of intermediate composition.

Controls on location of porphyry deposits within a volcanic belt are poorly understood. The occurrence of many active volcanoes along faults or at fault intersections suggests tectonic control of related porphyry copper mineralization. Intense shattering of the ore host rocks has been attributed to mineralization stoping (Locke 1926), pulsating magma movements (Perry 1961), fault and joint development related to strike-slip faults (Bryner 1968), and pressure release beneath a caldera (Taylor 1972)—for example, by explosive emission of nuée ardentes. The common development of both calderas and strike-slip faults toward the close of arc volcanism suggests that related porphyry copper deposits are emplaced late in the development of a magmatic arc.

Mercury deposits.—Deposits of cinnabar and minor amounts of quicksilver occur in the Philippines, Japan, and New Zealand, but are not known in the less complex arcs. Most deposits are in Cenozoic volcanic rocks and are located near either active or old volcanic centers.

Restriction of the known ore bodies to the more complex arcs is probably of genetic significance, but does not indicate whether the metal is magmatic or non-magmatic in origin. If the mercury is magma-derived, the presence of thick crust typical of complex arcs may be necessary to allow differentiation of a rising high-level pluton and concentration of volatiles. If the metal is sediment-derived, a thick stratigraphic prism including fine-grained sediments may be necessary to form a source from which mercury is expelled by magmatic heat (Moiseyev 1971). White et al. (1971) suggested that mercury, separated from less volatile metals in vapor-dominated reservoirs, could be deposited above boiling brine zones in which porphyry copper deposits develop. A mercury deposit in this position beneath a subaerial volcano is shown in Figure 2D.

Gold associated with granodioritic plutons.—Problems in relating the origins of some types of gold deposit to geological environments above a Benioff zone have been discussed by Sawkins (1972). Two of the most important settings for gold mineralization in island arcs are gold-quartz veins associated with granodioritic plutons, and gold in andesitic volcanics; the latter is discussed under Stage 5 below.

In the Solomon Islands, and probably in the western belt of the New Hebrides, small quantities of placer gold have been derived from lodes associated with the margins of dioritic or granodioritic plutons. Limited evidence suggests that these magmatic-hydrothermal deposits are emplaced at a deeper structural level than porphyry copper ore bodies and are related to nonporphyritic intrusions. Deposits in this type of environment are widespread in western North America, but the related plutons were probably emplaced mostly in an Andean-type igneous belt on a continental margin rather than in island arcs.

Pyrometasomatic deposits.—Pyrometasomatic deposits are common around plutons in the more complex arcs, such as Japan and the Philippines. However, deposits of this type are neither restricted to, nor particularly characteristic of, island arcs, and are, therefore, not considered here.

STAGE 5: ARC REVERSAL AND DEVELOPMENT OF NEW VOLCANIC ARC

GEOLOGICAL EVENTS

In many arcs, successive volcanic belts have not necessarily developed in the same place. Change in volcanic arc position can result from arc rifting and interarc basin development (Karig 1971a, 1971b; 1972). Thus, the relative positions of the active Tonga arc and the Lau-Colville inactive or remnant arc have been explained by late Cenozoic splitting and eastward migration

of the active volcanic belt (Karig 1972; Sclater et al. 1972). Similarly, the Mariana arc has probably migrated eastward away from the west Mariana remnant arc (Karig 1971a).

Change in position of a volcanic arc related to changes in Benioff zone inclination have been suggested in the Peruvian Andes (J. Cobbing, personal communication, 1972), but have yet to be convincingly demonstrated in an island arc. In Japan, where the distribution of Quaternary volcanoes coincides approximately with the Miocene "Green Tuff" volcanic belt, the two volcanic episodes possibly resulted from intermittent descent of lithsphere along a Benioff zone.

Changes in position of the volcanic arc, or arc reversal, related to changes in direction of dip of the Benioff zone (McKenzie 1969) shown in figures 1E and 2E may have occurred in the New Hebrides (Mitchell and Warden 1971), the New Ireland-Bougainville part of the Solomon Islands arc (Mitchell and Garson 1972), and Taiwan (Murphy 1972). Arc reversals probably result from attempted subduction of continental, island-arc, or oceanic island crust. Changes in Benioff zone dip may be related to emplacement of mélanges and consequent oceanward migration of the trench, to changes in rate of lithosphere descent, or to lithosphere drift relative to underlying deep mantle (Hyndman 1972).

Development of a new volcanic arc is probably preceded by associated trench formation. The old inactive trench becomes filled with sediments and rises isostatically as belts of thick folded and faulted flysch-type rocks, perhaps bordered on the arc side by mélanges and glaucophane schists. The Mentawai-Nicobar-Andaman Islands and the Indoburman Ranges could be interpreted as trench fill rather than mélange deposits.

The belt of flysch, glaucophane schists, and mélanges commonly lies parallel to uplifted and eroded rocks of an extinct volcanic arc, showing high-temperature metamorphism (see fig. 2F). These form the paired metamorphic belts (Miyashiro 1961) sometimes separated by a tectonic line along which major strike-slip movements may have occurred (Miyashiro 1972).

Block faulting of inactive arc segments can result in raised atolls, such as Rennell Island south of the main Solomons chain, and extensive raised carbonate platforms, as in Jamaica. Volcanic arc rocks lying between a younger active trench and volcanic arc can be elevated due to upward flexure of the arc plate above the downgoing plate (Fitch and Scholz 1971), resulting in tilted terraces, as in the western belt of the New Hebrides (Mitchell and Warden 1971), southwest Japan, and Eua in Tonga.

The younger volcanic arcs may develop on oceanic crust, on submarine sediments derived from an older arc (e.g., the active Central Chain in the New Hebrides), or on a subaerial or submerged erosion surface of older arc deposits and plutons (e.g., the Japanese "Green Tuff" succession). These younger volcanic successions may resemble either the subaqueous or subaerially erupted successions of Stages 2 or 3 described above.

The composition of these later volcanic rocks varies widely both within an arc and between different arcs. In many late Cenozoic volcanic arcs, the potash content increases with increasing depth to the Benioff zone (Kuno 1966; Hatherton and Dickinson 1969). In some arcs (e.g., Honshu), tholeiites nearest the trench pass laterally into calc-alkaline rocks and finally into shoshonites (Jakeš and Gill 1970). Variations from calc-alkaline to tholeiitic volcanism along the length of the arc are known in the Central Islands of the New Hebrides (Mitchell and Warden 1971) and also in the Lesser Antilles (Donnelly et al. 1971), where calc-alkaline rocks are associated with under-saturated basaltic lavas (Sigurdsson et al. 1973). Ignimbrites are common in the late Cenozoic calc-alkaline belt of some complex arcs

because the islands are larger and the lava more acidic than in arcs at earlier stages of development.

Figures 1E and 2E show arc reversal with "flipping" of a Benioff zone, and development of a related volcanic arc.

FORMATION AND EXPOSURE OF MINERAL DEPOSITS

The endogenous Kuroko-type massive sulfide and gold deposits described below may be exposed during the subsequent Stages 6 and 7 of arc development. Deposits formed in the older magmatic arc during Stages 2, 3, and 4 may be exposed during Stages 6 and 7, and also during tilting and faulting described above.

Kuroko-type massive sulfides.—Among the best-known massive sulfide ore bodies in a modern island arc are those at Kuroko in northeast Japan. They are all associated with predominantly clastic dacitic or more rarely andesitic volcanic rocks interpreted as shallow near-shore marine deposits. The host rock therefore differs considerably from that of both Cyprus-type and Besshitype deposits. Massive sulfides associated with clastic andesitic or more acidic rocks emplaced mostly in shallow water are here termed "Kuroko-type" deposits.

In the Kosaka deposits (Horikoshi 1969; Horikoshi and Sato 1970), the stratiform ore bodies are vertically layered with an upper zinc-rich layer of black (= Kuroko) ore and a lower zinc-poor yellow layer. The main minerals are pyrite, chalcopyrite, sphalerite, galena, and minerals of the tetrahedrite group. The ores have been interpreted as volcanic exhalative deposits formed by hydrothermal activity during the last stages of volcanism. This activity followed phreatic explosions which accompanied emplacement of dacite domes and flows and formed lenticular units of lithic dacitic fragments. Graded bedding in the upper levels of the ore bodies indicates a syngenetic sedimentary origin for at least part of the deposit.

Examples of Kuroko-type ore bodies are probably fairly common in orogenic belts now lying within continents. In the Archaean Keewatin lithofacies of Canada, Hutchinson et al. (1971) have stressed the association of massive pyrite base metal ores with felsic extrusives and pyroclastics. Possible examples are the Horne Mine (Sinclair 1971) and the Delbridge deposit in the Noranda area (Jenks 1971). It is just possible that certain deposits associated with earlier, more mafic, extrusives should be included within the Besshi-type, for example, deposits near Asmara in Ethiopia (Anon. 1971), and the pyrite-chalcopyrite-sphalerite massive sulfides in Precambrian rocks at Jerome, Arizona (Anderson and Nash 1972).

The Kosaka deposits lie in the "Green Tuff" belt of intense early Miocene volcanic activity in which andesitic and dacitic rocks predominate (Sugimura et al. 1963). This succession overlies a basement of Mesozoic age or older, and is bordered on the east by an older paired metamorphic belt. It therefore postdates the first period of island-arc volcanism in Honshu, although Kuroko-type deposits might also be expected within calc-alkaline rocks of the first episode of shallow marine volcanism of Stage 3; a possible example is the deposit at Undua in Fiji (H. Colley, personal communication, 1973).

Gold associated with andesites and monzonites.—Gold deposits, mostly associated with quartz veins, are common in thick successions of andesitic lavas and meta-sedimentary rocks in complex arcs. For example, in the Hauraki Peninsula in New Zealand, gold occurs in early Tertiary propylitized andesitic and dacitic flows overlain by Pliocene lavas (Lindgren 1933).

Host rocks in deposits of this type are commonly metamorphosed and folded but not necessarily cut by intrusions. The metamorphism indicates relatively deep burial and suggests that the succession was overlain by younger rocks, possibly of volcanic arc facies. Migration and concentration of gold together with quartz probably accompanied metamorphism and

deformation of the andesites (e.g., Helgeson and Garrels 1968). We therefore place the origin of gold occurring within andesites or meta-andesites in this stage of arc evolution.

Gold deposits at Vatukoula in Fiji occur in brecciated andesitic rocks of Pliocene age, closely associated with a caldera boundary fault. Infilling of the caldera with sedimentary rocks and andesites was followed by intrusion of trachyandesite and monzonite plugs. Tellurides and auriferous sulfide mineralization followed plug emplacement (Denholm 1967). The more basic Vatukoula rocks are island-arc alkali basalt or shoshonites (Dickinson et al. 1968; Gill 1970), and locally lie unconformably on older rocks of calc-alkaline or island-arc tholeiite composition which form much of the island. Hence they were erupted subsequent to the first major volcanic episode in Fiji.

Deposits broadly similar in structural setting and mineralogy to those at Vatukoula occur at Antamok and Acupan in the Philippines (Bryner 1969; Callow and Worley 1965).

Bauxite.—Economic deposits of bauxite on elevated limestones in island arcs are known in Jamaica and on Rennell Island (De Weisse 1970) south of the Solomon Chain; minor deposits occur in the Dominican Republic, Haiti, and the Lau Islands.

In Central Jamaica, large gibbsitic bauxite deposits occur in solution pockets, sinkholes, and troughs on a karstic surface. The limestone, of Oligicene and early Miocene age, overlies upper Eocene carbonates and accumulated after the final major volcanic episode in the island. The bauxite developed in post-Miocene time, following mid-Miocene faulting and uplift, in well-drained areas at elevations of 700–1,000 m. The hypothesis that the ore is a residual deposit resulting from weathering of several hundred feet of limestone (e.g., Hose 1963) is supported by trace element data (Sinclair 1967). Some authors (Zans 1954; Burns 1961) favored derivation from weathered volcanic rocks above the limestone, a hypothesis now coming back into favor.

Stratiform manganese.—Manganese deposits associated with raised limestones are largely restricted to recently elevated arcs in or near the tropics. These deposits are distinct from raised ocean floor manganese nodule deposits which are not currently economic. Economic deposits associated with raised limestones occur only in the New Hebrides and Cuba, although other types of occurrence are known, for example on Hanesavo island in the Solomon Islands (Grover et al. 1962).

In the New Hebrides, manganese occurs at or near the contact of raised late Cenozoic reef limestone with clastic volcanic rocks—as, for example, in Erromango, the Torres Islands, and Malekula. The Erromango deposits are probably syngenetic, but those on the other islands are considered to be largely or entirely epigenetic. The recently exploited Forari deposits on Efate Island, which mostly occur at the contact of laterites or limestones with underlying Pliocene volcaniclastic rocks. were possibly precipitated from solutions leached from volcanic rocks (Warden1970).

In Cuba, economic deposits of manganese are largely restricted to the southwest of Oriente Province. The ores occur near the top of an Upper Cretaceous to Middle Eocene thick marine volcaniclastic and sedimentary unit, and are concentrated within a few tens of meters of the contact between pyroclastic rocks and an overlying limestone member. Psilomelane, pyrolusite, and wad are the chief ore minerals. Simons and Straczek (1958) considered the deposits to be syngenetic and related to hot submarine springs.

STAGE 6: COLLISION OF ARCS

Reversal of arcs and loss of marginal basin crust along a younger Benioff zone can result in approach of the active and remnant arc and their eventual collision.

GEOLOGICAL EVENTS

Collision of arcs in the late Oligocene or early Miocene probably took place in New Caledonia (Dewey and Bird 1971; Karig 1972), where blueschist metamorphism accompanied southwestward thrusting of peridotites (Avias 1967) over older rocks including Permo-Triassic greywackes (Lillie and Brothers 1969). Derivation of these greywackes from the Australian continent presumably preceded late Mesozoic rifting and northeastward drift of the New Caledonia ridge. Attempted subduction of the ridge, perhaps beneath the Loyalty Islands, resulted in thrusting or "obduction" (Coleman 1971) of the Loyalty Island plate margin onto the New Caledonia plate.

In the Philippines, Tertiary arc collision is suggested by the discontinuous arcuate belt of layered ultrabasic rocks and diabase-gabbro dike swarms exposed on Palawan and western Luzon. These rocks were possibly emplaced during attempted subduction of the Palawan-western Luzon arcuate ridge along a Benioff zone which dipped east beneath an island arc, since fragmented by opening of the Sulu sea and by tectonic movements in western Luzon.

In the Solomon Islands, the ultrabasic rocks, basic lavas, and folded pelagic sediments of the northeastern Pacific Province (Coleman 1970) may be remnants of a late Mesozoic-early Cenozoic marginal basin thrust or obducted southwestward onto the Central Province during the Oligocene.

In the Oriente Province of Cuba, an ultrabasic complex interlayered with gabbroic rocks lies at the eastern end of a belt of northward-thrust ultrabasic slices of probable late Jurassic age (Meyerhoff and Hatten 1968). To the south, Cretaceous quartz diorite and granitic plutons (Khudoley 1967) were probably intruded beneath a volcanic arc. Upper mantle and ocean crust forming the layered complex were either emplaced in a mélange or obducted northward onto the Bahama Bank carbonate platform during southward subduction of a marginal basin. Emplacement of the layered complex may thus have involved collision of a Cretaceous island arc with the Bahama Bank continental margin (M. Itturalde-Vinent, personal communication, 1973).

Figures 1F and 2F show collision of the reversed arc with the previously rifted continental fragment. Obduction of ophiolites over continent-derived greywackes resembles that inferred for New Caledonia.

EMPLACEMENT AND EXPOSURE OF MINERAL DEPOSITS

Mineral deposits formed on the ocean floor, in oceanic crust, or in upper mantle are tectonically emplaced within ophiolites in island arcs; they occur in obducted slices, in mélanges, and possibly in fault-elevated blocks. Mineral deposits are not known from serpentinized ultrabasic rocks, perhaps mantle-derived, which are locally emplaced along strike-slip faults. Although only obducted ophiolites are emplaced during collisions, we consider here the emplacement of mineral deposits in any of these settings.

Cyprus-type massive sulfides.—Deposits of Cyprus-type could occur in ocean floor basalts emplaced either as obducted slices or within mélanges (Sillitoe 1972a). At present no deposits are known from mélanges; those in Newfoundland, for example, probably occur in obducted slices later deformed by continent-continent collision.

Podiform chromite.—The podiform chromites in upper mantle rocks in Cuba and the Philippines were probably emplaced together with their host rocks as obducted slices during collision of the island arcs with, respectively, a continental margin and another arc. Exposure of the ore followed erosion or tectonic removal of overlying ocean crustal rocks. Podiform chromite within upper mantle rocks could also occur in mélanges, or in fault slices or diapirs emplaced along major faults, but no deposits are known in these settings.

Nickel sulfides.—Nickel sulfides in the Philippines were presumably emplaced together with the chromite with which they

are associated. Like podiform chromites, they are likely to be exposed only at deep structural levels, probably in obducted ophiolites.

Formation of nickeliferous laterites.— In southeast Sulawesi, nickel-bearing laterites overlie partly serpentinized harzburgite and lherzolite possibly emplaced as mélange during early Tertiary westward descent of lithosphere (Hamilton 1972). The ore is best developed as nickeliferous serpentine in the lower level of the laterite profile (PT International Nickel Indonesia 1972), which presumably formed during late Cenozoic weathering. Economic deposits of nickeliferous laterites on rocks possibly emplaced in mélanges also occur on Obi and Gube Islands in the Moluccas (Anon. 1972).

In Burma, nickeliferous laterite deposits approaching economic grade overlie ultrabasic rocks in the early Tertiary flysch belt forming the Chin Hills (Gnau Cin Pau, personal communication, 1972). These and other elongate ultrabasic bodies exposed along the eastern margin of the Indoburman Ranges are probably related to eastward subduction of lithosphere prior to late Cenozoic sedimentation in the Central Valley of Burma.

The nickeliferous laterites of the Dominican Republic also probably belong to this environment (F. J. Sawkins, personal communication, 1973).

STAGE 7: DEVELOPMENT OF COMPLEX ARCS AND CHANGES IN MAGMA COMPOSITION WITH TIME

GEOLOGICAL EVENTS

Repeated arc riftings, marginal basin spreading, and arc reversals lead to development of successive magmatic arcs, flysch belts, and paired metamorphic belts (fig. 1G). Arc collisions and strike-slip fault movements result in juxtaposition of tectonic blocks of different age and lithology. An original simple arc thus becomes increasingly complex with magmatic and tectonic addition of younger arcs. For example, the late Cenozoic simple Scotia arc might progress through an arc reversal to the stage of the New Hebrides arc; collision with remnant arcs, obduction of ophiolites, and major strike-slip movements would result in an arc with the complexity of the Philippines. Alternatively, with intermittent loss of lithosphere and no reversal, an arc of Andaman-Nicobar type could eventually reach the complexity of Honshu Island in Japan.

As stated earlier, the development through time of a complex arc involves changes in magma composition. Jakeš and White (1972) considered that initially tholeiitic, predominantly basaltic, lavas are overlain by both tholeiites and minor calc-alkaline andesitic and dacitic rocks, and that finally tholeiitic, calc-alkaline, and shoshonitic rocks are erupted. Although such changes in composition with time have been described from Puerto Rico (Donnelly et al. 1971) and Viti Levu in Fiji (Gill 1970), they are evidently not universally found. In the Japanese "Green Tuff" belts, thick Miocene successions of largely dacitic and andesitic rocks (Sugimura et al. 1963) are overlain by Quaternary, predominantly tholeiitic, basalts. In the southern Kitakami Massif of eastern Honshu, late Palaeozoic and Cretaceous high-alumina basalts are overlain by Quaternary low-alkali tholeiites (Sugisaki and Tanaka 1971). In the new Britain-Schouten Islands, andesitic clastic rocks of probable early Tertiary age are cut by porphyries (Thompson and Fisher 1965) and overlain in the north by predominantly tholeiitic Quaternary lavas lacking contemporaneous calc-alkaline rocks (Jakeš and White 1969; Lowder and Carmichael 1970). Moreover, evidence that alkaline or shoshonitic rocks in arc environments are related to extensional tectonics typical of marginal basins (Martin and Piwinskii 1972) suggests that they are unrelated to this stage of arc evolution.

There are several possible controls on changes in composition of island-arc magma with time. Donnelly et al. (1971)

suggested that island-arc tholeiites result from partial melting of primitive upper mantle prior to development of a Benioff zone; they considered that later calc-alkaline rocks originate from partial melting of tholeiites metamorphosed to amphibolite or eclogite along a Benioff zone, and that as the arc develops descent of increasing volumes of arc-derived sediment would contaminate the tholeiitic layer. Armstrong (1971) suggested that descent of continent-derived ocean floor sediments and mixture with partially melting tholeiite, could explain the lead isotope ratios and high proportion of Pb, Ba, Th, K, Rb, and Cs in some calc-alkaline arc magmas.

The composition of arc magma may be related to rate of lithosphere descent (Sugisaki 1972) which could control the inclination of the Benioff zone (Luyendyk 1970). Benioff zones beneath active arcs erupting tholeiitic magma—such as the Marianas, Izu Islands, and Scotia arcs—mostly dip at more than 40°, while those beneath some arcs erupting predominantly calc-alkaline magma—such as the western Honshu arc and the Aleutians—are less steeply inclined. Marginal basin crust is not developing above the shallow dipping zones which underlie the Peruvian Andes and Central America; in the Marianas arc system, the generation of marginal or interarc basin crust may be dependent on the dip of the Benioff zone which possibly varies cyclically (Bracey and Ogden 1972).

There is a possibility that calc-alkaline magmas may be generated above a Benioff zone. Rise of volatiles from the descending plate into the zone of isotherm inversion above the descending cold slab of oceanic lithosphere could reduce melting points below the ambient temperature (McBirney 1969), facilitating partial melting of wet peridotite to produce liquids of calc-alkaline composition (Yoder 1969). Consequent magma compositions would be largely independent of the nature of the descending lithospheric plate, and would change with time due either to depletion in the low-temperature melting fraction or to addition of descending ocean crust material to the upper mantle (Arculus and Curran 1972). Magma composition may also be determined partly by the increasing thickness of crust as the arc develops (Hamilton 1972), which could control the degree of differentiation of, and partial melting around, rising magma.

The composition of plutons is related partly to the stage of arc development. Quartz diorite and granodiorite occur in relatively simple arcs with crust of moderate thickness (such as the Solomon Islands, Puerto Rico, and Fiji), but large bodies of alkali granite or adamellite have been described only from the complex Japanese arcs where they form batholiths, some of late Cretaceous age (e.g., Murakami 1970). The granites could either be highly differentiated products of partially melted mantle or they could have resulted from partial melting or anatexis of the lower part of the island-arc crust. A time difference of 40 m.y. between emplacement of diorites and granites has been demonstrated in a calc-alkaline association of Palaeozoic age at Yeoval in New South Wales (Gulson and Bofinger 1972).

FORMATION OF MINERAL DEPOSITS

Successive volcanic arcs developing in the same or adjacent localities may each be accompanied by formation of similar ore bodies—for example, Kuroko and perhaps Besshi-type massive sulfides, porphyry coppers, and mercury deposits.

As the thickness of the island-arc crust increases, emplacement of adamellitic and granitic plutons may be accompanied by mineralization associated with alkali and, particularly, soda granites, commonly believed to be largely restricted to continental crust. This mineralization includes deposits of the granitophile elements, sometimes considered to have originated in the crust together with palingenetic granitic plutons (Smirnov 1968) during the later stages of geosynclinal evolution (McCartney and Potter 1962).

Tin - tungsten - molybdenum - bismuth.— These are of economic importance only in Japan, occurring mainly around batholiths in southwest Honshu (Shunso 1971). However, tin ores are also known in the Philippines (Bryner 1969). Fluorine, invariably present with tin and tungsten deposits (e.g., Rub 1972), may originate in the upper mantle or lower crust, or it may be derived with other volatiles from downgoing oceanic crust at depths of 200–400 km (Mitchell and Garson 1972).

ANCIENT ARCS ON AND WITHIN CONTINENTS

Continued addition to an island arc of igneous rocks and of tectonically emplaced ophiolites results in development of crust with a thickness approaching that of continents. However, before an arc complex can grow to continental thickness it usually becomes attached to, and forms part of, an older continental mass. Tectonic, and less common sedimentary, accretion of arcs and related ore bodies to a continental margin have been discussed elsewhere (e.g., Dewey and Bird 1971; Mitchell and Garson 1972).

Island arc-continent accretion may eventually be followed by collision with another continent, resulting in further orogeny and deformation of the arc succession. Thrust movements exceeding 100 km during collision have been explained by crustal "flaking" whereby continental crust on the descending plate splits into an upper overriding and lower underriding slab (Oxburgh 1972). Late rifting of the continent over a spreading center may take place in a zone different from that of the collision, resulting in continental fragments each containing arc complexes lying between older shields.

Many orogenic belts of Palaeozoic and Proterozoic age within continents contain metamorphosed greenstone belts. Parts of these resemble in lithology and chemistry the successions in Cenozoic island arcs, with submarine volcanic rocks of calc-alkaline and island-arc tholeiite composition, and calc-alkaline plutons. Examples are the late Precambrian Harbour Main, Conception, and Holywood rock groups of the Avalon Peninsula Newfoundland (Hughes and Bruckner 1971).

Archaean shield areas also include low-grade metamorphic greenstone belts with volcanic rocks broadly resembling those of island arcs in composition and, in some cases, in lithology. Examples include the igneous rocks of the Slave Province in the Yellowknife area of Canada (Folinsbee et al. 1968), part of the Kalgoorlie System in Western Australia (Glikson 1970), and the Onverwacht Group in South Africa (Viljoen and Viljoen 1969). Despite minor chemical differences between some of these successions and modern island arcs (Glikson 1971; Jakeš and White 1971; Hart et al. 1970), many recent workers consider that tectonic accretion of island arcs to continental margins has continued for at least 3.5×10^9 years (e.g., Engel and Kelm 1972).

Although the formation of some types of massive sulfide deposits may be characteristic of certain periods of the earth's development (Hutchinson 1973), we consider that similarities between ancient and modern arc rocks together with similarities in associated mineral deposits suggest that ore-forming processes in arcs have changed little during the last 2×10^9 years.

CONCLUSIONS

Major mineral deposits in island arcs can be divided into three main groups according to their environment of formation: (1) deposits formed in magmatic arcs, (2) deposits formed in oceanic crust or upper mantle and tectonically emplaced within ophiolites, and (3) exogenous deposits formed in or on raised limestones and on ultrabasic rocks.

Magmatic arc deposits consist of stratiform massive sulfides and deposits related to plutons. Stratiform massive sulfides of Kuroko-type form in a shallow nearshore marine environment together with acidic volcaniclastic rocks; those of Besshi-type form in a deep-water environment together with sediments and minor amounts of intermediate to basic volcanic rocks; possibly massive sulfides also form in submarine island-arc tholeiites. Magmatic hydrothermal porphyry copper, gold, and some mercury deposits form around the upper margins of dioritic and granodioritic plutons emplaced beneath volcanoes. Gold mineralization also occurs around monzonitic intrusions and in andesites.

Deposits formed in or on oceanic crust comprise stratiform massive sulfides of Cyprus-type, and similar deposits may form on Hawaii-type oceanic volcanoes. Near the mantle-crust boundary, podiform chromites and possibly nickel sulfides develop. These are emplaced tectonically with the host rocks in mélanges or in obducted slices.

Exogenous deposits include bauxites developed on karstic raised limestones, stratiform manganese deposits formed near limestone-tuff contacts, and nickeliferous laterites developed on tectonically emplaced upper mantle rocks.

Ore deposits likely to be preserved in island arcs now within continents are magmatic arc deposits of deeply eroded porphyry copper and of gold and in some cases tin and tungsten deposits, and island-arc tholeiite-type, Besshi-type, and Kuroko-type massive sulfides; preserved deposits in ophiolites within arcs include chrome deposits, Cyprus-type massive sulfides, and possibly nickel sulfides.

Acknowledgments.—We thank Dr. H. G. Reading and Mr. E. Eadie of Oxford University, Dr. M. S. Garson of the Institute of Geological Sciences, London, and Dr. Frederick J. Sawkins for critically and constructively reviewing the manuscript. The first author is grateful to the Rio Tinto Zinc Co., Ltd. for permission to publish the paper.

REFERENCES CITED

ANDERSON, C. A., 1969, Massive sulfide deposits and volcanism: Econ. Geology, v. 64, p. 129–146.

——, and NASH, J. T., 1972, Geology of massive sulfide deposits at Jerome, Arizona—a reinterpretation: Econ. Geology, v. 67, p. 845–863.

ANON., 1970, Deep-sea drilling project: Leg 11: Geotimes, v. 15, p. 14–16.

—— 1971, Ethiopia Geol. Survey, 1970, Ann. Rept., p. 31–38.

—— 1972, Financial Times, October 10.

ARCULUS, R. J., and CURRAN, E. B., 1972, The genesis of the calc-alkaline rock suite: Earth and Planetary Sci. Letters, v. 15, p. 255–262.

ARMSTRONG, L. A., 1971, Isotopic and chemical constraints on models of magma genesis in volcanic arcs: Earth and Planetary Sci. Letters, v. 12, p. 137–142.

AVIAS, J., 1967, Overthrust structure of the main ultrabasic New Caledonian massives: Tectonophysics, v. 4, p. 531–541.

BAKER, P. E., 1968, Comparative volcanology and petrology of the Atlantic island arcs: Bull. volcanol., v. 32, p. 189–206.

BIQ, CHINGCHANG, 1971, Some aspects of post-Ordovician block tectonics in Taiwan, in Recent crustal movements: Royal Soc. New Zealand Bull., v. 9, p. 19–24.

BLAKE, D. H., and MIEZITIS, Y., 1966, Geology of Bougainville and Buka Islands, Territory of Papua and New Guinea: Australia Dept. Nat. Devel., Bur. Mineral Resources Geol. Geophys., Records No. 62.

BRACEY, D. R., and OGDEN, T. A., 1972, Southern Mariana arc: geophysical observations and hypothesis of evolution: Geol. Soc. America Bull., v. 83, p. 1509–1522.

BRYAN, W. B.; STICE, G. D.; and EWART, A., 1972, Geology, petrography and geochemistry of the volcanic islands of Tonga: Jour. Geophys. Research, v. 77, p. 1566–1585.

BRYNER, L., 1968, Notes on the geology of the porphyry copper deposits of the Philippines: Mineral Eng. Mag., v. 19, p. 12–23.

—— 1969, Ore deposits of the Philippines—an introduction to their geology: Econ. Geology, v. 64, p. 644–666.

BURNS, D. J., 1961, Some chemical aspects of bauxite genesis in Jamaica: Econ. Geology, v. 56, p. 1297–1303.

CALLOW, K. J., and WORLEY, B. W., 1965, The occurrence of telluride minerals at the Acupan Gold Mine, Mountain Province, Philippines: Econ. Geology, v. 60, p. 251–268.

CANN, J. R., 1969, Spilites from the Carlsberg Ridge Indian Ocean: Jour. Petrology, v. 10, p. 1–19.

—— 1970, New model for the structure of the ocean crust: Nature, v. 226, p. 928–930.

COATS, R. R., 1962, Magma type and crustal structure in the Aleutian arc: Australian Geophys. Union Geophys. Mon. 6, p. 92–109.

COBBING, J., 1972, Tectonic elements of Peru and the evolution of the Andes: Internat. Geol. Cong., 24th, Montreal 1972, Rept., p. 306–315.

COLEMAN, P. J., 1970, Geology of the Solomon and New Hebrides Islands, as part of the Melanesian re-entrant, Southwest Pacific: Pacific Sci., v. 24, p. 289–314.

COLEMAN, R. G., 1971, Plate tectonic emplacement of upper mantle periodotites along continental edges: Jour. Geophys. Research, v. 76, p. 1212–1222.

CRONAN, D. S.; VAN ANDEL, T. H.; HEATH, G. H.; DINKELMAN, M. G.; BENNETT, R. H.; BULENY, D.; CHARLESTON, S.; KANEPS, A.; RODOLFO, K. S.; and YEATS, R. S., 1972, Iron-rich basal sediments from the eastern equatorial Pacific: Leg 16, Deep-Sea Drilling Project: Science, v. 175, p. 61–63.

DAVIDSON, C. F., 1966, Some genetic relationships between ore deposits and evaporites: Inst. Mining and Metallurgy Trans., sec. B, v. 75, p. B216–225.

DENHOLM, L. S., 1967, Geological exploration for gold in the Tavua Basin, Viti Levu, Fiji: New Zealand Jour. Geology and Geophysics, v. 10, p. 1185–1186.

DESBOROUGH, G. A.; ANDERSON, A. T.; and WRIGHT, T. C., 1968, Mineralogy of sulfides from certain Hawaiian basalts: Econ. Geology, v. 63, p. 636–644.

DE WEISSE, G., 1970, Bauxite sur un atoll du Pacifique: Mineralium Deposita, v. 5, p. 181–183.

DEWEY, J. F., and BIRD, J. M., 1971, Origin and emplacement of the ophiolite suite: Appalachian ophiolites in Newfoundland: Jour Geophys. Research, v. 76, p. 3179–3206.

DICKEY, J. S., JR.; YODER, H. S.; and SCHAIRER, J. F., 1971, Chromium in silicate oxide systems: Carnegie Inst. Washington Year Book 70, p. 118–122.

DICKINSON, W. R., 1971, Clastic sedimentary sequences deposited in shelf, slope and trough settings between magmatic arcs and associated trenches: Pacific Geology, v. 3, p. 15–30.

——; RICHARD, M. J.; COULSON, F. I.; SMITH, J. G.; and LAWRENCE, R. L., 1968, Late Cenozoic shoshonitic lavas in northwestern Viti Levu, Fiji: Nature, v. 219, p. 148.

DONNELLY, T. W., 1964, Tectonic evolution of eastern Greater Antillean island arc: Am. Assoc. Petroleum Geologists Bull., v. 48, p. 680–696.

———; ROGERS, J. J.; PUSHKAR, P.; and ARMSTRONG, R. L., 1971, Chemical evolution of the igneous rocks of the eastern West Indies: an investigation of thorium, uranium, and potassium distributions, and lead and strontium isotopic ratios: Geol Soc. America Mem. 30, p. 181–224.

DUNHAM, K. C., 1972, Basic and applied geochemistry in search of ore: Inst. Mining and Metallurgy, Trans., sec. B, v. 81, p. 13–18.

ENGEL, A. E. J., and KELM, D. L., 1972, Pre-Permian global tectonics: a tectonic test: Geol. Soc. America Bull., v. 83, p. 2225–2340.

ERNST, W. G., 1972, Possible Permian oceanic crust and plate junction in Central Shikoku, Japan: Tectonophysics, v. 15, p. 233–239.

FITCH, T. J., and SCHOLZ, C. H., 1971, Mechanism of underthrusting in southwest Japan: a model of convergent plate interactions: Jour. Geophys. Research, v. 76, p. 7276–7292.

FLEMING, C. A., 1969, The Mesozoic of New Zealand: chapters in the history of the Circum-Pacific mobile belt: Geol. Soc. London Quart. Jour., v. 125, p. 125–170.

FOLINSBEE, R. E.; BAADSGAARD, H.; CUMMING, G. L.; and GREEN, D. C., 1968, A very ancient island arc, in KNOPOFF, L.; DRAKE, C. L.; and HART, P. J., eds., The crust and upper mantle of the Pacific area: Am. Geophys. Union Geophys. Mon. 12, p. 441–448.

GILL, J. B., 1970, Geochemistry of Viti Levu, Fiji, and its evolution as an island arc: Contr. Mineralogy and Petrology, v. 27, p. 179–203.

GLIKSON, A. Y., 1970, Geosynclinal evolution and geochemical affinities of early Pre-Cambrian systems: Tectonophysics, v. 9, p. 397–433.

——— 1971, Primitive Archaean element distribution patterns: chemical evidence and geotectonic significance: Earth and Planetary Sci. Letters, v. 12, p. 309–320.

GRIP, E., 1951, Geology of the sulfide deposits at Menstrask: Sveriges Geol. Undersokning, ser. C., no. 515: Stockholm, Norstedt and Soner, 52 p.

GROVER, J. C.; THOMPSON, R. B.; COLEMAN, P. J.; STANTON, R. L.; and BELL, J. D. et al., 1962, The British Solomon Islands geological record, v. 2, 1959–62: London, H.M.S.O., 208 p.

GUILD, P. W., 1947, Petrology and structure of the Moa district, Oriente Province, Cuba: Am. Geophys. Union Trans., v. 28, p. 218–246.

GULSON, B. L., and BOFINGER, V. M., 1972, Time differences within a calc-alkaline association: Contr. Mineralogy and Petrology, v. 36, p. 19–26.

HAMILTON, W., 1972, Tectonics of the Indonesian region: U.S. Geol. Survey Project Rept., Indonesian Inv. (IR) IND-20, 13 p.

HART, S. R.; BROOKS, C.; KROGH, T. E.; DAVIS, G. L.; and NAVA, D., 1970, Ancient and modern volcanic rocks: a trace element model: Earth and Planetary Sci. Letters, v. 10, p. 17–28.

———; GLASSLEY, W. E.; and KARIG, D. E., 1972, Basalts and sea floor spreading behind the Mariana island arc: Earth and Planetary Sci. Letters, v. 15, p. 12–18.

HATHERTON, T., and DICKINSON, W. R., 1969, The relationship between andesitic volcanism and seismicity in Indonesia, the Lesser Antilles and other island arcs: Jour. Geophys. Research, v. 74, p. 5301–5310.

HELGESON, H. C., and GARRELS, R. M., 1968, Hydrothermal transport and deposition of gold: Econ. Geology, v. 63, p. 622–635.

HESS, H. H., 1962, History of ocean basins, in Petrologic studies (Buddington volume): New York, Geol. Soc. America, p. 599–620.

HORIKOSHI, EI., 1969, Volcanic activity related to the formation of the Kuroko-type deposits in the Kosaka District, Japan: Mineralium Deposita, v. 4, p. 321–345.

———, and SATO, TAKEO, 1970, Volcanic activity and ore deposition in the Kosaka mine, in TATSUMI, T., ed., Volcanism and ore genesis: Tokyo, Tokyo Univ. Press, p. 181–195.

HOSE, H. R., 1963, Jamaica-type bauxites developed on limestones: Econ. Geology, v. 58, p. 62–69.

HUDSON, D. R., 1972, Evaluation of genetic models for Australian sulfide nickel deposits: Australasian Inst. Mining and Metallurgy Newcastle Conf., p. 59–68.

HUGHES, C. J., and BRUCKNER, W. D., 1971, Late Pre-Cambrian rocks of the eastern Avalon Peninsula Newfoundland—a volcanic island complex: Canadian Jour. Earth Sci., v. 8, p. 899–915.

HUTCHINSON, R. W., 1965, Genesis of Canadian massive sulfides reconsidered by comparison to Cyprus deposits: Canadian Inst. Mining and Metallurgy Trans., v. 68, p. 266–300.

———, 1973, Volcanogenic sulfide deposits and their metallogenic significance: Econ. Geology (in press).

———; RIDLER, R. H.; and SUFFEL, G. G., 1971, Metallogenic relations in the Abitibi Belt, Canada: a model for Archeaan Metallogeny: Canadian Mining and Metall. Bull., v. 64, p. 49–57.

HYNDMAN, R. D., 1972, Plate motions relative to the deep mantle and the development of subduction zones: Nature, v. 238, p. 263–265.

JAKEŠ, P., and GILL, J., 1970, Rare earth elements and the island arc tholeiitic series: Earth and Planetary Sci. Letters, v. 9, p. 17–28.

210

————, and WHITE, A. J. R., 1969, Structure of the Melanesian arcs and correlation with distribution of magma types: Tectonophysics, v. 8, p. 233–236.

————, — ———— 1971, Composition of island arcs and continental growth: Earth and Planetary Sci. Letters, v. 12, p. 224–230.

————, — ———— 1972, Major and trace element abundances in volcanic rocks of orogenic areas: Geol. Soc. America Bull., v. 83, p. 29–40.

JENKS, W. F., 1971, Tectonic transport of massive sulphide deposits in submarine volcanic and sedimentary host rocks: Econ. Geology, v. 66, p. 1215–1224.

JONES, W. G., 1966, Intraglacial volcanoes of southwest Iceland and their significance in the interpretation of the form of the marine basaltic volcanoes: Nature, v. 212, p. 586–588.

———— 1969, Pillow lavas as depth indicators: Am. Jour. Sci., v. 267, p. 181–195.

KANEHIRA, K., and TATSUMI, T., 1970, Bedded cupriferous iron sulphide deposits in Japan, a review, in TATSUMI, T., ed., Volcanism and ore genesis: Tokyo, Tokyo Univ. Press, p. 51–76.

KARIG, D. E., 1971a, Structural history of the Mariana island arc system: Geol. Soc. America Bull., v. 82, p. 323–344.

———— 1971b, Origin and development of marginal basins in the western Pacific: Jour. Geophys. Research, v. 76, p. 2542–2561.

———— 1972, Remnant arcs: Geol. Soc. America Bull., v. 83, p. 1057–1068.

KATO, T., 1937, Geology of ore deposits [in Japanese] (new ed.): Tokyo, Fuzambo.

KEAR, D., 1957, Erosional stages of volcanic cones as indicators of age: New Zealand Jour. Sci. and Technology, sec. B, v. 38, p. 671–682.

KHUDOLEY, K. M., 1967, Principal features of Cuban geology: Am. Assoc. Petroleum Geologists Bull., v. 51, p. 668–677.

KUNO, H., 1966, Lateral variation of basalt magma types across continental margins and island arcs: Bull. volcanol., v. 29, p. 195–222.

LANDIS, C. A., and BISHOP, D. G., 1972, Plate tectonics and regional stratigraphic-metamorphic relations in the southern part of the New Zealand geosyncline: Geol. Soc. America Bull., v. 83, p. 2267–2284.

LIDDY, J. C., 1972, Mineral deposits of the southwestern Pacific: Mining Mag. (London), March 1973, p. 197–203.

LILLIE, A. R., and BROTHERS, R. N., 1969, The Geology of New Caledonia: New Zealand Jour. Geography and Geophysics, v. 13, p. 145–183.

LINDGREN, W., 1933, Mineral deposits: New York, McGraw-Hill, 930 p.

LOCKE, A., 1926, The formation of certain ore bodies by mineralization stoping: Econ. Geology, v. 21, p. 431–453.

LOWDER, G. G., and CARMICHAEL, I. S., 1970, The volcanoes and caldera of Talasea, New Britain: geology and petrology: Geol. Soc. America Bull., v. 81, p. 17–38.

LUSK, J., 1969, Base metal zoning in the Heath Steele B-1 orebody, New Brunswick, Canada: Econ. Geology, v. 64, p. 509–518.

———— 1972, Examination of volcanic exhalative and biogenic origins for sulphur in the stratiform massive sulfide deposits of New Brunswick: Econ. Geology, v. 67, p. 169–183.

LUYENDYK, B. P., 1970, Dips of downgoing lithospheric plates beneath island arcs: Geol. Soc. America Bull., v. 81, p. 3411–3416.

McBIRNEY, A. R., 1969, Compositional variations in Cenozoic calc-alkaline suites of Central America, in McBIRNEY, A. R., ed., Proceedings of the Andesite Conference: Upper Mantle Proj., Sci. Rept. 16, Oregon State Bull. 65, p. 185–189.

McCARTNEY, W. D., and POTTER, R. F., 1962, Mineralisation as related to structural deformation, igneous activity and sedimentation in folded geosynclines: Canadian Mining Jour., v. 83, p. 83–87.

McKENZIE, D., 1969, Speculations on the causes and consequences of plate motions: Royal Astron. Soc., Geophys. Jour., v. 18, p. 1–32.

MACNAMARA, P. M., 1968, Rock types and mineralisation at Panguna porphyry copper prospect, Upper Kaverong Valley, Bougainville Island: Australasian Inst. Mining and Metallurgy Proc., v. 228, p. 71–79.

MALAHOFF, A., 1970, Gravity and magnetic studies of the New Hebrides Island arc: New Hebrides Condominium Geol. Survey Rept., 67 p.

————, and WOOLLARD, G. P., 1969, The New Hebrides Islands' gravity network, pt. 1, Final Rept.: Honolulu, Hawaii Inst. Geophysics, Univ. Hawaii, 26 p.

MALLICK, D. I. J., 1970, South Pentecost, in MALLICK, D. I. J., ed., Annual Report of the Geological Survey for the year 1968: New Hebrides Anglo-French Condominium, p. 22–27.

MARTIN, R. F., and PIWINSKII, A. J., 1972, Magmatism and tectonic setting: Jour. Geophys. Research, v. 77, p. 4966–4975.

MATSUDA, T., and UYEDA, S., 1971, On the Pacific-type orogeny and its model-extension of the paired belts concept and possible origin of marginal seas: Tectonophysics, v. 11, p. 5–27.

MEYERHOFF, A. A., and HATTEN, C. W., 1968, Diapiric structures in Central Cuba: Am. Assoc. Petroleum Geologists Mem. 8, p. 315–357.

MITCHELL, A. H. G., 1966, Geology of South Malekula: New Hebrides Condominium Geol. Survey Rept., 42 p.

———— 1970, Facies of an early Miocene volcanic arc, Malekula Island, New Hebrides: Sedimentology, v. 14, p. 201–243.

———— 1971, Geology of Northern Malekula:

New Hebrides Condominium Geol. Survey Regional Rept., 56 p.

———, and GARSON, M. S., 1972, Relationship of porphyry copper and circum-Pacific tin deposits to palaeo-Benioff zones: Inst. Mining and Metallurgy Trans., sec. B, v. 81, p. B10–B25.

———, and READING, H. G., 1971, Evolution of island arcs: Jour. Geology, v. 79, p. 253–284.

———, and WARDEN, A. J., 1971, Geological evolution of the New Hebrides Island arc: Geol. Soc. London Jour., v. 127, p. 501–529.

MIYASHIRO, A., 1961, Evolution of metamorphic belts: Jour. Petrology, v. 2, p. 277–331.

——— 1972, Metamorphism and related magmatism in plate tectonics: Am. Jour. Sci., v. 272, p. 629–656.

———; SHIDO, F.; and EWING, M., 1971, Metamorphism in the Mid-Atlantic Ridge near 24° and 30°N: Royal Soc. (London) Philos. Trans., v. A268, p. 589–603.

MOISEYEV, A. N., 1971, A non-magmatic source for mercury ore deposits: Econ. Geology, v. 66, p. 591–601.

MOORE, J. G., 1965, Petrology of deep-sea basalt near Hawaii: Am. Jour. Sci., v. 263, p. 40–52.

——— 1971, Relationship between subsidence and volcanic load, Hawaii: Bull. volcanol., v. 4, p. 562–576.

———, and FISKE, R. S., 1969, Volcanic substructure inferred from dredge samples and ocean-bottom photographs, Hawaii: Geol. Soc. America Bull., v. 80, p. 1191–1202.

MURAKAMI, N., 1970, An example of the mechanism of emplacement of the Chugoku Batholith—the Kuga Granites, southwest Japan: Pacific Geology, v. 3, p. 45–56.

MURPHY, R. W., 1972, The Manila Trench—West Taiwan foldbelt: a flipped subduction zone (Abs.): Regional Conf. Geology Southeast Asia, 1st, Geol. Soc. Malaysia.

OLIVER, J., and ISAACS, B., 1967, Deep earthquake zones, anomalous structures in the upper mantle, and the lithosphere: Jour. Geophys. Research, v. 72, p. 4259–4275.

OSMASTON, M. F., 1971, Genesis of ocean ridge median valleys and continental rift valleys: Tectonophysics, v. 11, p. 387–405.

OXBURGH, E. R., 1972, Flake tectonics and continental collision: Nature, v. 239, p. 202–204.

———, and TURCOTTE, D. L., 1971, Origin of paired metamorphic belts and crustal relation in island arc regions: Jour. Geophys. Research, v. 76, p. 1315–1327.

PARK, C. F., and MACDIARMID, R. A., 1964, Ore deposits: San Francisco, Freeman, 475 p.

PEARCE, J. A., and CANN, J. R., 1971, Ophiolite origin investigated by discriminant analysis using Ti, Zr and Y: Earth and Planetary Sci. Letters, v. 12, p. 339–349.

PEREIRA, J., and DIXON, C. J., 1971, Mineralisation and plate tectonics: Mineralium Deposita, v. 6, p. 404–405.

PERRY, V. D., 1961, The significance of mineralised breccia pipes: Mining Eng., v. 13, p. 367–376.

P. T. INTERNATIONAL NICKEL INDONESIA, 1972, Laterite deposits in the southeast arm of Sulawesi (Abs): Regional Conf. Geology Southeast Asia, 1st, Geol. Soc. Malaysia, p. 32.

ROBINSON, G. P., 1969, The geology of North Santo: New Hebrides Geol. Survey Rept., 77 p.

RUB, M. G., 1972, The role of the gaseous phase during the formation of ore-bearing magmatic complexes: Chemical Geology, v. 10, p. 89–98.

SANGSTER, D. F., 1968, Relative sulphur isotope abundances of ancient seas and strata-bound sulphide deposits: Geol. Assoc. Canada Proc., v. 19, p. 79–91.

SARASIN, P., 1901, Entwurf einer geografischen und geologischen Beschreibung der insel Celebes: Wiesbaden.

SAWKINS, F. J., 1972, Sulfide ore deposits in relation to plate tectonics: Jour. Geology, v. 80, p. 377–397.

SCHERMERHORN, L. J. G., 1970, The deposition of volcanics and pyritite in the Iberian pyrite belt: Mineralium Deposita, v. 5, p. 273–279.

SCLATER, J. G.; HAWKINS, J. W.; MAMMERICKX, J.; and CHASE, C. G., 1972, Crustal extension between the Tonga and Lau ridges: petrological and geophysical evidence: Geol. Soc. America Bull., v. 83, p. 505–518.

SHIRAKI, K., 1971, Metamorphic basement rocks of Yap Islands, Western Pacific: possible oceanic crust beneath an island arc: Earth and Planetary Sci. Letters, v. 13, p. 167–174.

SHUNSO, ISHIHARA, 1971, Major molybdenum deposits and related granitic rocks in Japan: Geol. Survey Japan, Rept. 239, 183 p.

SIGURDSSON, H.; BROWN, G. M.; TOMBLIN, J. F.; HOLLAND, J. G.; and ARCULUS, R. J., 1973, Strongly undersaturated magmas in the Lesser Antilles island arc: Earth and Planetary Sci. Letters, v. 18, p. 285–295.

SILLITOE, R. H., 1972a, Formation of certain massive sulphide deposits at sites of sea-floor spreading: Inst. Mining and Metallurgy Trans., sec. B, v. 81, p. B141–148.

———, 1972b, A plate tectonic model for the origin of porphyry copper deposits: Econ. Geology, v. 67, p. 184–197.

SIMONS, F. S., and STRACZEK, J. A., 1958, Geology of the manganese deposits of Cuba: U.S. Geol. Survey Bull. 1057, 289 p.

SINCLAIR, I. G. L., 1967, Bauxite genesis in Jamaica: new evidence from trace element distribution: Econ. Geology, v. 62, p. 482–486.

SINCLAIR, W. D., 1971, A volcanic origin for the No. 5 zone of the Horne Mine, Noranda, Quebec: Econ. Geology, v. 66, p. 1225–1231.

SMIRNOV, V. I., 1968, The sources of ore-forming fluids: Econ. Geology, v. 63, p. 380–389.

SMITHERINGALE, W. G., 1972, Low-potash Lush's Bight tholeiites: ancient oceanic crust in Newfoundland?: Canadian Jour. Earth Sci., v. 9, p. 574–588.

STANTON, R. L., 1960, General features of the conformable "pyritic" ore bodies: Canadian Inst. Mining and Metallurgy Trans., v. 63, p. 22–36.

———. 1972, Ore petrology: New York, McGraw-Hill, 713 p.

———, and BELL, J. D., 1969, Volcanic and associated rocks of the New Georgia Group, British Solomon Islands Protectorate: Overseas Geology and Mineral Resources, v. 10, p. 113–145.

STRONG, D. F., 1972, The importance of volcanic setting for base metal exploration in Central Newfoundland (Abs.): Canadian Mining and Metallurgy Bull., v. 65, p. 45.

SUGIMURA, A.; MATSUDA, T.; CHINZBI, K.; and NAKAMURA, K., 1963, Quantitative distribution of late Cenozoic volcanic materials in Japan: Bull. volcanol., v. 26, p. 125–140.

SUGISAKI, RYUICHI, 1972, Tectonic aspects of the Andesite Line: Nature, v. 240, p. 109–111.

———; MIZUTANI, S.; ADACH, M.; HATTORI, H.; and TANAKA, T., 1971, Rifting in the Japanese late Palaeozoic geosyncline: Nature, v. 233, p. 30–31.

———, and TANAKA, T., 1971, Magma types of volcanic rocks and crustal history in the Japanese pre-Cenozoic geosynclines: Tectonophysics, v. 12, p. 393–413.

TAYLOR, D., 1972, The liberation of minor elements from rocks during plutonic igneous cycles and their subsequent concentration to form workable ores, with particular reference to copper and tin: Geol. Soc. Malaysia, 5th Presidential Address, unpub. rept.

THAYER, T. P., 1964, Principal features and origin of podiform chromite deposits and some observations on the Guleman-Soriday district, Turkey: Econ. Geology, v. 59, p. 1497–1524.

——— 1959, Alpine-type sensu strictu (ophiolitic) peridotites: refractory residues from partial melting or igneous sediments?—a contribution to the discussion of the paper: "The origin of ultramafic and ultrabasic rocks," by P. J. Wyllie: Tectonophysics, v. 7, p. 511–516.

THOMPSON, G., and MELSON, W. G., 1972, The petrology of oceanic crust across fracture zones in the Atlantic Ocean: evidence of a new kind of sea-floor spreading: Jour. Geology, v. 80, p. 526–538.

THOMPSON, J. E., 1972, Evidence for a chemical discontinuity near the basalt-"andesite" transition in many anorogenic volcanic suites: Nature, v. 236, p. 106–110.

———, and FISHER, N. H., 1955, Mineral deposits of New Guinea and Papua and their tectonic setting: Commonwealth Mining and Metallurgy Cong., 8th Proc., A.N.Z. Preprint 129, p. 59.

VILJOEN, M. J., and VILJOEN, R. P., 1969, The geochemical evolution of granitic rocks of the Baberton region, in Upper Mantle Project: Geol. Soc. South Africa Spec. Pub. 2, 189 p.

VOGT, P. R., 1972, Evidence for global synchronism in mantle plume convection and possible significance for geology: Nature, v. 240, p. 338–342.

WARDEN, A. J., 1967, Geology of the Central Islands: New Hebrides Condominium Geol. Survey Rept. No. 5, 108 p.

———, 1970, Evolution of Aoba Caldera Volcano, New Hebrides: Bull. volcanol., v. 34, p. 107–140.

WHITE, D. E.; MUFFLER, L. J. P.; and TRUESDELL, A. H., 1971, Vapor-dominated hydrothermal systems compared with hot water systems: Econ. Geology, v. 66, p. 75–97.

WOOLLARD, G. P., and MALAHOFF, A., 1966, Magnetic measurements over the Hawaii Ridge and their volcanological implications: Bull. volcanol., v. 29, p. 725–760.

WRIGHT, T. L., and FISKE, R. S., 1971, Origin of the differentiated and hybrid lavas of Kilauea Volcano, Hawaii: Jour. Petrology, v. 12, p. 1–65.

YODER, H. S., 1969, Calc-alkaline andesites: experimental data bearing on the origin of their assumed origin, in McBIRNEY, A. R., ed., Proceedings of the Andesite Conference: Upper Mantle Proj., Sci. Rept. 16, Oregon State Bull. 65, p. 77–89.

ZACHRISSON, E., 1971, The structural setting of the Stekenjokk ore bodies, Central Swedish Calendonides: Econ. Geology, v. 66, p. 641–652.

ZANS, V. A., 1954, Bauxite resources of Jamaica and their development: Colonial Geology and Mineral Resources, v. 3, p. 307–333.

MASSIVE SULPHIDES AND PLATE TECTONICS

M. Solomon

SEVERAL authors have considered the genesis of various ores in terms of plate tectonic theory[1-6]. Most have discussed various possibilities for the origin of the ore solutions of massive sulphides and, with the exception of Guild[6], have stated or implied a fairly close genetic relationship between ore deposition and the processes of seafloor spreading and subduction. The nature of this relationship is, however, by no means clear and I suggest here that it may be very tenuous.

The deposits included in the term 'massive sulphide' are stratiform and wholly or largely composed of sulphides of iron with varying proportions of the sulphides of copper, lead and zinc, and minor gold and silver. Stratiform, massive oxide deposits (largely magnetite and/or haematite) are related ore types. The term includes 'volcanogenic massive sulphides'[7], but as used here it also covers very similar ores in sedimentary or sedimentary/volcanic successions (for example, Sullivan, Rammelsberg, Cobar, and Mount Isa).

Many geologists have discussed the possibility that massive sulphide deposits may be derived from heated seawater, or from meteoric water that has become an ore solution because of a hot water–rock interaction below the surface[8-11]. The principal requirements of this model are: a thermal anomaly sustained for sufficient time to precipitate an ore deposit; a system allowing long term downward penetration of water (probably to depths of several kilometres) and then upwards to the surface through major fractures; the presence of a suitable depositional site, which is commonly on the sea floor[7,11]. This geothermal model seems to account for the salient features of massive sulphides.

The principal host rocks to massive sulphides are volcanics of varying composition (mainly calc-alkaline), but some deposits occur in thick sediments (mainly shales and greywackes) with little or no evidence of volcanism (for example, Meggen, Rammelsberg, Cobar). There seems to be no evidence to support the statement[12] that massive sulphides are associated with mainly mafic volcanics. Deposits in volcanic successions do not show any correlation between the ore type (in terms of the Cu:Pb:Zn ratios) and the nature of the host rocks[12].

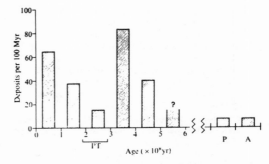

Fig. 2 Variation in the number of massive sulphide deposits per 100 Myr throughout the geological column. The Proterozoic (P) and Archaean (A) are each assumed to be 1,500 Myr long. PT, Permo-Triassic.

This seems to be true for all massive sulphides, except that where the host rocks are dominantly or entirely basaltic (as in Cyprus) the lead content is very low. If ore solutions derive their metal content through water–rock interactions several kilometres below the site of deposition, then correlations involving rock walls will be of little value. But a comparison between ore compositions and the nature of the rock successions for several kilometres below the deposits also showed no correlation, although the quality of the available data was poor. Copper and zinc are relatively abundant in most igneous and sedimentary rocks and the low degree of correlation between rock type and ore type is reflected in the geothermal model. Wide variations in the Cu:Zn ratio may be brought about by chemical factors during transport and at the site of deposition. Also predictable is the absence of lead ores in thick sequences of basic rocks, as these contain only a few parts per million (p.p.m.) of that metal. Another important feature of the host rocks is the lack of coeval plutons.

Massive sulphides tend to occur at distinct stratigraphic

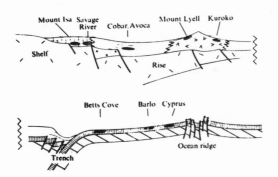

Fig. 1 Diagrammatic cross section of a continental margin to illustrate the range of tectonic environments of massive sulphides.

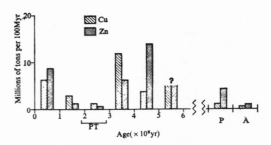

Fig. 3 Variation in the tonnages per 100 Myr of copper and zinc in massive sulphides through the geological column. The tonnages include production and reserves and are very approximate. A, Archaean; P, Proterozoic; PT, Permo-Triassic.

horizons[7,14] which may be unconformities (Mount Lyell), sediment lenses within volcanics (Rosebery, Captains Flat), tops of volcanic sequences (Rio Tinto, Cyprus), or at changes in the composition of volcanics (Noranda area) or the nature of the volcanism (dacite domes of northern Honshu). Thus, the timing of ore deposition does not, apparently, relate to any clearly recognisable, unique event but to periods of hydrothermal activity brought about by various agencies such as faulting, pauses in volcanism, and so on.

Phanerozoic massive sulphides occur in a wide variety of tectonic environments[15], particularly in environments associated with orogenic belts (Fig. 1). Most occur in calc-alkaline island arcs and Andean-type volcanic belts (Mount Morgan, Rosebery, Mount Lyell, Rio Tinto, Hopa-Murgul, Shakanai), but some occur in the flysch and shelf sequences of the continental margin (Cobar, Avoca?), or in oceanic crust in ridge and ocean floor environments (Betts Cove, Skouriotissa, Besshi?). In the Palaeozoic of south-eastern Australia most massive sulphides occur in Andean-type volcanics, and formed fairly late in the development of the volcanic belts during the 'paratectonic' phases of the main tectonic cycles[16], when the volcanics were undergoing faulting, folding and granitic intrusion. Massive sulphides commonly occur in groups aligned along major features that provide deep seated plumbing systems.

Sulphide-sulphur in massive sulphides is enriched in [34]S relative to the meteoritic standard and generally shows only small variations within deposits. The values for some deposits may reflect mixing and re-equilibration between sulphur in magmatic water rising from depth and sulphate in seawater[17,18]. The same results can be achieved by mixing sulphur derived from the leaching of rocks with sulphate-bearing seawater in geothermal conditions.

Unlike several ore-types associated, and apparently genetically linked, with plutonic rocks (for example, tin-wolfram veins, nickel sulphide deposits and porphyry coppers), massive sulphides are well represented throughout the geological column (Fig. 2). Figure 2 is not very reliable because there are little adequate data from many areas, particularly from the Cambrian. The main features include the high number of deposits in the Carboniferous and Devonian, and the low number in the Permo-Triassic and the Archaean. The Carboniferous high reflects the large number of deposits in the Iberian pyrite belt, and the low value for the Precambrian (and the Cambrian?) may well be a function of the low preservation rate of these rocks[19]. The plot of tonnages (production plus reserves) of zinc and copper every 100 Myr (Fig. 3), though based on limited data, seems to confirm the Archaean and Permo-Triassic lows, and indicates the existence of several very large Proterozoic deposits (Sullivan, McArthur River, Mount Isa).

The principal features of massive sulphides do not conflict with, and are to some extent predicted by, the geothermal model outlined. In particular the model accounts for the lack of related plutons, the poor correlation between host rock and ore type, and the wide range of tectonic environments in which the deposits occur. The grouping along faults and the S-isotope data and overall chemical composition, are also compatible with the model.

If massive sulphides are essentially of superficial origin, with the entire ore forming process extending no more than a few km below the surface, it is difficult to see why there should be any close relationship with plate tectonic processes and related magmatism. The tendency for ore to be concentrated in marine volcanic zones in orogenic belts may simply be because such areas are the most favourable for developing convective hydrothermal systems and providing plumbing systems and depositional sites. The Permo-Triassic low in the distribution graphs may reflect the relatively low output of volcanic rocks at this time[20], and the large size of the few Proterozoic deposits may be a function of relatively stable conditions in large basins Though the model seems to account for most of the data it does not necessarily explain the genesis of every deposit. The model needs to be tested on individual cases, possibly using isotope tracers[21].

M. SOLOMON

Department of Geology,
University of Tasmania,
Hobart, Australia 7005

Received December 17, 1973; revised April 9, 1974.

1. Sawkins, F. J., *J. Geol.*, **80**, 377 (1972).
2. Sillitoe, R. H., *Trans. Instn Min. Metall.*, **B81**, 141 (1972).
3. Mitchell, A. H. G., *Nature phys. Sci.*, **245**, 49 (1973).
4. Mitchell, A. H. G., and Garson, M. S., *Trans. Instn Min. Metall.*, **B81**, 10 (1972).
5. Tarling, D. H., *Nature*, **243**, 193 (1973).
6. Guild, P. W., *Proc. twenty-fourth int. geol. Congr.*, **4**, 17 (1972).
7. Sangster, D. F., *Can. geol. Surv.*, Paper 72-22 (1972).
8. Craig, H., in *Hot Brines and Recent Heavy Metal Deposits in the Red Sea* (edit. by Degens, E. T. and Ross, D. A.), (Springer-Verlag, Berlin, 1969).
9. Ellis, A. J., in *Geochemistry of Hydrothermal Ore Deposits* (edit. by Barnes, H. L.), (Holt, Rinehart and Winston, 1967).
10. Corliss, J. B., *J. geophys. Res.*, **76**, 8128 (1971).
11. Henley, R. W., *Trans. Instn Min. Metall.*, **B82**, 1 (1973).
12. Anderson, C. A., *Econ. Geol.*, **64**, 129 (1969).
13. Hutchinson, R. W., and Hodder, R. W., *Trans. Can. Inst. Min. Met.*, **75**, 16 (1972).
14. Stanton, R. L., *Econ. Geol.*, **50**, 681 (1955).
15. Mitchell, A. A. G., and Garson, M. S., *Trans. Instn Min. Metall.*, **B82**, 43 (1973).
16. Solomon, M., and Griffiths, J. R., *Geol. Soc. Aust.*, Spec. Pub. 4 (in the press).
17. Lusk, J., *Econ. Geol.*, **67**, 169 (1972).
18. Sasaki, A., *Geochem. J.*, **4**, 41 (1970).
19. Garrels, R. M., and Mackenzie, F. T., *Evolution of Sedimentary Rocks* (Norton, New York, 1971).
20. Ronov, A. B., *Sedimentology*, **10**, 25 (1968).
21. Cooper, J. A., and Richards, J. R., *Geochem. J.*, **3**, 1 (1969).

Editor's Comments
on Papers 26 Through 29

METALLOGENESIS AND PLATE TECTONICS
—LATER DEVELOPMENTS

As plate tectonic models themselves become more sophisticated, so will their application to metallogenic theory. Processes at plate margins differ in detail from place to place: as rates of sea-floor spreading change, for example, so also must rates of subduction at destructive plate margins. And the angle of subduction at such margins is known to vary because the dip of Benioff zones is not the same everywhere.

In Paper 26, Mitchell expands upon the theme introduced in Figure 5 and related text of Paper 18, and considers how metallogenesis can be controlled by the angle and rate of subduction. He concludes that while we can deduce how plate movements have been responsible for the distribution of *known* mineral deposits, we are still some way from applying plate tectonic models in a predictive way to the location of new deposits. Is this a realistic or an overly pessimistic view?

If we develop Mitchell's main theme further, there is no inherent reason why subduction zones should have the same angle of dip and rate of movement along their whole strike length. Sillitoe applies this reasoning to magmatism and metallogenesis in the Andes, where contrasted igneous rock assemblages and mineral deposits have been recognized for many years (Paper 27). He suggests that the location of mineral deposits is more likely to be controlled by the boundaries between such segments, which must reflect discontinuities in the underlying lithosphere, rather than by less fundamental lineament and fracture systems in the crust. Is it possible that some of the major lineaments believed to control ore emplacement in other parts of the world represent such tectonic boundaries?

The Mantle Plume Concept

Constructive plate margins are linear rift-like features, along which upper mantle material rises to form new oceanic lithosphere. Mantle plumes or hot spots are localized sources of deep mantle upwelling, which are believed to be responsible for much intra-plate magmatism. Since this magmatism is often associated with rifting, it has also been suggested that mantle plumes may initiate the crustal stretching and fracturing that lead to continental break-up and sea-floor spreading (e.g. Morgan, 1971, 1972; Burke and Dewey, 1973).

The plume concept is relevant to metallogenesis, for it is widely believed that plumes bring primitive—i.e. largely unfractionated—mantle material from depth into the lithosphere. Such material should therefore contain the higher concentrations of ore-forming elements than both the asthenosphere and lithosphere, and the magmatic products of plume activity should be enriched in these elements.

Both Iceland and the Azores have been identified as plume sites (e.g. Morgan, 1971, 1972; Schilling, 1973a), although the geochemical criteria for such an identification have been challenged in the case of Iceland: O'Hara (1973) maintains that the data can be explained just as convincingly by invoking fractional crystallization at high crustal levels (but see Schilling, 1973b). On the other hand, small amounts of mineralization identified in both places (Dmitriev et al., 1971; Jancovic, 1972) could be attributed to the rise of primitive mantle material beneath them.

Burke and Dewey (1973) suggest that mantle plumes rising beneath continental lithosphere give rise to broad uplifts which develop crustal rifts in three directions at about 120°. Where this doming is followed by continental break-up, two of the rifts commonly become spreading ridges, the third becomes inactive—the so-called failed arm of the triple junction. They go on to say (1973, p. 427):

> Mineralization in rifts and failed arms (e.g. the Keweenawan, Coppermine, Belt of Montana, Southern Alberta buried rift, Danish arm of the Jutland structure, Copperbelt of Zambia and Katanga) commonly consists of copper and other base metals normally deposited syngenetically as sulphides in volcanic deposits or sediments. The mineralization in the Benue trough . . . lead, zinc, and copper sulphides . . . in . . . anti-

217

> clines . . . is perhaps similar. . . . Red Sea and Salton Sea brines have
> often been compared to rift, or failed arm, mineralization.

Presumably they envisage failed arm mineralization to be similar to that of constructive plate margins; although not everyone would agree with their identification of the Red Sea and Salton Sea as failed arms.

Rifting is not always a necessary concomitant of mineralization related to mantle plumes beneath continental areas. Thus, Crockett and Mason (1968) suggested that the distribution of kimberlites and nickel deposits in southern Africa was controlled by fracture systems related to "foci of mantle disturbance," which could readily be re-interpreted as mantle plumes, even though they involved no continental movements. In a somewhat similar approach, Sillitoe (Paper 28) has suggested that the anorogenic granites and related tin mineralization in parts of Africa and South America were a consequence of plume activity and that the tin and associated ore metals are principally of mantle, rather than crustal, derivation.

Implicit in practically all the literature on this subject, is the assumption that mantle plumes are stationary with respect to moving lithosphere. On the other hand, if anorogenic continental granites (and any related mineralization) are really the product of plume activity (Paper 28), at least some of these plumes cannot be stationary at all, but *must* move with the lithosphere (Wright, 1973), especially if there has been repeated mineralization. It follows *either* that mantle plumes cannot originate beneath the asthenosphere as commonly supposed, for if they do they must be stationary; *or* that anorogenic granites need not be products of mantle plume activity.

It is of course also possible that plumes beneath continents may be of a different kind from those beneath oceans; and there is another alternative too, namely that mantle plumes do not exist at all, either beneath continents (e.g. Bailey, 1974, 1975) or beneath oceans (e.g. O'Hara, 1973; Anguita and Hernan, 1975)

The Transform Fault Setting

The great majority of transform faults are not plate boundaries, and since they must be deep fractures, they could provide suitable loci for mineralization, particularly at intersections with other fracture systems. Thus, Kutina (1969) has suggested that several of the important ore deposits of North America may be related to continental extensions of major east-west transform fractures in the Pacific plate, particularly where such extensions intersect an empirically determined "net" of northwest-southeast and northeast-southwest lineaments (compare also Kutina, 1972).

On a more local scale, Guilbert (1971) has related the Boleo copper deposit of Baja California to development of the Gulf of California, which is where the spreading axis of the East Pacific Rise becomes the San Andreas transform (transcurrent) fault system (Figure 2). The Boleo ores occur among shallow water sediments and calc-alkaline volcanics, the latter being normally a feature of destructive plate margins. Therefore, although the deposit appears to be situated upon a transform fracture, and possibly at an intersection with a spreading ridge segment, Guilbert acknowledges he cannot be sure that this was the main control on mineralization.

In Paper 29, on the other hand, Bignell shows that the distribution of brine pools in the Red Sea median valley may well be controlled by transform fault intersections; a conclusion which could have important implications for discovery of possible metal concentrations in other parts of the global ridge-rift system. There are mineralized sediments of more than one age in the Red Sea deeps, and Bignell also concludes that their deposition coincided with major sea-level changes during the Pleistocene. Somewhat unconvincingly, he discounts magmatic heat as a source of energy for driving the hydrothermal system: the brine pools lie over a spreading axis where the geothermal gradient must be higher than normal due to the rise of magma from the upper mantle along this accreting plate margin.

Although some transcurrent faults in continental regions may be landward extensions of oceanic transform faults, the majority are probably not. They are major fractures nonetheless, and there are plenty of exploration geologists who believe them to be important loci of mineralization, irrespective of whether or not they recognise any relationship to plate tectonics.

For example, in a paper discussing concepts in mineral exploration, which is remarkable (refreshing?) for its complete lack of any mention of plate tectonics, James (1972) lays great emphasis on the delineation of fracture systems as an exploration tool. He has this to say about transcurrent faults (= transform faults?) in continental regions (pages B139–B140):

> The recognition of large-scale transcurrent faulting in crystalline basements or shield areas may have a profound effect on mineral exploration. Most of the Canadian Shield is now covered by aeromagnetic surveys. . . . Compilation of these data . . . shows the pattern of transcurrent faulting in the shield not always evident from geologic mapping. . . . The recognition of major transcurrent faulting throughout the Canadian Shield . . . may fundamentally modify some of the concepts used in mineral exploration. First, some of these faults appear to have horizontal translations of the order of tens to hundreds of miles and must disrupt the crust to the crust-mantle interface. Secondly, shearing of this magnitude and extent is unlikely to be confined to the northern part of the North American continent. . . . it is likely to be a world-

wide phenomenon of crystalline basements so far unmapped or un-
recognised for lack of gewophysical data. . . . by using the idea of
rejuvenated large-scale faulting in . . . crystalline basements . . . as chan-
nelways for . . . metalliferous mineralization . . . it is possible to evolve
a unified concept of mineralization initiated about the crust-mantle
interface, but greatly modified by its journey outward through the
crust. . . . the massive belts of transcurrent faulting in Africa suggest the
presence of nickel sulphide mineralization as yet undetected because
of. . . . lack of regional geophysical data.

Another review paper by Guild (1977) recapitulates the theme of
earlier sections and considers some of the new developments introduced
in this one.

Metallogenic Belts and Angle of Dip of Benioff Zones

A. H. G. MITCHELL

Overseas Development Administration, c/o Overseas Division Institute of Geological Sciences, 5 Princes Gate, London SW7 1QN

The emplacement of metals and associated granites near Benioff zones can now be understood and interpreted; but the state of the art does not yet allow fruitful application of these relationships to predictions of the positions of as yet untapped mineral resources.

BENIOFF zone models have been used in the past two years to explain the presence of metallogenic belts above modern and ancient consuming plate margins[1]. For example, porphyry copper and Kuroko-type stratiform massive sulphides located on continental margins and in island arcs have been related to Benioff zones at a constant position and inclination[2-5], and deposits of the tin-tungsten-bismuth-fluorite association located near continental margins bordered by marginal basins and island arcs have been interpreted in terms of oceanward migration of Benioff zones[2].

Relationships of igneous, metallogenic and tectonic processes to changes in dip of the underlying Benioff zone have received relatively little attention. But the possible importance of these relationships is indicated by evidence that Benioff zone inclination may control marginal basin spreading and the consequent migration of island arcs. Behind most island arcs which overlie Benioff zones dipping at more than about 35° there has been either late Cainozoic extension of the marginal basin or, less commonly, major displacements along strike-slip faults. Behind the Marianas arc marginal basin spreading may be cyclic and related to changes in the inclination of the Benioff zone, spreading taking place only when the inclination is steep[6]. This suggests that development of marginal basins occurs only above Benioff zones with a steep or moderate dip.

I suggest here that rhyolitic rocks and granites with associated high level porphyry copper and deep level tin deposits emplaced near continental margins are related to shallow-dipping Benioff zones, and that andesites, tonalites and associated porphyry copper deposits emplaced on continental margins or in island arcs are related to Benioff zones with a steeper dip.

Quaternary Porphyry Copper Deposits

In the Peruvian Andes the youngest igneous rocks are Pliocene–Quaternary quartz porphyry stocks and ignimbrites lying in a belt east of the late Mesozoic and early Tertiary andesitic volcanic rocks and tonalitic batholiths[7] (Figs 1c and 2a_2). Porphyry copper and vein-type mineralization are associated with some of the porphyry stocks (E. J.

Cobbing, personal communication). Variations in composition of erupted rocks with depth to the Benioff zone near other plate margins[8] suggest that the origin of the Pliocene-Quaternary igneous belt in Peru is related to the present Benioff zone. This lies less than 200 km beneath the belt and approaches the surface in a submarine trench approximately 300 km to the south-west; seismic evidence indicates an angle of dip beneath the continent of less than 30° (ref. 9).

In Sumatra (Fig. 1a) Quaternary igneous rocks include large volumes of silicic pyroclastics and occur in and near an active strike-slip fault zone[10,11] in a belt parallel to the island's trend (Fig. 2b). Active volcanoes are restricted to the same belt and erupt similar silicic and intermediate pyroclastic rocks[12]. The Benioff zone lies not more than 200 km beneath the volcanoes[13], approaches the surface beneath a poorly defined trench 300 km to the south-west, and thus has an average north-easterly dip of 30° to 35°.

Late Mesozoic Tin Mineralization

Near the Pacific margin of Asia and Alaska and the south-western margin of South-east Asia there are belts of granites and acidic eruptive rocks, mostly late Mesozoic[14] (Fig. 1a). Associated with the upper parts of these granites are deposits of tin, tungsten and fluorite with minor bismuth and molybdenum, mostly emplaced at a depth of several kilometres. Examples are in the Burma-Thai Peninsula[15], in the Kiangsi belt of South China[16], in Korea, East China and eastern Siberia[17,18] and in the Seward Peninsula of Alaska[19]. If granodioritic and tonalitic plutons are the deep level equivalents of calc-alkaline andesitic volcanic rocks[20], then granitic rocks of eastern Asia and Alaska may be interpreted as the deep-level equivalents of the ignimbrites and acidic stocks of Peru and possibly of the ignimbrites of Sumatra.

The late Mesozoic granitic rocks of eastern Asia are mostly not underlain by present day Benioff zones, but are bordered on the ocean side either by marginal basins and island arcs or less commonly by calc-alkaline igneous belts along the continental margin[2] (Fig. 1a). The concept of marginal basin spreading[21] implies that the bordering island arcs were formerly on or close to the continental margin, and there is evidence that the Andaman and Nicobar islands were probably adjacent to the Burma-Thai Peninsula in the early Tertiary[22], that Japan lay adjacent to mainland Asia in the late Mesozoic[23], and that Borneo lay close to South China prior to late Mesozoic opening of the South China Sea[24]. The association between silicic igneous activity and an underlying shallow-dipping Benioff zone in the Peruvian Andes and Sumatra suggests that during the late Mesozoic to early Tertiary granitic rocks near the eastern and south-western margins of eastern Asia were emplaced above shallow-dipping Benioff zones (Fig. 2c_1). The zones probably approached the surface beneath a trench adjacent to the continental margin, prior to the development of marginal basins and island arcs, as has been suggested for the Cretaceous margin of Alaska[25].

The tin and related mineral deposits which occur in and around the Asian granite plutons were emplaced either during or subsequent to the final stages of igneous activity. In the Quaternary volcanic rocks of Sumatra and Peru, which can be interpreted as higher-level equivalents of the east Asian granites, the scarcity or absence of tin mineralization may reflect the shallow depth of erosion. But tin deposits might also be absent at depth either if granite

Asia initial rifting, probably along a fault zone (Fig. $2c_2$), was followed by growth of marginal basin crust and oceanward migration of the developing island arc (Fig. $2c_3$).

Development of Cainozoic island arcs, above Benioff zones at depths of 120 to 200 km which were mostly inclined at more than about 35°, resulted largely from calc-alkaline predominantly andesitic volcanism between 150 and 200 km from the trench, and island arc tholeiite volcanism nearer the trench[28]. Mineralization in the migrating arcs included magmatic-hydrothermal porphyry copper and gold deposits in or above dioritic and tonalitic stocks, and shallow-water

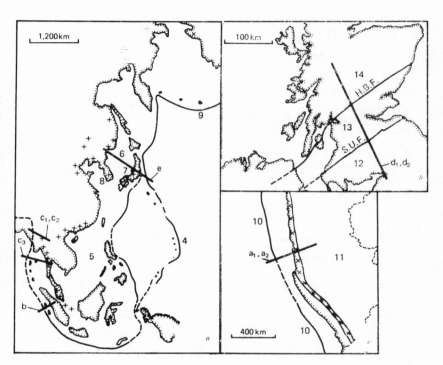

Fig. 1 Location of cross sections shown in Fig. 2. *a*, Eastern Asia: 1, Andaman Sea; 2, Sumatra; 3, Borneo; 4, Marianas Trench; 5, South China Sea; 6, Japan Sea; 7, Honshu I.; 8, Korea; 9, Aleutian Trench; +, tin-bearing granites; trenches (solid heavy line), and other probable plate boundaries (broken heavy line) from Fitch[13]. *b*, Scottish Caledonides: 12, Southern Uplands; 13, Midland Valley; 14, Grampian Highlands; SUF, Southern Uplands Fault; HBF, Highland Boundary Fault. *c*, Peruvian Andes: late Mesozoic to early Tertiary andesitic volcanics and coastal batholith shown by oblique crosses (from Cobbing and Pitcher[31]); 10, Peru-Chile Trench; 11, Eastern Andes and Craton of Peru.

batholiths have not yet been emplaced beneath the volcanic rocks, or if the mineralization is related to subsequent Benioff zone migration[3].

Inclination and Mineralization

Increase in inclination of the Benioff zones beneath eastern Asia, from shallow-dipping zones above which granites were emplaced during the late Mesozoic, to steeper-dipping zones above which Cainozoic marginal basins and island arcs developed, was probably related to decrease in rate of lithosphere descent[26]. This decrease may have resulted from a change in spreading direction with consequent oblique descent of lithosphere. Stress changes associated with the oblique descent could have caused strike-slip faulting through or near to the volcanic belt, as in Sumatra today[13] (Fig. 2b). Part of the Sumatra fault has been referred to as a volcano-tectonic rift zone, and it has been suggested that the Gulf of California opened along a similar rift zone between a belt of silicic igneous rocks on the continental margin and younger calc-alkaline andesitic rocks in Baja California, nearer the trench[27]. Similarly around the margins of eastern

Kuroko-type[2,4] and deep water Besshi-type[29] stratiform massive sulphides (Fig. $2c_3$).

The position of the rift zone and subsequent marginal basin relative to the granitic belt and younger andesitic or tonalitic belt of the island arc varies considerably between different arc systems, and is perhaps related to the inclination of the initially shallow-dipping Benioff zone. Beneath the Barren and Narcondam island in the Andaman Sea there may be only a narrow segment of original continental margin crust, and the belt of tin-bearing granites remains on the Burma-Thai Peninsula (Figs 1a and $2c_3$). But in the northeast of Honshu Island in Japan late Mesozoic tin-bearing granites are bordered and locally overlain by younger andesitic rocks[30] (Fig. 1e). These granites are probably part of the tin-bearing granite belt which extends through Korea, ripped off the East Asian mainland during early Tertiary rifting before development of the sea of Japan.

In the Andes the non-development of Cainozoic island arcs could be explained by early Cainozoic decrease in inclination of an initially steeply dipping Benioff zone. Thus in Peru the late Mesozoic to early Tertiary andesitic volcanic rocks and the tonalites and minor granite differentiates of the coastal batholith[31] (Fig. 1c) may have formed above slowly descending lithosphere along a Benioff zone with steep or moderate dip (Fig. 2a_1). Before a marginal basin could develop, decrease in dip of the Benioff zone, probably resulting from increase in rate of lithosphere descent associated with changing global spreading patterns, led to em-

are late Ordovician and Silurian, has been related to northward descent of lithosphere along a Benioff zone which approached the surface in a trench near the present site of the Southern Uplands[32-34] (Fig. 2d_1). Subsequent emplacement of the calc-alkaline lavas and plutons of the Midland Valley, closer to the trench, during the late Silurian and

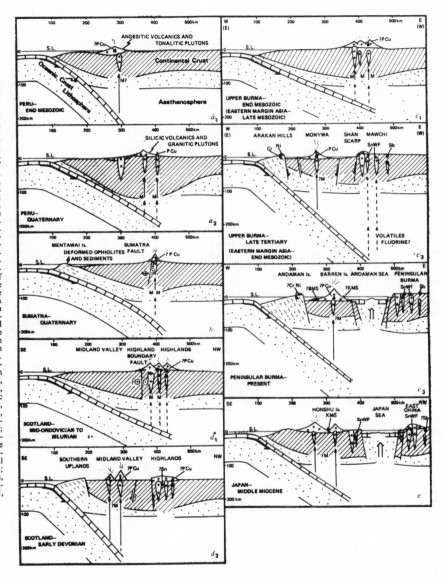

Fig. 2 Generalized diagrammatic cross sections through consuming plate margins. For explanation see text. a_1–a_2, Peruvian Andes; b, Sumatra; c_1–c_2–c_3, stages of development of northern part of Sunda arc represented by cross sections at different localities; c_1–c_2–e, stages of development of Asian Pacific margin; in e, pre-arc rift zone lies further from original trench position than in c_3. I emphasize the similarity of igneous activity and tectonic setting between present-day Andes (a_2), present-day Sumatra (b), late Mesozoic of East Asia margins (c_1) and late Ordovician and Silurian of Scotland (d_1); also the similarity between late Tertiary of upper Burma (c_2) and early Devonian of Scotland (d_2). Possible rise of fluorine and other volatiles above deep Benioff zone shown in c_2 is omitted from other cross sections. Crustal thickness is exaggerated ×2. Known mineral deposits: P Cu, porphyry copper; Cr, chromite; Ni, lateritic nickel; Sn W F, tin, tungsten, fluorite; Sb, antimony; BMS, Besshitype massive sulphide; KMS, Kuroko-type massive sulphide. Possible locations for mineral deposits: solid vertical arrows, rising magma; broken vertical arrows, rising volatiles or volatiles and magma; M, partial melting.

placement of the late Cainozoic acidic stocks and ignimbrites and inhibited rifting and island arc development (Fig. 2a_2).

Certain features of some older orogenic belts could be interpreted in terms of increase in dip of Benioff zones. In the British Caledonides (Fig. 1b), for example, emplacement in the Scottish Highlands of granitic plutons, many of which

Devonian could be explained by increase in inclination of the initially shallow-dipping Benioff zone (Fig. 2d_2). The granitic rocks of the highlands lack associated tin mineralization, and are mostly rather less acidic than tin bearing granites, but by analogy with the late Cainozoic stocks of Peru porphyry copper deposits may have been present

223

at structural levels higher than the present erosion surface. Similarities between the middle Palaeozoic development of Scotland (Fig. 2d) and the Cainozoic of Burma (Fig. 2c) will be considered in a paper in preparation.

Origin of Metals

It has been suggested that the metals in porphyry copper, some types of massive stratiform sulphide[3,4,35], and tin deposits[35] are derived from oceanic crust descending along the underlying Benioff zone. If calc-alkaline volcanic rocks result from partial melting of metamorphosed descending oceanic crust[36], then either the sedimentary or underlying basaltic layers could supply copper to these melts. But evidence that tonalitic plutons form by partial melting at the base of the continental or island arc crust[37,38] suggests that the metals in associated porphyry copper deposits are also derived from the crust. There is no obvious source of tin in either the downgoing oceanic crust and sediments or in the overlying mantle, and if granites are derived from partial melting of continental crust[39] it is probable that the associated tin, tungsten and bismuth also have a crustal source.

Extraction, transport, concentration and deposition of the metals are probably dependent largely on the nature and volume of volatiles[40] expelled from the descending crust along the Benioff zone. Thus deposition of tin could be contemporaneous with emplacement of fluorine-rich granite[41]. Alternatively, tin mineralization might take place later as a result of either increase in inclination of the Benioff zone or Benioff zone migration during island arc development; a rising stream of fluorine and other volatiles, derived from descending crust at depths exceeding 200 km, could then intersect bodies of already emplaced but still hot granite[2] (Fig. 2c_2). In West Malaysia, the large volumes of tin-bearing granites were possibly intruded intermittently throughout the Mesozoic above successive shallow-dipping Benioff zones, resulting in several heating events[42].

The occurrence of porphyry copper deposits, associated both with silicic volcanic rocks related to a shallow dipping Benioff zone 300 km from a submarine trench as in Peru, and also with andesitic volcanic rocks related to steeper Benioff zones as in the volcanic arc of Burma and in island arcs, suggests that two gradational types of porphyry deposit corresponding to the two settings might be recognized. There is some evidence that copper-molybdenum deposits are more common in the former setting and copper-gold in the latter. More speculatively, the location of antimony deposits in Burma[43] and China[44] mostly on the continental side of the tin-bearing granite belts suggests that antimony mineralization could be related to rise of volatiles and heat from Benioff zones at a depth of several hundred kilometres (Fig. 2c_2).

Mineral Exploration

I conclude that tin-tungsten-bismuth-fluorite deposits and associated granites are emplaced near continental margins above shallow-dipping Benioff zones. Porphyry copper deposits are emplaced both in andesitic or tonalitic rocks on continental margins and island arcs above steeply dipping Benioff zone, and also in silicic volcanic rocks above tin-bearing granites. The location of Mesozoic and early Tertiary tin deposits near continental margins mostly bordered by island arcs is related to postgranite steepening of the Benioff zones and resulting marginal basin development.

Benioff zone models cannot be applied successfully to mineral exploration until they can be used to predict, as opposed to interpret, the distribution of metallogenic belts, and until tectonic controls on magma and ore emplacement

can be related to large-scale plate movements. Evidence for the changing dip of Benioff zones indicates that attempts to relate metallogenic belts of different ages within an orogenic belt[45] to a Benioff zone at constant inclination are unlikely to be successful.

I thank Dr E. J. Cobbing for discussions and unpublished data, and Dr R. Garrisson, M. S. Garson and W. Johnson, Mr T. Marshall and Dr R. Pankhurst for helpful comments.

Received July 13, 1973.

[1] Pereira, J., and Dixon, C. J., *Mineralium Deposita*, **6**, 404 (1971).
[2] Mitchell, A. H. G., and Garson, M. S., *Trans. Instn Min. Metall.*, **B82**, 40 (1972).
[3] Sillitoe, R. H., *Econ. Geol.*, **67**, 184 (1973).
[4] Sawkins, F. J., *J. Geol.*, **80**, 377 (1972).
[5] Guild, P. W., in *International Geological Congress, Sect. 4, Mineral Deposits*, 17 (Montreal, 1972).
[6] Bracey, D. R., and Ogden, T. A., *Geol. Soc. Am. Bull.*, **83**, 1509 (1972).
[7] Cobbing, E. J., *Geological Map of the Western Cordillera of Northern Peru*, Overseas Development Administration (Ordnance Survey, London, 1972).
[8] Kuno, H., *Can. Geol. Surv. Paper*, 66–15, 317 (1966).
[9] James, D. E., *Geol. Soc. Am. Bull.*, **82**, 3325 (1971).
[10] Katili, J. A., *Geol. Rundsch.*, **59**, 581 (1970).
[11] Posavec, M. M., Taylor, D., van Leeuwen, T., and Spector, A., in *Regional Conference on the Geology of South-east Asia* (edit. by Haile, N. S.), abstract, 47 (Geological Society of Malaysia, 1972).
[12] Van Bemmelen, R. A., *Geology of Indonesia, General Geology*, **1A** (Government Printing Office, The Hague, 1949).
[13] Fitch, T. J., *J. geophys. Res.*, **77**, 4432 (1972).
[14] Matsumoto, T., *Pacif. Geol.*, **1**, 77 (1968).
[15] Gardner, L. S., and Smith, R. M., *Rep. Investig. Dept. Mineral Resources*, 10 (Bangkok, 1965).
[16] Hsu, K. C., and Ting, J., *Geol. Surv. China Mem.*, Ser. A, 17 (1943).
[17] Gallagher, G., *Mineral Resources of Korea*, **6A** (Mineral Branch, Industrial Mining Division, 650M/Korea, 1963).
[18] Sainsbury, C. L., *U.S. Geol. Surv. Bull.*, 1301 (1969).
[19] Sainsbury, C. L., *U.S. Geol. Surv. Bull.*, 1287 (1969).
[20] Hamilton, W., *Geol. Soc. Am. Bull.*, **80**, 2409 (1969).
[21] Karig, D. E., *J. geophys. Res.*, **76**, 2542 (1971).
[22] Rodolfo, K. S., *Geol. Soc. Am. Bull.*, **80**, 1203 (1969).
[23] Matsuda, T., and Uyeda, S., *Tectonophysics*, **11**, 5 (1971).
[24] Ben-Avraham, Z., and Uyeda, S., *Earth planet. Sci. Lett.*, **18**, 365 (1973).
[25] Moore, J. C., *Science, N.Y.*, **175**, 1103 (1972).
[26] Luyendyk, B. P., *Geol. Soc. Am. Bull.*, **81**, 3411 (1970).
[27] Karig, D. E., and Jensky, W., *Earth planet. Sci. Lett.*, **17**, 169 (1972).
[28] Jakes, P., and Gill, J. B., *Earth planet. Sci. Lett.*, **9**, 17 (1970).
[29] Mitchell, A. H. G., and Bell, J. D., *J. Geol.* (in the press).
[30] Sekine, Y., in *Geology and Mineral Resources of Japan*, second ed. (edit. by Saito, M., *et al.*), 141 (Geological Survey of Japan, 1960).
[31] Cobbing, E. J., and Pitcher, W. S., *Nature phys. Sci.*, **240**, 51 (1972).
[32] Dewey, F. J., and Pankhurst, R. J., *Trans. R. Soc. Edinb.*, **68**, 361 (1970).
[33] Dewey, J. F., *Scottish J. Geol.*, **7**, 219 (1971).
[34] Garson, M. S., and Plant, J., *Nature phys. Sci.*, **242**, 34 (1973).
[35] Halls, C., and Sillitoe, R. H., *Trans. Instn Min. Metall.*, **B82**, 44 (1973).
[36] Jakes, P., and White, A. J. R., *Geol. Soc. Am. Bull.*, **83**, 29 (1972).
[37] Jakes, P., and White, A. J. R., *Earth planet. Sci. Lett.*, **12**, 224 (1971).
[38] Brown, G. C., *Nature phys. Sci.*, **241**, 26 (1973).
[39] Stern, C. R., and Wyllie, P. J., *Earth planet. Sci. Lett.*, **18**, 163 (1973).
[40] Rub, M. G., *Chem. Geol.*, **10**, 89 (1972).
[41] Tischendorf, G., *Trans. Instn Min. Metall.*, **B82**, 9 (1973).
[42] Snelling, N. J., Bignell, J. D., and Harding, R. R., *Geologie Mijnb.*, **47**, 358 (1968).
[43] *Distribution of Mineral Deposits* (Directorate Geological Survey and Exploration, Rangoon, 1972).
[44] *Oxford Economic Atlas* (edit. by Jones, D. B.), fourth ed. (Oxford University Press, Oxford, 1972).
[45] Sillitoe, R. H., *Geol. Soc. Am. Bull.*, **83**, 813 (1972).

Tectonic segmentation of the Andes: implications for magmatism and metallogeny

Richard H. Sillitoe

Department of Mining Geology, Royal School of Mines, Imperial College of Science and Technology, London SW7 2BP, UK

The Andean orogen consists of a series of tectonic segments separated by transverse boundaries that reflect discontinuities on the underlying subduction zone. The characteristics of belts of magmatic rocks and the type, age and size of ore deposits in a series of longitudinal belts may change at tectonic boundaries.

THE central Andes provide a classic example of a volcano–plutonic orogen developed along a convergent plate margin[1]. The longitudinal continuity and transverse variability of characteristic volcano-plutonic orogens are commonly emphasised[2]. Although I accept that longitudinal continuity is dominant in such orogens, I stress here the potential significance of a fundamental, longitudinal segmentation of the central Andes and, by analogy, of other comparable orogenic belts above subduction zones.

In Chile a series of longitudinal, physiographic provinces (the Norte Grande, Norte Chico, Central Chile, and Southern Chile) has been recognised for many years. Some of the boundaries between the physiographic provinces must have a seismic significance because they coincide with positions at which the level of seismicity changes[5,6]. The Norte Grande and Central Chile, with longitudinal, fault-bounded valleys, bordered to the east by lines of recent volcanoes[3,4,7], contrast with the Norte Chico from which these features are absent, thereby demonstrating that the physiographic provinces also·exert a control over tectonism and calc-alkaline magmatism.

Similar, possibly more important, transverse boundaries which coincide with major features in the east-central Pacific, effectively define the northern and southern limits of the central Andes and mark a single, transverse division[8].

A study of intermediate and deep focus earthquakes in the Japanese arcs[9] revealed a series of offsets and changes in strike in the deep seismic zone, which is thus effectively divided into a number of 100–300 km long segments. These discontinuities can be correlated with surface geological features, such as: offsets or changes in the strike of the trench axis or of belts of volcanoes; and lines of volcanoes, prominent structures or topographic changes transverse to the arc[9]. It is proposed here that the well known longitudinal subdivisions of Chile, and the boundaries emphasised by Gansser[8], are comparable to the transverse geological features in Japan and are probably also coincident with discontinuities on the underlying deep seismic zone.

Using the criteria of Carr *et al.*[9], together with additional geological evidence, I have subdivided the central Andes into tectonic provinces (Fig. 1).

Definition of tectonic segments

Boundary 1 (Fig. 1) is known as the Amotape zone[7,10,11], and marks the northern limit of the central Andes[8]. The overall geology changes markedly at this point; to the north, a coastal belt of ophiolitic rocks appears[8,12], together with a line of recent stratovolcanoes associated with a longitudinal graben[7]. The boundary coincides with local transverse strikes[8] and faults[13], aligned with the Gulf of Guayaquil and the Carnegie Ridge to the west, and the Amazon depression, perhaps an old shield lineament[11], to the east.

Boundary 2 approximately coincides with the Huancabamba deflection[14] at a point where the Andes change their strike from north–west to north–north–east.

Transverse strikes have been noted on boundary 3 (ref. 10), which marks the southern limit of the Coastal Cordillera.

Boundary 4 coincides with the Pisco[14] or Abancay[15,16] deflection and has been proposed as a line of division in the central Andes[8]. It is marked by a major step in the coastline, which indicates the abrupt northward disappearance of the Precambrian formations of the Coastal Cordillera[17,18], and by the offshore Nazca Ridge, and a shallowing of the Peru–Chile trench[19]. On the boundary there is a marked northward narrowing and change in direction of the Eastern Cordillera, and north-east–south-west striking formations[16,20]. Important changes in the Mesozoic palaeogeography have also been observed[16].

Boundary 5 is marked by the northern limits of the belt of recent volcanoes and the Altiplano–Puna block.

Boundary 6 is characterised by a change in Andean strike, from north to north-west, and by a northward narrowing of the Eastern Cordillera, where it has been termed the Ichilo fault zone[21] or line[22], or the Arica Elbow line[23]. East of the Andean orogen the Precambrian basement is veneered by Cainozoic sediments to the north of the boundary, whereas to the south, Palaeozoic and Cretaceous formations are also present[24]. The

westward extension of the Arica Elbow line is less apparent but has been proposed[22,23,25].

Boundary 8 is so positioned because of the coincidence of a sinuous portion of the Peru–Chile trench[19] with the increase in width of the longitudinal valley (Fig. 1), the northern limit of the north-north-east trending Cordillera Domeyko[26], a westward step in the longitudinal belt of recent volcanoes, and an extension of Quaternary volcanoes in an irregular transverse belt extending as far as the Eastern Cordillera, especially in Sud Lipez in southernmost Bolivia.

Boundary 10, between the Norte Grande and the Norte Chico, coincides with the approximate southern limits of the belt of recent volcanoes, the longitudinal valley and the elevated

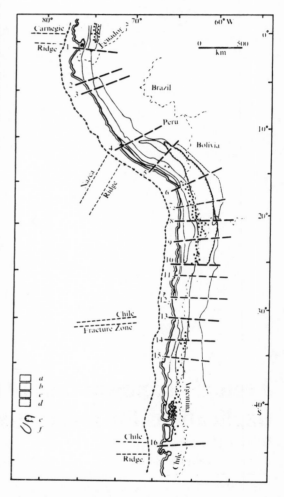

Fig. 1 The tectonic boundaries proposed for the central Andean volcano–plutonic orogen, in relation to certain physiographic features. *a–d*, metallogenic belts: *a*, iron; *b*, copper–(gold–molybdenum); *c*, copper–lead–zinc–silver; *d*, tin–(tungsten–silver); *e*, Altiplano–Puna; *f*, Longitudinal and Central Valleys; x, Location of recent volcanoes; +–+–+–, axis of Peru–Chile trench. Heavy dashed lines are the boundaries between the tectonic segments; numbered as referred to in the text; a subscript 2 denotes a boundary of only secondary importance.

Altiplano-Puna block[27] (Fig. 1). To the south there is a distinctive Cainozoic geomorphological development[28,29]. The boundary also seems to have controlled the southern limit of widespread volcanism in the Jurassic, and the northern limit of widespread submarine volcanism in the Lower Cretaceous[30].

Boundary 13 marks the junction between Central Chile and the Norte Chico. It limits the northernmost extent of the belt of recent volcanoes, of the Central Valley—a Palaeogene feature[31,32], of the belt of Upper Tertiary andesitic–basaltic volcanics, and of widespread Precambrian rocks and Palaeozoic intrusives in the Coastal Cordillera[26]. In Argentina, the boundary controls the southernmost extent of the Precordillera[33] and the Sierra de Córdoba. The boundary also lies at approximately the same latitude as the Chile fracture zone in the east-central Pacific (Fig. 1).

Boundary 16 is the southern limit of the central Andes[8] and of the Central Valley of Chile (Fig. 1), and lies at the same latitude as the spreading Chile Ridge. It also corresponds to the southern limit of volcanism in Upper Cretaceous, Eocene, late Tertiary and recent time[30].

Correlations with seismicity

In order to confirm whether or not the present seismic boundaries in the Andes coincide with the geologically defined boundaries suggested here, and in an attempt to recognise additional boundaries, a study of the latitudinal distribution of intermediate and deep focus earthquakes is necessary. Some information on seismic provinces is, however, available and correlates reasonably well with the tectonic segmentation proposed here. Gajardo and Lomnitz[5] defined provinces in Chile according to the levels of seismicity measured during a 16 yr period. Some of their boundaries correspond to boundaries 6, 8, 10, 14 and 15 (Fig. 1). Stauder[31] defined a similar series of boundaries between distinct seismic provinces, which correspond well to boundaries 6, 9, 10, 14 and 16 (Fig. 1). He also postulated that "the lithospheric slab" is apparently "segmented into a series of tongues that are absorbed independently". South of boundary 15 there is a shift of major seismicity to an offshore position[34,35]. Kelleher[36] studied the rupture pattern of the shallow part of the seismic zone beneath the Andes. He determined that the zone between boundaries 1 and 3 (Fig. 1) has been aseismic this century, whereas that between boundaries 3 and 5 has experienced large earthquakes, and that between boundaries 13 and 16 has ruptured about once each century along its entire length. He also recognised the intervals between boundaries 5 and 9 and 9 and 13 as distinct seismic provinces.

Nature of tectonic boundaries

As favoured by Carr *et al.*[9] for the Japanese arcs, I believe that the surface geological discontinuities in the central Andes probably correlate with subjacent seismic discontinuities, and reflect transverse boundary zones between separate segments of oceanic lithosphere which are subducted as individual units, perhaps even at different rates. The segmentation of the subduction zone is presumed to result from stresses created by the underthrusting of a cap-like slab of lithosphere, and the pattern produced may well be self perpetuating. Transform faults, which divide up the oceanic lithosphere into a series of strips, may also play a part in the production of the segments. Shield lineaments could also be involved at the boundaries at which changes in Andean strike are apparent[8], but they probably exert only a subsidiary influence.

The surface manifestations of the tectonic boundaries suggest that the magnitudes, and perhaps the histories, of their activity

are different. Furthermore, some boundaries (such as No. 13) seem to be fairly abrupt, and others (such as No. 10) more diffuse. It is not clear whether tectonic boundaries are relatively permanent features or whether they tend to migrate slowly or jump around with time. The fact that some of the present boundaries apparently occupied the same positions in early Cainozoic and even Mesozoic times, however, implies some degree of stability, or at least of repeated reactivation.

Although the discontinuities at the surface and on the subduction zone probably coincide, it does not necessarily mean that they are connected by major fault zones. Rather, I believe that the crustal segments overlying the various segments of the sinking slab have been subjected to similar, but clearly distinct, tectonic, magmatic, geomorphological, magmatic and metallogenic regimes, with zones of transition between them. Consequently, the wholesale extrapolation of geological histories along the lengths of volcano–plutonic orogens should be avoided.

Implications for magmatism

Across the central Andes, post-Triassic calc-alkaline magmas, of probable subduction zone origin[1,2,37,38], have been emplaced episodically upon and within the continental crust as a series of discrete pulses[30,39,40]. Each magmatic pulse has given rise to long, narrow belts of intrusive and extrusive rocks roughly parallel to the continental margin. Successive belts young eastwards in the western and central parts of the orogen[30,39,40], as is common in volcano–plutonic orogens[11], whereas little lateral migration of magmatism took place in a second locus of activity along the eastern margin of the Andes at latitudes 14°–24° S.

Radiometric age data[30,40,42], together with the mapped distribution of the magmatic belts both in the west and centre of the orogen and further east, support the concept that the belts are not longitudinally continuous. Therefore, magmatism in any particular tectonic segment can be expected to possess its own unique episodicity and spatial distribution, replaced by similar but clearly distinguishable regimes to the north and south. Clear discontinuities in belts of intrusive rocks have been recognised at boundaries 2, 3, 4, 5, 6, 7, 8, 10, 13 and 14, and volcanism, particularly recent activity, only occurs in certain segments at any one time.

Although magmatism in the North American Cordillera is locally episodic, it is essentially continuous if viewed over the full extent of the orogen[13]. Meso-Cainozoic intrusive epochs cannot be correlated throughout the orogen[11]. This may be simply explained if it is accepted that age patterns are only valid within individual tectonic segments.

Implications for metallogeny

The majority of the principal ore deposits in the central Andes have an inherent spatial and temporal relationship with the magmatic activity, and they are believed to derive their metal contents from the underlying subduction zone[15-17]. Thus, it might be expected that the regional metallogeny varies between segments. An examination of the distribution of post-Palaeozoic metallogenic belts in the central Andes (Fig. 1, and ref. 46) reveals a coastal belt rich in iron, bordered to the east by a copper-(gold–molybdenum) belt, a copper–lead–zinc–silver belt, and, in the central part of the region, a tin-(tungsten–silver) belt. When tectonic boundaries are crossed these metallogenic belts tend either to end or to undergo changes in their characteristics, including: second-order variations in their metal contents; changes in their widths; and changes in the ages, types and sizes of deposits. The tin belt is restricted to three segments enclosed by boundaries 5 and 8 (Fig. 1), and is subdivided into a narrow, northern segment dominated by vein-type tin–tungsten deposits related to Mesozoic and early Miocene batholiths, and a broader, southern part in which tin and tin–silver deposits of vein and porphyry[49] types are largely related to late Tertiary subvolcanic stocks[50-52]. Contact-metasomatic and vein type iron deposits are located in the coastal zone between boundaries 1 and 2, 4 and 6, and 9 and 12 (refs 30, 53, 54). The copper belt extends northwards from boundary 15 (ref. 53), but loses a good deal of its importance north of boundary 4. The polymetallic province is particularly enriched in silver north of boundary 1 (ref. 13), but is most important, despite its narrowness, from boundary 2 to boundary 4. From boundary 4 to boundary 5 the metal content is anomalous and lead, zinc and silver are overshadowed by copper and iron (Fig. 1, and ref. 54), and south of boundary 8 mineralisation is rather sparse and dominated by copper[55]. Within the copper belt the termination of long lines of distinctive ore types is also controlled by tectonic boundaries. Examples are the chalcopyrite–magnetite–actinolite veins that are known between boundaries 8 and 12 (ref. 53) and the disseminated, stratiform (manto-type) copper deposits in Jurassic volcanics from boundary 8 to boundary 10 (ref. 56). The porphyry copper deposits in the central Andes decrease in age eastwards[38] but, in addition, the major deposits change in age along the strike, so that Palaeocene deposits are dominant between boundaries 5 and 6 (refs 40 and 57), late Eocene–Oligocene deposits are dominant between boundaries 7 and 10, and late Miocene–Pliocene deposits are dominant between boundaries 12 and 14 (ref. 58). It is concluded that the definition of metallogenic segments is an important facet of the metallogenic characterisation of volcano–plutonic orogens and effectively explains the longitudinal variability of their metallogeny.

Mineral exploration

The mapping of metallogenic segmentation in both young and ancient volcano–plutonic orogens is clearly of paramount importance in the planning of mineral exploration programmes since metal contents, ore types and even economic importance change at tectonic boundaries.

Although a conclusive statement cannot be made at present, there is evidence from the central Andes that some major ore deposits tend to be located on or close to tectonic boundaries; the Chuquicamata and Río Blanco-Disputada porphyry copper deposits lie on boundaries 8 and 13, and the Oruro and Potosí porphyry tin deposits on boundaries 6 and 7. If magmas and their related metals are derived from the vicinity of the subduction zone, then this possibility seems to be more reasonable than the proposal that the locations of major ore deposits are controlled by megalineaments and their intersections (ref. 59 and 60), especially as lineaments in the central Andes are commonly at high angles to the tectonic boundaries[61-63] and are relatively superficial flaws in the upper continental crust.

Preparation of this article was commenced during tenure of a Shell postdoctoral Research Fellowship. Dr E. J. Cobbing read the manuscript.

Received March 13, 1974.

1. James, D. E., *Bull. geol. Soc. Am.*, **82**, 3325–3346 (1971).
2. Dickinson, W. R., *Rev. Geophys. Space Phys.*, **8**, 813–860 (1970).
3. Brüggen, J., *Fundamentos de la Geología de Chile* (Inst. Geogr. Milit., Santiago, 1950).
4. Muñoz Cristi, J., in *Handbook of South American Geology* (edit. by Jenks, W. F.), 187–214 (*Mem. geol. Soc. Am.*, **65**; 1956).
5. Gajardo, E., and Lomnitz, C., *Proc. 2nd World Conf. Earthquake Engng*, **3**, 1529–1540 (1960).
6. Lomnitz, C., *J. geophys. Res.*, **67**, 351–363 (1962).
7. Gerth, H., *Der Geologische Bau der südamerikanischen Kordillere* (Borntraeger, Berlin, 1955).
8. Gansser, A., *J. geol. Soc. Lond.*, **129**, 93–131 (1973).
9. Carr, M. J., Stoiber, R. E., and Drake, C. L., *Bull. geol. Soc. Am.*, **84**, 2917–2930 (1973).
10. Steinmann, G., *Geología de Perú* (Carl Winters Universität, Heidelberg, 1930).
11. de Loczy, L., *An. Acad. brasil. Ciénc*, **42**, 185–205 (1970).
12. Goossens, P. J., and Rose, W. I. jun, *Bull. geol. Soc. Am.*, **84**, 1043–1052 (1973).
13. Goossens, P. J., *Econ. Geol.*, **67**, 458–468 (1972).
14. Ham, C. K., and Herrera, L. J. jun, in *Backbone of the Americas* (edit. by Childs, O. E., and Beebe, B. W.), 47–61 (*Mem. Am. Ass. Petrol. Geol.*, **2**; 1963).
15. Jenks, W. F., in *Handbook of South American Geology* (edit. by Jenks, W. F.), 215–247 (*Mem. geol. Soc. Am.*, **65**; 1956).
16. Marocco, R., *Cah. ORSTOM, sér. Géol.*, **3**, 1, 45–58 (1971).
17. Bellido, E., *Bol. Serv. geol. Min., Lima*, **22** (1969).
18. Cobbing, E. J., and Pitcher, W. S., *Nature phys. Sci.*, **240**, 51–53 (1972).
19. Fisher, R. L., and Raitt, R. W., *Deep-Sea Res.*, **9**, 423–443 (1962).
20. Ruegg, W., *Bol. Soc. Geol. Perú*, **38**, 97–142 (1962).
21. Rod, E., *Bull. Am. Ass. Petrol. Geol.*, **44**, 107–108 (1960).
22. Radelli, L., *Trav. Lab. Geol. Grenoble*, **42**, 237–261 (1966).
23. Schlatter, L. E., and Nederlof, M. H., *Bol. Serv. geol. Bolivia*, **8** (1966).
24. Ahlfeld, F., *Geol. Rdsch.*, **59**, 1124–1140 (1970).
25. Sonnenberg, F. P., in *Backbone of the Americas* (edit. by Childs, O. E., and Beebe, B. W.), 36–46 (*Mem. Am. Ass. Petrol. Geol.*, **2**; 1963).
26. *Mapa Geológico de Chile*, 1:1.000,000 (Inst. Invest. Geol., Santiago, 1968).
27. Turner, J. C. M., *Geol. Rdsch.*, **59**, 1028–1063 (1970).
28. Sillitoe, R. H., Mortimer, C., and Clark, A. H., *Trans. Inst. Min. Metall.*, **B77**, 166–169 (1968).
29. Mortimer, C., *J. geol. Soc. Lond.*, **129**, 505–526 (1973).
30. Ruiz, F. C., Aguirre, L., Corvalán, J., Klohn, C., Klohn, E., and Levi, B., *Geología y Yacimientos Metalíferos de Chile* (Inst. Invest. Geol., Santiago, 1965).
31. Levi, B. D., Aguilar, A. M., and Fuenzalida, R. P., *Bol. Inst. Invest. Geol., Santiago*, **19** (1966).
32. Carter, W. D., and Aguirre, L., *Bull. geol. Soc. Am.*, **76**, 651–664 (1965).
33. Herrero-Ducloux, A., in *Backbone of the Americas* (edit. by Childs, O. E., and Beebe, B. W.), 16–28 (*Mem. Am. Ass. Petrol. Geol.*, **2**, 1963).
34. Stauder, W., *J. geophys. Res.*, **78**, 5033–5061 (1973).
35. Lomnitz, C., *Geol. Rdsch.*, **59**, 938–960 (1970).
36. Kelleher, J., *J. geophys. Res.*, **77**, 2087–2103 (1972); Kelleher, J., *et al.*, *J. geophys. Res.*, **78**, 2547–2585 (1973).
37. Dickinson, W. R., *Am. J. Sci.*, **272**, 551–576 (1972).
38. Sillitoe, R. H., *Econ. Geol.*, **67**, 184–197 (1972).
39. Farrar, E., Clark, A. H., Haynes, S. J., Quirt, G. S., Conn, H., and Zentilli, M., *Earth planet. Sci. Lett.*, **10**, 60–66 (1970).
40. Stewart, J. W., Evernden, J. F., and Snelling, N. J., *Bull. geol. Soc. Am.* (in the press).
41. Dickinson, W. R., *J. geophys. Res.*, **78**, 3376–3389 (1973).
42. Clark, A. H., and Farrar, E., *Econ. Geol.*, **68**, 102–106 (1973).
43. Gilluly, J., *Bull. geol. Soc. Am.*, **84**, 499–514 (1973).
44. Lanphere, M. A., and Reed, B. L., *Bull. geol. Soc. Am.*, **84**, 3773–3782 (1973).
45. Sillitoe, R. H., *Bull. geol. Soc. Am.*, **83**, 813–818 (1972).
46. Sillitoe, R. H., *Geol. Assoc. Canada Spec. Pap.* (in the press).
47. Sawkins, F. J., *J. Geol.*, **80**, 377–397 (1972).
48. Petersen, U., *Geol. Rdsch.*, **59**, 834–897 (1970).
49. Sillitoe, R. H., and Halls, C., submitted for publication.
50. Ahlfeld, F., and Schneider-Scherbina, A., *Los Yacimientos Minerales y de Hidrocarburos de Bolivia* (Depto. Nac. Geol. Bol. 5, La Paz, 1964).
51. Ahlfeld, F., *Miner. Deposita*, **2**, 291–311 (1967).
52. Turneaure, F. S., *Econ. Geol.*, **66**, 215–225 (1971).
53. Ruiz, F. C., and Ericksen, G. E., *Econ. Geol.*, **57**, 91–106 (1962).
54. Bellido, E., and de Montreuil, L., *Aspectos Generales de la Metalogenia del Perú* (Serv. Geol. Min., Geol. Econ. 1, Lima, 1972).
55. Angelelli, V., Fernandez Lima, J. C., Herrera, A., and Aristarain, L., *Descripción del mapa Metalogenético de la República Argentina. Minerales Metalíferos* (Direc. Nac. Geol. Min., Buenos Aires, 1970)
56. Ruiz, C., Aguilar, A., Egert, E., Espinosa, W., Peebles, F., Quezada, R., and Serrano, M., in *Proc. IMA-IAGOD Meetings '70, IAGOD Vol.* (edit. by Takeuchi, T.), 252–260 (Soc. Mining Geol. Japan Spec. Issue 3, Tokyo, 1971).
57. Laughlin, A. W., Damon, P. E., and Watson, B. N., *Econ. Geol.*, **63**, 166–168 (1968).
58. Quirt, S., Clark, A. H., Farrar, E., and Sillitoe, R. H., *Geol. Soc. Am. Absts. with Prog.*, **3**, 7, 676–677 (1971).
59. Mayo, E. B., *Min. Engng.*, **10**, 1169–1175 (1958).
60. Thomas, N. A., *Geol. Rdsch.*, **59**, 1013–1027 (1970).
61. Segerstrom, K., *Prof. Pap. US geol. Surv.*, **700-D**, 10–17 (1971).
62. United Nations, *Fotolineamientos y Mineralización en el Noroeste Argentina* (Exploración Minera de la Region Noroeste, Argentina, inf. téc., New York, 1973).

Tin mineralisation above mantle hot spots

MANY important tin belts of Phanerozoic age, including those in Bolivia, Mexico, Alaska, Malaysia-Indonesia, eastern Australia and western Europe, are believed to have formed above subduction zones at convergent plate margins[1,2]. If the late Tertiary mineralisation in the southern part of the Eastern Cordillera of Bolivia is representative of this type, then the mineralisation occurred above the deepest parts of subduction zones with the tin perhaps related to partial fusion of subducted oceanic lithosphere[1].

In addition to tin deposits at active or inactive convergent plate margins, and Precambrian tin-bearing pegmatites (for example, Congo and Western Australia), there is a third genetic type of economically important tin deposit, which is discussed here. This type formed away from lithospheric plate boundaries, in intraplate environments, and is represented by major deposits on the Jos Plateau in Nigeria[3,4] and in Rondônia and adjoining states of southern Amazonia, Brazil[5,6]. Smaller deposits occur in the Aïr and Zinder areas of Niger[7], at Mayo Darlé in Cameroun[8], in the Hoggar massif of southern Algeria[9] and its extension into the Adrar des Iforas in Mali, at Sabaloka in northern Sudan[10], in south-eastern Egypt[11], in the Tibesti area of Chad, in southern Morocco, and in the Damaraland province of Southwest Africa[11,12] (Fig. 1). Tin deposits in Transbaikalia and Mongolia are also of this type (N. Varlamoff, personal communication, 1973).

All of these deposits are associated with small, alkaline and peralkaline granite plutons of anorogenic character—the Younger Granites of Africa—that cut discordantly through the basement which is commonly of Precambrian age. The important Nigerian[3,4] and Rondônian[5,6] examples are typical and occur in circular or oval volcanic-plutonic complexes 5 to 20 km wide. These are characterised by early rhyolitic volcanics that are succeeded by multiple granitic intrusives, the emplacement of which was commonly controlled by ring fracturing and cauldron subsidence. Basic and intermediate rocks are also present at some centres. The granites and their porphyritic equivalents are rich in fluorine, containing accessory fluorite and topaz. Mineralisation is in the form of cassiterite accompanied by topaz, lithia mica and some wolframite in greisens and swarms of quartz veins in the roof zones of biotite granites. The complexes in certain areas, such as Hoggar[13] and Transbaikalia-Mongolia, lack ring structures and may represent a deeper level of emplacement. Although mineralogically similar, the tin deposits in Southwest Africa are somewhat different since many occur in pegmatites; at one locality, however, a stockwork of quartz veins is mined[11,12].

The tin-bearing, subvolcanic complexes occur in variously oriented lines, several of which constitute elongated belts, approximately 300 to 400 km long, in Nigeria, Rondônia, South West Africa (Fig. 2) and elsewhere. Progressive age trends along the lines of complexes have been claimed[14,15] but are not strongly supported by the available radiometric data[16,17] (Fig. 2).

Radiometric dating has demonstrated emplacement of the stanniferous complexes over a protracted time interval: late Precambrian (around 1,000 m.y. ago) in Rondônia[18,19], Cambrian in the Hoggar[9] and at Sabaloka in Sudan[10], Carboniferous in Aïr[20], Jurassic in Nigeria[21,15], Jurassic and Cretaceous in South West Africa[22], Cretaceous in Transbaikalia-Mongolia (N. Varlamoff, personal communication, 1973), and Eocene at Mayo Darlé in Cameroun[8].

The concept of hot spots or mantle plumes was introduced by Morgan[23,24] after the earlier ideas of Wilson[25]. Hot spots are considered to be manifestations of solid-state convection in the lower mantle, and consist of vertical columns or plumes of ascending primitive mantle material. Within the asthenosphere the plumes spread laterally[23,24], producing, in some

cases, rifting and consequent separation of plates of continental lithosphere[26]. The fixed mantle plumes are considered to 'burn' through lithospheric plates, thus giving rise to localised igneous activity. At the outset, when the lithosphere is continental, this activity is represented by alkaline magmatism including alkaline granitic, subvolcanic complexes[14,15,26], whereas once oceanic crust has been generated the plumes give rise to basaltic oceanic island chains or to aseismic ridges[23,24,25].

I therefore propose that the tin-bearing, alkaline granitic complexes in Nigeria, Rondônia and elsewhere were generated above mantle hot spots during a number of brief intervals since the late Precambrian. In all cases the effect of the hot spots seems to have been limited to domal uplift and the emplacement of linear belts of complexes, and activity apparently ceased before the inception of the rifting stage proposed by Burke and Dewey[26]. In many cases the plumes sought out fundamental breaks in the continental lithosphere and these now seem to control the location and linear array of the stanniferous complexes[16,17,20,27]. The extents of the tin-bearing belts (Fig. 2) are presumed to reflect the dimensions of the hot spots that produced them and not to indicate widespread migration of magmatism resulting from differential motion between the plumes and the overlying lithosphere[14,15].

Economic concentrations of tin are traditionally thought to have been derived from the continental crust (see ref. 28). The concept of tin mineralisation above rising plumes of mantle material suggests, however, that the mantle may be a possible source for the tin and accompanying tungsten and fluorine. A source in the mantle is difficult to prove because extraction from continental lithosphere, by magmas or magmatic heat provided by rising plume material, cannot be easily discounted. Initial Sr^{87}/Sr^{86} ratios of 0.721 and 0.718 determined for samples of granites from complexes in Nigeria[29] and Rondônia[18] suggest that significant amounts of the granitic rocks associated with the tin deposits have a crustal source, a hypothesis also supported by other workers[26]. Much lower initial Sr isotope ratios have, however, been obtained for some granitic units of the White Mountain

FIG. 1 Areas of tin-bearing, alkaline granitic complexes plotted on a pre-drift reconstruction of Africa and South America. The tin-bearing complex in Cameroun was, however, emplaced after the inception of spreading in the South Atlantic in the early Cretaceous[34].

plume trace in New Hampshire[30], possibly indicating a subcrustal origin for their parent magmas. In Nigeria, tin potentially available for extraction, occurs in pegmatites in the Precambrian basement[31]. The abundance of boron, beryllium and tantalum in these deposits and the virtual absence of these elements from the Jurassic tin deposits have been cited as evidence against the hypothesis of remobilisation of Precambrian tin[27], unless some mechanism for element segregation is proposed. The rather widespread occurrence of tin in association with anorogenic, alkaline granitic complexes and, perhaps more significantly, its presence along the full length of belts of complexes, also tend to decrease the likelihood of a source in remobilised Precambrian deposits. Furthermore, the absence of significant amounts of the more common base metals from these complexes seems to preclude anatexis or assimilation of unmineralised Precambrian crustal rocks as an origin for the tin. Although some of the granitic rocks are derived at least in part from the continental crust, the evidence, although far from unequivocal, seems to favour a source in the mantle for the tin (compare with ref. 27).

Support for a mantle derivation of lithophile elements at hot spots is provided by evidence from two localities at which continental crust is absent. Blocks dredged from the axial trough of the mid-Atlantic Ridge in the vicinity of the Azores hot spot show evidence of hydrothermal attack and have high contents of several heavy metals, including tin[32]. At the Iceland hot spot, granitic rocks derived from parent magmas of mantle origin contain concentrations of molybdenum[33], an element which has a geochemical affinity to tin and which occurs with it in some of the complexes (for example, Nigeria[4] and Sudan[10]).

The generation of tin mineralisation above some hot spots and not above others (White Mountain magma series, western Scotland Tertiary province) might reflect either an inhomogeneous distribution of tin (or fluorine) in the mantle or

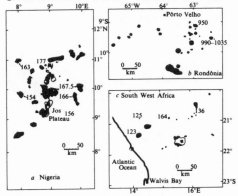

FIG. 2 Linear distribution of tin-bearing, alkaline granitic complexes within elongate belts in *a*, Nigeria; *b*, Rondônia; and *c*, South West Africa. Radiometric ages (numbers) in Nigeria from Jacobson *et al.*[21] and van Breemen and Bowden[15]; in Rondônia from Priem *et al.*[18]; and in South West Africa from Siedner and Miller[22].

the propagation of mantle plumes, each with a different tin (or fluorine) content, from several levels in the mantle.

The concept of tin mineralisation related to granitic complexes of alkaline type generated in intraplate environments, and unrelated to arc-trench systems, should be useful in the search for ore. In poorly exposed regions, such as Rondônia,

the observation that hot-spot activity commonly gives rise to several lines of complexes within an elongated belt (Fig. 2) could facilitate the delimitation of areas for more detailed exploration. Tin deposits might also accompany felsic differentiates at some hot spots in the ocean basins.

Discussions with Drs J. W. Stewart and N. Varlamoff and review of a manuscript by Drs K. Burke, Stewart and Varlamoff are gratefully acknowledged. The paper was prepared during tenure of a Shell Postdoctoral Research Fellowship.

RICHARD H. SILLITOE

Department of Mining Geology,
Royal School of Mines,
Imperial College of Science and Technology,
Prince Consort Road,
London SW7 2BP, UK

Received November 20, 1973; revised January 29, 1974.

1 Sillitoe, R. H., *Bull. geol. Soc. Am.*, **83**, 813 (1972).
2 Mitchell, A. H. G., and Garson, M. S., *Trans. Inst. Mining Metall.*, B81, B10 (1972).
3 Jacobson, R. R. E., MacLeod, W. N., and Black, R., *Mem. geol. Soc. Lond.*, 1 (1958).
4 MacLeod, W. N., Turner, D. C., and Wright, E. P., *Bull. geol. Surv. Nigeria*, **32**, 1 (1971).
5 Kloosterman, J. B., in *A Technical Conference on Tin*, **2**, 383 (International Tin Council, London, 1967).
6 Kloosterman, J. B., in *A Second Technical Conference on Tin*, 1 (edit. by Fox, W.), 197 (International Tin Council and Department of Mineral Resources, Government of Thailand, Bangkok, 1969).
7 Raulais, M., *C. r. hebd. Séanc. Acad. Sci. Paris*, **223**, 96 (1946).
8 Gazel, J., Lasserre, M., Limasset, J.-C., and Vachette, M., *C. r. Somm. Séanc. Acad. Sci. Paris*, **256**, 2875 (1963).
9 Boissonnas, J., Leutwein, F., and Sonet, J., *C. r. Somm. Séanc. Soc. géol. France*, **7**, 251 (1970).
10 Almond, D. C., *Geol. Mag.*, **104**, 1 (1967).
11 de Kun, N., *The Mineral Resources of Africa* (Elsevier, Amsterdam, 1965).
12 Pelletier, R. A., *Mineral Resources of South-Central Africa* (Oxford University Press, Cape Town, 1964).
13 Boissonnas, J., *C. r. hebd. Séanc. Acad. Sci. Paris*, **250**, 4016 (1960).
14 Rhodes, R. C., *Tectonophysics*, **12**, 111 (1971).
15 van Breemen, O., and Bowden, P., *Nature phys. Sci.*, **242**, 9 (1973).
16 Wright, J. B., *Nature*, **244**, 565 (1973).
17 Marsh, J. S., *Earth planet. Sci. Lett.*, **18**, 317 (1973).
18 Priem, H. N. A., Boelrijk, N. A. I. M., Hebeda, E. H., Verdurmen, E. A. Th., Verschure, R. H., and Bon, E. H., *Bull. geol. Soc. Am.*, **82**, 1095 (1971).
19 Cordani, U. G., Melcher, G. C., and de Almeida, F. F. M., *Can. J. Earth Sci.*, **5**, 629 (1968).
20 Black, R., and Girod, M., in *African Magmatism and Tectonics* (edit. by Clifford, T. N., and Gass, I. G.), 185 (Oliver and Boyd, Edinburgh, 1970).
21 Jacobson, R. R. E., Snelling, N. J., and Truswell, J. F., *Overseas Geol. Miner. Resour.*, **9**, 168 (1964).
22 Siedner, G., and Miller, J. A., *Earth planet. Sci. Lett.*, **4**, 451 (1968).
23 Morgan, W. J., *Nature*, **230**, 42 (1971).
24 Morgan, W. J., *Bull. Am. Ass. Petrol. Geol.*, **56**, 203 (1972).
25 Wilson, J. T., *Phil. Trans. R. Soc.*, A258, 145 (1965).
26 Burke, K., and Dewey, J. F., *J. Geol.*, **81**, 406 (1973).
27 Wright, J. B., *Econ. Geol.*, **65**, 945 (1970).
28 Stoll, W. C., *Mining Mag.*, **112**, 22, 90 (1965).
29 Bowden, P., and van Breemen, O., in *African Geology* (edit. by Dessauvagie, T. F. J., and Whiteman, A. J.), 105 (University of Ibadan, Nigeria, 1972).
30 Foland, K. A., Quinn, A. W., and Giletti, B. J., *Am. J. Sci.*, **270**, 321 (1971).
31 Jacobson, R. R. E., and Webb, J. S., *Bull. geol. Surv. Nigeria*, 17 (1946).
32 Dmitriev, L., Barsukov, V., and Udintsev, G., in *Proc. of IMA-IAGOD Meetings '70, IAGOD Vol.* (edit. by Takeuchi, T.), 65 (Soc. Mining Geol. Japan Spec. Issue 3, Tokyo, 1971).
33 Jancovic, S., *Rep. twenty-fourth int. geol. Congr., Montreal, Canada*, **4**, 326 (1972).
34 Le Pichon, X., and Hayes, D. E., *J. geophys. Res.*, **76**, 6283 (1971).

29

Reprinted from *Trans. Instn Min. Metall.* 84:B1–B3, B5–B6 (1975)

Timing, distribution and origin of submarine mineralization in the Red Sea

R. D. Bignell B.Sc., Associate Member
Applied Geochemistry Research Group, Imperial College, London

553.068.24(267.5)

Synopsis

The location of brine pools and metalliferous sediments along the median valley of the Red Sea is related to the distribution of probable transform faults within the Red Sea inferred from morphology, continental fracture lines and bathymetric data. The timing of brine activity (mineralization) within the Red Sea appears to be related to glacio-eustatic changes in sea-level. It is suggested that elevations in sea-level at the end of glacial periods promote the movement of sea water along fault zones and result in discharge of brines within the median valley.

The heat provided by volcanic activity on the floor of the median valley is considered to be unnecessary to heat the brines and to promote brine circulation; moreover, models for the origin of these deposits that suggest brines are discharged due to volcanic heat fail to explain the distribution and episodic nature of brine activity.

The discovery and detailed investigation of hot brines and metalliferous sediments in the Atlantis II Deep of the Red Sea, during the mid 1960s, revealed a potentially exploitable orebody in the process of formation.[1,2] Following this discovery, investigations were made to find additional deposits of this type. The central trough of the Red Sea was surveyed by the Applied Geochemistry Research Group aboard the *Nereus* in 1970 and 1971, and by German scientists aboard the M. V. *Wando River* in 1969 and the M. V. *Valdivia* in 1971 and 1972.[3] A geochemical investigation of cores provided by these subsequent surveys has revealed the variable nature and widespread distribution of metalliferous sediments in the Red Sea.[4]

In this paper an attempt is made to explain the distribution of these deposits in relation to known structure in the Red Sea region, and to interpret the timing of mineralization (brine activity) in relation to the recent sedimentary history of the Red Sea. The controls on the timing and location of mineralization are considered to be important factors in the genesis of these deposits. Detailed knowledge of the distribution and genesis of the brine pools and metalliferous sediments will, it is hoped, provide guidelines for future exploration of these and related deposits formed in rifting environments.[5]

Geotectonic position of deposits

Relation to active spreading

The Red Sea metalliferous sediments consist predominantly of oxides, sulphides and silicates rich in Fe and Mn, with high concentrations of Zn, Cu, Pb, Cd and Ag. These deposits were among the first to be recognized as showing a direct relationship between ore formation and sea-floor spreading. The formation of metalliferous sediments along both mature[6] and embryonic[7] sections of the world's active ridge system is well documented. In none of these cases, however, has mineralization been found to the extent that it occurs in the Red Sea.

The Red Sea deposits are forming at a time when sea-floor spreading has only recently restarted after an unusually long quiescent period.[8] During this pause a

Manuscript first received by the Institution of Mining and Metallurgy on 4 September, 1974; revised manuscript received on 10 December, 1974. Paper published in February, 1975.

thick (up to 5-km) sequence of evaporites was deposited.[9] The renewal of sea-floor spreading has rifted apart, and faulted, these evaporites—a situation that would appear to be most favourable to the formation of dense brines by the leaching of fractured salt formations by sea water.

Despite the presence of these favourable conditions along the length of the Red Sea, however, the sites of mineralization have a limited distribution. They are located in the northern and central Red Sea along those parts of the axis of spreading where the median valley is best developed. In these areas the tectonic situation has resulted in a deep central rift, divided into separate deeps in which the brines accumulate. In the southern Red Sea, where the median valley is poorly developed, consisting of graben-horst structures running parallel with the axis of spreading, no mineralization has yet been found.

Relation to structure

The distribution of brine pools and metalliferous sediments along the median valley is shown in Fig. 1. The deposits are frequently located opposite points where changes in the trend of the coastline take place or where the Red Sea narrows, or where there are sharp changes in the trend of the bathymetry of the median valley. The latter are most noticeable as offsets in the position and trend of the steep sides of the median valley. They run parallel to the axis of spreading, and it is suggested that these offsets result from transform faulting associated with sea-floor spreading.

The suggested location of faulting is, in part, supported by the work of Garson and Krs,[10] who have mapped the positions of the major fracture lines along parts of the coastal areas of the Red Sea and have suggested these fracture lines are transformed faults. The extension of these fractures below the Red Sea is frequently indicated by sharp changes in the bathymetry. Along the sections of the Red Sea coasts from about 23° to 24°N, and south of about 21°N, insufficient geological data were available to enable the distribution of faults to be determined.[10] The presence of faults in these areas may, however, be inferred from sharp changes in the bathymetry of the Red Sea and in the trend of the coastlines.

The location of the deposits appears to be closely related to positions where transform faults intersect the median valley. This relationship suggests that the sites of brine discharge (metal deposition) in the Red Sea are controlled by transform faulting.

Time of deposition in relation to sedimentation

Recent history of sedimentation in the Red Sea

The history of sedimentation in the Red Sea for the past 100 000 years has been compiled from micropalaeontological, radiocarbon and oxygen isotope data (see Berggren[11] and references cited). These data show that at least five periods of lowered sea-level and increased salinity have occurred as a result of glacio-eustatic movements during the past 85 000 years (Fig. 2).[12] Each period was characterized by a steady lowering of sea-level and increase in salinity when the Red Sea was a closed basin, followed by an abrupt rise

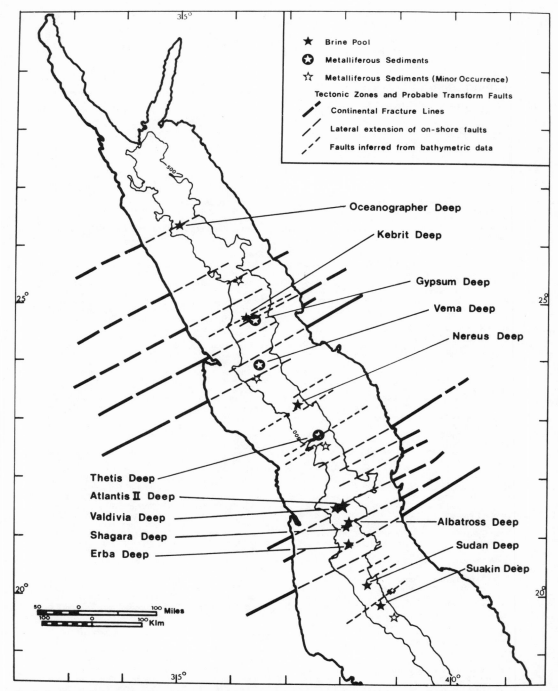

Fig. 1 Location of brine pools and metalliferous sediments in relation to distribution of tectonic zones and probable transform faults in Red Sea

in sea-level and decrease in salinity when connexion between the Red Sea and Indian Ocean was re-established.

These events in the history of the Red Sea are reflected by conspicuous changes in the nature of the sediments deposited at the time. Periods of open circulation are characterized by an abundance of micro-

fossils, mostly foraminifera and pteropods, and during the periods of increasing salinity the abundance of fossils decreased sharply. At the end of each glacial-salinity cycle slight elevations in temperature caused aragonite to be precipitated from the highly saline waters, resulting in the deposition of lithified aragonitic layers.

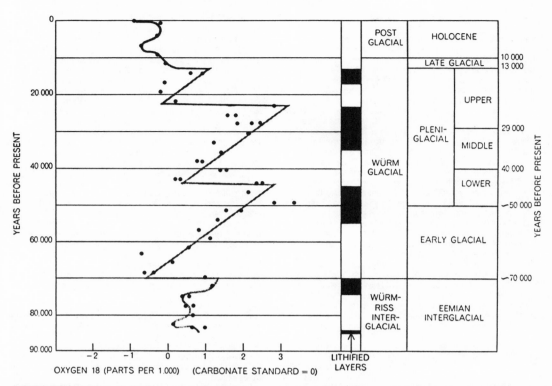

SALINITY CHANGES in the Red Sea over the past 70.000 years are reflected in changes in the amount of the isotope oxygen 18 incorporated in the fossils of foraminifera found at different depths in bottom sediments. Four slow rises to peak salinity are evident, followed by four abrupt declines. Stony layers of microfossils coincide with periods of high salinity. The cycles appear to be controlled by a glacial mechanism. The isotope values shown represent deviation with respect to ocean-water standard in parts per 1 000.

Fig. 2　Late Pleistocene–Holocene salinity changes in the Red Sea. After Degens and Ross[12]

[*Editor's Note:* Material has been omitted at this point.]

Discussion

Influence of structure on location of deposits

The distribution of brine pools and metalliferous sediments is related to the tectonic situation along the axis of spreading. Sites of mineralization are found along sections of the spreading axis where the median valley is best developed (deepest) and where it forms a number of isolated deeps. The extent to which the median valley has developed appears to be a function of the rifting caused by the recent phase of spreading modified by salt flowage into the rift.[9,17] The formation of isolated deeps, and, thus, favourable sites for brine accumulation, is a result of transform faulting due to changes in the trend of the median valley in relation to the direction of spreading. Transform faulting and mineralization are most common along sections of the Red Sea where the direction of spreading and the trend of the median valley converge. In these areas magnetic anomalies suggest that spreading occurs along a number of short ridge segments offset by transform faults (R. Searle, personal communication). Alternatively, where the axis of the median valley and the direction of spreading are at right angles, as in the southern Red Sea, there are fewer transform faults, and linear features have developed parallel to the axis of spreading.

It has been suggested that the Red Sea brines evolve from sea water sinking through the floor of the Red Sea and dissolving salts from the underlying strata, before discharging on to the floor of the median valley.[18,19] The strata flanking the median valley, however, consist predominantly of Miocene evaporites, which have a very poor permeability and would not allow the movement and discharge of brines.[20] The strong association between sites of brines discharge and transform faulting suggests that these faults may provide a flow path for the brines through the evaporites.

Role of sea-level changes in promoting brine activity

During glacial periods the worldwide lowering of sea-level isolated the Red Sea from the Indian Ocean. With the loss of inflowing water from the Indian Ocean, the level of water in the Red Sea would become considerably lower due to loss of water by evaporation. Olausson[21] assumed an evaporation rate during glacial periods of 2 m per year and calculated that 90 per cent of the Red Sea would become dry after 500 years. Thus, during the glacial periods considerable changes in sea-level took place in the Red Sea.

These changes in the amount of water in the Red Sea would be expected to have an effect on the isostatic equilibrium. Assuming an average rock density of $2 \cdot 5$ below the Red Sea and a density of 1 for sea water, a change in sea-level of 500 m would result in a 200-m change in the level of the sea-floor. According to Girdler,[22] the Red Sea is in isostatic equilibrium at present, so, since the abrupt rise in sea-level at the end of the Würm glaciation, isostatic readjustment has taken place. During this readjustment movement may occur along active fracture lines, such as transform faults, allowing sea water along the faults to sink into the strata beneath the floor of the Red Sea. Investigations of the sediments obtained during Leg 23B of the D.S.D.P. have shown that these strata contain sufficient concentrations of available metals to form the deposits.[31] The rise in sea-level also means that a greater length of the fault is submerged, and sea water would be seeping into the fault zone over a wider area. Thus, the larger amount of water flowing into the fault zone will make the hydraulic gradient along it greater, resulting in an increase in discharge from it.

Relationship to volcanism

Most of the controversy surrounding the genesis of the Red Sea deposits results from their location along a zone of active volcanism, which has led to numerous suggestions that the deposits have a volcanic origin.

Isotope data for sulphur,[23] strontium[24] and lead[25] suggest that these elements in the Atlantis II Deep could have been derived from volcanic rocks. The rocks involved may be either young basic volcanics associated with the sea-floor spreading, or water-laid volcanic tuffs interbedded with the Miocene evaporites.[9] There is now considerable evidence that sea water circulating through volcanic rocks will remove ore metals.[6] Although the Red Sea deposits are spatially connected with submarine volcanism and geothermal anomalies, no direct supply of magmatic liquids has yet been proved. Extensive sampling of sediments in areas of recent volcanism, such as the islands Jebel Tair and Zubair, and in Volcano Deep, show no major occurrences of metalliferous sediments.

It has been suggested that the heat provided by volcanic activity in the axial zone of the Red Sea serves both to heat the brines and promote brine circulation.[26] Recent evidence would suggest that the geothermal gradient alone would be sufficient to heat the brines. The results of the D.S.D.P. within the Red Sea show that the geothermal gradient in the sediments on either side of the median valley is about $0 \cdot 1°C/m$.[28] It has been estimated that the temperature of the incoming brine in Atlantis II Deep, discharging at a depth of 2000 m, is about 104°C.[19] To raise the temperature of the brines to this level it is thus only necessary for them to sink to a depth of about 800 m within the flanking sediments.

Several models for the formation of ore deposits have suggested that brines discharge as a result of the heat supplied by volcanic activity from below the floor of the median valley.[26,27] Such models, however, do not account for either the distribution of the deposits in relation to faulting or the episodic nature of brine discharge.

In addition to the brine pools and metalliferous sediments in the deeps on the floor of the median valley, a number of metalliferous sediment layers, and the brines in Valdivia, Oceanographer and Albatross Deeps, occur well above the floor of the median valley. These occurrences may result from the seepage of brines at the intersection of transform faults and the steep sides of the median valley.

Conclusions and implications for mineral exploration

The metalliferous sediments in the southern part of the Atlantis II Deep contain 150 000 000–200 000 000 tons of dry salt-free mud, with an average content of 5 per cent Zn and 1 per cent Cu.[29] Although this is the only deposit at present of economic interest in the Red Sea, the search for similar deposits has revealed the widespread nature of metalliferous sediment deposition. The most significant feature of the distribution of brine pools and metalliferous sediments in the Red Sea is their occurrence along the extensions of known continental fractures. It is suggested that sites of brine discharge and metalliferous sediment deposition are related to transform faults intercepting the deep central rift. A consideration of the distribution of probable transform faults indicates a number of potential areas for the discovery of metalliferous sediments.

The onset of past periods of brine activity appears frequently to have occurred shortly after the return to normal oceanic conditions following periods when the Red Sea was isolated. It is suggested that the major rises in sea-level at these times promote brine activity. Of the areas with major metalliferous sediment deposits, only Atlantis II Deep remains active. It is thus concluded that we are observing the end of an active period of metalliferous sediment deposition within the Red Sea, initiated by the influx of sea water at the end of the last glaciation. A compilation of micropalaeontological, radiocarbon dating and oxygen isotope data suggests that four glacio-eustatic changes in sea-level have occurred during the past 70 000 years (Fig. 2),[2,11] so three periods of major metalliferous sediment deposition may occur at depth. In Atlantis II Deep seismic profiles have been interpreted as suggesting that at least 100 m of sediment may be present.[30] The deposition of a sedimentary sequence of this thickness over recently formed oceanic crust would require the high rates of deposition of the metalliferous sediments. Because of the inadequacy of present coring techniques, and the thickness of the major metalliferous deposits, where major deposits have formed, only the most recent cycle of deposition has been observed.

Acknowledgement

This paper is a development of ideas expressed by the author at the symposium on metallogeny and plate tectonics, St. Johns, Newfoundland, 1974. The research was supervised by Dr. D. S. Cronan and supported by a grant from the Natural Environment Research Council. The author wishes to thank Dr. H. Bäcker of Preussag, A.G., for providing sediment samples, Mr. S. Ali for undertaking micropalaeontological investigations, and Dr. M. Garson for providing a map of the continental fracture pattern in the Red Sea area.

References

1. **Miller A. R.** *et al.* Hot brines and recent iron deposits in deeps of the Red Sea. *Geochim. cosmochim. Acta*, **30**, 1966, 341–59.
2. **Degens E. T. and Ross D. A.** eds. *Hot brines and recent heavy metal deposits in the Red Sea* (New York, etc.: Springer Verlag, 1969), 600 p.
3. **Bäcker H. and Schoell M.** New deeps with brines and metalliferous sediments in the Red Sea. *Nature, phys. Sci.*, **240**, 1972, 153–8.
4. **Bignell R. D. Cronan D. S. and Tooms J. S.** Red Sea metalliferous brine precipitates. *In Metallogeny and plate tectonics. Spec. Publ. geol. Soc. Can.*, 1974, in press.
5. **Russell M. J.** Structural controls of base metal mineralization in Ireland in relation to continental drift. *Trans. Instn Min. Metall. (Sect. B : Appl. earth sci.)*, **77**, 1968, B117–28.
6. **Dymond J.** *et al.* Origin of metalliferous sediments from the Pacific oceans. *Bull. geol. Soc. Am.*, **84**, 1973, 3355–72.
7. **Degens E. T.** *et al.* Microcrystalline sphalerite in resin globules suspended in Lake Kivu, East Africa. *Mineral. Deposita*, **7**, 1972, 1–12.
8. **Girdler R. W. and Styles P.** Two stage Red Sea floor spreading. *Nature, Lond.*, **247**, 1974, 7–11.
9. **Ross D. A. and Schlee J.** Shallow structure and geologic development of the southern Red Sea. *Bull. geol. Soc. Am.*, **84**, 1973, 3827–48.
10. **Garson M. S. and Krs M.** Geophysical and geological evidence of the relationship of Red Sea transverse tectonics to ancient fractures. In preparation.
11. **Berggren W. A.** Micropaleontologic investigations of Red Sea cores—summation and synthesis of results. Reference 2, 329–35.
12. **Degens E. T. and Ross D. A.** The Red Sea hot brines. *Scient. Am.*, **222**, April 1970, 32–42.
13. **Bäcker H. and Richter H.** Die rezente hydrothermal-sedimentäre Lagerstätte Atlantis-II-Tief im Roten Meer. *Geol. Rdsch.*, **62**, 1973, 697–741.
14. **Risch H.** Mikropaleontologische Untersuchungen im Roten Meer und Golf von Aden. *Valdivia, Wissenschaftliche Ergebnisse*, 1974, 1–13. (BfB, Hannover, internal publication)
15. **Bignell R. D.** *et al.* An additional location of metalliferous sediments in the Red Sea. *Nature, Lond.*, **248**, 1974, 127–8.
16. **Berggren W. A. and Boersma A.** Late Pleistocene and Holocene planktonic foraminifera from the Red Sea. Reference 2, 282–98.
17. **Girdler R. W. and Whitmarsh R. B.** Miocene evaporites in Red Sea cores, their relevance to the problem of the width and age of oceanic crust beneath the Red Sea. *Initial Rep. Deep Sea Drilling Project*, **23**, 1974, 913–21.
18. **Craig H.** Geochemistry and origin of the Red Sea brines. Reference 2, 208–42.
19. **Ross D. A.** Red Sea hot brine area: revisited. *Science*, **175**, 1972, 1455–7.
20. **Manheim F. T. Dwight L. and Belastock R. A.** Porosity, density, grain density, and related physical properties of sediments from the Red Sea drill cores. *Initial Rep. Deep Sea Drilling Project*, **23**, 1974, 887–908.
21. **Olausson E.** Quaternary correlations and the geochemistry of oozes. In *The micropalaeontology of the oceans* Funnell B. M. and Riedel W. R. eds. (Cambridge: The University Press, 1971), 375–92.
22. **Girdler R. W.** The Red Sea—a geophysical background. Reference 2, 38–58.
23. **Kaplan I. R. Sweeney R. E. and Nissenbaum A.** Sulfur isotope studies on Red Sea geothermal brines and sediments. Reference 2, 474–98.
24. **Carwile R. H. and Faure G.** Strontium isotope ratios and base metal content in a core from the Atlantis II Deep, Red Sea. *Chem. Geol.*, **8**, 1971, 15–23.
25. **Delevaux M. H. and Doe B. R.** Preliminary report on uranium, thorium and lead contents and lead isotopic composition in sediment samples from the Red Sea. *Initial Rep. Deep Sea Drilling Project*, **23**, 1974, 943–6.
26. **White D. E.** Environments of generation of some base-metal ore deposits. *Econ. Geol.*, **63**, 1968, 301–35.
27. **Henley R. W.** Contribution to discussion. *Trans. Instn Min. Metall. (Sect. B : Appl. earth sci.)*, **80**, 1971, B59.
28. **Girdler R. W. Erickson A. J. and Von Herzen R.** Downhole temperature and shipboard thermal conductivity measurements aboard D/V *Glomar Challenger* in the Red Sea. *Initial Rep. Deep Sea Drilling Project*, **23**, 1974, 879–86.
29. **Amann H. Bäcker H. and Blissenbach E.** Metalliferous muds of the marine environment. Paper no. 1759 presented at Offshore technology conference, Houston, 1973.
30. **Whitmarsh R. B.** *et al.* Site 226. *Initial Rep. Deep Sea Drilling Project*, **23**, 1974, 595–600.
31. **Bignell R. D.** The geochemistry of metalliferous brine precipitates and other sediments from the Red Sea. Unpublished thesis, Imperial College, London, 1975.

Editor's Comments
on Papers 30 Through 34

METALLOGENIC AND PLATE TECTONICS
—REGIONAL STUDIES

This section surveys the application of plate tectonic models to metallogenesis in specific provinces. It may represent a somewhat artificial subdivision, for some earlier papers could have come under this heading. However, they were judged more appropriate in other sections, in that they illustrate a principle rather than an actual application.

These regional studies are particularly striking for the range of opinion about the relevance of plate tectonics to metallogenesis and mineral exploration: from qualified acceptance, through scepticism to outright rejection. This is partly because several quite different models are available for many provinces—and even where a single model is generally agreed, most people hold that it lacks sufficient precision to be used as a prospecting tool.

Western America

Along the Pacific margin of North and South America, there is no dispute about the general plate tectonic setting: eastward subduction of Pacific ocean floor beneath the continents. But until much more is known about changes with time of such features as the rate, angle of inclination, position, and number of subduction zones, plate tectonics can be of little direct use in planning exploration programs. Thus, in their comprehensive review of porphyry coppers in the Pacific Northwest, Field et al. (Paper 30) pay little more than lip service to plate tectonic concepts, as their concluding paragraphs show.

Lowell (1974) is sceptical of any association at all between mineralization and plate movements in the southwestern porphyry province of North America, while Noble (Paper 31) is even more emphatic that the whole huge metallogenic province of western North America is unrelated to plate tectonics. This paper is notable for three other major conclusions: first, metal deposits of different type show no significant systematic distribution pattern (compare Paper 19); second, the ore metals originated in the upper mantle; and third, that the broad association in space and time between magmatic rocks and ore deposits is due to structural rather than genetic controls.

One possible answer to these arguments is that the plate tectonic model is not yet sophisticated enough to explain these complex relationships. If this province is considered in its circum-Pacific context, some association between porphyry mineralization and subduction zones seems inescapable. Moreover, Noble acknowledges (Paper 31) that vast amounts of heat were needed to generate the great volume of magmas which provided the pathways for mineralization. In continental margin orogenic belts, is there a better heat source than an underlying subduction zone?

In South America, too, plate tectonics can provide a general explanation for metallogenesis, but fails when it comes to details. Thus, in a regional review of South American copper deposits, Hollister (1974, page 48) finds that: "the erratic distribution of porphyry copper deposits in time and space in the [South American] plate needs some fuller explanation than is now offered by Sillitoe's model," (Paper 17 in this volume).

Hollister also records that in general porphyry coppers of disseminated stockwork type lie near known faults; whereas breccia pipe deposits usually do not. The faults are often of transverse type, and descriptions by Hollister (1974) Goossens and Hollister (1973), and Hollister and Sirvas (1974) suggest that two of these might coincide with Boundaries 1 and 2 (or 3) of Figure 1 in Paper 27. Perhaps a more consistent distribution pattern could emerge if the subduction zone beneath western South America were treated as a segmented system (Paper 27) rather than as a continuous belt.

Hercynian Belt of Europe

There are several plate tectonic models for the broad and complex Hercynian (Variscan) orogenic belt of western Europe, and a number of these relate directly to mineralization in two of the best-known metallogenic provinces: the Iberian pyrite belt of volcanic-sedimentary origin and the tin-tungsten deposits associated with K-rich granites.

There is, however, remarkably little agreement about these models. For the massive sulphide deposits of Rio Tinto type in Spain and Portugal, Bernard and Soler (1971) drew comparisons with the Japanese Kuroko deposits (see Paper 24), and suggested a *southerly* dipping subduction zone beneath southern Iberia. Carvalho (1972) argued that the subduction zone was inclined *northeastwards* and correlated the changing pattern of mineralization in that direction with increasing depth to the Benioff zone (Compare Figure 1 in Paper 19).

Schermerhorn (1975) has provided a useful review of these and other models, but questions whether plate tectonics played any part in the development of the Hercynian belt as a whole, let alone the mineralization related to it (see also Schermerhorn, 1974, and compare Williams et al., 1975)

Plate tectonic models for the tin-tungsten mineralisation related to the Hercynian granites further north are if anything even more varied than those for the pyrite belt; one might be excused for wondering if Mitchell and Bromley[1] (Papers 32 and 33) are writing about the same subject (compare also Badham and Halls, 1975).

Caledonian Orogenic Belt

Here again several plate tectonic interpretations are available, but so far only very few attempts have been made to link them in any detailed way with metallogenesis; while metallogenic studies have commonly tended to take little note of plate tectonic concepts. Thus, a recent review of mineralization in the British Caledonides by Wheatley (1971), notable for its wholly "traditional" approach, drew criticism from Arculus and Smith (1972), who felt that a plate tectonic interpretation would provide a better insight into the origins of the mineralization (see also Rea, 1972).

On the other side of the Atlantic, however, Strong (Paper 34) has

---

[1] For discussion of Mitchell's and Bromley's papers (Papers 32 and 33) see Inst. Mining and Metallurgy Trans., B, **84**:69 (1975).

recognized three main types of plate tectonic setting for the mineral deposits of Newfoundland. He claims to have identified criteria upon which to base further exploration, in particular by relating the zonal pattern of broadly defined mineral belts in the Appalachians to the inclination of an ancient subduction zone. This reconstruction seems to provide the exception to the general rule that in most regions, plate tectonic models cannot yet offer sufficiently precise criteria for the exploration geologist to use in finding new mineral deposits (see also Hammond, 1975).

Mediterranean Region

The first attempt specifically to relate metallogenesis with plate tectonics in this highly complex region was made by Dixon and Pereira (1974), who acknowledge that their's is a preliminary study only. They see the Tethys as having been "an ocean full of islands" from Mesozoic time onwards, and they trace the progressive collision and suturing together of these islands, during closure of the Tethyan ocean. Much of the mineralization is therefore located at plate (or miniplate) margins, which is broadly the conclusion reached by Barnes (1973) in a brief discussion on the origin of mineralized ophiolite complexes in the eastern Mediterranean/Black Sea region.

An alternative model of Mediterranean evolution is presented by Evans (1975), to explain the scarcity of subduction-related ore deposits—e.g. porphyry coppers—a feature which Dixon and Pereira (1974) found difficult to explain. Evans suggests that the main sense of plate movements in the "Alpine geosyncline" was transcurrent rather than convergent.

Other Regions

Two other accounts on the relationship between plate tectonics and metallogenesis in specific regions deserve mention: From part of the Solomon Islands group, G. R. Taylor (1974) has described Cyprus-type massive sulphide deposits which occur in "part of an obducted ophiolite suite marginal to a now defunct trench system." The distribution of metalliferous deposits of South Korea is outlined by Shin (1974), whose attempt to place them in a plate tectonic setting is somewhat unconvincing, if only because he appears to group together ore deposits of several ages, ranging from Precambrian to Mesozoic.

The papers in this section show that it will be some time before plate tectonic models for particular metallogenic provinces will be of much use to exploration geologists. Even when there is general agreement about a single model, it seems to be too generalized to enable specific target areas to be located (as in western America). More commonly there is such a range of conflicting models (e.g. the Hercynian belt), that we are obviously still at the stage of trying to explain in plate tectonic terms why ore deposits occur where they do, which is a long way from being able to predict where others may be found. Despite the occasional exception (e.g. Paper 34) and the optimistic tone of Hammond's (1975) review, many people would be tempted to agree with Mitchell's comment (written communication, 1974):

> It's difficult to separate papers relating mineralization to plate tectonics from those discussing mineralization in terms of modern environments. All the papers on pre-Cenozoic orogens are really the latter and whether they draw a Benioff zone underneath or not doesn't really matter.

240

30

Reprinted from *Trans. Soc. Min. Eng. AIME* **225**:20, 22 (1974)

Porphyry Copper-Molybdenum Deposits of the Pacific Northwest

by Cyrus W. Field, Michael B. Jones, and Wayne R. Bruce

[*Editor's Note:* In the original, material precedes this excerpt.]

Summary and Conclusions

Porphyry copper-molybdenum deposits of the Pacific Northwest exhibit many overall similarities to their generally younger counterparts in the continental U.S. with respect to size, geometry, plutonic host rocks, types and assemblages of ore and gangue minerals, geochemical and mineralogic zonations, and sulfur isotope compositions. Principal differences include: (1) the common association with plutons of batholithic size and with eugeosynclinal country rocks having an imprint of weak (greenschist) regional metamorphism; (2) the large apparent range in ages (18-210 m.y.) and older (120-210 m.y.) groups of deposits; (3) generally lower average grades of ore (0.60% Cu, 0.28% MoS$_2$, and less) for producing mines; (4) the relatively minor importance of secondarily enriched ore; (5) dominance of fracture-controlled sulfides in governing the distribution of ore; (6) rather large bornite:chalcopyrite ratios (unity and larger) for several deposits; and (7) the lack of significant lead-zinc-silver deposits fringing porphyry ore. Radiometric (K-Ar) data indicate near-contemporaneity of plutonic and hydrothermal events. Broad temporal similarities between the plutonic host rocks and marine country rocks for the older deposits suggest that the emplacement of magma and metallization may have occurred beneath the ocean floor. The mineral deposits are typically associated with porphyritic intrusions, breccias, and potassic alteration. Exploration programs routinely employ both geochemical (stream sediment, soil, and rock) and geophysical (IP) surveys with considerable success. Porphyry-type deposits are found in nearly all provinces within this geologically complex and diverse mobile tectonic belt. In terms of plate tectonic theory, the Northern Cordillera has evolved since Precambrian time with attendant processes of sedimentation, plutonism, volcanism, metamorphism, and deformation related to subduction of oceanic crust beneath the continental margin during active convergence of the Pacific and North American plates. Complexity has resulted from repeated episodes of convergence, multiple and migrating zones of subduction, and transcurrent faulting. Available chemical and isotopic data suggest a deep-seated source for magmas, sulfur, and water but they are insufficient on which to estimate the relative contributions from mantle versus continental and oceanic crustal sources.

Continuing field investigations supported by laboratory research, improvements and revisions in the interpretation of analytical data, and further developments in the concept of plate tectonics, and in conjunction with similar studies of South America and the Southwest Pacific, undoubtedly will contribute to a more precise geologic knowledge of the Pacific Northwest and of the genesis of mineral deposits in this vast and complex region. Unfortunately, we have not attempted to integrate other economically and genetically important types of deposits in this review. Probably significant, and yet another distinguishing feature of the older porphyry-type deposits of this region, is an apparent association with occurrences of stratiform volcanogenic sulfides. This relationship has been discussed by Hutchinson and Hodder[95] and Northcote and Muller[96] and is entirely consistent with both the temporal coincidence of plutonism-volcanism and the island arc regime of pre-Late Cretaceous magmatism. Additional data are needed relating to (1) possible associations between porphyry-type and volcanogenic ore-forming processes, (2) possible contamination of volcanic, plutonic, and hydrothermal systems by waters of oceanic and meteoric origin, and (3) the apparent enigma whereby subvolcanic porphyry-type deposits of widely differing ages are found in close regional proximity throughout the Cordillera.

REFERENCES

[95] Hutchison, R.W., and Hodder, R.W., "Possible Tectonic and Metallogenic Relationships Between Porphyry Copper and Massive Sulphide Deposits," *Transactions*, Canadian Institute of Mining & Metallurgy, Vol. 75, 1972, pp. 16-22.
[96] Northcote, K.E., and Muller, J.E., "Volcanism, Plutonism, and Mineralization, Vancouver Island," *Canadian Mining & Metallurgical Bulletin*, Vol. 65, No. 726, 1972, pp. 49-57.

31

Reprinted from *Mineral. Deposita* **9**:1-5, 7, 8, 10, 14-17, 23-25 (1974)

Metal Provinces and Metal Finding in the Western United States

J. A. NOBLE

Pasadena, California, U. S. A.

The metal provinces are believed to outline a primitive or precrustal heterogeneity of metal distribution in the upper mantle, and the well established space and time relationship between ores and intrusive bodies is a structural rather than a genetic control. The distribution of the metal deposits of the western United States in space and time is governed by some as yet unknown deep-seated source of heat; it cannot, however, be related to current theories of 'the new global tectonics'. Although we lack a full understanding of the processes of ore deposition, we can use the assembled data in an empirical way to guide exploration. This is outlined for the porphyry copper deposits of the southern Arizona Copper Quadrilateral. Most of the deposits are of Laramide age; one is Jurassic. Post-Laramide deformations, mainly Basin and Range orogeny, removed part of the deposits and buried the others; these are now being found where they are emerging at the eroding edges of the basins as the basin fill is being removed by erosion.

Es wird angenommen, daß Metallprovinzen eine primitive oder vor der Bildung der Erdkruste bestehende Ungleichmäßigkeit der Metallverteilung im oberen Mantel abzeichnen und daß die deutliche Zeit- und Raumabhängigkeit zwischen Erzen und Intrusivgesteinskörpern weniger genetisch als vielmehr strukturell bedingt ist. Die Verteilung der Erzlagerstätten der westlichen Vereinigten Staaten auf Zeit und Raum wird durch einige noch ungeklärte, tiefgehende Wärmequellen verursacht. Sie kann jedoch nicht in Beziehung zu den modernen Theorien der „Neuen globalen Tektonik" gesetzt werden. Obwohl uns noch die volle Kenntnis der Bildungsprozesse von Erzlagerstätten fehlt, können wir die vorliegenden Daten empirisch als Hilfsmittel für die Exploration verwenden. Dies wird am Beispiel der porphyrischen Kupfererze des Kupfervierecks im südlichen Arizona gezeigt. Die meisten der Lagerstätten haben dort laramisches Alter; eine ist jurassisch. Nachlaramische Verstellungen, insbesondere im Zusammenhang mit „Basin and Range"-Bildungen, bewirkten die Abtragung eines Teils der Lagerstätten und die Überdeckung des anderen Teils. Diese wird nun dort gefunden, wo die Erosion an den Becken den Untergrund freigibt.

Introduction

An earlier paper (NOBLE 1970), described the methods used in constructing metal province maps for the western United States, and from the maps drew some conclusions regarding the provenance and emplacement of the metals in the ore deposits. Subsequent studies now permit adding to these data and conclusions. The techniques used in assembling and plotting the data, set forth in the earlier paper, will not be repeated. Two factors are, however, worth mentioning, because they also bear on the new studies. The plotting of metal mines and districts on the work sheets is essentially to scale; the mines and districts on the 1:250,000 scale are not appreciably exaggerated in size. In addition, the coverage is complete; every mine with recorded production of any of the common metals is shown. Undoubtedly some mines were missed, either from inadequate search or lack of record, but the percentage of the total that was missed is certainly quite small. This means that boundaries of the outside limits of commercial amounts of metals can be drawn with confidence, and blank areas are real, except where there is very extensive and nearly complete cover by younger formations; these are discussed. These conclusions are justified because the western United States is, I think, the most extensively studied large area of important mineralization in the world.

New deposits are being found, but they do not change the province boundaries; this result is discussed below.

Figure 1, the map of combined metal provinces, is a composite of the distributions of all the important metals, together with the platform areas (Great Plains and Colorado Plateau) and the thick piles of Tertiary volcanic rocks. The distribution of the metal provinces is nearly parallel to the continental margin to a distance of about 200 mi inland, then the pattern diverges into northeast and east-northeast trends. Five separate metal provinces are peripheral to the Colorado Plateau, and four of these are immediately adjacent to piles of Tertiary volcanic rocks also peripheral to the Plateau. Northeast and east-northeast trend lines, important in controlling metal provinces, are not

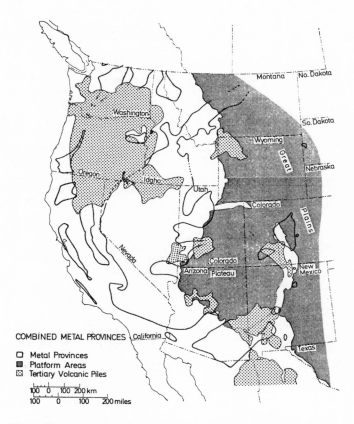

Fig. 1. Combined metal provinces in the western United States. Adapted from NOBLE (1970)

important crustal tectonic features in the western United States. Lines of intrusions trend northeast in the Front Range Mineral Belt of Colorado and in the so-called "porphyry belts" (ANDERSON 1951) of central Idaho, extending into western Montana, and those in Colorado may be related to faulting of Precambrian age (TWETO and SIMS 1963), but the relationship is not obvious. The linear northeast-trending boundaries of the well-defined quadrilateral in southern Arizona have no crustal expression.

The metal province maps and the tabulated data show that the ore-forming process has resulted not only in a high degree of concentration of the metals but also in a remarkable separation of the metals. Table 1 gives a summary of the

Table 1. *Distribution of high metal concentrations*

Copper

I.   So. Arizona Quadrilateral — 71%
II. Bingham — 10%
III. Butte — 8%
IV. Northern Arizona — 4%
V.   Ely — 3%

Molybdenum

I.   Climax-Urad — 66%
II. Questa — 31%

Lead and Zinc

I.   Cœur d'Alene — 30%
II. P. C. — Bingham — Tintic — 26%
III. Butte — 15%
IV. Colorado Front Range — 9%

Gold

I.   Black Hills — 24%
II. M. L. — G. V. — Alleghany — 13%
III. Cripple Creek — 9%

Silver

I.   Cœur d'Alene — 22%
II. P. C. — Bingham — Tintic — 17%
III. Butte — 15%
IV. Colorado Front Range — 10%
V.   Comstock — 8%

Tungsten

I.   Pine Creek — 62%

Mercury

I.   Calif. Coast Ranges — 68%

Distribution of total values

Copper	58.6%
Molybdenum	19.4%
Lead (and Zinc)	8.3%
Gold	6.4%
Silver	5.9%
Tungsten	0.5%
Mercury	0.3%
Other	0.8%

data, and Figure 2 gives the distributions. Of the 7 metals in sufficient abundance to outline and tabulate, only lead and silver stay together in the high concentrations, and in one important silver area, Comstock, there is almost no lead. All the other high concentrations scatter widely over the map.

In the earlier paper, an attempt was made to generalize the space distribution of the different metals, but I now believe that no significant pattern is present in the western United States. The first appearance of the separate metal provinces, going eastward from the coast, is in the order: mercury, copper, gold, silver, lead and zinc, tungsten, molybdenum. This is not the order of the greatest concentrations of the metals, however, as is clearly shown by Figure 2. On that map, the high concentrations of copper form a central north-trending belt; a lead-silver belt crosses the copper belt diagonally, but most is to the east of the copper; molybdenum lies to the east of these, just along the edge of the Great Plains; whereas mercury lies to the west near the Pacific Coast. Gold is scattered all the way across the map, and silver of the Comstock district is anomalous. On the other hand, the distribution of metals in areas rather than in high concentrations is largely random; only mercury is mainly restricted to a position near the Pacific Coast.

As a working hypothesis, it was concluded in the earlier paper that the sources of the metals that form the ore deposits could not be in the crust, but had to be in the upper mantle, and that the patterns of metal provinces reflected a primitive or precrustal heterogeneity of metals in the upper mantle. The metal provinces make coherent patterns, but these, except along the continental border, are essentially independent of crustal tectonic features. The remarkable

1*

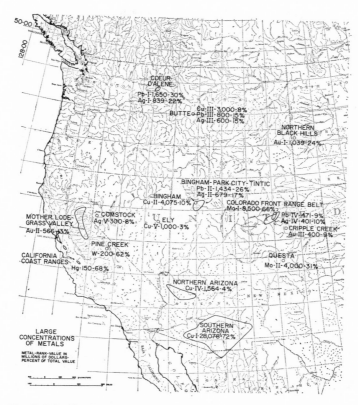

Fig. 2. Distribution of large concentrations of metals in the western United States (see Table 1). From NOBLE (1970); some new mines in the southern Arizona Copper Quadrilateral not included, compare Fig. 11

degree of high concentration of metals, and at the same time their separation, probably cannot be explained by crustal processes alone. The simple, nearly linear, shapes of the metal province maps presumably represent a deep-seated control existing in either the lowermost crust or the upper mantle. The space and time relationships between magmatic processes and metallization are considered to be structural, not genetic. Evolution of magma and its rise toward the surface of the earth gave opportunity for the escape of metal-bearing complexes distributed in patterns throughout the upper mantle. These factors were required for the formation of an ore deposit: a source of metals; a process for their escape, generally in the wake of a rising mass of magma; and a site of deposition in the upper crust.

Support for the concept that deep-seated, subcrustal, tectonic features may control metal deposition comes from a map of the compressional (Pn) velocities in the upper mantle (Fig. 3), which shows a broad pattern very much like that of the combined metal province map (Fig. 1): parallel to the coast for about 200 mi, then diverging at a high angle, northeast, eastnortheast, and east, farther inland. Even in detail there are some intriguing correlations between the two maps, all the more remarkable considering that one map is based on geophysical data recorded currently whereas the other is a summation of results of metallization that occurred over most of Mesozoic and Tertiary time.

The earlier paper made little or no reference to three interesting but poorly defined aspects of

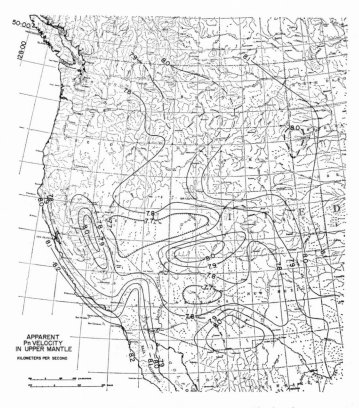

Fig. 3. Apparent P_n velocity in the upper mantle in the western United States. Adapted from ARCHAMBEAU *et al.* (1969)

the metallization process: the ages of the deposits; their relationship, if any, to theories of plate tectonics; and the overall problems of metal finding.

[*Editor's Note:* Material has been omitted at this point.]

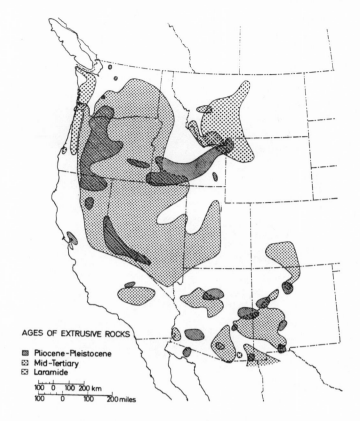

Fig. 4. Distribution of ages of extrusive rocks in the western United States

247

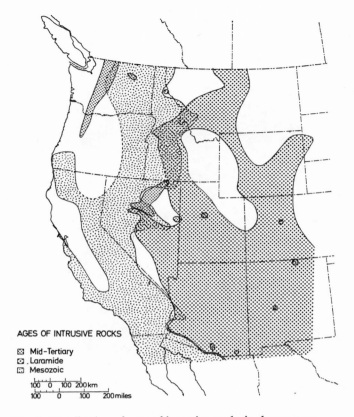

Fig. 5. Distribution of ages of intrusive rocks in the western
United States

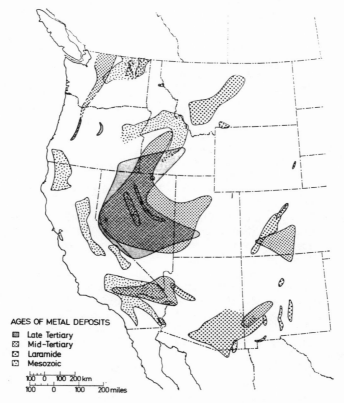

Fig. 6. Distribution of ages of metal deposits in the western
United States

Metal Provinces and Theories of Plate Tectonics

Is there a relationship, in the western United States, between metal provinces and theories of plate tectonics? To answer this question requires defining these two concepts.

Metal Provinces

Discussion and descriptions of metallogenic provinces extend over a long period of time, but definitions are hard to find. The concept is old, but the specific term came into general use in the early part of this century; perhaps SPURR (1923, p. 431—485) made first clearcut distinction between epochs and provinces, although he preferred to call them metallographic provinces. Usage in North America to-day (for example, TURNEAURE 1955) makes no reference to epochs in defining metallogenic provinces. The European usage on the other hand (for example, PETRASCHECK 1965) prefers to restrict the metallogenic province in both space and time. I find that I agree with Spurr's conclusion that provinces and epochs are different concepts. What I have called "metal provinces" and what he called "metallographic provinces" are essentially identical, but they may not be what some to-day prefer to call "metallogenic provinces". Perhaps the restrictive terms are misleading and the more general term "metal province" covers all that we want to say. A metal province is a part of the earth in which certain metals or combinations of metals are found, in distinction from other areas with no metals or with other metals or combinations of metals. Given sufficient information, these perhaps can be subdivided; the porphyry copper deposits can be outlined within a copper province. The subdivisions will depend on the operation of processes, such as specific types of host rock, fracturing, and depth at time of metallization in the case of porphyry copper deposits, but the metal province itself depends on a source of metals at depth, and this, I think, is independent of crustal processes. The concept of age or epoch may or may not apply to the subdivisions. In the southern Arizona Copper Quadrilateral, 20 porphyry coppers are Laramide but one is Jurassic; on the other hand, in the western United States, 3 porphyry coppers are Mesozoic, 24 are Laramide, and 1 is Tertiary. The formation of porphyry coppers was not confined, in the western United States, either to specific metal provinces or to specific epochs.

Theories of Plate Tectonics

Theories of plate tectonics are manifold and in process of growth. They originate with the concept of continental drift, first stated explicitly by WEGENER (1912), mainly in the manner of the geometric fitting of broad continental outlines. On the basis of much new oceanographic and geophysical, particularly paleomagnetic, information, the concept was reappraised by HESS (1962), who suggested that mantle material rising on mid-ocean ridges spread laterally (sea-floor spreading) as new oceanic crust, rafting the continents with it. Geophysical studies had showed that patterns of paleomagnetism supported, in fact demanded, that the continents had broken up and moved outward from one or perhaps two primordial supercontinents, beginning in Triassic or Jurassic time. Meanwhile, further magnetic studies of natural remanent magnetism in volcanic rocks demonstrated that worldwide reversals of magnetic polarity had occurred and could be traced back in time, and the patterns of reversals could be correlated with "stripes" of magnetic anomalies arranged symmetrically on each side of the mid-ocean ridge crests. These results not only supported the concept of continental drift continuing over a long period of time but also gave additional information on the character of the oceanic crust. This was found, in the North Pacific and eventually in many other places, to be broken up into blocks bounded by the mid-ocean ridge crests and transverse fractures showing enormous displacements, some measured in hundreds of kilometers. Additional and independent support for the concept of continental drift comes from the oceanic drilling. In none of the large oceanic basins is there sediment older than Jurassic, and in most basins the oldest sediments are Cretaceous, from which it is concluded that the basins first came into existence at about the time of the beginning of continental drift, in early Mesozoic time.

The concept of subduction zones is a correlary of the concept of continental drift motivated by sea-floor spreading. Mantle material rising along the crest of the mid-ocean ridge becomes part of the oceanic crust and is an increment to the crust; but since there is no overall expansion of the earth's crust, equivalent amounts of crust must be "consumed" elsewhere. This introduces the concept of subduction zones, whose presence in a few designated places had in fact been suggested some years before by BENIOFF (1954, 1955) on the basis of seismic data. Vast areas of oceanic crust, moving away from the axis of the mid-ocean rise, on encountering the stationary continental mass are thought to slide under the continent on an inward-dipping surface, the "Benioff zone". An oceanic trench at the toe of the continental slope, parallel to the continental margin, marks the line at which the oceanic plate bends down to pass under the continental block. The areas studied by Benioff were thought to be essentially plane surfaces dipping at about 33 degrees to 300 km vertical depth, and at about 60 degrees to 700 km vertical depth. Several similar areas have since been described, mainly around the Pacific Ocean. Still unexplained, however, is how the inward-dipping plane surface on the Chilean coast, one of Benioff's examples, produces the longest straight line on the surface of the globe. The subduction zone concept seems to be the weakest link in the chain of reasoning leading to a unified plate tectonic theory (DIETZ 1970). There does not seem to be enough evidence of subduction going on to-day to account for the amount of mid-ocean spreading that seems to be in progress.

The plate concept is derived from studies of the magnetic patterns within the blocks of oceanic crust formed by the mid-ocean crest lines and the transverse fracture zones. Some boundaries of the blocks formed by fracture zones show strong seismic activity, suggesting that motion is going on at the present time. The concept that the ocean basins could be made up of a multitude of smaller blocks or plates having diverse motions gave another opening for development of new theory, which is still under way. Moreover, if the oceanic crust is made up of a mosaic of plates, each with its own motion, why not the continents as well? Boundaries between plates having differing motions are seismic belts and possible fracture zones leading into the mantle, up which mantle material can rise. Not much observational control on plate outlines is at present available, and there is great lack of unanimity in drawing the plate boundaries and the paths of motion. Most of the paths of motion are believed to be horizontal, but subduction may take place where a moving plate encounters a fixed continental mass or an island arc.

The foregoing brief summary of plate tectonic theory, designed to test the correlation of those theories with the concept of metal provinces, is perhaps the shortest summary on record. More complete discussions are by DICKINSON (1970; 1971); KNOPOFF (1969a); MCKENZIE (1969); SMITH (1970); and WILSON (1970).

Relationship Between Metal Provinces and Plate Tectonic.

Continental drift, the best documented portion of the plate tectonic theories, seems to have had no influence on the pattern of metal distribution or of igneous activity. Metal distribution (Figs. 1 and 6) shows a westward-facing arc near and parallel to the continental margin and an eastward-facing arc far inland, with some diagonal, mostly northeast-trending, belts between them. The western arc is Mesozoic, the eastern arc Laramide, and the central portion mostly mid-Tertiary. Neither the distribution pattern nor the age relationships support the concept of the splitting up and drifting apart of a primordial continent beginning in early Mesozoic time. But if the plane of translation was well below the Mohorovičic discontinuity (KNOPOFF 1969b), then crust and upper mantle were not separated, and the metal patterns, which I believe are derived from the upper mantle, would be undisturbed by the drifting. Similarly, the age patterns of extrusive and intrusive rocks (Figs. 4 and 5) seem to have no control attributable to continental drift. The extrusive rocks (Fig. 4) have a crude symmetry with oldest outside and youngest inside, but the pattern of Mesozoic rocks is not shown because of inadequate mapping; it may be different. The age pattern of intrusive rocks (Fig. 5) has Mesozoic near the coast, Laramide inland; but mid-Tertiary in the center overlaps both earlier groups. A straight-line progression might be interpreted as evidence of drifting if the continent was moving over a deep source of magma, but the position of the mid-Tertiary intrusions is aberrant.

Testing relationships between distribution and age patterns and theories of subduction zones presents problems, because many different subduction models have been proposed (COATS 1962; HAMILTON 1969; DEWEY and BIRD 1970; DIETZ 1970; ATWATER 1970; JAMES 1971; SILLITOE 1972a, b). Unless there are complicating factors, it seems probable that there should be a progression in age and position inward from the continental margin (DEWEY and BIRD 1970; JAMES 1972; SILLITOE 1972a, b). Just such a pattern has been shown by Clark and associates (CLARK and ZENTILLI 1972) in northern Chile. The maps of age patterns for the western United States (Figs. 4, 5, and 6) were constructed to test this possibility, and in my opinion they fail, for the most part, to show this relationship. The age distributions of extrusive rocks (Fig. 4), although incomplete because of lack of sampling, is roughly concentric, with older units on the outside; in detailed work in the Great Basin, however, ARMSTRONG et al. (1969) have a concentric pattern of older units inside, younger outside. The age pattern of intrusive rocks (Fig. 5) and of metal districts (Fig. 6) show oldest units on the west, intermediate ages on the east, and youngest toward the center, again a roughly concentric pattern.

Complicating factors may in fact exist; one is the possibility that the continental block, moving away from the Mid-Atlantic Rise, has overridden the East Pacific Rise, beginning probably in Miocene time. The possible complexities are probably too great to analyze, but the most direct test for the location of the axis of the rise is a linear pattern of abnormally high heat flow, and no such pattern is found on the heat flow maps (ROY et al. 1968). Most of the area covered by Laramide orogeny shows higher than normal heat flow, greater than 1.5 microcalories per centimeter square per second, but the highest values, in excess of 3.0 units, are randomly distributed across the map. Studies of the motions of plates in this area (ATWATER and MENARD 1969; ATWATER 1970; CHASE et al. 1970) use patterns of magnetic anomalies in the oceanic crust, and are largely expressed in terms of horizontal movements, with only small amounts of subduction, limited in time. Few attempts have been made, I believe, to extend the studies across the continental block. Ultramafic rocks, which might be thought to mark boundaries between plates, occur in a long sinuous belt near the coast, from Washington to southern California, but elsewhere are too rare to be significant. Almost all studies of plate motions bordering the Pacific Coast of the United States assume large amounts of right-lateral displacement on the San Andreas fault, but the validity of this assumption I question.

Subcrustal motion in horizontal directions could perhaps be used to explain the wave pattern of the Basin and Range orogeny (GILLULY 1963), but among other objections there is serious question if the crust and upper mantle possess sufficient fluidity to produce bow waves around the advancing prow of the Colorado Plateau. A more detailed study of the position of the base of the crust (Mohorovičić discontinuity) might in fact show a wave pattern on this surface comparable to that at the top of the crust, in which case we might accept a wave motion as a reasonable mechanism. The effects of the Basin and Range orogeny were limited in time, however, in most of the western United States to Miocene and Pliocene time, and their correlation with any plate motions being discussed at present probably is impossible.

Conclusion

These studies do not contradict the general conclusions of the hypotheses of continental drift, sea-floor spreading, and plate tectonics, but they indicate to me that the process of formation of ore deposits in the western United States has no relation to those hypotheses and is not governed by them. The initiation of sea-floor spreading in the Atlantic Ocean and the first great surge of batholithic intrusion in western United States, both in early Mesozoic time, were nearly synchronous and may have been related, but the nature of the relationship certainly is not clear. Some metallization accompanied this orogeny, but unless very large amounts have been completely destroyed by erosion, which seems unlikely, the succeeding Laramide orogeny and volcanism produced much greater amounts of metallization, without any clear time relationship to global tectonics. The simultaneous concentration and separation of metals, on an arcuate zone reaching to distances of about 1000 mi (1600 km) from the continental margin, cannot be fitted to any global tectonic model known to me. Large forces, or large amounts of heat, are required to explain the volcanism and metal formation in the western United States, and it seems likely that these were contained within the limits of the maps I have been describing; we need to look down, not to one side for the sources (GILLULY 1970). It has been noted (DAMON and MAUGER 1966; ARMSTRONG 1970) that volcanism on both large and small scale tends to begin with a great surge and then taper off gradually. It has also been noted above that volcanism shows some tendency to contract toward a center; in a broad way, older units surround younger units in the age distribution maps. These two generalizations seem to point to the sporadic and rather sudden generation at great depths of large amounts of heat by mechanisms for which at present we probably have no adequate explanations. The rise and emplacement of the metals to form ore deposits is probably part of this process, and these likewise are not yet explained.

Acknowledgements

Dr. PAUL E. DAMON kindly furnished data from which I constructed the age distribution maps. The substance of this paper was used in a Society of Economic Geologists Visiting Lecture Series in 1972 at McGill University, Montreal, Quebec; Montana State University, Bozeman; Queen's University, Kingston, Ontario; The University of Georgia, Athens; and the University of Utah, Salt Lake City. I am indebted to the staffs and students of these schools for many discussions. Parts of the paper were also presented at a symposium on correlation between metallogenic and geochemical provinces, Leoben, Austria, and at The Institute of Geological Sciences, London. Again I am indebted to the participants for discussions. Dr. JOHN C. RUCKMICK read and criticized the final manuscript.

References

ANDERSON, A. L.: Metalliferous epochs in Idaho. Econ. Geol. 46, 592—607 (1951).

ANDERSON, C. A: Arizona and adjacent New Mexico. In: Ore deposits in the United States 1933—67. A. I. M. E Graton-Sales Vol., p. 1163—1190 (1968).

ARCHAMBEAU, C. D., FLINN, E. A., LAMBERT, D. G.: Fine structure of the Upper Mantle. J. Geophys. Res. 74, 5825—5865 (1969).

ARMSTRONG, R. L.: Geochronology of Tertiary igneous rocks, eastern Nevada and vicinity, U.S.A. Geochim. Cosmochim. Acta 34, 203—232 (1970).

— EKREN, E. B., McKEE, E. H., NOBLE, D. C.: Space-time relations of Cenozoic silicic volcanism in the Great Basin of the western United States. Am. J. Sci. 267, 478—490 (1969).

ATWATER, T.: Implications of plate tectonics for the Cenozoic tectonic evolution of western North America. Geol. Soc. Am. Bull. 81, 3513—3536 (1970).

— MENARD, H. W.: Magnetic lineations in the northeast Pacific. Earth Planetary Sci. Letters 7, 445—450 (1969).

BATEMAN, P. C., CLARK, L. D., HUBER, N. K., MOORE, J. G., RINEHART, C. D.: The Sierra Nevada batholith — a synthesis of recent work across the central part. U.S. Geol. Surv. Prof. Paper 414—D, 46 p. (1963).

BENIOFF, H.: Orogenesis and deep crustal structure — additional evidence from seismology. Geol. Soc. Am. Bull. 65, 385—400 (1954).

— Seismic evidence for crustal structure and tectonic activity. Geol. Soc. Am. Spec. Papers 62, 61—74 (1955).

CHADWICK, R. A.: Belts of eruptive centers in the Absaroka-Gallatin volcanic province, Wyoming-Montana. Geol. Soc. Am. Bull. 81, 267—274 (1970).

— Volcanism in Montana. Northwest Geol. 1, 1—20 (1971).

CHASE, G. C., MENARD, H. W., LARSON, R. L., SHARMAN, F. G., III, SMITH, S. M.: History of sea-floor spreading west of Baja California. Geol. Soc. Am. Bull. 81, 491—498 (1970).

CLARK, A. H., ZENTILLI, M.: The evolution of a metallogenic province at a consuming plate margin: The Andes between latitudes 26° and 29° South. Can. Min. Met. Bull. 65, 37 (abstr.) (1972).

COATS, R. R.: Magma type and crustal structure in the Aleutian Arc. In: The crust of the Pacific Basin (ed. MACDONALD, G. A., and KUNO, H.). Am. Geophys. Union Geophys. Mon. 6, 92—109 (1962).

COHEE, G. V.: Compiler, Tectonic map of the United States. U.S. Geol. Surv. 1962.

DAMON, P. E.: Correlation and chronology of ore deposits and volcanic rocks. Atom. En. Comm., Annual Progress Rept., Contract AT (11—1)—689, 73 p. (1968) plus appendices.

— The relationship between late Cenozoic volcanism and tectonism and orogenic-epeirogenic periodicity. In: The late Cenozoic Glacial Ages (ed. TUREKIAN, K. K.), p. 15—35. New York: John Wiley and Sons 1971.

— MAUGER, R. L.: Epeirogeny-orogeny viewed from the Basin and Range Province. Am. Inst. Min. Eng. Trans. 235, 99—111 (1966).

DANE, C. H., BACHMAN, G. O.: Compilers, geologic map of New Mexico. U. S. Geol. Surv. 1965.

DEWEY, J. F., BIRD, J. M.: Mountain belts and the new global tectonics. J. Geophys. Res. 75, 2625—2647 (1970).

DICKINSON, W. R.: Global tectonics. Science 168, 1250—1259 (1970).

— Plate tectonics in geologic history. Science 174, 107—113 (1971).

— GRANTZ, A.: Indicated cumulative offsets along the San Andreas fault in the California Coast Ranges. In: Proceedings of conference on geologic problems of the San Andreas fault system (ed. DICKINSON, W. R., and GRANTZ, A.). Stanford Univ. Pubs. Geol. Sci. 11, 117—120 (1968).

DIETZ, R. S.: Continents and ocean basins. In: The megatectonics of continents and oceans (ed. JOHNSON, H., and SMITH, B. L.), p. 24—46. New Brunswick, N. J.: Rutgers Univ. Press 1970.

EASTWOOD, R. L.: A geochemical-petrological study of mid-Tertiary volcanism in parts of Pima and Pinal Counties, Arizona. Ph. D. thesis, Univ. Arizona, 211 p., 1970.

GILLULY, J.: The tectonic evolution of the western United States. Quart. J. Geol. Soc. London 119, 133—174 (1963).

— Chronology of tectonic movements in the western United States. Am. J. Sci. **265**, 306—331 (1967).

— Crustal deformation in the western United States. In: The megatectonics of continents and oceans (ed. JOHNSON, H., and SMITH, B. L.), p. 47—73. New Brunswick, N. J.: Rutgers Univ. Press 1970.

HAMILTON, W.: Mesozoic California and the underflow of Pacific mantle. Geol. Soc. Am. Bull. **80**, 2409—2430 (1969).

HESS, H. H.: History of ocean basins. Geol. Soc. Am. Petrologic Studies (Buddington Vol.), p. 599—620 (1962).

HILL, M. L., DIBLEE, T. W., Jr.: San Andreas, Garlock, and Big Pine faults, California. Geol. Soc. Am. Bull. **64**, 443—458 (1953).

— HOBSON, H. D.: Possible post-Cretaceous slip on the San Andreas fault zone. In: Proceedings of conference on geologic problems of the San Andreas fault system: (ed. DICKINSON, W. R., and GRANTZ, A.). Stanford Univ. Pubs. Geol. Sci. **11**, 123—129 (1968).

HOLMES, A.: A revised time scale. Edinburgh Geol. Soc. Trans. **17**, 183—216 (1959).

JAMES, D. E.: Plate tectonic model for the evolution of the central Andes. Geol. Soc. Am. Bull. **82**, 3325—3346 (1971).

KING, P. B.: Compiler, Tectonic map of North America. U.S. Geol. Surv. 1969.

KNOPOFF, L.: Continental drift and convection. In: The Earth's crust and upper mantle (ed. HART, P. J.). Am. Geophys. Union Geophys. Mon. **13**, 683—689 (1969a).

— The upper mantle of the earth. Science **163**, 1277—1287 (1969 b).

KULP, J. L.: Geologic time scale. Science **133**, 1105—1114 (1961).

LIVINGSTON, D. E., MAUGER, R. L., DAMON, P. E.: Geochronology of the emplacement, enrichment, and preservation of Arizona porphyry copper deposits. Econ. Geol. **63**, 30—36 (1968).

LOWELL, J. D., GUILBERT, J. M.: Lateral and vertical alteration-mineralization in porphyry ore deposits. Econ. Geol. **65**, 373—408 (1970).

MACKIN, J. H.: Structural significance of Tertiary volcanic rocks in southwestern Utah. Am. J. Sci. **258**, 81—131 (1960).

MCKEE, E. H., NOBLE, D. C., SILBERMAN, M. L.: Middle Miocene hiatus in volcanic activity in the Great Basin area of the western United States. Earth Planetary Sci. Letters **8**, 93—96 (1970).

MCKENZIE, D. P.: Speculations on the consequences and causes of plate motions. Geophys. J. **18**, 1—32 (1969).

NASON, R. D.: San Andreas fault at Cape Mendocino. In: Proceedings of conference on geologic problems of the San Andreas fault system: (ed. DICKINSON, W. R., and GRANTZ, A.).

Stanford Univ. Pubs. Geol. Sci. **11**, 231—241 (1968).

NIELSEN, R. L.: Hypogene texture and mineral zoning in a copper-bearing granodiorite porphyry stock, Santa Rita, New Mexico. Econ. Geol. **63**, 37—50 (1968).

NOBLE, J. A.: Metal provinces of the western United States. Geol. Soc. Am. Bull. **81**, 1607—1624 (1970).

PAKISER, L. C., ZIETZ, I.: Transcontinental crustal and upper-mantle structure. Rev. Geophys. **3**, 505—520 (1965).

PETRASCHECK, W. E.: Typical features of metallogenic provinces. Econ. Geol. **60**, 1620—1634 (1965).

ROBERTS, R. J., RADTKE, A. S., COATS, R. R.: Gold-bearing deposits of north-central Nevada and southwestern Idaho, with a section on periods of plutonism in north-central Nevada by SILBERMAN, M. L., and MCKEE, E. H. Econ. Geol. **66**, 14—33 (1971).

ROY, R., DECKER, E. R., BLACKWELL, D., BIRCH, F.: Heat flow in the United States. J. Geophys. Res. **73**, 5207—5221 (1968).

SILLITOE, R. H.: A plate tectonic model for the origin of porphyry copper deposits. Econ. Geol. **67**, 184—197 (1972a).

— Relation of metal provinces in western America to subduction of oceanic lithosphere. Geol. Soc. Am. Bull. **83**, 813—817 (1972b).

SMITH, B. L.: Foreword. In: The megatectonics of continents and oceans (ed. JOHNSON, H., and SMITH, B. L.). New Brunswick, N. J.: Rutgers Univ. Press. p. vii—xii, 1970.

SMITH, J. G., MCKEE, E. H., TATLOCK, D. B., MARVIN, F. R.: Mesozoic granitic rocks in northwestern Nevada: A link between the Sierra Nevada and Idaho batholiths. Geol. Soc. Am. Bull. **82**, 2933—2944 (1971).

SPURR, J. E.: The ore magmas, 915 p. New York: McGraw-Hill 1923.

STEWART, J. H.: Basin and Range structure: A system of horsts and grabens produced by deep-seated extension. Geol. Soc. Am. Bull. **82**, 1019—1044 (1971).

TITLEY, S. R.: Pre-ore environment of southwestern North American porphyry copper deposits. 24th Int. Geol. Congr. Proc., Montreal, Canada, section **IV**, p. 252—260 (1972).

TURNEAURE, F. S.: Metallogenic provinces and epochs. Econ. Geol. **50th Ann.**, 38—98 (1955).

TWETO, O., SIMS, P. K.: Precambrian ancestry of the Colorado mineral belt. Geol. Soc. Am. Bull. **74**, 991—1014 (1963).

WATERS, A. C.: Basalt magma types and their tectonic associations: Pacific Northwest of the United States. In: The crust of the Pacific Basin (ed. MACDONALD, G. A., and KUNO, H.). Am. Geophys. Union Geophys. Mon. **6**, 158—170 (1962).

WEGENER, A.: Die Entstehung der Kontinente. Petermanns Mitteilungen p. 185—195, 253—256, 305—309 (1912).

WILSON, E. D., MOORE, R. T., COOPER, J. R.: Compilers, geologic map of Arizona. Ariz. Bur. Mines U.S. Geol. Surv. 1969.

WILSON, J. T.: Science of the earth. In: The megatectonics of continents and oceans. (ed. JOHNSON, H., and SMITH, B. L.), p. 253—268. New Brunswick, N. J.: Rutgers Univ. Press 1970.

Received December 5, 1972

JAMES A. NOBLE
1475 East California Boulevard, Pasadena, California 91106, U.S.A.

32

Reprinted from *Trans. Instn Min. Metall.* 84:B95–B97 (1974)

Southwest England granites:

magmatism and tin mineralization

in a post-collision tectonic setting

A. H. G. Mitchell M.A., D.Phil., M.I.M.M.
Department of Geology, Oxford University, Oxford (at
present U.N.D.P., P.O. Box 650, Rangoon, Burma)

551.24 : 552.3 : 553.45(423.5/.7)

Tin deposits and associated granitic rocks, particularly those of late Mesozoic to early Cenozoic age in the Andes and Southeast Asia, have recently been interpreted as having been emplaced above a subducting oceanic plate inclined beneath continental crust.[1,2,3] The tin-bearing granites of southwest England cannot be satisfactorily explained either by this model or by mantle plumes,[23] and evidence given below suggests that they were related to subduction, but were emplaced following continent–continent collision.

The Hercynian orogeny in southwest England has been interpreted as an Andean-type continental margin above a northward-dipping Benioff zone,[4,5] and as a continental collision belt following northward[6] and both northward and southward[7] subduction; development of a marginal basin in the orogenic belt above a northward-dipping Benioff zone has also been considered.[8] A difficulty with these interpretations is that calc-alkaline volcanic rocks of Devonian and Carboniferous age indicative of the inferred northward-dipping Benioff zone are largely absent in Devon and Cornwall. In this note analogies with Cenozoic arc systems and collision belts (Table 1) are used to indicate that the

Table 1 Rock units and tectonic features in late Mesozoic and Cenozoic arc systems and collision belts comparable with those in southwest England

Southwest England	Cenozoic analogy
Tectonically emplaced Gramscatho Beds	Barbados; Indo-Burman Ranges; Kodiak Shelf, Alaska
Lizard Thrust and 'wildflysch'	Naga metamorphic complex and Miocene wildflysch; Indo-Burman Ranges
Bathyal lull facies and volcanic rocks	Mid-ocean ridge; Ninety East Ridge
Crackington Formation	Bengal Fan deposits
Bude Formation	Upper Tertiary sediments of Assam and Siwaliks
North-directed overthrusts	Himalayan thrusts south of Indus suture
Post-tectonic tin-granites	Badrinath Granite, Himalayas
Permian alkaline volcanics	Rome Province, Italy
Inferred Carboniferous sediments, English Channel	Tibetan Basin; upper Tertiary molasse, Burma

relationships of rock units and the presence of granites can most simply be explained by southward subduction of ocean floor or marginal basin crust (the Rheic Ocean)[9] followed by continental collision. The stratigraphy and facies relationships are based on recently published summaries.[7,8]

The thick succession of middle Devonian turbidites that form the Gramscatho Beds in south Cornwall[8] can be interpreted as ocean-floor sediments derived partly from a volcanic source area to the south; comparable sediments in a similar setting occur on the Aleutian abyssal plain.[10] The start of southward to southeastward subduction of the ocean floor beneath the southern continent in the middle Devonian probably resulted in tectonic thinning of the crust beneath the continental margin above the Benioff zone and in underthrusting of ocean crust beneath continental crust, sedimentary rocks and ophiolites of the overriding plate, forming the incipient Lizard Thrust[11] with Caledonoid trend (Fig. 1(a)).

Continued southward subduction was accompanied by underthrusting of successive wedges of Gramscatho turbidites above the subduction zone, as suggested by Burne,[7] in a manner analogous to that described from Cenozoic sediments on the Barbados Ridge.[12,22] The resulting thick pile of flysch-type deposits with north to northwestward-facing folds preserved in south Cornwall resembles the deformed flysch belt of the Kodiak Shelf in Alaska[13] and the Indo-Burman Ranges.[14] With further northward movement of the Lizard Thrust, blocks from the advancing thrust sheet broke off and accumulated above the deformed Gramscatho Beds, forming the 'wildflysch' of the upper Devonian Veryan Series and Gidley Wells Beds.[8] Comparable Cenozoic events are the westward thrusting of the Naga Metamorphic Complex and Kanpetlet Schists in Burma over the Indo-Burman Ranges flysch, resulting in Miocene wildflysch.[14] Subduction with tectonic growth and uplift of the Gramscatho Beds continued into the early Carboniferous, forming the mid-Devonian to Namurian orogeny in south Cornwall.[8]

Subduction and partial closure of the Rheic Ocean resulted in the approach of either the mid-ocean ridge or a non-spreading oceanic ridge analogous to the Indian Ocean Ninety East Ridge to the subduction zone. The crustal rocks of the ridge, like the Gramscatho Beds flysch, were, in turn, underthrust from the north or northwest by oceanic crust, and now form the bathyal lull facies with alkaline and tholeiitic lavas of south Devon[8] (Fig. 1(b)).

In north Devon, on the passive southern margin of the northern continent, paralic sedimentation continued from the middle Devonian to the upper Carboniferous.[7] To the south, turbidites of the middle Carboniferous Crackington Formation[7] were deposited on the closing ocean floor north of the underthrust rise, in an analogous setting to the Cenozoic Bengal Delta fan deposits.[15] With continued subduction, partial closure of the basin and approach of the northern continent to the subduction zone a change from marine turbidite to non-marine sedimentation forming the Bude Formation took place.[7] The Bude Formation thus occupies an analogous tectonic position to the late Tertiary molasse of Assam and the southern Himalayas.[16]

With continued southward movement the northern continent underthrust the sedimentary units of south Devon and south Cornwall, and south-dipping thrust planes of Hercynian trend developed progressively further to the north, in north Devon, south Wales and southern Ireland.[11] These thrusts were synthetic to the direction of underthrusting and analogous to those on the underthrusting Indian plate in the Himalayas (Fig. 1(c)).

Emplacement of granites and associated mineralization in southwest England post-dated tectonic activity, and took place when southward underthrusting had almost ceased. Decrease in rate of subduction is commonly accompanied by increase in inclination of the Benioff zone;[17] hence, by the time subduction had ceased at the close of the Carboniferous, a sub-vertical zone of oceanic crust of the subducting plate probably underlay southwest England. The location and age of the granites suggest that they were emplaced above

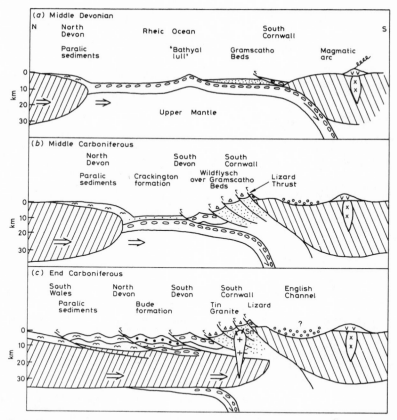

Fig. 1 Diagrammatic cross-sections through Hercynian arc system and collision belt of southwest England: (*a*) middle Devonian; (*b*) middle Carboniferous; and (*c*) end Carboniferous

this zone, and presumably related to rise of hot magma or volatiles from the zone at depth. The granites occupy a similar position with respect to the orogen to the post-tectonic tourmaline-bearing two-mica granites of the Himalayas.[16]

The Permian potash-rich lavas of Devonshire, sometimes used as evidence for northward subduction,[18] are probably post-collision rocks not directly related to subduction processes, analogous in tectonic setting to the Cenozoic alkaline rocks of the Rome Province in Italy.[19] The English Channel, south of the collision belt of Devon and Cornwall, occupies a comparable position to the Tertiary Basin of Burma and the Tibetan basin, and might be underlain by a thick Carboniferous clastic succession.

The tin mineralization and the geochemical characteristics of the granites were probably directly related to the tectonic setting during magma emplacement. If the sedimentary host rocks of Devon and Cornwall were not underthrust by the northern continent, generation of the tin-bearing granites must have taken place within the buried sedimentary succession, within the underlying oceanic crust and upper mantle, or within subducted lithosphere, and continental crust was not involved. If, however, continental crust underthrust the sedimentary host rocks, as shown in Fig. 1(*c*), the granites and associated metals were either generated within or penetrated this crust.

The tin in southwest England could be explained if abnormal concentrations of the metal were present in the underthrust southern margin of the continent,[20] and were remobilized, transported and further concentrated by the rising magma. Alternatively, in the absence of

evidence for this pre-Hercynian concentration of metal, it can be argued that the mineralization is characteristic of most granites generated in a continental collision tectonic setting. Differences in composition between Caledonian and Variscan granites of Western Europe, and inferred differences in the geothermal gradient,[21] can perhaps be explained if most of the Variscan granites were emplaced during or subsequent to continental collision and the Caledonian granites were mostly emplaced adjacent to the continental margin in the more widely recognized pre-collision type of tectonic setting.[1,2,3] Although commonly exposed within sedimentary host rocks, the composition of granites and mineral deposits similar to those of Devon and Cornwall may be dependent on the presence of underlying older underthrust continental crust.

References

1. **Mitchell A. H. G.** and **Garson M. S.** Relationship of porphyry copper and circum-Pacific tin deposits to palaeo-Benioff zones. *Trans. Instn Min. Metall. (Sect. B: Appl. earth sci.)*, **81**, 1972, B10–25.

2. **Sillitoe R H.** Relation of metal provinces in western America to subduction of oceanic lithosphere. *Bull. geol. Soc. Am.*, **83**, 1972, 813–7.

3. **Mitchell A. H. G.** Metallogenic belts and angle of dip of Benioff zones. *Nature, phys. Sci.*, **245**, Sept. 24 1973, 49–52.

4. **Nicolas A.** Was the Hercynian orogenic belt of Europe of the Andean type? *Nature*, **236**, 1972, 221–3.

5. **Floyd P. A.** The tectonic environment of South-West England: contributions to the discussion of a paper by **Floyd P. A.** [*Proc. Geol. Ass.*, **83**, 1972, 385–404] *Proc. Geol. Ass.*, **84**, 1973, 243–7.

6. **Burrett C. F.** Plate tectonics and the Hercynian orogeny. *Nature*, **239**, 1972, 155–7.

7. **Burne R. V.** Palaeogeography of south west England and Hercynian continental collision. *Nature, phys. Sci.*, **241**, 1973, 129–31.

8. **Johnson G. A. L. Reading H. and Hall A.** The tectonic environment of South-West England: contributions to the discussion of a paper by **Floyd P. A.** [*Proc. Geol. Ass.*, **83**, 1972, 385–404] *Proc. Geol. Ass.*, **84**, 1973, 237–47.

9. **McKerrow W. S. and Ziegler A. M.** Palaeozoic oceans. *Nature, phys. Sci.*, **240**, 1972, 92–4.

10. **Hamilton E. L.** Marine geology of the Aleutian Abyssal Plain. *Marine Geol.*, **14**, 1973, 295–325.

11. **Anderson J. G. C. and Owen T. R.** *The structure of the British Isles* (Oxford, etc.: Pergamon Press, 1968), 162 p.

12. **Westbrook G. K. Bott M. H. P. and Peacock J. H.** Lesser Antilles subduction zone in the vicinity of Barbados. *Nature. phys. Sci.*, **244**, 1973, 118–20.

13. **Moore J. C.** Cretaceous continental margin sedimentation, southwestern Alaska. *Bull. geol. Soc. Am.*, **84**, 1973, 595–614.

14. **Brunnschweiler R. O.** On the geology of the Indoburman Ranges. *J. geol. Soc. Aust.*, **13**, 1966, 137–94.

15. **Curray J. R. and Moore D. G.** Growth of the Bengal Deep-Sea Fan and denudation in the Himalayas. *Bull. geol. Soc. Am.*, **82**, 1971, 563–72.

16. **Gansser A.** *Geology of the Himalayas* (London: Interscience, 1964), 289 p.

17. **Luyendyk B. P.** Dips of downgoing lithospheric plates beneath island arcs. *Bull. geol. Soc. Am.*, **81**, 1970, 3411–6.

18. **Cosgrove M. E.** The geochemistry of the potassium-rich Permian volcanic rocks of Devonshire, England. *Contr. Mineral. Petrol.*, **36**, 1972, 155–70.

19. **Appleton J. D.** Petrogenesis of potassium-rich lavas from the Roccamonfina volcano, Roman Region, Italy. *J. Petrology*, **13**, 1972, 425–56.

20. **Hall A.** The tectonic environment of South-West England: contributions to the discussion of a paper by **Floyd P. A.** [*Proc. Geol. Ass.*, **83**, 1972, 385–404] *Proc. Geol. Ass.*, **84**, 1973, 243.

21. **Hall A.** Regional geochemical variation in the Caledonian and Variscan granites of western Europe. In *24th Int. Geol. Congr.*, *Sect. 2* (Montreal: The Congress, 1972), 171–80.

22. **Mitchell A. H. G.** Flysch ophiolite successions: polarity indicators in arc and collision-type orogens. *Nature*, **248**, 1974, 747–9.

23. **Sillitoe R. H.** Tin mineralisation above mantle hot spots. *Nature*, **248**, 1974, 497–9.

33

Copyright © 1975 by The Institution of Mining and Metallurgy

Reprinted from *Trans. Instn Min. Metall.* 84:B28–B30 (1975)

Tin mineralization of Western Europe : is it related to crustal subduction?

A. V. Bromley Ph.D., F.G.S.
Camborne School of Mines, Camborne, Cornwall

551.24 : 552.3 : 553.45(4–15)

In a recent note Mitchell[1] proposed that the tin-bearing granites of southwest Britain were generated above a steeply dipping, southward-inclined subduction zone, following continental collision in late Carboniferous times. Mitchell's hypothesis involves the emplacement of the passive margin of a northern continent beneath the Devonian–Carboniferous sedimentary pile of south Devon and Cornwall, with concomitant northward overthrusting by the southern continent. During late Carboniferous times the active subduction zone was overridden and sealed by the advance of the northern continent (Mitchell, Fig. 1). Mitchell has suggested that tin-bearing granites were generated above the sealed subduction zone, either by fusion of the underthrust continental crust, which may have carried anomalous concentrations of tin, or in the absence of abnormal tin concentrations as a result of granite magma generation in a continental collision tectonic regime.

The writer believes Mitchell's hypothesis to be untenable—for the following reasons : (1) there is no evidence that the pre-Hercynian (Precambrian–Lower Palaeozoic) basement of Britain contains anomalous tin concentrations ; (2) there is no evidence that granites generated in a continental collision tectonic setting (e.g. Alpine–Himalayan system) carry anomalous concentrations of tin as a general rule : tin mineralization is associated with convergent plate margins of the cordilleran type,[2,3] in island arc–trench regimes, and is also characteristic of certain granite complexes in intraplate environments, as is Nigeria, South West Africa, Rondônia and elsewhere ;[4] and (3) any hypothesis which seeks to explain the origin of tin-bearing granites in southwest Britain, in terms of magma generation above a subduction zone, must also account for the distribution of Hercynian tin-bearing granites in Germany, Brittany, the Massif Central and the Iberian Peninsula. If the tin mineralization is related to a Hercynian subduction zone, that zone cannot be located beneath south Cornwall because, regardless of its polarity, it could not be responsible for the generation of tin-bearing magmas in southwest Britain *and* on the southern side of the oceanic suture. Conversely, if an Hercynian subduction zone is located beneath south Cornwall, the tin-bearing granites must have been generated by some process other than subduction.

The Hercynian orogenic belt of southwest Britain has been interpreted by other authors as a cordilleran-type continental margin above a northward-dipping subduction zone,[5] as a continental collision belt with northward subduction[6] and in terms of both northward and southward subduction.[7] Floyd,[8] Reading[9] and Hall[10] have suggested that a northward-dipping subduction zone lay to the south of the Massif Central in the Tethyan region—a distance of 700 km from the south Cornwall area, even if no allowance is made for crustal shortening. Although Floyd's hypothesis has the advantage of placing all the tin-bearing granites within the same crustal plate, the distance between a Tethyan subduction zone and southwest Britain seems rather excessive. Furthermore, palaeogeographical considerations demand that a mid-European ocean (Rheic Ocean) separated northern and south-central Europe at least as early as Ordovician times[11] and that this ocean was only eliminated at the end of the Carboniferous period.[6,12] The most probable location for this oceanic suture is between the Rheno-Hercynian zone and the Saxo-Thuringian zone of the European Hercynides.[12,13,14] Aubouin[14] proposed that the external Saxo-Thuringian zone, characterized in middle Europe by powerful ophiolite volcanism, extends westward beneath more recent rocks to re-emerge in the Lizard area of south Cornwall. It is here that most convincing evidence of Hercynian subduction is preserved.

In south Cornwall the Veryan Series,[15,16] of characteristic wildflysch facies, contains Ordovician quartzite and Devonian limestones with shelly faunas of southern affinities, thick sequences of Devonian pillow lava and acid–intermediate volcanic rocks of uncertain provenance. The Veryan rocks are thrust northwards over autochthonous middle Devonian Gramscatho Beds and are, in turn, overthrust by the mafic–ultramafic rocks of the Lizard Complex.[17] The Lizard Complex peridotite has previously been interpreted as an horizontal thrust sheet emplaced at low temperature[18] and as a hot diapiric intrusion,[19] though Thayer[20] claimed that the Lizard rocks constitute a true ophiolite complex. In the southeastern part of the Lizard Complex there is a clearly displayed northward (and probably upward) succession from peridotite (lherzolite and dunite) into foliated and commonly brecciated gabbros via a complex transition zone in which innumerable dykes of gabbro, gabbro pegmatite and troctolite penetrate the peridotite and where ultrabasic xenoliths are abundant in the gabbros. The gabbros, in turn, pass northwards into the root zone of a sheeted dyke complex.[21] Such sequences are now widely recognized as constituting fragments of ancient oceanic crust[22,23] and, in association with wildflysch deposits, are regarded as providing important evidence in the location of former subduction zones. The apparently conflicting evidence for both high- and low-temperature emplacement of the Lizard peridotite is resolved in terms of the ocean crust model. Ophiolite complexes are generated under the high-temperature conditions of active spreading axes and may be emplaced tectonically at accreting continental margins. They may preserve evidence of both environments.[24]

If the Iberian Peninsula is restored to the position it occupied before the opening of the Bay of Biscay,[25] the Hercynian structures of the Massif Central and Brittany appear continuous with those of northern Spain and Portugal, forming a tight arc. With this restoration in mind it is tempting to extend the German–Lizard subduction zone into southern Portugal, between the Ossa Morena zone (with ophiolite volcanism and ultrabasic intrusions) and the south Portugal zone (with Carboniferous rocks of Culm facies).[26] Even on this restoration (Fig. 1) it is not clear if the apparent distribution of tin-bearing granites reflects their original configuration, since it is uncertain whether the Brittany–Iberian arc is primary (curvature of the subduction zone) or secondary and related to subsequent deformation.

Since the advent of the plate tectonics hypothesis it has become fashionable to invoke crystal subduction to explain a variety of igneous phenomena. The first requirement of subduction-driven igneous action is that the igneous rocks should be generated within or above the subduction zone and subduction zones can only be inclined in one direction at a time. If the tin-bearing granites of Hercynian Europe were generated by the same mechanism and/or derived from a similar source, then their distribution seems to exclude the subduction hypothesis. Selective fusion by radioactive heating in a tectonically thickened sialic crust following continental collision could account for the simultaneous generation of granite magma on both sides of the subduction zone, but it does not go far towards explaining why they should be tin-enriched. Alternatively, the granites may

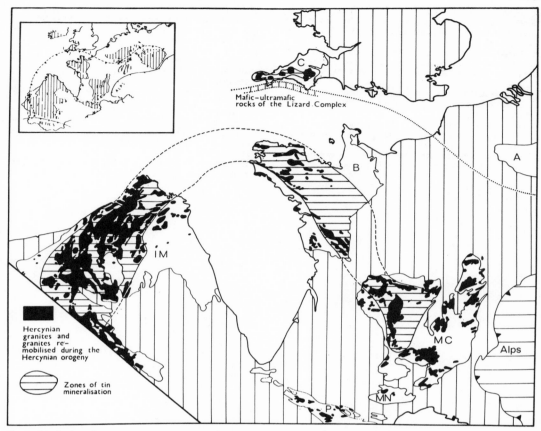

Fig. 1 Distribution of tin mineralization in western Europe (C, Cornubia; B, Brittany; A, Ardennes; IM, Iberian Meseta; MC, Massif Central; MN, Montagne Noire; P, Pyrenees). Inset map shows suggested position of Hercynian oceanic suture in western and central Europe (1, Rheno-Hercynian zone; south Portugal zone in Iberian Peninsula; 2, Saxo-Thuringian zone; Ossa Morena zone in the Iberian Peninsula; 3, Moldanubian zone; Galicia–Castille and Alcudia–eastern Lusitania zone in Iberian Peninsula; Hercynian Massifs, vertical lines)

have been generated above fixed mantle plumes or 'hot spots' and their location with respect to the Hercynian orogen might even be casual rather than causative. Evidence for mantle derivation of a number of characteristically lithophile elements, including tin, has been provided from the Azores hot spot[27] and from Iceland.[28] In this context it may be significant that recent strontium isotope data[29] from granitic rocks in southwest Britain suggest derivation from the mantle rather than from the sialic crust.

References

1. **Mitchell A. H. G.** Southwest England granites: magmatism and tin mineralization in a post-collision tectonic setting. *Trans. Instn Min. Metall. (Sect. B: Appl. earth sci.)*, **83**, 1974, B95–7.

2. **Sillitoe R. H.** Relation of metal provinces in western America to subduction of oceanic lithosphere. *Bull. geol. Soc. Am.*, **83**, 1972, 813–7.

3. **Mitchell A. H. G. and Garson M. S.** Relationship of porphyry copper and circum-Pacific tin deposits to palaeo-Benioff zones. *Trans. Instn Min. Metall. (Sect. B: Appl. earth sci)*, **81**, 1972, B10–25.

4. **Sillitoe R. H.** Tin mineralisation above mantle hot spots. *Nature, Lond.*, **248**, 1974, 497–9.

5. **Nicolas A.** Was the Hercynian orogenic belt of Europe of the Andean type? *Nature, Lond.*, **236**, 1972, 221–3.

6. **Burne R. V.** Palaeogeography of south west England and Hercynian continental collision. *Nature, phys. Sci.*, **241**, 1973, 129–31.

7. **Burrett C. F.** Plate tectonics and the Hercynian orogeny. *Nature, Lond.*, **239**, 1972, 155–7.

8. **Floyd P. A.** Geochemistry, origin and tectonic environment of basic and acidic rocks of Cornubia, England. *Proc. Geol. Ass.*, **83**, 1972, 385–404.

9. **Reading H.** Contribution to the discussion of reference 8. *Proc. Geol. Ass.*, **84**, 1973, 239–42.

10. **Hall A.** Contribution to discussion of reference 8. *Proc. Geol. Ass.*, **84**, 1973, 243.

11. **Whittington H. B. and Hughes C. P.** Ordovician geography and fauna provinces deduced from trilobite distribution. *Phil. Trans. R. Soc.*, **B263**, 1972, 235–78.

12. **Johnson G. A. L.** Closing of the Carboniferous sea in western Europe. In *Implications of continental drift to the earth sciences* Tarling D. H. and Runcorn S. K. eds (London: Acadamic Press, 1973), vol. 2, 843–50.

13. **Kossmat F.** Gliederung des varistischen Gebirgebaues. *Abh. sächs geol. Landesamts*, **1**, 1927, 1–39.

14. **Aubouin J.** *Developments in geotectonics 1, geosynclines* (Amsterdam: Elsevier, 1965), 335 p.

15. **Sadler P. M.** An interpretation of new stratigraphic evidence from South Cornwall. *Proc. Ussher Soc.*, **2**, 1973, 535–50.

16. **Hendriks E. M. L.** Facies variations in relation to tectonic evolution in Cornwall. *Trans. R. geol. Soc. Cornwall*, **20**, 1968, 114–51.

17. **Flett J. S.** Geology of the Lizard and Meneage. *Mem. geol. Surv. Gt Br.* 359, 1947, 208 p.

18. **Sanders L. D.** Structural observations on the South-East Lizard. *Geol. Mag.*, **92**, 1955, 231–40.

19. **Green D. H.** The petrogenesis of the high-temperature peridotite intrusion in the Lizard area, Cornwall. *J. Petrology,* **5,** 1964, 134–88.

20. **Thayer T. P.** Chemical and structural relations of ultramafic and feldspathic rocks in Alpine intrusive complexes. In *Ultramafic and related rocks* **Wyllie P. J. ed.** (New York: Wiley, 1967), 222–39.

21. **Bromley A. V.** The sequence of emplacement of basic dykes in the Lizard Complex, South Cornwall (abstract). *Proc. Ussher Soc.,* **2,** 1973, 508.

22. **Dewey J. F. and Bird J. M.** Origin and emplacement of the ophiolite suite: Apalachian ophiolites in Newfoundland. *J. geophys. Res.,* **76,** 1971, 3179–206.

23. **Le Pichon X. Francheteau J. and Bonnin J.** *Developments in geotectonics 6, plate tectonics* (Amsterdam: Elsevier, 1973), 300 p.

24. **Bromley A. V.** A new interpretation of the Lizard Complex, South Cornwall. In preparation.

25. **Bullard E. Everett J. E. and Smith A. G.** The fit of the continents around the Atlantic. In *A symposium on continental drift* **Blackett P. M. S. Bullard E. and Runcorn S. K. eds.** *Phil. Trans. R. Soc.,* **A258,** 1965, 41–51.

26. **Lotze F.** Zur Gliederung der Varisziden der Iberischen Meseta. *Geotekt. Forsch.* no. 6, 1945, 79–92.

27. **Dmitriev L. Barsukov V. and Udintsev G.** Rift-zones of the ocean and the problem of ore-formation. In *Proc. IMA–IAGOD meetings, '70, IAGOD volume* **Takeuchi Y. ed.** (Tokyo: Society of Mining Geologists of Japan, 1971), 65–9. (*Spec. Issue* no. 3)

28. **Jankovic S.** The origin of base-metal mineralization on the Mid-Atlantic ridge (based upon the pattern of Iceland). In *Rep. 24th Int. geol. Congr. Montreal 1972* (Montreal: The Congress, 1972), **4,** 326–34.

29. **Harding R. R. and Hawkes J. R.** The Rb : Sr age and K/Rb ratios of samples from the St. Austell granite, Cornwall. *Rep. Inst. geol. Sci.* no. 71/6, 1971, 10 p.

34

Reprinted from *Trans. Soc. Min. Eng. AIME* **256**:121–128 (1974)

Plate Tectonic Setting of Appalachian-Caledonian Mineral Deposits as Indicated by Newfoundland Examples

by D. F. Strong

Most Newfoundland mineral deposits can be clearly classified as within rocks formed either as accreting plate margins (ophiolitic pyrite-chalcopyrite massive sulfides such as Betts Cove, Whalesback, York Harbour, chromite at Lewis Hills, and asbestos at Baie Verte), at consuming plate margins (Kuroko-type polymetallic massive sulfides such as Buchans and Pilleys Island, porphyry-type Mo-Cu at Rencontre, fluorite at St. Lawrence), or within plates (zinc deposits at Daniels Harbour, sedimentary iron ores at Wabana, manganese in Conception Bay.) As these metallogenic patterns can be convincingly extrapolated southward into the Appalachians and northward to the Caledonides, they provide an important means of eliminating and selecting areas for particular exploration targets. Some criteria are described for the recognition of these environments of differing economic potential.

The plate tectonics concept can be simply stated as follows,:[1]

1) The outer earth can be divided into three layers, namely, the rigid *lithosphere*, i.e., the crust and uppermost mantle to depths up to 100 km; the *asthenosphere*, with effectively no strength and extending to depths of several hundred kilometers; and the rigid *mesosphere*, the lower remaining portion of the mantle.

2) The lithosphere is divisible into a small number of thin blocks or plates which are in relative movement at the margins.

3) Plate margins, the principal zones of tectonic activity within the earth, can be described as *accreting* where plates separate and new crust is being created, e.g., at the ocean ridges; *consuming* where one plate is destroyed by underthrusting beneath another, e.g., at island arcs; and *transform faults* where one plate slides past another with a strike slip motion.

Two important extensions to this concept are especially important to economic geologists, viz.,

4) These tectonic zones or plate boundaries are characterized by unique assemblages of igneous, sedimentary, and metamorphic rocks.

5) Certain types of economic mineral deposits are uniquely formed in specific plate regimes.

This last point has recently been discussed in a general way by a number of authors[2-10] and their ideas are summarized in Fig. 1 and a more detailed discussion is given by Strong.[11] These ideas have been applied to parts of the Appalachian-Caledonian mineral deposits[11,12] and have been useful in the investigation of specific deposits.[13]

Geology and Mineral Deposits of Newfoundland

The island of Newfoundland is divisible into three main tectonic zones, namely, Precambrian-Paleozoic platforms on the east and west and a Paleozoic "mobile" or "eugeosynclinal" zone in the center.[14] These three zones are further divisible into eight subzones on the basis of differences in stratigraphy and structural

D. F. STRONG is with Dept. of Geology, Memorial University of Newfoundland, St. Johns, Nfld., Canada. SME Preprint 731320, SME Fall Meeting, Pittsburgh, Pa., September 1973. Manuscript, July 30, 1973. Discussion of this paper, submitted in duplicate prior to Sep. 15, 1974, will appear in SME Transactions, December 1974, and in AIME Transactions, 1974, Vol. 256.

styles.[15] The tectonostratigraphic divisions and associated mineral deposits are outlined in Fig. 2, and the following discussion (based on Strong[11]) summarizes the main features of these zones within the context of plate tectonic models. More detailed geological descriptions and references are given by Williams, et al.[15]

Precambrian: The Western Platform (Fig. 2) is dominated by an elongate block of Grenville basement making up the Long Range Mountains, and smaller areas of Grenville basement found in the Indian Head Range in the southern part of zone A. The only reliably documented example of Precambrian rocks in the central mobile belt is the gneissic basement to the Fleur de Lys rocks of the Burlington Peninsula recently discovered by M. F. DeWit (personal communication, 1972). Other areas of possibly Precambrian amphibolitic and granitic rocks have been suggested in Notre Dame Bay,[16,17] but these have now been reinterpreted as having younger origins.[18,19] Precambrian gneissic basement is found in the Gander Lake metamorphic belt (zone "G", Fig. 2) which marks the boundary between the Central Mobile Belt and the eastern parts of the Avalon platform.[20] The only mineral deposits recognized as Pre-Hadrynian are the anorthosite-ilmenite occurrences in the Indian Head Range, and these are probably not directly related to plate tectonics.

Late Hadrynian: The late Hadrynian (Fig. 3) was a period of extension and thinning of the older continental crust. This resulted in extrusion of tholeiitic plateau lavas[21,22] along north-trending fissures presently marked by diabase dike swarms cutting the Grenville basement of zone A, with thick wedges of clastic sediments deposited on the continental margins now represented by zones C and G. Syntectonic beryllium- and molybdenum-bearing granitoid rocks intrude gneissic basement in zone G. The Avalon Platform is dominated by large volumes of late Hadrynian volcanic rocks, primarily a bimodal basalt-rhyolite assemblage with volcanically derived sediments, which were interpreted by Hughes and Brueckner[23] as an island arc assemblage, and by Papezik[24] and Schenk[25] as produced in a rifting environment. Strong, et al.,[26] on the basis of an eastward increase in K_2O of granitoid rocks, drew an analogy between the Avalon and Gander Lake Belt and Cordilleran-type continental margins, which also explains the gneissic and granitoid rocks of the Gander Lake Belt, with the bimodal assemblage forming in a Basin and

263

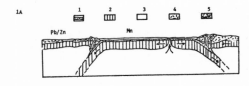

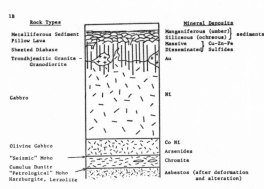

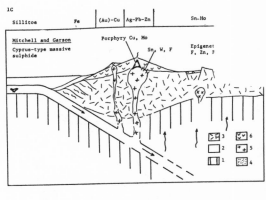

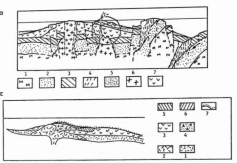

Fig. 1A—An idealized model of different plate tectonic regimes based on a cross section of the Pacific Ocean (after Dewey and Bird[28]). 1, Oceanic crust; 2, lithosphere; 3, asthenosphere; 4, continental crust; 5, intrusive and extrusive igneous rocks produced above subduction zones. A particular plate may contain a variety of metallogenic environments at its margins, as shown in B, C, and D. Other deposits show no relation to the tectonic processes taking place at plate margins; for example, the ferromanganese nodules and encrustations from the ocean floors are probably the most widely distributed deposits which are independent of plate boundary tectonics; lead-zinc deposits of the Mississippi Valley type are perhaps the most valuable group of intra-plate deposits being mined today; and numerous other types of intra-plate deposits which apparently result from processes other than plate tectonics, such as those of stratiform mafic intrusions, meteorite impact craters, titanium in anorthosites, kimberlitic and carbonatitic intrusions, coal deposits, etc.

Fig. 1B—An idealized cross section through the oceanic crust based on[54-66] with stratigraphic location of mineral deposits. Oceanic crust is created at ocean ridges by partial melting of the upper mantle. Chemical sediments in some areas are enriched in copper, iron, manganese, and other metals, the Red Sea Brines and sediments of the East Pacific Rise being the two best documented examples, and Cyprus-type massive sulfides being ancient examples. Along with the massive sulfide deposits, ophiolites also characteristically contain deposits of chromite (in peridotite, mainly cumulus dunite), nickel (in gabbro), and asbestos (in serpentinized periodite).

Fig. 1C—Mineral deposits of Cordilleran type consuming plate margins. Zonal distribution of mineral deposits across subduction zones according to Sillitoe[8] and Mitchell and Garson.[9] 1, Upper mantle; 2, oceanic crust; 3, continental crust; 4, clastic sediments; 5, granitoid intrusives; 6, volcanic rocks.

Figs. 1D and E—Geological setting of island arc (Kuroko-type) massive sulfide deposits. (D) Generalized island arc model (after Mitchell and Reading.[47]) 1, Upper mantle; 2, oceanic layer 3, gabbro and sheeted diabase; 3, oceanic layer 2, tholeiitic pillow lavas; 4, melange deposits and amphibolites; 5, volcanic turbidites; 6, granitoid intrusives; 7, calc-alkaline volcanic rocks. (E) Typical Kuroko deposits (after Horikoshi[68]). 1, Dacite lava dome; 2, andesitic lava; 3, explosion breccia; 4, siliceous ore (pyrite); 5, black ore (Zn, Pb, Cu); 6, yellow ore (Cu-py); 7, barite, chert, gypsum. Note that these deposits are fundamentally different from the simple Cu-Fe (±Zn) Cyprus-type deposits (Fig. 1B) in being polymetallic and occurring in late stage acidic volcanics of the calc-alkaline suite.

Range type of volcanic environment to the east. Copper and molybdenum occurrences are found on the Avalon Platform within the Holyrood granite, which also produced pyrophyllite deposits around its margins. The lead, zinc, silver, and tin occurrences of zone G are insufficiently known to say whether they have a syngenetic or hydrothermal origin. Any oceanic crust possibly produced at this time was either destroyed by subduction or has not yet been recognized.

Cambrian: The Cambrian of the Western Platform (Fig. 3) commences with arkosic sediments in northern Newfoundland, succeeded by eastward-thickening Cambrian to Lower Ordovician limestones and dolomites, orthoquartzites, and shales, reflecting relatively stable shallow water deposition. Some silicic volcanic rocks, such as the Cape St. John Group of zone C suggest that some calc-alkaline volcanism was taking place at this time. However there is little conclusive evidence as to the

direction of dip of any subduction zone, and two alternative hypotheses predominate. Church and Stevens[27] suggest that in the Lower Ordovician a southeast-dipping subduction zone is necessary to permit emplacement (westward obduction) of the ophiolites of west Newfoundland. It is possible that such a situation prevailed earlier on in the Cambrian, accounting for the Cape St. John Group, as suggested in Fig. 3B. Dewey and Bird[28] and Kennedy[29] suggest that a Cambrian marginal ocean basin existed in zone C with a westward-dipping subduction zone, and that the copper, asbestos, and chromium deposits of zone C were formed in this marginal basin, according to a model approximately as shown in Fig. 3C. The Rambler Cu-Fe-Zn-Au-Pb deposits show conflicting evidence of both Cyprus-type[30] and Kuroko-type origin, and require further study before they can be reliably characterized.

The eastern margin of the Central Mobile Belt (zone G) was characterized by a period of sedimentation in the early Cambrian and deformation and granitic in-

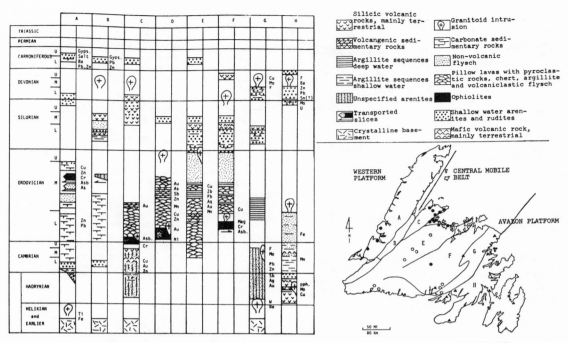

Fig. 2—Tectonic-stratigraphic zones of Newfoundland (after Williams, et al[18]) and main mineral occurrences (after Fogwill[90]).

trusion in the late Cambrian. It is not understood whether these sediments are trench deposits (cf. Fig. 3A) or were formed at a more stable Atlantic-type continental margin (as shown in Fig. 3B). The later tectonic-intrusive period might have resulted from sub-

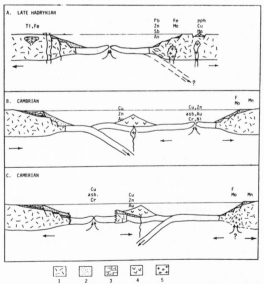

Fig. 3—Plate tectonic models for Newfoundland in the late Hadrynian (A) and the Cambrian compatible with models proposed by Church and Stevens[27] (B), or Dewey and Bird,[28] and Kennedy[29] (C). 1, Continental crust; 2, clastic sediments; 3, limestone; 4, volcanic rocks; 5, granitoid intrusive rocks; unshaded area enclosed by black line indicates oceanic crust.

duction (e.g., as shown in Fig. 3A) or from intersection of the continental margin and oceanic ridge (Fig. 3C) as suggested for production of the granites of western North America by Kistler et al.[31]

The Cambrian of the Central Mobile Belt is represented by few fossiliferous rocks (e.g., 32) and some of the ore-bearing rocks there could be of any age from Cambrian to Ordovician. These rocks are mostly mafic pillow lavas and volcaniclastic sediments with minor limestone lenses. The lowermost rocks contain sheeted dikes and are chemically similar to ocean-ridge basalts, and are thus interpreted as representative of oceanic crust.[33-35] These contain the typical Cyprus-type pyrite-chalcopyrite deposits such as Betts Cove, Tilt Cove, Whalesback, and Little Bay.[13] The allochthonous Bay of Islands ophiolites, transported westward during the Lower Ordovician,[36] are thought to have been derived from this central Newfoundland oceanic terrain. They likewise contain typical Cyprus-type deposits (e.g., York Harbour) as well as chromite and asbestos.

The Cambrian of the Avalon Platform commences with deposition of orthoquartzites, now being quarried for silica. These are overlain by Cambrian shales and minor limestones containing algal structures which suggest a shallow water origin. Low-grade manganese deposits occur within these shales, being most important in rocks of the Middle Cambrian.

Ordovician: The Ordovician was marked by continued deposition of reef limestones on the Western Platform, oceanic and island arc volcanism in the Central Mobile Belt, and clastic sedimentation on the Avalon Platform, with ophiolite obduction (zone A) and deformation on the western (zones C and D) and eastern (zone F) margins of the Central Mobile Belt. As described for Fig. 3, there is no conclusive evidence for direction of dip of any subduction zone, and it could have been as shown in either Fig. 4A[28,29] or Fig. 4B.[27] Likewise the

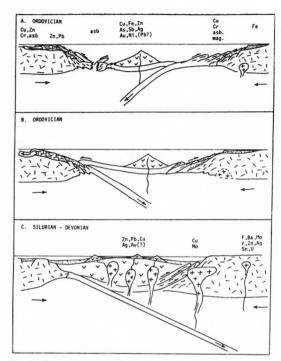

Fig. 4—Plate tectonic models for Newfoundland in the Ordovician after Dewey and Bird[28] and Kennedy[29] (A), or Church and Stevens[27] (B), and in the Silurian (C). Symbols as for Fig. 3.

eastern margin of the mobile belt is poorly understood, although Ordovician melange deposits in zones E and F[30,37] suggest the existence of a trench at this time.

The Mississippi-Valley-type zinc deposits, best known in the Daniel's Harbour area, occur within Lower Ordovician parts of the zone A limestone sequence, in the St. George Formation at a karstic unconformity with the Middle Ordovician Table Head Formation.[38] In the Central Mobile Belt the ophiolitic rocks are generally overlain by a thick sequence of volcanic flysch and pillow lavas which are interpreted as island arc deposits.[18,19] They contain a variety of polymetallic deposits ranging from the Cu-As-Sb-Au deposits of the Moreton' Harbor Area[19,39] to the typical Kuroko-type Cu-Zn-Pb-Ag deposits of Pilleys Island.[11,40] The large Buchans deposits[41] also fit within this latter category, although there is some doubt as to whether the Buchans deposits are Ordovician or Silurian in age.

The Cambrian sediments of the Avalon Platform are conformably overlain by thicker and sandier Lower Ordovician rocks which contain the Clinton-type oolitic hematite beds which were mined on Bell Island in Conception Bay.

Silurian (Fig. 4): The Silurian was a time of predominantly shallow marine and redbed alluvial fan type sedimentation and minor mainly subaerial calc-alkaline volcanism throughout the central and western parts of Newfoundland with no record being left on the Avalon Platform. This represents the period when ocean basins of the Central Mobile Belt were closing and consolidating. The red sandstones and acid and basic volcanics of the Micmac Formation on the Burlington Peninsula, the Springdale Group in the western Central Mobile Belt (zone E), and the Botwood Group of the eastern Cen-

tral Mobile Belt (zone E) are the best examples of these rocks. There are no important mineral deposits known to occur within these rocks unless, as mentioned previously, the Buchans and Pilleys Island Kuroko-type deposits are of Silurian age.

Some of the Newfoundland granitoid rocks are of Silurian age but no mineral deposits are known for these rocks.

Devonian: The Devonian period (Fig. 4C) is represented by shallow-water sediments on the Western and Avalon Platforms and only minor intrusive bodies on the eastern parts of the Avalon Platform, but in the Central Mobile Belt and the Gander Lake Belt (zone G) it was a period of extensive granitoid intrusive activity. The only known extrusive Devonian igneous rocks in Newfoundland are a small area of rhyolite flows in the center and southwest of the island, and these apparently are not mineralized.

The Devonian granitoid bodies represent a number of types, ranging from calc-alkaline zoned plutons in the Central mobile Belt (e.g., Mount Peyton and Hodges Hill plutons), microcline-megacrystic granites-quartz monzonites common to the Gander Lake Belt (e.g., the Ackley City Batholith), and the granites *sensu stricto* which occur as small discordant plutons of the Gander Lake Belt on the Avalon Platform (e.g., St. Lawrence).

Copper and molybdenum mineralization is common in the quartz monzonite plutons of the Gander Lake Belt in Fortune Bay, while fluorite (with minor Ba/Sn/Zn/Pb mineralization) is found in the discordant plutons of the Avalon Platform, the large fluorite deposits long mined from peralkaline granite at St. Lawrence being the best example. This eastward variation in type of granitoid mineralization is reminiscent of the zoning described by Mitchell and Garson[9] for Southeast Asia. However, little base metal exploration has been done in the granitic terranes of Newfoundland, so such interpretations of metal zoning are very tentative.

Carboniferous: The Carboniferous period (Fig. 5) is represented in Newfoundland by two sedimentary basins in the western part of the island, filled with shallow water conglomerates and sandstones with minor limestones and evaporites. The evaporites are being quarried for anhydrite and gypsum at Flat Bay and are now being investigated for barite potential.[42] The presence of base metal mineralization in similar rocks both to the north and south, e.g., the Walton deposits of Nova Scotia and the Carboniferous deposits of Ireland, lend some hope that similar deposits might be found in the Newfoundland Carboniferous.[43]

The only post-Carboniferous rocks known in Newfoundland are lamprophyre dikes (Triassic to Jurassic) and Pleistocene glacial deposits.

Applications Outside of Newfoundland

Given this apparently systematic plate-tectonic control of Newfoundland mineral deposits, there follows an attempt to crudely apply the previous interpretations to deposits elsewhere in the Appalachian-Caledonian mountain system. Because of the author's limited personal experience in studying such deposits outside of Newfoundland, the following discussions are strongly based on a literature survey. It is especially dependent on the compilations by previous workers who made metallogenic interpretations of the Appalachians based on tectonic cycles[44,45] or mineralization gradients.[46]

With regard to the Caledonides of Norway and Britain, some close similarities are readily apparent. Fig. 6 summarizes evidence supporting the plate-tectonic model for the development of Newfoundland and shows

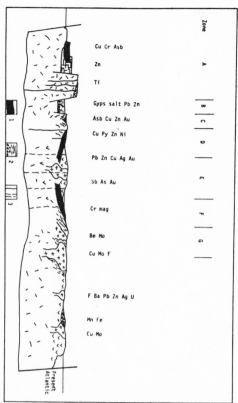

Fig. 5—Schematic cross section across Newfoundland showing the distribution of mineral deposits. 1, Ophiolite; 2, carboniferous basin sediments; 3, solid lines—Hadrynian diabase dikes; dashed lines—post Carboniferous lamprophyre dikes. All other symbols as for Fig. 3.

it to be compatible with models of the Norwegian Caledonides modified from Gale and Vokes[12] and the British Caledonides;[47,48] their similarity to the Newfoundland interpretation is obvious and striking, with regard to both the general geology and the mineral deposits. If the Carboniferous deposits are included, such comparisons appear to be strengthened even further.[48]

Comparisons with the Appalachians (Figs. 7 and 8) are more readily made because of the abundant recent work on both tectonic[15,16,49] and metallogenic interpretations.[45,46] Nevertheless it is still difficult to get the specific information on host rocks which is essential to classify a deposit, i.e., it is no longer adequate to merely know that a deposit is in volcanic rocks, or even just that it's in basic volcanic rocks, but we should know whether we are dealing with a tholeiitic or calc-alkaline suite. Two features of Appalachian stratigraphy appear to be outstandingly reliable in extending Newfoundland models southward, namely the Cambro-Ordovician platformal carbonates and the western ultramafic belt (the ophiolites).

The Cambro-Ordovician platformal limestones extend from Norway and Greenland through Scotland and Newfoundland[50] to Alabama. Furthermore they consistently contain the Mississippi Valley-type zinc (with or without lead) deposits at a Lower-Middle Ordovician paleokarst unconformity (e.g., the Jefferson City zinc deposit at the Shady dolomite and Knox limestone contacts in Tennessee; the Daniels Harbour zinc deposit at the St. George-Table Head contact in Newfoundland). Similar deposits are present, but less common, lower down in the sequence, e.g., the Lower Cambrian Austenville-Ivanhoe lead-zinc deposit of Virginia.

Ultramafic rocks have long been known to form a discontinuous belt throughout the Appalachians, and they are now generally interpreted as allochthonous masses thrust (obducted) westward onto a stable platform from an ocean basin lying to the east. This relation is clearly shown in Newfoundland as described previously, where the Bay of Islands and Hare Bay ophiolites were obducted onto the Western platform from the Central Mobile Belt. Deposits of chromite, asbestos, and copper are best known in the Canadian ex-

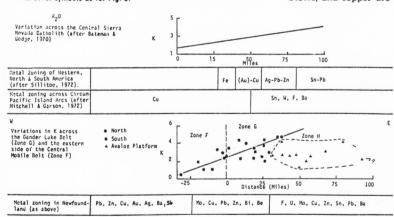

Fig. 6—Summary of evidence supporting plate-tectonic model for the development of Newfoundland, with data for analogous areas. This model is compatible with the available chemical data for Norway[18] and Britain.[61,62] The K data are averages for granitoid plutons.

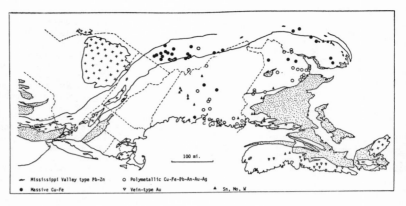

Fig. 7—Summary of postulated Paleozoic mineral zones of the Northern Appalachians. The mineral zones are drawn on the basis of data compiled by Potter[45] and from Ref. 63.

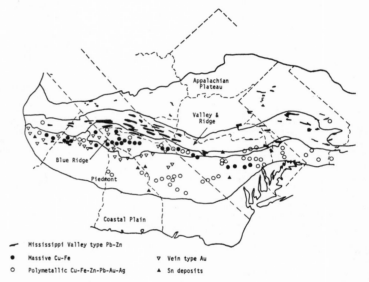

Fig. 8—Postulated Paleozoic mineral zones of the Southern Appalachians, based on data from Ref. 64.

amples, and this may be due to the U.S. examples being more intensely deformed and dismembered with critical parts of the sequence missing or isolated. It is suggested that a thorough investigation of Appalachian ophiolites is now warranted, bearing in mind the ophiolite stratigraphy outlined in Fig. 8.

Although there is insufficient data to outline the distribution of the nonultramafic portions of the ophiolite suite, there does appear to be some association of the massive pyrite-chalcopyrite deposits (Cyprus-type) with the ophiolites in both the Inner Piedmont and Blue Ridge.[51] There is likewise a great shortage of specific chemical data to outline areas of calc-alkaline volcanic rocks, although one gets the impression from the association of acid volcanics and pyroclastic sediments that most such rocks are found in the Piedmont of the southern Appalachians and the eastern volcanic belts of the northern Appalachians. This is supported by the occurrence of polymetallic Cu-Pb-Zn±Ag±Au deposits (Kuroko-type) in these areas. It is possible that many of the small disseminated deposits of these areas represent the so-called "footwall ore" deposited beneath the Kuroko-type massive sulfides of Bathurst, N.B., and Buchans, Nfld. The same may be said for many of the gold deposits of the inner Piedmont, since the fact that

many of them are found in metasediments and associated with granitic intrusives may not necessarily mean that they are nonvolcanogenic (e.g., see Ref. 52).

Tin deposits in the southern Appalachians (e.g., the Alabama tin belt and Irish Creek, Va.) tend to be distributed to the east of the other mineral deposits and this is also true for the tungsten of the Hamme district, North Carolina, in line with the Newfoundland model. The same can generally be said for molybdenum, although in Maine Mo-Cu deposits occur both at Catheart Mt. and Salley Mt. in the western part of the state, while the Catherine Hill and Cooper deposits are found in the east where expected.[53]

These generalizations, if valid, imply that not only are plate tectonic concepts useful in explaining Appalachian mineral deposits, but also that a specific model holds for the whole Appalachian-Caledonian system. In particular, the distributions of rocks, chemical variations, and metal zoning in Norway, Britain, and Newfoundland suggest that a southeastward-dipping subduction zone under an Andean-type continental margin to the east can account for most features. Metal zoning in the Appalachians conforms to this pattern, thus suggesting that some revision of previous models[16,49] will be necessary. Such revision is not the purpose of this paper, but it can be stated that such revised models with east-ward-subduction are more readily applicable to the available data.

The implications for mineral exploration are profound, since if the models are valid and reinforced by new data, then discoveries in, say, Norway could create a ripple of activity all the way to Alabama. Examples that are currently available compel a thorough investigation of Appalachian ophiolites to the south of Quebec, of the shelf carbonates in western Newfoundland, of the St. Lawrence type fluorite-bearing peralkaline granites wherever found. It has to be emphasized, however, that much of the evidence required for these interpretations has a chemical basis, and many more data are required. Nevertheless, plate tectonic concepts emphasize more strongly than ever that economic geologists working in the Appalachians cannot do so in isolation from developments either to the north or south.

Conclusion

There is gross zonal arrangement of Appalachian mineral deposits in linear belts, with Mississippi Valley-type zinc-lead deposits on the west followed eastward by Cyprus-type (ophiolitic) copper-iron massive sulfides followed in turn by gold deposits and overlapping polymetallic zones of Cu-Pb-Zn-Ag-Au which are in turn followed by tin, tungsten, and molybdenum deposits. This pattern corresponds to that produced during processes of plate tectonic evolution culminating in an Andean-type continental margin with an eastward-dipping subduction zone, according to models developed (independently of the metals zoning) in Newfoundland, Britain, and Norway.

Acknowledgments

I am grateful to E. R. W. Neale, W. G. Smitheringale, H. R. Peters, and K. D. Collerson for criticism of the manuscript, to them and colleagues and students at Memorial University for stimulating discussion, and to the National Research Council of Canada Grant A7975 and The Geological Survey of Canada Research Agreement 1135-D13-4-18/72 for financial support.

References and Bibliography

[1] Isacks, B., Oliver, J., and Sykes, L.R., "Seismology and the New Global Tectonics," *Journal of Geophysical Research*, Vol. 73, 1968, pp. 5855-5899.

[2] Pereira, J., and Dixon, C.J., "Mineralization and Plate Tectonics," *Mineralium Deposita*, Vol. 6, 1971, pp. 404-405.

[3] Sillitoe, R.H., "Relation of Metal Provinces in Western America to Subduction of Oceanic Lithosphere," *Geological Society of America Bulletin*, Vol. 83, 1972a, pp. 813-818.

[4] Sillitoe, R.H., "A Plate Tectonic Model for the Origin of Porphyry Copper Deposits," *Economic Geology*, Vol. 67, 1972b, pp. 184-197.

[5] Sillitoe, R.H., "Formation of Certain Massive Sulphide Deposits at Sea-Floor Spreading," *Transactions*, Institution of Mining and Metallurgy, Vol. 81, 1972c, pp. B141-B148.

[6] Guild, P.W., "Metallogeny: a Key to Exploration," *Mining Engineering*, Vol. 23, 1971, pp. 69-72.

[7] Guild, P.W., "Massive Sulphides vs. Porphyry Deposits in Their Global Tectonic Settings," *Proceedings*, Joint Mtg. MMIJ-AIME, Tokyo, No. G13, May 1972a.

[8] Guild, P.W., "Metallogeny and the New Global Tectonics," *International Geological Congress*, 24th Session, Vol. 4, 1972b, pp. 17-24.

[9] Mitchell, A.H.G., and Garson, M.S., "Relationship of Porphyry Copper and Circum-Pacific Tin Deposits to Paleo-Benioff Zones," *Transactions/Sec. B*, Institution of Mining and Metallurgy, Vol. 81, 1972, pp. B10-B25.

[10] Sawkins, F.J., "Sulphide Ore Deposits in Relation to Plate Tectonics," *Journal of Geology*, Vol. 80, 1972, pp. 377-397.

[11] Strong, D.F., "Plate Tectonic Setting of Newfoundland Mineral Deposits," *Geoscience Canada*, Vol. 1, 1974, pp. 21-31.

[12] Gale, G.H., and Vokes, F.M., "Norwegian Paleozoic Volcanogene Sulphide Deposits in the Light of a Plate Tectonic Model," unpublished.

[13] Upadhyay, H.D., and Strong, D.F., "Geological Setting of the Betts Cove Copper Deposits, Newfoundland: An Example of Ophiolite Mineralization," *Economic Geology*, Vol. 68, 1973, pp. 161-167.

[14] Williams, H., "The Appalachians in Northern Newfoundland—a Two Sided Symmetrical System," *American Journal of Science*, Vol. 262, 1964, pp. 1137-1158.

[15] Williams, H., Kennedy, M.J., and Neale, E.R.W., "The Appalachian Structural Province, Sp-11, Geological Assoc. of Canada, 1972, pp. 181-262.

[16] Bird, J.M., and Dewey, J.F., "Lithosphere Plate—Continental Margin Tectonics and the Evolution of the Appalachian Orogen," *Geological Society of America Bulletin*, Vol. 81, 1970, pp. 1031-1060.

[17] Kennedy, M.J., and DeGrace, J.R., "Structural Sequence and Its Relationship to Sulphide Mineralization in the Orodovician Lush's Bright Group of Western Notre Dame Bay, Newfoundland," *Canadian Mining & Metallurgical Bulletin*, Vol. 65, 1972, pp. 300-308.

[18] Kean, B.F., "Stratigraphy, Petrology and Geochemistry of Volcanic Rocks of Long Island, Newfoundland," MSC. Thesis, Memorial University of Newfoundland, 1973, 155 pp.

[19] Strong, D.F., and Payne, J.G., "Early Paleozoic Volcanism and Metamorphism of Moreton's Harbour Area, Newfoundland," *Canadian Journal of Earth Sciences*, Vol. 10, 1973, pp. 1363-1379.

[20] Kennedy, M.J., and McGonigal, M.H., "The Gander Lake and Davidsville Groups of Northeastern Newfoundland: New Data and Geotectonic Implications," *Canadian Journal of Earth Sciences*, Vol. 9, 1972, pp. 452-459.

[21] Strong, D.F., and Williams, H., "Early Paleozoic Flood Basalts of Northwest Newfoundland: Their Petrology and Tectonic Significance," SP-25, Geological Assoc. of Canada, 1971, pp. 43-44.

[22] Strong, D.F., "Plateau Lavas and Diabase Dikes of Northwestern Newfoundland," *Geological Magazine*, 1974, in press.

[23] Hughes, C.J., and Brueckner, W.D., "Late Precambrian Rocks of Eastern Avalon Peninsula, Nfld.—A Volcanic Island Comlex," *Canadian Journal of Earth Sciences*, Vol. 8, 1971, pp. 899-915.

[24] Papezik, V.S., "Petrochemistry of Volcanic Rocks of the Harbour Main Group, Avalon Peninsula, Newfoundland," *Canadian Journal of Earth Sciences*, Vol. 1, 1970, pp. 1485-1498.

[25] Schenk, P.E., "Southeastern Atlantic Canada, Northwestern Africa, and Continental Drift," *Canadian Journal of Earth Sciences*, Vol. 8, 1971, pp. 1218-1251.

[26] Strong, D.F., Dickson, W.L., and O'Driscoll, C.F., "Geochemistry of Eastern Newfoundland Granitoid Rocks," unpublished report, Newfoundland Mineral Research Div., 1973, 121 pp.

[27] Church, W.R., and Stevens, R.K., Early Paleozoic Ophiolite Complexes of Newfoundland Appalachians as Mantle-Ocean Crust Sequences," *Journal of Geophysical Research*, Vol. 76, 1971, pp. 1460-1466.

[28] Dewey, J.F., and Bird, J.M., "Origin and Emplacement of Ophiolite Suite: Appalachian Ophiolites in Newfoundland," *Journal of Geophysical Research*, Vol. 76, 1971, pp. 3179-3207.

[29] Kennedy, M.J., "The Relationship between Pre-Ordovician Polyphase Deformation and Ophiolite Obduction in the Newfoundland Appalachian System," Northeastern Section 8th Annual Meeting, Geological Society of America, Abstracts 5, 1973, p. 183.

[30] Gale, G.H., and Roberts, D., "Paleogeographical Implications of Greenstone Petrochemistry in the Southern Norwegian Caledonides," *Nature-Physical Science*, Vol. 238, No. 82, 1972, pp. 60-61

[31] Kistler, R.W., Evernden, J.F., and Shaw, H.R., "Sierra Nevada Plutonic Cycle: Part I, Origin of Composite Granite Batholiths," *Geological Society of America Bulletin*, Vol. 82, 1971, pp. 853-868.

[32] Kay, M., and Eldridge, N., "Cambrian Trilobites in Central Newfoundland Volcanic Belt," *Geological Magazine*, No. 105, 1968, pp. 372-377.

[33] Upadhyay, H.D., Dewey, J.F., and Neale, E.R.W., "The Betts Cove Ophiolite Complex, Newfoundland: Appalachian Oceanic Crust and Mantle," *Proceedings, Geological Assoc. of Canada*, Vol. 24, 1971, pp. 27-34.

[34] Smitheringale, W.G., "Low Potash Lush's Bight Tholeites: Ancient Oceanic Crust in Newfoundland?," *Canadian Journal of Earth Sciences*, Vol. 9, 1972, pp. 574-588.

[35] Strong, D.F., "Sheeted Diabases of Central Newfoundland: New Evidence for Ordovician Sea-Floor Spreading," *Nature*, Vol. 235, 1972, pp. 102-104.

[36] Stevens, R.K., "Cambro-Ordovician Flysch Sedimentation and Tectonics in West Newfoundland and Their Possible Bearing on a Proto-Atlantic Ocean," SP-7, Geological Assoc. of Canada, 1970, pp. 165-177.

[37] Kay, M., "Dunnage Melange and Lower Paleozoic Deformation in Northeastern Newfoundland," *24th International Geological Congress*, Sec. 3, pp. 122-133.

[38] Collins, J.A., and Smith, L., "Sphalerite as Related to the Tectonic Movements, Deposition, Diagenesis and Karstification of a Carbonate Platform," *24th International Geological Congress*, Sec. 6, 1972, pp. 209-216.

[39] Gibbons, R.V., and Papezik, V.S., "Volcanic Rocks and Arsenopyrite Veins of Morton's Harbour Area, Notre Dame Bay, Nfld.," *Proceedings, Geological Assoc. of Canada*, Vol. 22, 1970, pp. 1-10.

[40] Strong, D.F., and Peters, H.R., "The Importance of Volcanic Setting for Base Metal Exploration in Central Newfoundland," Abstract, Annual Meeting, Canadian Institute of Mining and Metallurgy, Ottawa, Apr. 1972.

[41] Swanson, E.A., and Brown, R.L., "Geology of Buchans Orebodies," *Canadian Mining & Metallurgical Bulletin*, Vol. 55, 1962, pp. 618-626.

[42] McArthur, J.D., "Barite and Celestite Deposits in Newfoundland," unpub. report, Mineral Research Div., 1973, 21 pp.

[43] Russell, M.J., "North-South Geofractures in Scotland and Ireland, *Scottish Journal of Geology*, Vol. 8, 1971, pp. 75-84.

[44] McCartney, W.D., and Potter, R.R., "Mineralization as Related to Structural Deformation, Igneous Activity and Sedimentation in Folded Geosynclines," *Canadian Mining Journal*, Vol. 83, 1962, pp. 83-87.

[45] Potter, R.R., "Metallogeny and Characteristics of Sulphide Deposits in the Appalachian Region," 1972 Convention, Canadian Institute of Mining & Metallurgy.

[46] Gabelman, J.W., "Metallotectonic Zoning in the North American Appalachians Region," *23rd International Geological Congress*, Vol. 7, 1968, pp. 17-33.

[47] Mitchell, A.H., and Reading, H.G., "Evolution of Island Arcs," *Journal of Geology*, Vol. 79, 1971, pp. 253-284.

[48] Church, W.R., and Gayer, R.A., "The Ballantrae Ophiolite," *Geological Magazine*, in press, 1973.

[49] Hatcher, R.D., "Developmental Model for the Southern Appalachians," *Geological Society of America Bulletin*, Vol. 83, 1972, pp. 2735-2760.

[50] Swett, K., and Smit, D.E., "Cambro-Ordovician Shelf Sedimentation of Western Newfoundland, Northwest Scotland and Central East Greenland," *24th International Geological Congress*, Sec. 6, 1972, pp. 33-41.

[51] Kinkel, A.R., Jr., Feitler, S.A., and Hobbs, R.G., "Copper and Sulfur," *Mineral Resources of the Appalachian Region*, U.S. Geological Survey, Professional Paper No. 580, 1968, pp. 377-384.

[52] Ridler, R.H., "Relationship of Mineralization to Volcanic Stratigraphy in the Kirkland Lake-Larder Lake Area, Ontario," *Proceedings, Geological Assoc. of Canada*, Vol. 21, 1970, pp. 33-42.

[53] Young, R.S., "Mineral Exploration and Development in Maine," *Ore Deposits of the Unitd States, 1933-1967*, Vol. 1, J. D. Ridge, ed., AIME, New York, 1968, pp. 125-139.

[54] Moores, E.M., and Vine, F.J., "The Troodos Massif, Cyprus, and Other Ophiolites as Oceanic Crust: Evaluation and Implications," *Proceedings*, Royal Society of London, Vol. 268, 1970, pp. 443-466.

[55] Church, W.R., "Ophiolite: Its Definition, Origin as Oceanic Crust, and Mode of Emplacement in Orogenic Belts, with Special Reference to the Appalachians," *The Ancient Oceanic Lithosphere*, Vol. 42, Earth Science Branch, Canada, 1972, pp. 71-85.

[56] Malpas, J., "A Restored Section of Oceanic Crust and Mantle in Western Newfoundland," Abstracts 5, Northeastern Sec. 8th Annual Meeting, Geological Society of America, 1973, p. 191.

[57] Degens, E.T., and Ross, D.A., *Hot Brines and Recent Heavy Metal Deposits in the Red Sea. A Geochemical and Geophysical Account*," Springer-Verlag, Berlin, 1969, 600 pp.

[58] Bostrom, K., and Peterson, M.N.A., "Precipitates from Hydrothermal Exhalations on the East Pacific Rise", *Economic Geology*, Vol. 61, 1966, pp. 1258-1265.

[59] Horikoshi, E., "Volcanic Activity Related to the Formation of the Kuroko-Type Deposits in the Kosaka District, Japan," *Mineralium Deposita*, Vol. 4, 1969, pp. 321-345.

[60] Fogwill, W.D., "Mineral Deposits and Prospecting Environments of Newfoundland," Information Circular No. 14, Newfoundland Dept. of Mines, 45 pp., (1970).

[61] Fitton, J.G., and Hughes, D.J., "Volcanism and Plate Tectonics in the British Ordovician," *Earth and Planet Science Letters*, Vol. 8, 1970, pp. 223-228.

[62] Thorpe, R.S., "Possible Subduction Zone for Two Precambrian Calc-Alkaline Plutonic Complexes From Southern Britain," *Geological Society of America Bulletin*, Vol. 83, 1972, pp. 3663-3668.

[63] Ridge, J.D. ed., "Ore deposits of the United States, 1933-1967," Vol. 1, 1968, AIME, New York, 991 pp.

[64] *Mineral Resources of the Appalachian Region*, U.S. Geological Survey, Professional Paper 580, 1968, 492 pp.

Editor's Comments
on Paper 35

35   WATSON
Influence of Crustal Evolution on Ore Deposition

PLATE TECTONICS AND THE SOURCE OF ORE METALS

To what extent has the application of plate tectonic concepts to metallogenesis affected ideas on the ultimate origin of ore metals? We already know that biological, sedimentary, hydrothermal, and magmatic processes variously act to recycle and concentrate elements within and upon the continental crust into economic concentrations. But have those metals been there from the earliest stages of crustal evolution or has the transfer of metals from mantle to crust been a continuing process throughout geological time?

Compilations of age versus abundance data have been made for different types of metal deposits (Pereira and Dixon, 1965; Petrascheck, 1974; see also Paper 35). They indicate that while significant concentrations of some metals can be found in rocks of all ages—e.g. gold, uranium, chromium, copper—others are found chiefly in rocks of less than about 1000 Ma age—e.g. lead, zinc, tin, tungsten, molybdenum. One possible conclusion from these data is that the processes that form ore deposits of these "younger" metals did not evolve until about 1000 Ma ago (compare Paper 23), although it has also been argued that the structural setting of these ores is such that deposits older than about 1000 Ma have been largely eroded away.

The existence of metallogenic provinces characterised by ore deposits of similar metals but different ages certainly points to long-lived crustal inhomogeneities (e.g. Papers 4 and 5; see also Krauskopf, 1971; Routhier et al., 1973), though it has become more appropriate to think of these inhomogeneities as lithospheric rather than simply crustal. This is not merely a semantic quibble: as the continental crust has evolved it has almost certainly become thicker; as progressively deeper levels are exposed by erosion, the only way that such thickening can occur is by a form of sialic underplating (e.g. Holland and Lambert, 1975). The necessary material could only come from the mantle, and would bring granitophile metals with it. Repeated mineralizing epochs are thus attributable to remobilization of higher than average metal concentrations anywhere within the 100 km thick combined crust/upper mantle column. Sialic underplating is one way of transferring metals from mantle to continental crust. Some metals can also be transferred by sea floor spreading and subduction, copper being a notable example (e.g. Dunham, 1972). There is no *positive* evidence that granitophile elements cannot be added in this way—lead, for example. So-called copper-lead lines have been drawn through metallogenic provinces of younger orogenic belts (e.g. Laznicka and Wilson, 1972), separating outer (oceanic) zones with relatively high copper:lead ratios, from inner (continental) zones with relatively low ratios.

The copper is therefore believed to have an oceanic (basaltic) source, the lead a continental (granitic) one. In plate tectonic terms, however, such lines are equally consistent with remobilization of lead from oceanic crust at greater depths down the continental margin subduction zone (Paper 19; see also Wright and McCurry, 1973a). Isotopic measurements on the oxygen, sulphur, strontium and lead of intrusives and ore deposits in western North America, for example (Jensen, 1971; Lange and Cheney, 1971) has not unequivocally discriminated between a mantle and crustal source for the metals. A kind of compromise is offered by Petrascheck (1972, 1973a, 1974) who suggests that mantle-derived heat and volatiles (especially H_2O, CO_2, H_2S, HCl) rise through the continental crust, collecting and concentrating granitophile elements (Zn, Pb, Sn, W, Au). This resembles Mitchell and Garson's view of volatile fluorides rising from a subduction zone and remobilizing tin in the continental crust (Paper 18).

In the debate about crustal versus mantle origin for the granitophile elements, therefore, the alternative you choose remains in the end a matter of opinion—the evidence is not yet conclusive enough to permit a positive selection between them.

Several of these points are recapitulated in more detail by Watson in Paper 35, who supports the view that metallogenic processes have evolved with time and that concentrations of granitophile elements are mainly of crustal origin.*

*For a discussion of Watson's paper (Paper 35), see *Inst. Mining Metallurgy Trans.*, B, **83**:39–41 (1974).

Reprinted from *Trans. Instn Min. Metall.* 82:B107–B113 (1973)

Influence of crustal evolution on ore deposition*

Janet Watson B.Sc., Ph.D.
Department of Geology, Imperial College, London

551.2/.3 : 553.21

Synopsis
The record of mineralization through almost four billion years of geological time is examined in relation to changes in the structure and behaviour of the earth's crust. The earliest stage of earth history was one in which the bulk of the crust was hot, weak and mobile. Most ore deposits older than 2·7 b.y. derived their material directly or through only one intermediate stage from sources in the mantle; they are deficient in those metals which are easily dispersed under unstable conditions. The evolution of the first stable and relatively cool cratonic crustal units at about 2·7 b.y. was followed by an increase in the variety of ore deposits. The first deposits concentrated by sedimentary processes date from this time, as do the first important concentrations of mobile elements such as uranium. During subsequent stages of earth history new mineralization was related both to the influx of magmas from the mantle and to the recycling of metals already dispersed in the crust. The transference of such metals as copper, lead and zinc to high levels in the crust was commonly followed by the appearance of concentrations of the same metals in sedimentary sequences of the adjacent cratons. The most constant styles of mineralization have been those resulting from the rise of partial melts generated in the mantle. The most variable have been those which involve concentration by sedimentary processes, which have been progressively influenced by the effects of organic evolution.

The record of mineralization
The abundant evidence relating to the distribution of ore deposits in time and in space provides many indications that the dominant styles of mineralization have not remained constant through the successive stages of earth history. The long-term variations revealed by the record of mineralization attain a particular interest when they are set against the geological evidence of changes in the behaviour and architecture of the earth's crust as a whole which has emerged during the last couple of decades. The object of the present paper is to explore the possible relationships between long-term variations in the style of ore deposition and the evolution of the crust over a time-span of three to four billion years.

At the present day the greater part of the crust is relatively cool, strong and rigid, forming a number of stable plates which move as units in response to movements in the mantle. The plate boundaries, at which new crustal material is added and surplus material is consumed, are tectonically mobile, are characterized by high geothermal gradients and are mechanically weak. Similar contrasts between mobile belts and stable units—which include the cratonic regions of the continental crust—can be recognized at every stage of earth history throughout Phanerozoic and Proterozoic times. Rocks dating from the first billion years of earth history, however, are those characteristic of unstable regimes, and their predominance indicates that the greater part of the early crust was hot, weak and unstable. A major factor in the evolution of the crust appears to have been the emergence, from 2·8 b.y. onwards, of stable massifs in which geothermal gradients were relatively low, and the complementary restriction of many manifestations of tectonic mobility and igneous activity to well defined mobile belts.

It is against this background of crustal evolution that the records of ore deposition are considered. Although quantitative assessments on the lines pioneered by Pereira and Dixon[1] have not been made, nevertheless, the outlines of a pattern which makes sense in the context of crustal history emerge from the data.

Archaean stage (3·8–2·6 b.y.)
The first well documented stage of earth history, lasting well in excess of a billion years, was characterized by widespread mobility. A skin of granitic material broadly comparable with the continental crust of later eras had been formed in some places before the start of this stage. During the Archaean, basic and ultrabasic volcanic rocks, derived from partial melting of the mantle, were repeatedly erupted on to or between the primitive continental rafts and subsided under their own weight to form the infolded tracts commonly called greenstone belts. Granitic material mobilized from the preexisting basement, or differentiated from the mantle, formed buoyant domes and massifs, which tended to rise progressively between the greenstone belts. The hot residuum of the rising granites remained in the lower crust in the form of gneisses, granulites and other high-grade metamorphic rocks. Steep geothermal gradients and chronic instability promoted repeated deformation and metamorphism to such an extent that few rocks remained unmodified.

The occurrence of clastic sediments in the upper parts of many greenstone-belt successions shows that land masses of high relief were subject to erosion, especially near zones of volcanism. Nevertheless, it seems doubtful whether the early granitic crust had the strength to support large land areas standing high above the primitive ocean floors after the manner of the present continents.

The characteristic ore deposits appear to have derived their principal metals directly or by not more than one intervening stage from sources in the mantle, their formation being related to the two dominant geological processes of the early Archaean crust—basic magmatism and granite formation.

The oldest deposits of economic interest are those containing chromium, nickel or copper, which are associated with early basic, ultrabasic or anorthositic intrusions in the greenstone belts (Fig. 1). An unconfirmed age of 3·6 b.y. for the chromite-layered anorthosite of Fiskenaesset in southwest Greenland[2] suggests that ore deposition of this type began within a couple of hundred million years of the formation of the oldest known rocks.[3] At Selukwe in Rhodesia chromite deposits associated with an ultramafic complex predate granites of about 2·9 b.y. in age,[4] and nickel-bearing ores at Kambalda in Western Australia are associated with similar complexes probably little older than 2·7 b.y.

Chromite is the distinctive ore mineral of early layered anorthosites. Chromite or nickel and/or copper sulphides characterize early ultramafic bodies. All these minerals, once segregated from their parent magmas by igneous processes, seem to have remained essentially stable during subsequent phases of deformation and metamorphism. The chromite layers in the Fiskenaesset anorthosite, for example, retained their identity through repeated folding and metamorphism to granulite facies,[5] and chromite masses at Selukwe have similarly survived inversion and low-grade metamorphism.[6]

Manuscript first received by the Institution of Mining and Metallurgy on 29 January, 1973; revised manuscript received on 18 May, 1973. Paper published in August, 1973.

*The substance of this paper formed the Presidential Address to Section C at the Annual Meeting of the British Association held in Leicester in September, 1972.

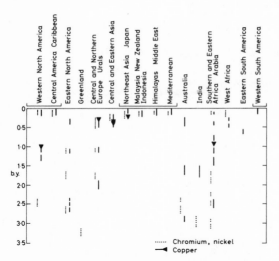

Fig. 1   Deposits characterized by chromium or nickel and copper plotted in terms of age and geographical location. Short vertical lines in Figs. 1–3 indicate approximate time of formation of deposits; inverted triangles show deposits formed mainly by agents of sedimentation and diagenesis; regions which belong to late Phanerozoic mobile belts are enclosed by square brackets

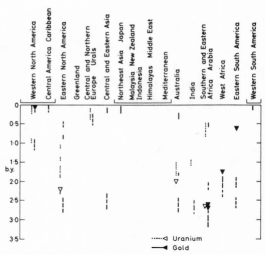

Fig. 2   Deposits characterized by gold and uranium

These very early deposits represent materials segregated from magmas originating in the mantle which took on essentially their present form immediately after their arrival in the crust. The influence of specifically crustal processes on their formation was minimal. It is not surprising, therefore, that rather similar concentrations of chromium, nickel, copper or platinum have been formed in association with mantle-derived intrusive complexes in a variety of crustal settings (Fig. 1). The relative importance of the Archaean deposits is mainly an expression of the scale of Archaean igneous activity fed from sources in the mantle.

The second major group of ore deposits characteristic of Archaean provinces is that dominated by gold, sometimes in association with silver, copper or zinc. Most of these deposits are located in greenstone belts, where they post-date much of the deformation and metamorphism and are generally related, spatially and in time, to granites emplaced in or on the margins of the belts. These relationships formerly led to the inference that gold and associated metals had been introduced into the crust by the rising granites. Recent studies in Canada[7] and in South Africa,[8] however, support MacGregor's early suggestion[9] that mineralization was a two-stage affair, the metals being initially introduced during the formation of greenstone belts but subsequently concentrated into workable deposits by the distinctively crustal processes connected with the mobilization and emplacement of granites.

In the Barberton area of South Africa and in the Archaean provinces of Rhodesia and Tanzania—where greenstone belts well in excess of 3 b.y. in age occur—mineralization began in association with granites dating from at least 2·9 b.y. In the Superior and Slave provinces of Canada and in Western Australia mineralization dates mainly from 2·7 to 2·6 b.y.[7,10,11] In West Africa, where Archaean styles of activity lingered to an exceptionally late date, the Eburnian granites, with which mineralization was connected, range in age down to about 1·8 b.y. Mineralization of the same general style spanned more than a billion years (Fig. 2), coming to an end when the granite–greenstone belt provinces were finally stabilized. The broad mobile belts formed during the succeeding

Proterozoic stages are almost devoid of gold deposits of this type, even where mantle-derived basic volcanic rocks are associated with voluminous granites. There seem to have been no further major influxes of gold from the mantle until late Phanerozoic times (Fig. 2).

Aftermath of the Archaean stage
The Archaean stage, characterized by very widespread mobility, petered out over a long transitional period, beginning about 2·8 b.y. ago, during which portions of the continental crust cooled and became more stable. This period—the first episode involving stabilization of large parts of the crust—appears to have established the pattern of geological activities for the remainder of earth history.[12] Its long-term effects on mineralization were shown by the emergence of styles of activity adapted to the newly defined cratonic regimes. A more immediate effect was the accumulation of a unique series of deposits, formed by the recycling of material derived from deposits of the types already discussed.

In several continents the eroded remnants of Archaean granites and greenstone belts are overlain unconformably by sedimentary successions, which began to accumulate soon after the onset of stabilization. Gold and/or uranium is concentrated towards the base of these successions in deposits which, although they have suffered considerable diagenetic or hydrothermal changes, have some of the attributes of placers. Many geologists consider that the gold and uranium of the Witwatersrand System, by far the most spectacular of the ore-bearing formations, were initially introduced by rivers flowing into the Witwatersrand Basin in detrital grains, which were mechanically segregated on account of their high density. The uranium-bearing Huronian deposits at Blind River and other localities in Ontario, the northern Australian deposits at Rum Jungle and the gold-bearing Tarkwaian sediments of West Africa all have relationships broadly similar to those of the Witwatersrand reefs (see, for example, Roscoe[13] and Bowie[14]).

These concentrations, which together supply most of the world's current production of gold and a fair proportion of uranium, represent the first major suite of ore deposits to derive their substance from material recycled at the surface of the earth, largely by erosion and deposition. The oldest deposits in the Dominion Reef Formation of the Witwatersrand date from 2·8 to 2·7 b.y.; the youngest, in the Tarkwaian of West Africa, from about 1·7 b.y. The occurrence of detrital uraninite, pyrite and other minerals which oxidize readily on weathering is in accord with the current view that the Archaean atmosphere was almost devoid of oxygen.

Sources for the gold were available not far from the deposits, since the host successions flank granite–greenstone belt provinces characterized by gold mineralization of the kind already discussed. Uranium mineralization, however, is not recorded on a significant scale in these provinces (Fig. 2). It must be supposed that any detrital uranium minerals were derived from concentrations formed in parts of the granite–greenstone-belt provinces subsequently removed by erosion. This inference is supported by age-determinations of 3·1 b.y. for thorian uraninite and monazite from the Dominion Reef.[15] It seems reasonable to suppose that uranium in the Archaean crust would have settled only in sites near the surface. The relatively high proportions of uranium in the first-formed strata of the Witwatersrand Basin and the preponderance of gold at higher levels in the succession[8] are in keeping with such an inference. That upward migrations of uranium did take place during Archaean times is suggested independently by geochemical studies of the residual granulites and gneisses formed at deep levels in the Archaean crust. In Greenland, Scotland and Norway rocks of these types contain far less than the crustal average amounts of uranium and thorium and are thought to have been impoverished in these elements during high-grade metamorphism.[16,17]

The gold–uranium deposits formed in the aftermath of the Archaean stage were laid down with a comparatively short time-lag after the stabilization of the provinces from which their substance was derived; and although gold-bearing Archaean provinces have been exposed to erosion at many later times, there has been no further episode on a comparable scale. This time-link illustrates what appears to be a general point—major ore deposits which derived their substance from the erosion of mineralized areas have tended to accumulate comparatively soon after the initial episodes responsible for the concentration of the relevant metals in the upper crust. The appearance of numerous gold–uranium deposits soon after the stabilization of the first cratonic massifs suggests that stabilization had made possible the emergence of larger land masses than those which could be supported by the weak early crust, and, hence, had exposed the potential source areas to the forces of erosion.

Late Archaean and early Proterozoic stage (ca 2·6–1·6 b.y.)

When activity in the granite–greenstone belt provinces came to a close, systems of mobile belts characterized by persistent instability and high geothermal gradients had been differentiated from a number of cooler continental cratons, usually less than 1000 km in diameter. The igneous and sedimentary assemblages formed under the newly defined cratonic regimes include types which were of little importance in the older Archaean provinces, and a corresponding diversification marks the record of mineralization at this stage.

Basic igneous activity in cratons

The cratons formed towards the end of the Archaean stage were invaded, at various times between 2·6 and 1·9 b.y., by basic magmas derived from partial melting of the mantle. Both dyke swarms of regional extent and large layered intrusions were emplaced as a result of this magmatic activity. The event was comparable in scale with the cratonic basic magmatism of two later periods (ca 1·4–0·9 b.y. and ca 0·3–0·0 b.y.), which appears to have been genetically related to extension and fracturing of the cratons.

The principal ore deposits associated with cratonic basic magmatism of late Archaean and earliest Proterozoic times are the deposits of chromium, platinum, nickel and iron–titanium, which occur as magmatic segregations in the large differentiated intrusions. The Great Dyke of Rhodesia (ca 2·6 b.y.), the Bushveld complex of South Africa (ca 1·95 b.y.) and the Petsamo complex in the north of the Baltic shield are among the hosts of these segregations (Fig. 1).

It is worth mentioning that copper mineralization, which was a feature of the two younger basic episodes in the cratons, is of little importance, perhaps because few plateau-lavas (other than the Ventersdorp lavas of the Witwatersrand Basin) have survived.

The Sudbury intrusive complex, which lies close to the eastern border of the principal Archaean craton in Canada, and which is dated at about 1·7 b.y., seems to have no precise analogues among intrusions of other ages. Dietz' invocation of meteor impact[18] and associated suggestions of extraterrestrial sources for the nickel sulphide ores have gained many adherents, and although such a *deus ex machina* seems to raise more problems than it solves, it is difficult to find a place for the complex in the context of crustal evolution.

Kimberlites in cratons

The distinctive pipes and dykes of kimberlite which are the primary sources of diamonds are concentrated almost exclusively in cratonic areas of the continental crust. There seems to be general agreement that kimberlite is, at least in part, derived from the mantle well below the levels from which most basaltic partial melts are generated—the formation of diamonds at a minimum depth of some 200 km being favoured on experimental grounds by some authors.[19] The transmission to the crust of the minute mobile fractions from which kimberlite is derived seems to have been most feasible when and where the upper levels of the mantle were too cool to permit large-scale partial melting, and the appearance of kimberlites, therefore, suggests the existence of a low geothermal gradient. With these considerations in mind, it is perhaps surprising that the first diamondiferous kimberlites made their appearance so early. The kimberlite which supplies Premier mine near Pretoria in South Africa is firmly dated at more than 1·2 b.y., and is probably as much as 1·7 b.y. in age.[20] The occurrence of detrital diamonds in the Roraima Formation of the Guyana shield, which was deposited more than 1·7 b.y. ago, suggests that kimberlites were exposed to erosion in South America even before this date. Cratonic carbonatite–alkaline complexes make their appearance in the geological record at about 1·7 b.y.

Parts of the crust and mantle had, therefore, cooled sufficiently to allow the generation of diamondiferous kimberlites within a billion years of the stabilization of the first cratons. The vast majority of known kimberlites, however, were emplaced during the late Palaeozoic and Mesozoic events connected with the disruption of Laurasia and Gondwanaland. Despite the enormous time-gap of 1·5 b.y. between the first kimberlites and the main swarms, most Phanerozoic kimberlites, especially the diamondiferous ones, are crowded into the Archaean massifs, which represent the first cratons to stabilize.[21] Their preference for long-established cratons, such as those of Tanzania and the Transvaal in Africa and the Anabar massif in Siberia, emphasizes the special characters of these first-formed stable massifs; that such special characters may be traced back to abnormalities in the underlying mantle is suggested not only by the style of igneous activity in them but also by the fact that they tend to form swells in the present continental structure.

Sedimentary deposits of cratons

The formation of relatively strong continental cratons, it was suggested above, facilitated the emergence of moderate-size land masses in which crystalline rocks were soon unroofed by erosion. The metals released by

275

this means became available for recycling at the earth's surface at about the time at which stable environments favourable to the processes of sedimentary differentiation became more extensive. It was in this setting that the gold–uranium deposits already referred to were formed.

The unique banded iron formations of the Lake Superior and Labrador regions of North America, the Hamersley Range of Western Australia, the Ukraine of eastern Europe and several other localities represent segregations formed by chemical or biochemical means. The bulk of these formations consists of rhythmically banded cherts carrying magnetite or hematite. Terrigenous detritus is almost lacking and the purity of the sediments, together with their remarkable lateral consistency,[22] indicate that they accumulated in tranquil conditions.

Banded iron formations of this type were deposited over a short time-range: they came in at about $2 \cdot 5$ b.y., preceded only by thinner jaspery iron formations in certain Archaean volcanic assemblages, and disappeared at about $1 \cdot 9$ b.y. Their limited time-span has been attributed to the effects of organic evolution.[23] The precipitation of iron oxides is seen as a side-effect of the release of oxygen during photosynthesis by primitive organisms incapable of tolerating free oxygen. The subsequent evolution of organisms with a more sophisticated metabolism removed the need for iron as an oxygen acceptor and put an end to the precipitation of banded iron formations. The build-up of oxygen in the atmosphere after $2 \cdot 0$ b.y. is attested by the appearance of red beds and evaporites in the stratigraphical record from about $1 \cdot 7$ b.y. onward.

Deposits of mobile belts

The ore deposits formed in mobile provinces during the late Archaean and early Proterozoic stage were conspicuously different from those of the old granite–greenstone belt terrains. Provinces which consist mainly of regenerated Archaean material tend to be poorly supplied with contemporaneous mineral deposits. The Ubendian belt of East Africa, the

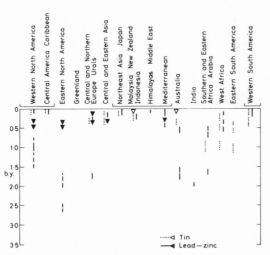

Fig. 3   Deposits characterized by lead and/or zinc and tin

Limpopo belt of southern Africa and the Churchill province of Canada, all located at the margins of gold-bearing Archaean terrains and containing reworked greenstone belts, carry only minor amounts of gold. Such gold as survived (for example, at Flin-Flon) in the Churchill province is generally located in areas which largely escaped the effects of regeneration.[24]

Reworking at the level exposed appears to have dispersed earlier concentrations.

In provinces which incorporate early Proterozoic cover successions, copper, lead, zinc and silver are important.

The copper sulphide ores of the Svecofennide belt in the Baltic shield are associated with, though not always located in, slightly metamorphosed basic volcanic rocks. The lead–zinc (copper) ores of Mount Isa and Broken Hill in Australia also lie in the vicinity of supracrustal groups that include volcanic rocks. These deposits, dating from $1 \cdot 8$ to $1 \cdot 6$ b.y., are of kinds scarcely represented during the Archaean stages of mineralization (Fig. 3). Zinc is indeed an important component of ores in the Abitibi greenstone belt ($2 \cdot 6$ b.y.), but the metal is seldom prominent in older greenstone belts, and where it does occur it is rarely associated with significant amounts of lead. The delayed appearance of the typical lead–zinc association may be attributed to the scarcity, during the early stage, of the relatively cool and stable environments which provided favoured sites.

It seems probable that the mobile Archaean crust simply proved inhospitable to mineralization of this type. The many small hydrothermal uranium deposits in the Churchill province of Canada (*ca* $1 \cdot 8$ b.y.) also suggest that conditions suitable for the fixing of fugitive metals had been attained in early Proterozoic times. That such elements shifted repeatedly to new sites is shown by the enormous range of ages obtained from ore minerals in areas of uranium mineralization.[14]

Later Proterozoic and Palaeozoic stages ($1 \cdot 6$–$0 \cdot 3$ b.y.)

After a terminal phase of abundant granite production many of the early Proterozoic mobile belts became quiescent about $1 \cdot 8$ b.y. ago. The stabilized parts of these belts were welded to older massifs to produce the larger cratons which seem to have dominated the continental crust in mid-Proterozoic times. Extension and fracturing associated with the second major episode of cratonic basic magmatism took place in these cratons over the period $1 \cdot 4$–$0 \cdot 9$ b.y. and were accompanied by the development of a new network of mobile belts. The pattern of structures blocked out at this time provided the tectonic framework within which the Atlantic and Indian Oceans were subsequently developed.

The later Proterozoic stage of crustal development saw not only the initiation of a global arrangement of structures recognizably, though remotely, related to the present-day arrangement but also the beginnings of styles of mineralization with a decidedly modern aspect. It will be convenient to concentrate on these new developments rather than to attempt a general survey of mineralization to late Palaeozoic times. Such deposits as the copper ores associated with volcanic rocks in the Caledonian mobile belt of Scandinavia have so much in common with those formed in similar settings in earlier times that comment is superfluous.

Beginnings of tin mineralization

Although cassiterite occurs as an accessory mineral in pegmatites dated at $2 \cdot 6$–$2 \cdot 7$ b.y. in West Africa and Rhodesia, forms detrital grains in the Dominion Reef Formation of South Africa ($2 \cdot 8$–$2 \cdot 7$ b.y.) and is associated with the Bushveld granites (*ca* $1 \cdot 9$ b.y.), tin mineralization of economic importance was largely confined to younger periods (Fig. 3). About a billion years ago mineralization in two rather different settings marked the widespread incoming of tin, tungsten and their associates to the record. The Karagwe–Ankole or Kibaran mobile belt of Africa, which reached the terminal stages of orogenic activity at this time, was characterized by abundant granites and by metamorphism of a low-pressure facies series usually attributed to the operation of a high geothermal gradient. Tin–tungsten deposits are associated with late orogenic (0·9 b.y.) granites at various localities from Uganda to South West Africa. In the Rondônia district of western Brazil[25,26] tin deposits are clustered near a group of high-level anorogenic granites dated at about 0·9 b.y. and emplaced in a basement of granite and gneiss.

These tin-bearing assemblages share many of the characteristic features of younger tin deposits. The Karagwe–Ankole belt, like the tin-bearing Hercynian belt of western Europe,[27,28] was a zone in which a steep geothermal gradient developed; it was not subjected to a major influx of mantle-derived igneous material, and some, at least, of the mineralized granites appear to represent remobilized basement granites. The granites of Rondônia have much in common with the Mesozoic Younger Granites of Nigeria, which are strongly fractionated high-temperature intrusions emplaced in a granitic or gneissose basement. These relationships render it unlikely that the appearance of concentrations of tin, tungsten and allied metals about a billion years ago was connected directly with the influx of material from the mantle.

The well known evidence of repeated mineralization in tin provinces such as those of Nigeria and western South America shows that later phases were commonly richer than the earlier.[29,30] As Schuiling has emphasized in another connexion,[31] rich deposits tend to occur where fields of more than one age overlap. With the remarkably late start to effective tin mineralization, this evidence suggests that recycling within the crust was an inefficient process, which worked best in terrains previously enriched in the metals.

Cratonic basic magmatism and copper mineralization

The exceptionally widespread emplacement of basic dykes and of larger differentiated basic intrusions which took place between about 1·4 and 0·9 b.y. was associated with the extrusion of plateau basalts and, more locally, with rifting of the crust. Magmatic ore deposits of various kinds are associated with the mantle-derived intrusive bodies—for example, the Fe–Ti deposits of the Duluth intrusion in the northern U.S.A. The plateau lavas to which the regional dyke swarms served as feeders are associated with copper deposits of varying importance: those of the Keweenawan lavas of Lake Superior, located mainly in permeable amygdular or conglomeratic horizons, are a major source of the metal, those of Arctic Canada (Coppermine River lavas) and of East Africa (the Bukoban lavas and their equivalents)[32] are scarcely economic. These occurrences, together with the fact that related dykes extend far beyond the present limits of the lava plateau, suggest that copper was transferred to the upper parts of the crust from sources in the mantle on a considerable scale during the period 1·4–0·9 b.y.

It can hardly be a coincidence that copper deposits of sedimentary origin enter the geological record in force at about the same time, occurring in Africa, in Siberia and in North America (Fig. 1). The cupriferous formations from which the huge ore deposits of Zambia, Katanga and the adjacent territories were formed lie in the lower part of the Katangan System, whose basal members[33] represent a molasse dating from the terminal stages of the Karagwe–Ankolean orogeny at or soon after 0·9 b.y. These formations fringe the southern border of the large late Proterozoic craton on which the Bukoban volcanics were extruded about 0·9 b.y. ago. In North America cupriferous sediments of little economic importance characterize horizons in the Proterozoic Belt Supergroup deposited at the western fringe of the craton on which the Keweenawan lavas were extruded between 1·2 and 1·0 b.y. ago.[34] On the assumption that the early deposits of basins fringing the late Proterozoic cratons were derived largely from erosion of the cratons themselves, it seems reasonable to suppose that the copper they contain was directly or indirectly supplied by the recently erupted lavas. Once again, recycling of a metal by surface agencies followed closely the delivery of this metal to the upper parts of the crust.

Lead–zinc in cratonic environments

Towards the end of the Palaeozoic era, and to some extent from Cambrian times onward, lead–zinc concentrations began to appear in the sedimentary cover successions of the cratons, most notably in North America, where the huge deposits of Mississippi Valley type accumulated (Fig. 3).

Sedimentation and diagenesis appear to have been responsible for the initial concentration of lead and zinc in these environments, though substantial redistribution by circulating fluids preceded the formation of the ore deposits themselves. The richest concentrations are in the North American craton, which was flanked on the west by the terrain of Proterozoic lead–zinc mineralization[35] now incorporated in the Cordilleran orogenic belt. The preferred host rocks in the cratonic sequences—limestones, dolomites and associated evaporites—are types whose abundance increased notably over late Proterozoic times with the establishment of an oxygen-rich atmosphere and evolutionary explosion of the Metazoa. The concentration in a sedimentary environment of lead and zinc supplied to the cratonic basin by the erosion of a mineralized terrain was, no doubt, favoured by the abundance of the appropriate host rocks in the Palaeozoic succession. It is interesting that favourable successions in cratonic areas, such as western Europe and northwest Africa, which were not flanked by zones of extensive mineralization, carry smaller concentrations (as in Britain) or are barren.

The latest stage (<0·3 b.y.)

The latest stage of crustal history has been dominated by the fragmentation of large continental cratons and the consequent growth of the Atlantic and Indian Oceans. Disruption and stretching of the cratons were accompanied by the eruption of plateau basalts and the emplacement of mantle-derived dyke swarms and differentiated intrusions, continuing to the present day. Deposits of copper, nickel and platinum—for example, in the Triassic Siberian traps—are broadly comparable with those associated with the products of earlier cratonic basic magmatism (Fig. 1). The initiation of major fracture systems in the cratons coincided in time with the emplacement of kimberlites, especially in the Archaean portions of the cratons. This event, although not new in kind, was on a quite different scale from previous phases of kimberlite emplacement. It was responsible for the influx of most of the world's diamonds and was followed after a short time-lag by the formation of diamondiferous placers, such as those of South West Africa.

In the mobile belts at the junctions of advancing crustal plates a remarkable range of ore deposits is displayed. Some are evidently derived from sources in the mantle, directly or at only one remove. Most of the characteristic deposits associated with the ultrabasic–basic ophiolite suite—massive sulphide copper ores, such as those of Cyprus, and nickel ores of such regions as New Caledonia—fall in this category. Mineralization related to the ophiolitic assemblages—regarded by many geologists[36] as fragments of oceanic crust produced from sources in the mantle at mid-oceanic ridges and emplaced more or less passively within orogenic fold belts—had a good deal in common with the initial mineralization in the Archaean greenstone belts and was on a scale which had not been attained since Archaean times (Fig. 1). The Tertiary gold–sulphide deposits have more complex relationships not unlike those of the porphyry coppers. Both were connected spatially and in time with granites or high-level, acid–intermediate intrusions emplaced in island arcs or mountain belts above active Benioff zones. The

277

metals themselves may have originated in the mantle-derived, oceanic crustal plates overridden at Benioff zones, the acid magmatism serving mainly as an agent of transport and segregation.[37] In style and scale the gold mineralization bears some resemblance to that of the Archaean provinces (Fig. 2).

In other respects mineralization in the mobile belts reflects crustal influences very different from those which prevailed in Archaean times. The diversity of metals represented reflects the maturity of the continental crust. In the Cordilleran and Andean sectors alone copper, lead, zinc, silver, gold, uranium, tin and mercury are all of major importance at some locality. Evidence that the crust was already well stocked with many of these metals makes it natural to infer that recycling of metals by fluids circulating in the upper crust played a major part in mineralization. Such cannibalistic activity had no real counterpart in the immature Archaean crust. Finally, the variety of metals represented is increased by the short time-lapse since the final stages of mineralization in that near-surface deposits of fugitive elements, such as arsenic, antimony, and mercury, have not yet been dispersed by erosion. If precedent is anything to go by, the metals driven up in the mobile belts during Tertiary orogenic phases should be transferred over the next hundred million years to concentrations in cratonic sedimentary environments.

Retrospect

When examined in the light of the present geological situation, the history of mineralization falls into perspective (Fig. 4). Two principal themes may be distinguished—that connected with the transfer of metals from the mantle to the crust and that connected with the redistribution of metals within the crust itself.

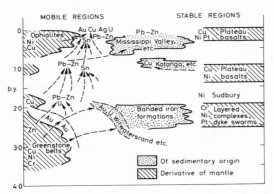

Fig. 4 Summary diagram illustrating possible relationships of ore deposits formed during successive stages of crustal history

Although the most consistent styles of mineralization have been those related to the influx of metals from sources in the mantle, these processes have not continued at a uniform rate. In mobile belts a long Proterozoic period of relative quiescence intervened between the abundant Archaean magmatism of the greenstone belts and widespread formation of Phanerozoic ophiolite assemblages. The accessions of copper, nickel, gold and other metals indirectly or directly associated with the Phanerozoic phases of ophiolite formation were on a scale unmatched since Archaean times; these contrasts suggest that the global tectonics of Proterozoic eras conformed to patterns unlike that of the Phanerozoic model of plate tectonics. In the cratons the major phases of basic magmatism and attendant mineralization coincided with periods of fracturing and crustal extension, the most recent being linked also with continental drift and sea-floor

spreading. The emplacement of diamondiferous kimberlites and of alkaline complexes with rare-earth concentrations took place over much the same periods.

The deposits mustered by crustal agencies, more particularly by agencies working at the surface, show conspicuous variations through times which may be attributable to several causes. Deposits of the first billion years show deficiencies which reflect both the immaturity and the instability of the crust. The scarcity of tin and its associates may be put down to the fact that the preliminary recycling necessary for the concentration of these metals had not been completed. The scarcity of lead and uranium seems more likely to have stemmed from the perennial crustal instability which promoted the dispersal rather than the concentration of fugitive elements. The increased importance of uranium and lead–zinc mineralization during the stages which followed the evolution of the first continental cratons illustrates the importance of the appearance of cooler and more stable environments where concentrations of mobile elements might escape subsequent dispersal. The observed changes in tectonic regime appear to provide a more acceptable explanation of variations in the style of mineralization within the crust than do the variations in depth of erosion favoured by Laznicka.[38]

The first suites of ore deposits formed mainly by agents of sedimentation and diagenesis followed rapidly on the emergence of the first land masses in which large mineralized terrains were exposed to erosion. From $2 \cdot 8$ b.y. onward metals which had been transferred to the upper crust by igneous or hydrothermal activity tended to find their way after a short time-lag into the adjacent sedimentary basins. Such metals were concentrated in ore deposits mainly in cratonic environments, which favoured the sorting and re-working of sedimentary materials. The deposits so formed show long-term variations directly or indirectly related to the effects of organic evolution. Some unusual features of the Archaean deposits characterized by gold and/or uranium were connected with anoxygenic conditions of weathering. The early Proterozoic banded iron formations record biologically controlled processes of oxidation; and the majority of younger lead–zinc deposits appear to have depended for their formation on the presence of host rocks formed only in an oxygenic atmosphere. The most recent of all surface concentrations of metals—the dumps of our industrial cities—have been formed by the activities of the newly evolved human species.

Table 1   Incoming of new styles of mineralization

b.y.	
$0 \cdot 0$	First deposits formed by human agencies
$0 \cdot 5$	First major Pb–Zn deposits of Mississippi Valley type
$1 \cdot 0$	First major sedimentary copper ores
$1 \cdot 5$	First important tin mineralization
	First evaporites
	First red beds
$2 \cdot 0$	First diamonds
	First major banded iron formations
$2 \cdot 5$	First stable cratons
	First deposits of cratonic layered igneous complexes (Cr, Pt)
$3 \cdot 0$	First ores formed by sedimentary processes (Au, U)
$3 \cdot 5$	First gold mineralization in greenstone belts
	First deposits of layered igneous complexes (Cr)

The record of mineralization suggests that a substantial proportion of the metals now fixed in ore deposits had entered the crust well before the end of the Archaean stage. Since these metals have been repeatedly recycled in or at the surface of the crust, the possibility arises that they may have been selectively concentrated in certain crustal environments. It seems probable that such mobile metals as uranium, lead, zinc and mercury, recent concentrations of which are located predominantly in the upper and outer parts of mobile belts or on the cratons, are now recycled mainly by agents operating in the upper crust and are returned to depth only where the collision of crustal plates has forced rock masses bodily down towards the mantle. Such a restriction in the scope of recycling could result in a long-term and large-scale geochemical differentiation in the crust. As Table 1 suggests, there is no reason to suppose that the evolution of mineralizing processes has yet come to an end.

References

1. **Pereira J. and Dixon C. J.** Evolutionary trends in ore deposition. *Trans. Instn Min. Metall.*, **74**, 1965, 505–27.
2. **Evenson M. M. Murthy V. R. and Windley B. F.** In preparation.
3. **Black L. P. *et al.*** Isotopic dating of very early Precambrian amphibolite facies gneisses from the Gothaab district, west Greenland. *Earth planet. Sci. Lett.*, **12**, 1971, 245–59.
4. **Wilson J. F.** The Rhodesian Archaean craton—an essay in cratonic evolution. *Phil. Trans. R. Soc.*, **273A**, 1973, 389–411.
5. **Windley B. F.** Anorthosites of southern west Greenland. *Mem. Am. Ass. Petrol. Geol.*, no. 12, 1969, 899–915.
6. **Stowe C. W.** Intersecting fold trends in the Rhodesian basement complex south and west of Selukwe. *Geol. Soc. S. Afr. Annexure to (Trans.)*, **71**, 1968, 53–77.
7. **Goodwin A. M.** Metallogenic patterns and evolution of the Canadian shield. *Spec. Publs geol. Soc. Aust.*, **3**, 1971, 157–74.
8. **Viljoen R. P. Saager R. and Viljoen M. J.** Some thoughts on the origin and processes responsible for the concentration of gold in the early Precambrian of Southern Africa. *Mineral. Deposita*, **5**, 1970, 164–80.
9. **MacGregor A. M.** The primary source of gold. *S. Afr. J. Sci.*, **47**, 1951, 157–61.
10. **Green D. C. Baadsgaard H. and Cumming G. L.** Geochronology of the Yellowknife area, Northwest Territories, Canada. *Can. J. Earth Sci.*, **5**, 1968, 725–35.
11. **Arriens P. A.** The Archaean geochronology of Australia. *Spec. Publs geol. Soc. Aust.*, **3**, 1971, 11–24.
12. **Sutton J.** Some developments in the crust. *Spec. Publs geol. Soc. Aust.*, **3**, 1971, 1–10.
13. **Roscoe S. M.** Huronian rocks and uraniferous conglomerates in the Canadian shield. *Pap. geol. Surv. Can.* 68–40, 1969, 205 p.
14. **Bowie S. H. U.** Some geological concepts for consideration in the search for uranium provinces and major uranium deposits. In *Uranium exploration geology* (Vienna: International Atomic Energy Agency, 1970), 285–98.
15. **Nicolaysen L. O. Burger A. J. and Liebenberg W. R.** Evidence for the extreme age of certain minerals from the Dominion Reef conglomerates and underlying granite in the western Transvaal. *Geochim. cosmochim. Acta*, **26**, 1962, 15–23.
16. **Heier K. S.** Geochemistry of granulite facies rocks and problems of their origin. *Phil. Trans. R. Soc.*, **273A**, 1973, 429–42.
17. **Lambert, I. B.** The composition and evolution of the deep continental crust. *Spec. Publs geol. Soc. Aust.*, **3**, 1971, 419–28.
18. **Dietz R. S.** Sudbury structure as an astrobleme. *J. Geol.*, **72**, 1964, 412–34.
19. **Kennedy G. C. and Nordlie B. E.** The genesis of diamond deposits. *Econ. Geol.*, **63**, 1968, 495–503.
20. **Allsopp H. L. Burger A. J. and Zyl C, van.** A minimum age for the Premier kimberlite pipe yielded by biotite Rb–Sr measurements with related galena isotopic data. *Earth planet. Sci. Lett.*, **3**, 1967, 161–6.
21. **Clifford T. N.** Tectono-metallogenetic units and metallogenic provinces of Africa. *Earth planet. Sci. Lett.*, **1**, 1966, 421–34.
22. **Trendall A. F. and Blockley J. G.** The iron formations of the Precambrian Hamersley Group, Western Australia. *Bull. geol. Surv. West. Aust.* 119, 1970, 366 p.
23. **Cloud P. E. Jr.** Atmospheric and hydrospheric evolution on the primitive earth. *Science, N.Y.*, **160**, 1968, 729–36.
24. **Davidson A.** The Churchill province. In *Variations in tectonic styles in Canada* **R. A. Price and R. J. W. Douglas eds.** (Toronto: Geological Association of Canada, 1972), 381–433. (*Spec. Pap. geol. Ass. Can.* 11)
25. **Kloosterman J. B.** Granites and rhyolites of São Lourenço: a volcano-plutonic complex in southern Amazonia. *Engenh. Miner. Metal.*, **44**, 1966, 169–71.
26. **Priem H. N. A. *et al.*** Granitic complexes and associated tin mineralizations of 'Grenville' age in Rondônia, western Brazil. *Bull. geol. Soc. Am.*, **82**, 1971, 1095–102.
27. **Zwart H. J.** The duality of orogenic belts. *Geologie Mijnb.*, **46**, 1967, 283–309.
28. **Hall A.** The relationship between geothermal gradient and the composition of granitic magmas in orogenic belts. *Contr. Mineral. Petrol.*, **32**, 1971, 186–92.
29. **Wright J. B.** Controls of mineralization in the older and younger tin fields of Nigeria. *Econ. Geol.*, **65**, 1970, 945–51.
30. **Turneaure F. S.** The Bolivian tin–silver province. *Econ. Geol.*, **66**, 1971, 216–25.
31. **Schuiling R. D.** Tin belts on the continents around the Atlantic Ocean. *Econ. Geol.*, **62**, 1967, 540–50.
32. **Harris J. F.** Summary of the geology of Tanganyika, Part IV: Economic geology. *Mem. geol. Surv. Dep. Tanganyika* no. 1, 1961, 143 p.
33. **Cahen L. and Lepersonne J.** The Precambrian of the Congo, Rwanda and Burundi. In *The Precambrian, vol. 3* **Rankama K. ed.** (New York, etc.: Interscience, 1967), 143–290.
34. **Harrison J. E.** Precambrian Belt basin of northwestern United States. *Bull. geol. Soc. Am.*, **83**, 1972, 1215–40.
35. **Zartman R. E. and Stacey J. S.** Lead isotopes and mineralization ages in Belt Supergroup rocks, northwestern Montana and northern Idaho. *Econ. Geol.*, **66**, 1971, 849–60.
36. **Sillitoe R. H.** Formation of certain massive sulphide deposits at sites of sea-floor spreading. *Trans. Instn Min. Metall.* (*Sect. B: Appl. earth sci.*), **81**, 1972, B141–8.
37. **Mitchell A. H. G. and Garson M. S.** Relationship of porphyry copper and circum-Pacific tin deposits to palaeo-Benioff zones. *Trans. Instn Min. Metall.* (*Sect. B: Appl. earth sci.*), **81**, 1972, B10–25.
38. **Laznicka P.** Development of nonferrous metal deposits in geologic time. *Can. J. Earth Sci.*, **10**, 1973, 18–25.

Editor's Comments
on Papers 36 Through 45

PLATE TECTONICS AND SEDIMENTARY MINERAL
DEPOSITS

The development of many sedimentary mineral deposits—e.g. coal, phosphorites, evaporites of Stassfurt type, placers—is controlled chiefly by climatic and geomorphological factors. The paleo-distribution of climatically controlled sediments has been cited as evidence both to support continental drift, and in more recent attempts to refute it, especially by Meyerhoff and his co-workers. The abstract of one of Meyerhoff's papers is reproduced as Paper 36, one of several in the same vein (e.g. Meyerhoff and Teichert, 1971; Meyerhoff et al, 1972; Meyerhoff and Meyerhoff, 1972a,b). Meyerhoff is more than usually sceptical about palaeomagnetic data, but his own thesis has not met with wide acceptance—though he has prosecuted it with such vigor that one might almost suppose him to have been acting as devil's advocate. Meyerhoff's findings could be accommodated within the framework of continental drift and plate tectonics, if continental displacements have had a dominant latitudinal component, the subordinate meridional movements being restricted to a fairly limited spread of latitudes. Drewry et al. (1974) appear to have reached a similar conclusion (p. 531): "Phanerozoic latitude variations have generally not been large enough to create a present-day distribution of Phanerozoic sediments that provides unambiguous evidence for or against continental drift" (see also Schopf, 1973; Gordon, 1975).

Plate tectonics has in general relatively little to do with the formation of sedimentary mineral deposits, except insofar as plate movements have determined the palaeo-distributions of land and sea with respect to climatic belts, so to a great extent this is merely the theme of an earlier section in a new guise (but see the section on Petroleum and Plate Tectonics). This point is apparent from Paper 37 by Blissenbach and Fellerer, which also reviews the more direct influence of plate tectonic processes on the development of certain other sedimentary environments. There is some overlap with material in previous sections, particularly in the discussion of Red Sea-type metalliferous muds, and in the suggestion that subduction of metal-rich sediments and manganese nodules might contribute to continental ore deposits.

In Paper 38, Bardossy quite clearly recapitulates the theme of Papers 4 through 10, providing a well-developed argument for the paleo-distribution of bauxite deposits in terms of continental drift, plate tectonics as such being largely irrelevant in this context. The author uses the distribution of Permo-Carboniferous bauxites, moreover, as evidence to reconstruct parts of eastern Asia, where paleomagnetic data for this period are meagre. He also attempts a correlation between types of bauxite (lateritic versus karstic) and tectonic setting, but it is certainly possible

281

to question his final point that the rate of bauxitization has increased through geological time. The superficial nature of bauxite deposits ensures that they are highly susceptible to erosion, so it may even be surprising that so much Paleozoic bauxite is in fact preserved.

The provision of graben structures suitable for the accumulation of evaporite deposits is explored for the margins of Africa and South America by Burke in Paper 39. The presence of thick evaporites along the margins of the Red Sea, and the later Pleistocene history of this region (Paper 29), both suggest a generally similar origin for the salt deposits there. Climate also clearly played an important part in the development of these deposits, if the assumptions outlined in Paper 39 are valid.

PreCambrian Banded Iron-Formations (BIF)

Iron-formations composed of thinly bedded chert and iron minerals which contain at least 15 per cent iron are probably the most abundant chemically precipitated sedimentary rocks known. They occur in a wide variety of geological environments and because of the diversity in chemical properties of their elemental constituents are highly sensitive indicators of the depositional environments in which they formed. . . . Because there are relatively few examples of cherty iron-formation in rocks of Mesozoic age or younger and apparently no modern examples exist where banded cherty iron sediments are forming today, we have no complete contemporary model or guide to the geological parameters affecting the origin of these special sediments (Gross, 1973).

In the second part of a paper concerned with the evolution of Precambrian crust, Goodwin (Paper 40) provides a summary classification and description of cherty iron-formations and reviews their global distribution in Gondwanaland, Laurasia and Pangaea.

Climatic conditions in the Precambrian were probably an important factor in the origin of iron-formation sediments (e.g. Cloud, 1973; Eugster and Chou, 1973; Garrels et al., 1973; Holland, 1973), but development of the basins in which they accumulated, and the mechanisms of supply and concentration of metals, were not necessarily the result of processes peculiar to Precambrian times. Gross (1973), for example, has compared the sulphides in Algoma-type iron-formations (see Paper 40) with modern metalliferous sediments in the Red Sea and suggests they may have had a similar volcanic/hydrothermal origin "centred along major faults or tectonic features in the crust." Goodwin (1973) has compared the development or iron-formation basins with that of modern interarc basins in the Western Pacific, each basin being a centre of crustal spreading with accumulation of oceanic volcanics in the interior and arc-type volcanics at the margins:

It is proposed that the basins represent surface expressions of thermal plumes . . . originating in the deep mantle or core. . . . near-surface Archaean plumes caused local horizontal migration of thin Archaean lithospheric plates . . . . (which) behaved (similarly) to modern plates; in contact with sialic crust the plates were underthrust along subduction zones with . . . development of arc-type volcanics . . . . Volcanic exhalations and . . . clastics were deposited mainly within the basins in the form of chemical (iron-formation) and clastic sediments. . . . thermal plumes were the controlling feature in development of Archaean basins. An isolated plume produced an isolated basin. . . . Close-packed plumes produced close-packed basins. A linear alignment of close-packed plumes produced a linear array of basins which coalesced to produce a volcanic-rich belt, the . . . greenstone belt in Archaean crust.

Petroleum and Plate Tectonics

As indicated in Paper 37, the evolution of petroleum basins is fundamentally related to plate movements, so plate tectonics *is* important in development of this type of sedimentary accumulation. Obvious examples are basins along continental margins, which generally represent past or present constructive or destructive plate boundaries. Plate tectonic concepts have brought greatly increased understanding of the structure and evolution of continental margin basins, and hence of their petroleum potential (Hedberg, 1970; Henderson, 1973; Beck and Lehner, 1974; Fischer, 1975; Kinsman, 1975; Curray, 1975).

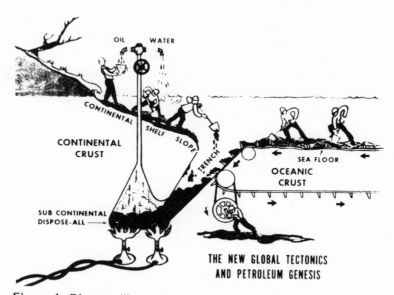

Figure 4 Diagram illustrating an early view of how plate tectonics might control the development of petroleum accumulations. Source: Hedberg, 1970, Fig. 32.

Petroleum basins are not of course confined to continental margins, but are found in many different settings. Klemme's classification of basin types in terms of plate tectonics is reproduced in Paper 41 (compare McDowell, 1971), followed by Tarling's more general review of the factors governing petroleum accumulation (Paper 42). Irving et al. have quantified some of these factors in Paper 43, with particular reference to the paleolatitude distribution of oil fields, in an attempt to explain why most of the world's known petroleum reserves occur in restricted areas in strata covering only a short span of geological time.

In later papers, Klemme (1972, 1975a,b) developed the point made in Papers 41 and 42, that the geothermal gradient is an important factor in controlling both the conversion of organic debris into petroleum, and its subsequent migration from source rocks into reservoirs.

If the continental margins of opening ocean basins are obvious sites for development of large petroleum basins, then sediment-filled rifts which may represent incipient or abortive spreading systems—so-called "failed arms"—could also be important for petroleum accumulations, particularly since heat flow in such basins is likely to be higher than average. Burke and Dewey (1973, page 428), state:

> Major fluid hydrocarbon accumulations in failed arms are very widespread. The northern North Sea, Gippsland Basin, Gulf of Suez, Niger delta, Maracaibo Basin, and Sirte Basin open into oceans but the older Anadarko Basin is linked to a collision orogen. Many young failed arms have not yet been fully explored for hydrocarbons.

This proposition has been explored in some detail for the North Sea by Whiteman et al., whose concluding paragraphs and Figures 1, 2, and 3 form Paper 44 (see also Whiteman et al., 1975). In Paper 45, Schlanger and Combs examine how structure and heat flow might determine the petroleum potential of small oceanic basins that form behind some destructive plate margins.

36

Reprinted from *J. Geol.* **78**:1 (1970)

CONTINENTAL DRIFT: IMPLICATIONS OF PALEOMAGNETIC STUDIES, METEOROLOGY, PHYSICAL OCEANOGRAPHY, AND CLIMATOLOGY[1]

A. A. MEYERHOFF

American Association of Petroleum Geologists, Tulsa, Oklahoma

ABSTRACT

Paleomagnetic methods apparently have little or no resolving power for studies of global tectonics. As the number of ancient pole determinations increases, the scatter of ancient pole positions has become so great, even from single localities and geologic provinces, that the circles of error for paleopoles from rocks of each age are wider than the Atlantic Ocean. From this fact it is apparent that paleomagnetic pole determinations cannot be used to show relative movements between continents. The use of paleoclimatology to interpret earth history in general—and global tectonics in particular—has not been successful because its practitioners have not utilized the results of present-day meteorology, physical oceanography, and climatology. The writer used marine evaporite, coal, desert eolian sandstone, and tillite deposits as paleoclimate indicators. From a plot of marine evaporite, coal, and proved desert eolian deposits on world maps—one for each age from the Proterozoic through the Miocene—a consistent geometric pattern emerged. Addition of proved tillite deposits to the maps brought out a second, but time-related, pattern. The geometric pattern on the maps shows that, since Devonian time, the major coal deposits of each age consistently are in two belts which extend to high latitudes. Equatorward from the two coal belts are two—or at times only one—evaporite belts. The widths of the coal and evaporite belts fluctuate through time. When the evaporite belts have the greatest latitudinal spread (up to 125°), the earth must have been very warm; these are termed "evaporite maxima." The coal belts are very narrow during evaporite maxima. When the coal belts are broadest, the evaporite belts are restricted (about 60°) and glaciation is most common. These periods are called "glacial maxima." The great fluctuation of the widths of these belts illustrates the second, or time-related pattern: specifically, the average world temperature has changed considerably during earth history, and the changes are episodic; ultimately a periodicity may be established. The climate changes do not appear to be related to orogeny. In the Northern Hemisphere, the northern limit of evaporite deposition through time fluctuates from about lat 40° N. to lat 83° N. In the Southern Hemisphere, the fluctuation is slight, ranging from lat 20° S. to lat 40° S. This northward "offset" of evaporite zones is similar to that of the present northward offset of the meteorological equator. The largest areas of desert eolian deposits are in two circumglobal belts: one extends from Spanish Sahara to India, and from Southern California to the Rio Grande Valley; the other crosses South America from Peru and Chile to southern Brazil, extends across southern Africa, and continues across most of Australia. The two belts where the known desert eolian sandstones are concentrated coincide with today's horse-latitude, or desert, belts. It was found that 95 percent of all ancient evaporites, Proterozoic to the present, are in areas which *today* receive less than 100 cm of annual rainfall. Thus the evaporites are in areas which today have the driest air currents. The significance of this discovery is that the planetary wind- and ocean-current circulation patterns have undergone little or no important change since middle Proterozoic time. This is not possible unless the positions of the rotational pole, ocean basins, and continents also have been the same for the past 800–1,000 m.y. An independent test of these statements is provided by a study of the tectonic history of the Arctic and North Atlantic Oceans and the relation of this history to evaporite deposition in Eurasia and North America. This study demonstrated that the Gulf Stream and North Atlantic Drift have been in existence since middle Proterozoic time. Thus the Atlantic and Arctic basins appear to be very ancient features. In view of the results presented here, modern concepts of continental drift (the new global tectonics) and polar wandering require complete reevaluation unless some other explanation can be found for (1) the evaporite- and coal-distribution patterns and (2) the causal relations of these patterns to the modern planetary ocean- and wind-circulation system, and to the present rotational pole.

37

Reprinted from *Geol. Rundsch.* 62:814–840 (1973)

Continental drift and the origin of certain mineral deposits

By ERICH BLISSENBACH and RAINER FELLERER, Hannover [*])

Introduction

Continental drift as conceived by WEGENER (1912) has become widely accepted, in principles. Sea-floor spreading and plate tectonics are considered by a majority of investigators as the mechanism involved. (These and related hypotheses are summarized and discussed by VINE & HESS, 1970; WILSON, 1970, and COULOMB, 1972, in the German literature by KRAUS, 1971.)

The causes of drift are believed to rest in convection cells within the earth's mantle as recently described by MORGAN (1972) and NELSON & TEMPLE (1972). The new global tectonics have not remained without criticism: MEYERHOFF & MEYERHOFF (1972) have summarized the arguments against drift and, in particular, against sea-floor spreading and plate tectonics. KEITH (1972) has even advanced the hypothesis of an opposite model implying a converging sea floor. In spite of the present authors' general acceptance of drift, no specific cause for it is supported in the following, and composition of crust and mantle as well as isostasy remain unconsidered. Furthermore, mineral deposits described below need not form exclusively in the course of continental drift but may originate elsewhere and at other times under a favorable interaction of genetic prerequisites.

We assume that trends for the origin of certain mineral deposits evolve in the course of continental drift. This was suggested by SCHNEIDER (1971), more specifically for the hydrocarbon potential by McDOWELL (1971) and for the metalliferous muds of the Red Sea type by BLISSENBACH (1972). The origin of porphyritic copper ore was explained in connection with a late stage of drift by SILLITOE (1972). (SCHULING, 1967 tried to delineate tin belts on a reconstructed Pangaea supercontinent prior to its breakup and RUNNELS, 1970 expects such trend and analyses to help in forecasting the prospects in areas insufficiently known, such as the Antarctic. Evidence for large ore-mineral provinces prior to rifting and subsequent drift has recently been forwarded by PETRASCHEK, 1972). We commence by presenting a theoretical model of continental breakup and subsequent drift together with typical sedimentation and attempt to offer examples from the present and the geological past. This is followed by an account what mineral trends prevail during certain stages. For this purpose it is sub-divided into e a r l y and a d v a n c e d g r a b e n s t a g e s and early, ad-vanced and l a t e d r i f t s t a g e s, respectively.

The authors are indebted to PREUSSAG AG for permitting this publication and grate-fully acknowledge the advice of their colleagues.

Drift and sedimentation model

Continental drift begins with the e a r l y g r a b e n s t a g e exhibiting the characteristics of tensional faulting (Fig. 1). This is commonly preceded by regional upwarping as evidenced along the East African rift (interpreted by NELSON & TEMPLE, 1972 as resulting from a secondary convection cell) or the Upper Rhine graben (ILLIES, 1970); the Basin-Range province of western North America believed to be related to the underlying East Pacific rise (MENARD, 1964) may represent another example for an early graben stage. According to ELDERS et al. (1972) a compilation of geothermal, geodetic and tectonic data and trends suggests that the continental crust along the Imperial Valley and the Gulf of California is rifted apart. Along subparallel faults central blocks subside with their marginal flanks rising. In the course of continued widening a central rift or rift-in-rift may develop. Grabens are commonly offset laterally and major system often composed of several graben units arranged en echelon. Through further opening double or multiple grabens may thus form leaving in between major horst blocks (e. g. Danakil Alps of the Ethiopian Afar triangle). It is a plausible assumption that remnants of continental crust, such as the islands of Madagascar or Ceylon or the Rockall plateau, have become separated from a mother continent in this way. Continued opening of the East African rift would lead to a similar separation (DIETZ & HOLDEN, 1970, p. 38).

Photographs of the Mars surface taken by Mariner 9 reveal graben-like structures comparable in size and extension to the East African rift (HAMMOND, 1972). This could indicate the universal nature of such features, at least on celestial bodies of similar composition and subjected to similar forces.

The lowlands within the graben system receive such coarse to fine-clastic sediments as they form depending mainly on climate and relief. Fanglomerates are typical along the major fault escarpments. In more arid regions such es part of the Basin-Range province and of the East African rift redbeds, arkosic deposits and impure evaporites of playa lakes are widespread. The principal faults extend to great depth, possible the base of the crust, where magma chambers exist or magma is generated by the processes leading to the updoming and rifting. The sediments of the early graben stage are therefore commonly associated with volcanic rocks. In the inital phase of rifting stratified basalt attaining a composite thickness of several hundreds or thousands of meters may erupt as in the areas of the present Ethiopian and the Colorado plateaux. The lower Miocene conglo-merates of the Suez graben and the Triassic Newark series of the North American Atlantic coast with the intercalated diabases are examples for sequences typical for the early graben stage of former geological times. Along lines of weakness

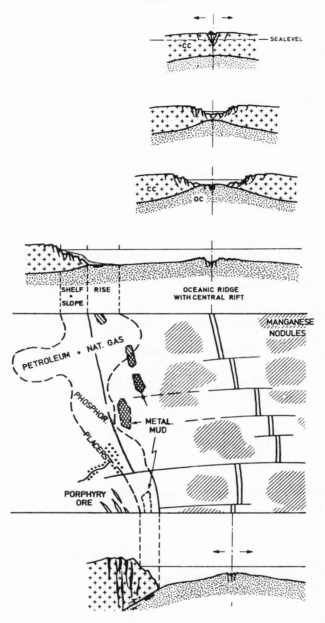

Fig. 1. Continental drift and mineral deposits. From top to bottom: (idealized cross-section through upper crust at) early graben stage, advanced graben stage, early drift stage, advanced drift stage with (plan view of) occurrences of principal mineral deposits (simplified), late drift stage (with generation of porphyry ore, modified after SILLITOE, 1972) (no scale) cc continental crust; oc oceanic crust.

carbonatites tend to be emplaced, in the opinion of Petrascheck (1972), who attributes the Mozambique occurrences to the rifting and formation of the East African continental margin and explains the Kaiserstuhl as similar feature within the Rhine graben.

The a d v a n c e d g r a b e n s t a g e is reached when opening and subsidence have developed sufficiently to let the sea enter (Fig. 1). Due to the predominantly subparallel graben margins, narrow elongated branches of the sea form, e. g. the Red Sea before the present wider opening (and possibly the southern-most segment of the East African rift in about 50 million years, according to Dietz & Holden, 1970, p. 38). Evaporites, mainly rock salt, are the most typical sediments, especially in low latitudes. The deposits are commonly pure and attain great thickness. The Miocene series of the Red Sea with salt up to 5,000 meters thick (Lowell & Genik, 1972) is an outstanding example. Many other grabens such as the Upper Rhine graben contain evaporitic sequences from this stage. In recent years numerous occurrences of evaporite formations dating back to the embryonic opening of the North Atlantic in Mesozoic times and now flanking the continental margins have become known (Pautot et al., 1970; Rona, 1970; Schneider & Johnson, 1970). Landwards the chemical deposits interfinger with clastic sediments from the nearby land or they end abruptly, as often in tropical seas, at reefs forming effective sediment-trapping dams. Pelagic deposition prevails outside the areas of clastic and chemical sedimentation. Where the bottom morphology does not permit the exchange of oxygen-depleted water, deposits rich in organic remains may accumulate. Schneider & Johnson (1970) report sapropelitic shale from the Mesozoic sequence underlying the western North Atlantic and follow a strong environmental analogy between the Cretaceous Atlantic and the Holocene Red Sea where similar sediments were encountered. At site 105, located in the western North Atlantic and drilled in the course of the Deep Sea Drilling Project, a 100 meters section of black clay rich in carbonaceous matter (some of it burnable) of lower Cretaceous age was penetrated (Hollister et al., 1972).

According to our definition the (advanced) graben stage ends and the e a r l y d r i f t s t a g e begins when the continental crust is torn apart (Fig. 1). Shelves form along the two sides of the central rift and undergo development through sedimentation and erosion, tectonic changes such as folding and block faulting or large-scale gravitational displacement. These processes have been described comprehensively (e. g. Heezen et al., 1959; Dietz, 1964; Stanley, 1969; Emery et al., 1970; and Schott, 1970).

The emplacement of basalt from the upper mantle in the central rift to form new oceanic crust is believed to be of paramount importance for the scope of our investigations. The Red Sea (for the latest several ten thousands of years and probably the near geological future) is regarded as the type state of early drift and the sediments deposited are therefore taken as typical for this stage.

By continued separation of the continents the a d v a n c e d d r i f t s t a g e is reached (Fig. 1). The Atlantic Ocean may serve as example. In the course of drifting the central rift gradually develops into an oceanic ridge. While the shelves undergo the changes indicated above, new units, the continental rises may form at the shelves' foot, huge fans such as the Ganges cone. Sediment transported beyond, mostly by turbidity currents, builds up the vast abyssal plains (turbidite-fill stage of SCHNEIDER, 1971; HORN et al., 1972). Pelagic sedimentation (clays and oozes) predominates over the ocean bottom with local abundance of concretions of various compositions.

The type of deposits depends on parameters such as water temperature, salinity, currents and variables such as the supply of material through volcanic activity or the carbonate compensation level. Horizontal and vertical sequences of facies types have been demonstrated by FRAKES & KEMP (1972). Contrary to the low sedimentation rate of 0.1 centimeter per 1,000 years over much of the ocean floor, there is more rapid and less uniform deposition in the central rift of the oceanic ridge and its surroundings. Ponded sediments described from the central North Atlantic by VAN ANDEL & KOMAR (1969) appear to be typical there.

In the course of continued drifting crustal plates may eventually collide thereby ending the late drift stage. This can imply the subduction of oceanic crust and the overthrust of continental crust (Fig. 1) as much of the western margins of the American continents is interpreted; or, in the case of two colliding continents, the compression and uplift of the sedimentary wedge in between as the Himalaya range may have originated.

The subdivision into graben and drift stages does not require their time-space continuity: different stages of graben and/or drift development may exist along major systems. The East African rift serves as prominent example. It extends over 80 degrees latitude, possibly from the Kalahari desert (REEVES, 1972) through East Africa, the Red Sea and the Gulf of Aden with connection to the Carlsberg ridge; through a western branch across the Mediterranean to the Upper Rhine and eventually the Oslo graben (ILLIES, 1970). The northernmost segment underwent its major movement in Permian (RAMBERG & SMITHSON, 1971), the Rine graben in Tertiary as also most of the other parts of the system (although it is not sufficiently known to what extent these movements occurred along older tectonic lines). The northern and the southern parts exhibit smaller amounts of movement, not exceeding the early or advanced graben stage while in the central part (Red Sea, Afar triangle, Gulf of Aden) transitions between the graben and the early and even the advanced drift stages exist. (A different degree of development may be due to the individual rotation of the participating plates, in the opinion of ALLAN, 1970).

The differentiated opening of the Atlantic can be taken as an example for the development of graben and drift stages. The North Atlantic underwent the early graben stage in Triassic, the advanced graben and the early drift stages in Jurassic. The advanced drift stage was reached in lower Cretaceous times when South America and Africa were still in the early graben stage (Fig. 2) (for a summary of recent findings on the opening of the South Atlantic, see BEURLEN, 1972).

The assumed development leading to the present position of Madagascar is another example. According to TARLING (1972) Madagascar was probably separated from Mozambique in such a way that a "bulge" of Karroo lavas and later sediments underwent initial fracturing in Jurassic. The early and advanced graben stages were followed by crustal separation, the early drift, in mid-Cretaceous.

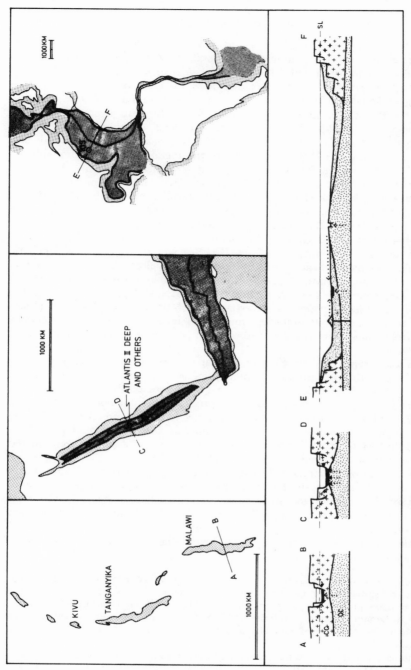

Fig. 2. Originating sites of metalliferous mud. Above, left: East African rift (western branch) (modified after Girdler & Sowerbutts, 1970) — precursory (minor) phase. Center: Red Sea — mature (major) phase. Right: Atlantic (at late Cretaceous times) (105 = site 105, leg 11 of JOIDES drilling program) — subsequent (minor)phase. Darker shading — oceanic crust from drifting. Below: cross-section models (not to scale) of above 3 phases. Black — metalliferous mud. Dotted arrows — schematic ways of feeder solutions. cc continental crust; oc oceanic crust.

Trends for mineral accumulations and economic aspects

In the course of volcanic activities and mostly postvolcanic processes of the early graben stage commercial mineral deposits may be expected to accumulate. The water of Lake Kivu of East Africa, an example of the early graben stage, is enriched in zinc as described by DEGENS et al. (1972). The generating of sphalerite globules is still in progress. In Lake Malawi, another example for this stage, iron silicates were found forming over an extensive bottom area (Prof. G. MÜLLER, Heidelberg, personal communication) [1]) comparable to the minerals formed presently in the Red Sea metalliferous muds and reported by BISCHOFF (1972). In these cases the metal supply can probably be attributed to the cycling of water under the influence of positive heat flow anomalies as known from the Red Sea, Lake Kivu and also proved by VON HERZEN & VACQUIER (1967) for Lake Malawi. DEGENS et al. (1971) report that except for one value other heat flow measurements on Lake Tanganyika bottom were normal. This and the absence of mineralizations, and the opposite combination, high heat flow and mineral enrichments in Lake Kivu and Lake Malawi sediments, and in the central rift of the Red Sea would support such a genetic relationship. Anomalous heat flow and mineralizations also coincide in the Salton Sea area of western North America, an early graben system (ELDERS et al., 1972). A connection between high heat flow and metal precipitates over the East Pacific rise results from the investigations of BOSTROM & PETERSON (1966) although these authors attribute the metal enrichments to ascending solutions derived from magmatic sources rather than to the supply through water cycling.

The Malawi type of iron enrichment can be regarded as the precursory (minor) phase of the precipitation of metalliferous muds culminating in the Red Sea type deposition (Fig. 2). The Miocene Pb-Zn ores of the Um Gheig area, Egypt, are believed to have originated as precipitates from hot brines (ANWAR et al., 1972). The confirmation of a similar origin of the ore deposits known from the opposite coastal strip in Saudi Arabia, at Um Laji, would point to a large distribution of metalliferous muds of the precursory phase accumulated during the early or advanced graben stages of the Red Sea.

The utilization of precipitate deposits of the advanced graben stage (mainly rock salt and gypsum) is still small. Sulfur forming above salt domes has been produced from the shelf since 1960 and had a share in the total production of 12% by 1969. The sulfur is mined by liquifying it with hot steam (Frash method), a process which may eventually be employed, with essential modifications, for the lode mining from below the ocean floor in the more distant future.

One group of mineral deposits of commercial interest occurring on the seabed or within the subsurface does not originate in connection with continental drift and therefore remains outside the scope of this presentation. It includes offshore placers, sources of rutile and ilmenite, cassiterite, gold, platinum and diamonds valued together at more than 200 million DM annually. Some mineral deposits mined below the sea

---

[1]) Preliminary finding of a campaign carried out by Prof. G. MÜLLER and team from Heidelberg University in May-June 1972, in cooperation with PREUSSAG AG, Hannover.

floor are merely extensions of land occurrences as several coal seams and iron ores. The underwater barite mining off Castle Island, Southeast Alaska, is similar in principles but shows already a way to marine lode mining, at least in shallow water (Smith, 1972).

A second group of deposits or potentials, phosphorite and organic mud, and the offshore hydrocarbons (the latter estimated to be worth 40 billion DM annually or about 90% of the value of mineral resources recovered from the offshore) do not necessarily accumulate in the course of drift but may do so and are therefore mentioned only briefly in the following.

The third and last group comprises metalliferous mud and manganese nodules. They are more exclusively associated with stages of continental drift and are therefore discussed in more detail.

For a general description of sea-bed mineral deposits, reference is made to Seibold (1970).

The Upper Rhine and the Gulf of Suez are examples of advanced graben stages from which sedimentary fills p e t r o l e u m and n a t u r a l g a s are produced. The sections of the offshore and coastal Gulf of Guinea having produced in 1972 about 110 million tons of petroleum also originated partly in the advanced graben stage of late Mesozoic times. The role of increased heat flow observed within most graben structures for generating hydrocarbon fields has been emphasized (Klemme, 1972). There are many untested areas in the world having passed this stage of development. The prospects for finding new petroleum provinces through research in this direction are therefore rated as encouraging. Oil and gas may continue to be generated following the breakup of the continental crust and the beginning of the e a r l y d r i f t s t a g e. Their origin is generally understood to depend on the interaction of source and reservoir rocks, and tectonics. The frequent occurrence of organic-rich sediments has already been mentioned. Litoral sands and, in tropical seas, porous reefs are common marginal facies and well suited as reservoirs. Movements within the continental margins (folding, block faulting, gravitational displacement) are often recorded. Oil and gas may also accumulate atectonically by hydrocarbons migrating from the lutites of deeper locations upwards into the porous sandstone or reefal limestone closer to the shore. The petroleum potential of the continental margins has been rated by Boigk & Porth (1970), and Hedberg (1970). The chances of the continental rise in particular were discussed by Emery (1969). A possible influence of continental drift and rifting for the development of special basins with petroleum potentials is assumed by McDowell (1971) which hypothesis could specifically point to new prospects.

The annual petroleum production from offshore areas presently amounts to about 450 million tons, close to 20% of the total. Until 1985 the offshore share is estimated to rise to about 2 billion tons annually or 35—40% of the total (Cnexo, 1972). The estimate that approximately 40% of all petroleum reserves occur under the sea may already be outdated in the light of new findings: the discovery of oil shows in sediments of the deep Gulf of Mexico (under 3,600 meters of water) and, more recently, of migrated hydrocarbons in deposits of the Pacific (Shatsky rise, water depth at drilling site about 4,300 meters) (McIver, 1972). Shell's exploration manager describes the state of knowledge and the resulting outlook as follows: Attractive hydrocarbon prospects do not end or even diminish at the edge of continental shelves. It is forecast that even in this century the deeper oceans will satisfy an important part of the world's hunger for energy (Beck, 1972).

Oil and gas may therefore be generated over a wide range of time and environments, from fills of the advanced graben stage to shelf and rise deposits
of the early drift and, although not sufficiently understood, to deep-sea sediments of the advanced and the later drift stages.

Phosphorite and organic mud (Fig. 1) commonly occurring together
on the shelf and its slope are not exploited commercially; they may, lowever,
develop into future mineral reserves. Both types of deposits form preferably under
the influence of mineral-rich upwelling water. Phosphorite (comprising ooze,
sand- to larger-size particles and crusts) is precipitated in many shelf areas, e. g.
along the West and East coast of both Americas, off the Asian continent and
South and West Africa. The deposits may become commercial in future where
a high demand for phosphate fertilizers coincides with a lack of high-grade land
occurrences of such minerals. Offshore phosphorite as the Californian deposits
may contain up to 30% P_2O_5 and is then comparable to land mines. The phosphorite off California has been estimated to amount to 1 billion tons with possibly
one tenth exploitable, according to EMERY (1960). (More recent investigators
warn against over-estimating the potential: SORENSON & MEAD (1969) report
that 85% of the samples taken during an extensive survey off the Californian
coast run less than 1% P_2O_5, the average being 3.3%).

Other precipitate deposits, including glauconite sands and ferro-manganese nodules,
may occur similarly on the shelf but do not appear economically attractive in the near
future. Oolithic rocks having formed in shallow water are important iron ores (Jurassic
Minette of Lorraine or the Silurian Clinton of the USA) but originate preferably in
lakes and along the shores of inland seas from metal solutions derived from weathering
and are therefore not considered as mineral deposits generated in the course of continental drift.

Organic mud, in fossil state known as black shale, may become a future
source of one or more of the heavy metals U, Th, Mo, V, Ni and Zn if enriched
beyond the presently known contents. The mud forms by a complex interaction
of metal supply from sea water, biological activity and metal concentrations in
organisms, release following decomposition, oxygen-depletion and reducing bottom conditions, and the eventual precipitation of metal compounds, especially
with humic complexes and adsorption by plant remains. For an account on the
possible economic importance as a source of uranium, reference is made to
MAUCHER (1962). Organic mud occurs on the shelf (approxim. 100—200 meters
of depth), several tens of kilometers from the shore, for example along the
western coasts of Africa and South America in shallow elongated basins, containing from a few to more than 10 meters of this sediment.

An interestang occurrence is off the coast of South West Africa (AVILOW &
GERSHANOVICH, 1970; CALVERT & PRICE, 1971; BATURIN, 1971; BATURIN et al.,
1971). The average content of U in these sediments is 10 ppm, within the phosphate intercalations enriched to 100 ppm, in diagenetically altered vertebra remains to 670 ppm with the interstitial water containing 0.65 ppm of uranium.
Organic carbon is present with 25%, at maximum, Ni up to 445 ppm and Zn up
to 337 ppm. In general the sediment is a diatomaceous ooze. Fossil occurrences
of organic muds include: the Cambrian Kolm shales of Sweden (averaging 0.03%
U, in nodules up to 0.4%) temporarily exploited and the Devonian Chattanooga
shale of Tennessee (averaging 0.006% U, in lenses up to 0.7%).

The humic subfacies of organic muds may grade oceanwards into a sapro-pelitic subfacies known from the stratigraphic column as oil shale and regarded as huge potential reserve. It has been announced recently that the US Govern-ment expects the shale-oil production from deposits in Colorado, Wyoming and Utah to start by 1976 with an annual rate of 2.5 million tons of oil, rising to about 50 million tons in 1985 (ANONYMOUS, 1972).

With the e a r l y d r i f t s t a g e optimum prerequisites for the origin of m e -t a l l i f e r o u s m u d prevail, the sediment first encountered in the Atlantis II deep of the Red Sea (Fig. 1). The mineral deposit is regarded as representing the mature (major) phase of this type sedimentation (Fig. 2). The occurrence and its geoscientific setting have been described in numerous special studies edited by DEGENS & ROSS (1969) and FALCON et al. (1970): deeps along the central rift contain hot and highly saturated brines (up to about 60° C and 30% NaCl). In the Atlantic II deep with an area of approxim. 50 square kilometers, stratified multicolored sediments with a thickness of several tens of meters rest on basalt, below the brine and about 2,000 meters of water. They consist of alternating manganites, iron oxides, the sulfides of Fe, Cu and Zn and intercalations of anhydrite or montmorillonite. Thin beds may exhibit concentrations of up to 20% Zn (of the saltfree dry substance). However, the overall contents of the value metals, Cu and Zn, do not exceed 1 and 5%, respectively. Cd, Ag, Pb and Au are present in minor amounts. Previous investigators have estimated the metal value of the uppermost 10 meters to be US $ 2.5 billion not allowing for production and processing losses. Scientific institutes and the industry are following up the assessment of this mineral occurrence (BÄCKER & SCHOELL, 1972). (The technical aspect of a production method has already been the sub-ject of various studies, e. g. BOES & BADE, 1971).

The metalliferous mud is believed to have accumulated by the following processes: water sinking into the fissured sea bottom becomes heated (especially under the increased heat flow), leaches salt from the flanking evaporite for-mations and metal ions from mineral-rich basalt aquifers or sediments high in Zn, Cu, V and Mo as recently encountered in the vicinity of the brine deeps (DSDP, 1972 b). The enriched brine is ejected again in pockets of the sea floor. Upon contact with the sea water the metallic compounds are precipitated as oxides and sulfides depending on the prevailing chemical environment. In addition exhalations associated with the submarine volcanism are likely to contribute metallic and other substances. This type of sedimentation has been going on for several ten thousands of years and also continues et present (BREWER et al., 1971).

The state of knowledge will probably be much improved following the eva-luation of data and material from the recently completed leg 23 of the DSDP in the Red Sea and from the two cruises of R/V VALDIVIA in 1971 and 1972 [2]). The latter campaigns had included, among the objectives, an economic evalua-tion of the metalliferous mud within the Atlantis II deep and the search for similar deposits elsewhere in the Red Sea and in the Gulf of Aden. Several new deeps with or without brines and with or without metal-enriched sediments were discovered in the Red Sea but none was encountered in the Gulf of

[2]) Supported by the Ministry of Education and Science of the Federal Republic of Germany.

Aden (BÄCKER & SCHOELL, 1972). This finding supports the hypothesis that the Red Sea is presently in a mature phase of originating metalliferous mud, the early drift stage (BLISSENBACH, 1972) (Fig. 2), whereas this state has already passed if it was ever present in the more developed Gulf of Aden (which could be regarded as in the advanced drift stage, according to our definition). The preferential generation of the mud during the early drift stage is easily explained by an optimum interaction of the prerequisites of anomalous heat flow promoting water cycling and metal deposition, volcanic exhalations etc. taking place in a narrow zone, in particular in sea-floor pockets formed by rifting and cross-faulting.

Mineral concentrations are less readily expected during a stage when the ocean has considerably widened leading to a disseminating effect. The central rift of a widely opened ocean is not, in our opinion, the zone to search for economic deposits, contrary to the belief of some investigators (e. g. WENK, 1969, p. 84—85). The scarce information from sampling such oceanic ridges is contradictory: There are no trends for interesting mineral accumulations, according to data by HOROWITZ (1970), but enrichments of Fe, Mn, Cu, Cr, Ni and Pb on the East Pacific rise as reported by BOSTROM & PETERSON, 1966. In the opinion of PETRASCHECK (1972) basaltic magmas of the oceanic crust as observed on Iceland are commonly poor in ore metals.

The role of the salt and a necessity of nearby evaporite formations as prerequisite are not quite understood. It is obvious, however, that a brine increases the solving potential for metals.

BONATTI et al. (1972) report the occurrence of a Fe-Mn-Ba deposit formed in the Pleistocene in the (then submerged) Northern Afar rift of Ethiopia, a branch of the Red Sea system. The deposits contain again nontronite (as presently forming in Lake Malawi and the Red Sea) and may be regarded as example of metalliferous deposit intermediate between the precursory and the mature phase. In the central Afar depression, near Dalol, a drill core of halite with inclusions of copper minerals was recovered (Dr. W. Thormann, formerly Ministry of Mines Addis Ababa, personal communication).

The Atlantis II brine contains traces of methane possibly derived from the flanking Tertiary, in particular where oil and gas are believed to be generated. (Shows have also been encountered by petroleum exploratory wells drilled along the eastern and western onshore and offshore areas of the Red Sea.) In our opinion it has not been sufficiently investigated whether connate water from the hydrocarbon forming or compaction fluids ejected from the evaporite formations contribute to the Red Sea brines.

At this point a brief account on the observed mineral supply by marine volcanism appears appropriate: several submerged volcanoes are found to emit gaseous exhalations and thermal springs, predominantly in the course of postvolcanic activities. Springs along the volcanoes of Matupi Harbor of New Britain, SW Pacific, contain Fe, Mn and Zn in concentrations similar to the Red Sea brines (FERGUSON & LAMBERT, 1972) and appear to represent sea water circulated through hot Quaternary volcanics thereby becoming enriched in leached metals under the influence of volcanic gasses. Gas emissions and powerful jets of mineralized water have been described from the submerged Banu Wuhu volcano of the island chain north of Celebes by ZELENOV (1964) with precipitations of colloidal sulfur and iron-manganese muds. They also contain: V and Sr (tenths of a per cent) Mo, Cu, Zn, Ba (hundredths of a per cent) and Ni, Co, Zr, Pb, Sn, Ge, Ga, Y and Yb (thousandths of a per cent). Exhalations in the shallow water around Vulcano of the western Mediterranean lead to a cementation of sands and tuffs by pyrite and marcasite, often in the form of framboid globules (HONNOREZ, 1969; SCHULZ, 1970). Such minerals are also formed in other recent sedimentary or volcanic deposits (e. g. Pauzhetka, Kamchatka Peninsula) and occur in ancient sediments, according to those investigators, such as the Rammelsberg (Germany), Chattanooga (U.S.A.), and Mount Isa (Australia) ores. In the submerged caldera of Santorini volcano of the eastern Mediterranean, iron-rich springs lead to the formation of a ferruginous layer with raised contents in Pb (about 0.03%) and Zn

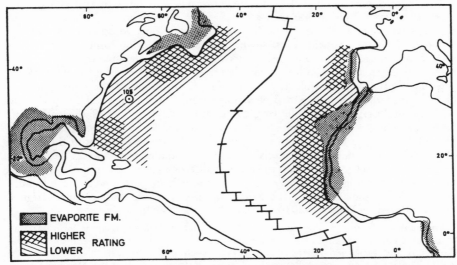

Fig. 3. North Atlantic — hypothetical zoning of areas prospective for metalliferous mud (Evaporite Formations after SCHNEIDER & JOHNSON, 1970).

(about 0.014%) (BUTUZOVA, 1969). The uppermost layer indicates a high precipitation rate of 30—35 centimeters per 1,000 years (Pacific deep-sea rates 0.1 centimeter or less). The mud is believed to correspond in origin to the iron ores of the Devonian Lahn-Dill prototype (Germany). In the case of an Antarctic volcano ELDERFIELD (1972) has demonstrated changes in the chemical composition of sea water due to the volcanic mineral discharge.

These observations have proved that exhalations of varying compositions in time and space discharging into the surrounding sea may contribute to generate metal-rich deposits. Although volcanoes need not occur exclusively in connection with continental drift they are nevertheless commonly associated with such processes and related crustal adjustment.

The section penetrated at site 105 of leg 11 of the DSDP, located in the western North Atlantic (Fig. 3), contains 50 meters of probably late Cretaceous multicolored sediments with abundant Fe and Mn oxide minerals and local sphalerite and pyrite concentrations; in one layer 3.6% Zn were analyzed. In the underlying section a thin veinlet of native Cu was encountered (HOLLISTER et al., 1972, LANCELOT et al., 1972). The multicolored deposits show a striking

53 Geologische Rundschau, Bd. 62

resemblance to the metalliferous mud of the Red Sea and suggest similar processes although no evidence for brines has been found. Other multicolored beds were not encountered in the drill holes of the vicinity although they penetrated the respective stratigraphic interval. This could be explained by the existence of a paleopocket holding the metalliferous deposits (comparable to the present Red Sea deeps) at site 105. A core obtained during leg 14 from the eastern Atlantic shows similar colorful banding and may as well be related in origin (Dr. J. D. HATHAWAY, personal communication).

It is tempting to assume that the metalliferous sediments of site 105 formed when the North Atlantic was in a state of early drift comparable to the present Red Sea. However, this cannot be the case as the respective deposits do not rest on newly generated oceanic crust (as in the Red Sea) but on top of a 350 meters section of Jurassic-Cretaceous sediments (overlying basalt) suggesting that they were laid down 60—70 million years later. A speculative reconstruction of the site-105 position at that time is offered in fig. 2. Forming of the metal-enriched deposits may then be understood as a subsequent (minor) phase of the precipitation of metalliferous mud, following the mature (major) phase of the early drift stage (present Red Sea) which, in turn, was preceded by a precursory (minor) phase (in Holocene East African rift). Although the metal contents in the section of site 105 are far below commercial interest, they might nevertheless be indicate for higher concentrations nearby. (The metalliferous sediments and the mineralization of the section drilled may also have formed, in part, under the influence of volcanism associated with the Bermuda rise, as could be suspected from a section offered by HEEZEN & HOLLISTER, 1971, fig. 13.6, p. 546).

The continued study of the conditions under which metalliferous muds form has lead to the recognition of an increasing number of ore deposits dating back to the pre-Cambrian and explained partly or entirely by such processes. These occurrences include: The Pb-Ag-Zn ores of Bawdwin (Birma) (HANNAK, 1972) or the Massive Sulfide deposits of Jerome, Arizona (ANDERSON & NASH, 1972). The Rammelsberg (Germany) Chattanooga (U.S.A.) and Mount Isa (Australia) ores have already been suspected previously as belonging to this type of origin.

The embryonic North Atlantic of Jurassic-Cretaceous times probably exhibited the ideal prerequisites for generating metalliferous muds: special deeps as resulting from variable drift; concentrated thermal and exhalative effects in a narrow central zone within an overall trough with closed northern and southern ends, and evaporite formations flanking the central rift. The prerequisites as having evolved from the above discussion permit to draw some preliminary conclusions as to the discovery chances of such deposits of the geological past. Fig. 3 is an attempt to tentatively define areas of a different degree of prospect for the North Atlantic. As will be seen the prospective areas are theoretically restricted to the continental margins and adjoining zones. The better rating is attributed to the areas off the known evaporite formations having originated during the embryonic opening of the ocean. In general, the older the deposits are, the deeper and the more indurated they tend to occur, more likely to be covered by a thick overburden making the utilization increasingly problematic to prohibitive. On the other hand even Paleozoic sediments are known not to have become lithified in certain areas and overburden once covering metalliferous deposits could have been removed by erosion or by mass displacement. RONA

(1971) has drawn the attention to the intersection of fracture zones of the ocean basins with continental margins as logical sites for mineral exploration. Indications as to variable sea-floor spreading as reported by Taylor et al. (1971) may serve as additional criteria.

With regard to a recovery of metals from lithified deposits, it has been speculated that atomic explosions could be used to shatter a mineralization zone prior to employ leaching processes. The estimated costs fo a well, a small-scale nuclear device and the removal of radioactive byproducts on land do not appear prohibitive and are comparable to those of a shallow petroleum wildcat or a production well of medium depth. The prospects of applying the sophisticated production technique of the pretroleum industry to underground deposits of solid minerals have recently been described by Lorbach (1972). According to him, the development of an ore deposit by such methods, including the drilling of injection and production wells, chemical leaching possibly aided by bacteria, and preceded by conventional or nuclear blasting, raises the hope for an efficient, low-cost exploitation. Prospecting for the potential of the metalliferous muds appears at first sight extremely complicated but it is maintained that the prerequisites for the origin of petroleum are not simpler nor less (theoretically) contradictory (e. g. source rocks from non-aerated, reservoirs from aerated environments). Future studies should first screen the vast amount of data available, in the search for prospective areas, before special sampling programs are justified.

In the course of continued movement of the continents, the metalliferous mud deposits of a central rift become split. Although a generally symmetrical separation is to be expected, this will nevertheless be of a higher order of magnitude than the dimensions of a mineral occurrence of a few kilometers. A metalliferous deposit on one side of the drift system need not have a twin counterpart on the other side although this may hold true for the general mineralization trend.

In the course of the a d v a n c e d   d r i f t vast abyssal plains form and become the environment of another major mineral potential, the ferro-manganese nodules, or m a n g a n e s e   n o d u l e s, for thort.

They have been known since the Challenger expedition 1873—76. Murray and Renard (1891) presented them as a scientifically highly interesting phenomenon. "Die im Stillen Ozean auf dem Meeresgrunde vorkommenden Manganknollen" (Gümbel, 1878) have almost been forgotten for the following nearly 90 years, except for scarce references and comparison to terrestrial deposits; they, however, began to attract world-wide interest in the sixties because of their important contents of the value minerals Cu, Ni, Co (Mero, 1965). Ore nodules have meanwhile become the objective of intensified prospecting and the key for designing new discovery tools and utilization methods as evidenced by almost 1000 publications on this subject. Beyond this information, economic interests and state organizations, particularly in the United States, Japan, USSR, Canada, France and the Federal Republic of Germany, command extensive data which have not become available for obvious reasons.

It would be too early to attempt a comprehensive description of the highly complex problems involving manganese nodules. Many preliminary findings will have to be secured by further data, numerous parameters are not sufficiently understood in their mutual relation and other functions can well be assumed but remain to be proved. Subject to future modifications a mode of origin can already be advanced, concerning the economically interesting varieties of nodules — only these are considered in the following — whereby the overall large-scale tectonic setting, an environmental situation of the advanced drift, forms the prerequisite and the framework for nodule generation.

Concentrations of manganese and iron represent a world-wide phenomenon of the exogenic cycle. The elements range in the scale of geochemical abundance with 0.07% at place No. 13 and almost 5% at No. 4, respectively and, therefore, become available in large amounts during the decomposition of exposed plutonic rocks. On the other hand the two elements react sensitively to changes in the chemo-physical environment resulting in changes of their valence and correspondingly in solution or deposition. In general, both elements react similarly but within narrow margins differently enough to lead to spatial, mineralogical separation. Enrichments in these metals, whereby quality and quantity are essentially a function of the length of time and the constancy of environment are due to local or periodic changes of the redox potential of manganese- and iron-enriched waters circulating in the soils under specific climates, to organisms or to anorganic precipitation through the admission of oxygen, to the decomposition of complex compounds and protective colloids and, to a smaller part, to evaporation.

The widely varying environmental positions of hydroxidic metal enrichments are shown in fig. 4. Among the strictly terrestric manganese deposits only the so-called residual ores of hot alternating climates attain economic importance. Their metals are not only derived directly from the underground (manganese caps of the oxidation zone) but also transported and deposited by ground water.

The occurrences of Ghana, Cuba, Haiti, Costa Rica, East India, Brazil and the deeply metamorphized deposits of Postmasburg also belong to this category. The giant deposits of lateritic iron ores of Mayari/Cuba and Conakry/Guinea (regarded because of the Ni-contents as competitors for the deep-sea manganese nodules) are to be mentioned in this connection as well.

In the temperate to cold and humid climate ferro-manganese ores generated in the terrestric limnic surroundings at times have become of importance in Northern Europe and Canada. Some lake occurrences in the USA, Canada, Scandinavia and Eastern Europe are comparable in origin and facies to the marine manganese nodules (HARRISS & TOUP, 1970; SCHOETTLE & FRIEDMAN, 1971; TERASMAE, 1971; ROSSMANN et al., 1972; CRONAN & THOMAS, 1972; VARENT-SOV, 1971; HOCKINGS et al., 1969).

In most cases the derival of metals is easily explained by surface and subsurface water transport from hinterland sources, complex anorganic-organic, partly rhythmic precipitation in eutrophic environments as hydrogels and/or by halmyrolysis of limnic muds or clastic deposits. Manganese nodules of this type are characterized, contrary to deep-sea nodules, by a widely varying Fe/Mn ratio and a low rate of value metals (Cu, Ni, Co). Their age is postglacial and, therefore, geologically young; from their size of several centimeters a high rate of growth follows.

Nodules of the shallow sea in the fiords of British Columbia and Scotland (GRILL et al., 1968; CALVERT & PRICE, 1970) and the smaller concretions of the Baltic and the Black Sea (MANHEIM, 1965; SEVASTJANOV & VOLKOV, 1967, SCHTERENBERG, 1971) have a closely related origin.

The two largest manganese deposits of the continents, Tschiaturi and Nikopol/USSR, represent a type of their own having originated in agitated and aerated sea water leading to the formation of oolites. Similar to the deep-sea manganese nodules, their deposition requires a tectonic setting of large-scale stable conditions over long periods of time: low relief energy on a deeply weathered continent, little clastic sedimentation

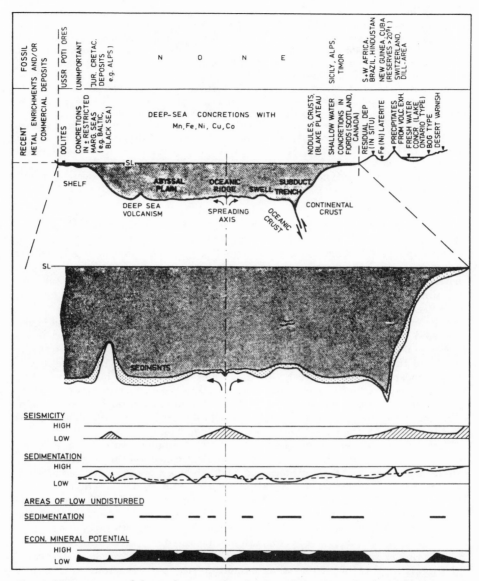

Fig. 4. Manganese nodules: schematic distribution of type enrichments of manganese on continents and ocean bottom (above), and nodule occurrences along latitudinal section through central Pacific (theoretical, highly simplified) and environmental relations (below) explanation to sedimentation: broken line — primary sedimentation; solid line — total sedimentation = primary plus secondary (e.g. slumping minus erosion) vertical scales greatly exaggerated.

and carbonate precipitation, restricted but well aerated shallow branches of the sea in 40—80 km distance from the shore with a subsiding tendency in which basins the protective colloids and humic substances become decomposed and the manganese precipitated. Their contents in Cu, Ni and Co are always low, in the range of ppm, comparable to those of fresh-water and shallow-marine nodules. Such depositional conditions have prevailed in times of world-wide transgressions such as in Silurian, Devonian, Jurassic, Upper Cretaceous and Lower Tertiary. Recent occurrences of this type and of large extent have not been observed. Fossil manganese nodules are rare and known only from the Jurassic of the Alps and Sicily (GARRISON & FISCHER, 1969; JENKYNS, 1970; GERMANN, 1971 among others), from the Cretaceous of Timor and Borneo (MOLENGRAAFF, 1922; EL WAKEEL & RILEY, 1961; AUDLEY-CHARLES, 1972) as well as from South Dakota (GRUNWALD, 1964). Some more deposits have been described but it remains uncertain whether they can be classified under this type. The value metals are always one order of magnitude lower than in recent deep-sea concretions and, based on numerous analyses, this can already be taken as criterion for a non-abyssal origin.

Genuine deep-sea nodules are hardly expected on the continents in fossil state as oceanic crust and its sedimentary cover are rarely incorporated into continents; if this happens it is commonly accompanied by metamorphic and palingenetic processes. In addition manganese nodules are highly instable concretions and are, therefore, not likely to be preserved: under the reducing conditions already prevailing under a cover of several centimeters or meters of sediment they tend to be dissolved again.

The amount of nodules occurring on the Pacific bottom alone has been estimated to be 10^{11} tons, at minimum, exceeding by far all known land deposits of Mn (4,000 $\times$), Ni (1,500 $\times$) and Cu (15 $\times$). Distributed evenly over the bottom of the Pacific 1 km² would hold only 56 tons. Actually the nodules tend to be concentrated on relatively small portions of the ocean bottom, whereby the occurrence of the economically attractive ore-type concretions depends on genetic prerequisites prevailing there (fig. 4). Within such areas 10^5 tons or more of nodules may be present on 1 km² representing, at a rate of 3% of combined Ni, Cu and Co, an amount of 3,000 tons of these metals.

These potentially economic deposits at the sediment-water interface are "two-dimensional" and directly "visible". Their prospection and exploration are less problematic than their exploitation and metal extraction. The deposits are characterized by a number of typical parameters. The origin of nodules and the interaction of individual factors leading to the formation of "good" deposits are still the subject of contraversal discussions. This applies to the derival of the metals, the transport and precipitation mechanisms, the growth processes such as rate of growth, total age, reasons for different facies etc.

The primary supply of manganese from the continents and submarine volcanic exhalations is plausible. None alone would suffice but numerous observations indicate an upgrading effect on nodule occurrences directly through volcanic activity. The deposition of manganese in the oxygen-rich bottom water and consequently the growth of nodules is probably effected anorganically through oxidation; bacterial action may have a modifying influence (in particular in the cases of fresh-water and shallow-marine deposits) but is not expected to play a major role at greater depths.

The actual growth of nodules is nowadays generally explained by the following process: low-soluble IV-Mn within a reducing environment (with organic substances and at increased pressure readily present below the sediment-water interface) becomes mobilized as II-Mn and transported upward to and beyond the interface by diffusion or compaction currents (e. g. Grasshoff, 1970). It is then oxidized and precipitated again. Ni and Cu from the seawater are preferably adsorbed by the micro- to crypto-crystallized hydrated Mn-mineral phases. Co prefers the (mostly X-ray — amorphous) iron hydrates. Ni from the numerous enclosed cosmic spherules is negligible, according to Finkelmann (1972).

Such a genetic process, the supply of Mn mostly from the underlying sediments, is also in accordance with other findings which could not be satisfactorily explained otherwise. This includes the observation that the nodules occur predominantly at the sediment surface although their rate of growth may be one order of magnitude lower than the clay deposition. There is no evidence that the nodules would be moved to the surface by the action of bottom-dwelling burrowing organisms. Even in areas without nodules there is still a Mn enrichment in the uppermost sediments. The assumed origin is also supported by the presence of nodules several tens of centimeters or, much more rarely, a few meters below the sediment top. They have probably been buried by young slumping or turbidity currents and commonly exhibit typical solution effects.

Once the nodule genesis started, the individuals (mostly with smooth upper and rough porous lower surface) continue to grow, possibly interrupted by occasional periods of solving trends, until they contact each other and thus form aggregates. There are characteristic values for the vertical growth, the socalled critical diameter, for different deposits. Upon reaching this statistical figure the nodule facies changes from spherical to flat oval and discoid. Encrustations may thus form representing a final stage of nodule genesis as the supply of Mn from below is then interrupted. Such "inactive" crusts are being covered by sediments in the course of time and may become dissolved again upon reaching a certain depth within the sediment.

Many bottom photos combined with sampling show that nodule growth is already partially prevented by thin caps of sediments as resulting from sudden rapid deposition (e. g. by turbidity currents): such nodules exhibit a dish shape. This process can be thought to continue with the marginal sediment-free zone of the nodule still growing until the central clay cap is completely enclosed. This would be a plausible explanation for the frequent obervation of an excentric "clay nucleus" within nodules and would not suggest a derival of the Mn exclusively from the sea water.

The old age of several million years and the corresponding low rate of growth of approxim. $0.1 \, cm/10^5$ years is another characteristic of economically interesting nodules. Such a rate is equivalent to only a few molecular layers per day. Age determinations after various methods are in good accordance with paleontological, paleomagnetic and geological findings, including the distribution of cosmic spherules.

Concretions with a high rate of growth exhibit a distinctly lower content of Ni, Cu and Co. Encrustations of manganese around objects of known age (deep-sea cables, oyster shells and scrap) show this clearly. Postglacial deposits of the lacrustrine, palustrine and shallow-marine environment and concretions around the roots of plants in lateritic soils are similar. Although Ni is present in much larger percentages than in sea water, the deposits contain less then deep-sea nodules. Similarly artificially grown nodules (experiments of Hawaii University) contain only traces of the interesting value metals in spite of ample supply. It follows that the manganese and iron mineral phases can only adsorb small quantities of other metals per unit time.

The prerequisites for nodule genesis at "slow-motion" speed, growing by a factor of 10^4—10^5 more slowly than poor-grade nodules, are therefore low sedimentation rate, highly quantitative separation of iron and manganese, limited supply of manganese and/or precipitation, constancy of environment over long periods of time and, for reasons of size, a high age of the deposits. The term of sedimentation rate is complex. Biogenic terrigenous deposition represents primary sedimentation, turbidity currents or slumpings a secondary one. Submarine erosion may locally interact. Apart from the resulting total sedimentation, the type of deposit appears to play an important role; the siliceous biogenic ooze, as present in a meridional belt of the North Pacific with its high porosity is believed to favor nodule generation (HORN et al., 1972) [3]). Volcanic rocks could also have a positive influence whereas carbonatic biogenic sediments of the equatorial belt do not favor the growth of ore-type nodules. Turbidite deposits, most hemi-pelagites, ice-rafted sediments and terrigenous and volcanic silts are also excluded as favorable substratum.

The quality of nodule occurrences is predominantly characterized by total deposition, facies, sediment thickness and derived chemo-physical parameters such as redox potential, diffusion speed and sediment chemistry.

Since the concretions are not only in contact with the underlying sediment but also with the water, the surrounding body of sea water must fulfil certain prerequisites, e. g. its volume must be large enough to guarantee uniform conditions over long periods of time. Bottom temperature should not exceed $10°$ C; in most nodule areas it is between 0 and $4°$ C. There must be sufficient oxygen, favored by the low water temperature, for oxidizing the manganese. The relatively warm Red Sea is devoid of nodules and may serve as example. Rare crusts begin to appear in the eastern Gulf of Aden where bottom water is colder than $10°$ C and nodule fields are eventually present in the Somali basin. The Mediterranean with a bottom temperature above $13°$ C is also devoid of nodules. Lower oxygen contents or high consumption by organisms or used for organic decomposition are therefore expected to have a negative effect on nodule growth.

The search for environments suited for originating high-quality nodules is directed to the major world oceans because only there can the prerequisites and their optimum interaction be expected: environmental constancy over large space and long time, extremely low deposition rate, possibly the additional supply of metals through exhalations or other volcanic products in connection with sea-floor spreading. The influence of depth to the chemistry and mineralogy of the nodules is not to be neglected (GLASBY, 1972). The 5,000—6,000 meters deep abyssal basins and plains, tectonically quiet, relatively old and stable, not more than slightly undulated, as normally present on both sides of the oceanic ridges

---

[3]) Note added in proof: more suggestions as to genetic relations between nodules and substrata are forwarded in a recent article by D. R. HORN, B. M. HORN & M. D. DELACH: Ocean manganese nodules metal values and mining sites. Techn. report No. 4, NSF GX 33616, 1973.

(but at a safe distance from them) and away from transverse fracture zone are thus the preferential prospection sites (fig. 4). These stable areas resembling in certain aspects the continental cratons (representing counterparts) and bounded by tectonic zones of global extent are the frame for the precipitation of nodules at the sea water-sediment interface and leading to a huge ore potential whose utilization begins to form a world-wide priority task of many states and private industries. The approximate and general distribution of deep-sea manganese nodules and, in particular, their economic variety is tentatively shown in fig. 5.

Sillitoe (1972 a) has advanced the hypothesis that porphyry copper deposits

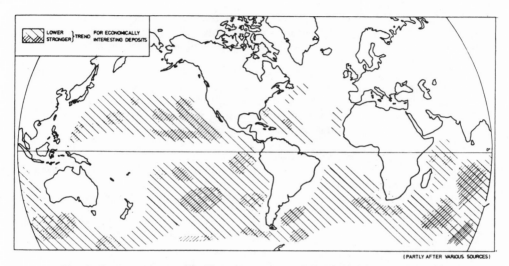

Fig. 5. Sea areas prospective for manganese nodules (simplified, schematic).

(accounting for more than half of the World's production) may originate from subducted and melted crust and pelagic sediments in the course of l a t e d r i f t (fig. 1). It is plausible to assume that the incorporation of metalliferous deposits into the melting zone would easily account for the episodic occurrence of metal-rich melts. The same author has also forwarded the suggestion that many cupri-ferous massive sulphide deposits associated with ophiolite complexes were generated in ocean basins at the sites of sea-floor spreading, the ocean rises. There is considerable evidence (such as in the Cyprus Troodos complex) for the recognition of layer 1, 2 and 3 rocks of the former oceanic crust; the massive stratiform sulphide deposits are presumed to have been formed initially as accumulations of brine-soaked sulphide mud in depressions on the sea floor (Sillitoe, 1972 b).

It has been suggested that the closing of the ocean separating northern and southern Europe in earlier Paleozoic times and leading to the collision of the continents in the Carboniferous mark the Asturic orogenic phase in Europe and the Allegheny phase in the adjoining America (Burrett, 1972). The volcanism accompanying the subduction zones may then be thought to have contributed to the supply of metal solutions of the Rammelsberg and Meggen ore deposits of Germany. These occurrences could therefore represent a late phase of generating metalliferous mud following the precursory, the mature and the subsequent phases described above (fig. 2). Following the model of OXBURGH (1972), such development could be favored by the collision of irregularly shaped continental margins leaving, between overthrust pieces of continental crust, pockets of old ocean floor as ideal basins for deposition.

Conclusion

Continental drift is though to begin with the e a r l y and the a d v a n c e d g r a b e n   s t a g e s. After continental breakup the e a r l y   d r i f t and the a d v a n c e d   d r i f t   s t a g e s follow. The development may be closed during a l a t e   d r i f t   s t a g e.

H y d r o c a r b o n s do not necessarily form in connection with continental drift although several oil provinces appear to be related to such processes (e. g. graben stages: Tertiary of Upper Rhine and Gulf of Suez, Mesozoic of West African margin; rifting and structural control of basins: Argentine). Recently discovered oil shows within sediments of the deep ocean suggest that large portions of it will probably become prospective, contrary to an earlier opinion.

P h o s p h o r i t e forming on shelf and slope has no economic importance at present. O r g a n i c   o o z e s occurring in upwelling shelf areas (e. g. West coasts of South America and Africa) and containing U and other heavy metals may represent a future potential where enriched beyond the presently known concentrations.

M e t a l l i f e r o u s   m u d s containing as value metals mostly Zn, Cu, Pb, Ag (apart from Fe, Mn, Ba) become deposited in various phases, a precursory (Lakes Malawi, Kivu) and a subsequent one (Cretaceous of DSDP site 105, western N. Atlantic) with a mature phase (Red Sea) in between coinciding with the e a r l y   d r i f t   s t a g e. The genetic prerequisites are then at optimum (heat-flow anomalies acting along narrow strips of central rifts promoting water cycling and leaching, possibly admixture of volcanic exhalations, bold relief of central rift with depositional pockets, likelihood of flanking evaporates favoring brines and metal transport). Several deeps within the Red Sea central rift hold metalliferous muds. Only the 50 km² Atlantis II deep contains volumes and grades of metals worth economic interest (averaging 5% Zn and 1% Cu in dry, saltfree substance). It appears likely that metalliferous muds of good grade and large amounts became deposited in former geological times when an opening ocean passed the early drift stage. In spite of technical difficulties in recovering them, a new and possibly large ore potential is in sight, possibly stimulating their development through improved techniques and innovated technologies.

The major type minerals of economic interest and possibly great commercial importance of the a d v a n c e d   d r i f t   s t a g e, in the abyssal plains with minimum sedimentation and away from tectonic activities, are represented by the m a n g a n e s e   n o d u l e s. Occurrences of the northern Pacific contain 3% or more of Cu, Ni and Co (apart from Mn and Fe) over very large areas (compared to land mines) with nodules spaced densely enough to offer the outlook of utilization once low-cost methods of lifting them and extracting the value metals are developed. The nodules are believed to be formed predominantly through the ascending mobilization of Mn from the underlying sediments and its subsequent precipitation to already existing ones or other nuclei whereas Ni, Cu and Co are adsorbed from the sea water by the Mn and Fe hydrates. High-grade nodules are believed to require low rates of growth. On the other hand geologically young nodule deposits of comparatively rapid growth contain the value metals Ni, Cu and Co in percentages about one order of magnitude lower. Deep-sea manganese nodules are not likely to be preserved in fossil state but tend to be dissolved under a sediment cover.

Deposits originated as metalliferous muds becoming conveyed via subduction trenches into the melting zones during a l a t e d r i f t s t a g e may be a source for p o r p h y r y o r e s.

Bibliography

Anderson, C. A., & Nash, J. T.: Geology of the Massive Sulfide deposits at Jerome, Arizona — a reinterpretation. — Econ. Geology, 67, no. 7, 845, 1972.

Anonymous: U.S. sees 1 million b/d of shale oil by 1985. — The Oil and Gas Journal, **70**, no. 38, 52, 1972.

Anwar, Y. M., El-Mahdy, O. R., & El-Dahhar, M. A.: Geology and origin of the lead-zinc ores of Um Gheig area, Egypt. — Econ. Geology, 67, no. 7, Abstracts, 1002, 1972.

Audley-Charles, M. G.: Cretaceous deep-sea manganese nodules on Timor: implications for tectonics and olistostrome development. — Nat. Physic. Sci., **240**, 137—138, 1972.

Avilov, I. K., & Gershanovich, D. Ye.: Investigation of the relief and bottom deposits on the Southwest African Shelf. — Oceanology (USSR), **10**, no. 2, 229—232, 1970.

Bäcker, H., & Schoell, M.: New deeps with brines and metalliferous sediments in the Red Sea. — Nat. Physic. Sci., **240**, no. 103, 153—158, 1972.

Baturin, G. N.: Uranium in oceanic ooze solutions of the southeastern Atlantic. — Dokl. Earth Sci. Sect. **198**, 224—226, 1971.

Baturin, G. N., Kochenov, A. V., & Senin, Yu. M.: Uranium concentration in recent ocean sediments in zones of rising current. — Geochem. Intern., **8**, no. 2, 281—286, 1971.

Beck, R. H.: A geologist examines deepsea formations for possible oil clues. — Offshore, **32**, no. 7, 1972.

Beurlen, K.: Die Kreide Nordost-Brasiliens — ein Beitrag zur Geschichte des Atlantischen Ozeans. — Nachr. Dtsch. Geol. Ges., no. 5, 44—47, 1972.

Bischoff, J. L.: A ferroan nontronite from the Red Sea geothermal system. — Clays and Clay Minerals, **20**, 217—223, 1972.

Blissenbach, E.: Continental drift and metalliferous sediments. — Oceanology International 72, Brighton, England. Conference papers, 412—416, 1972.

Boes, C. H., & Bade, P. G. A.: A system for the recovery of heavy metal sediments from the Red Sea deeps. — Underwater Journal, 220—228, 1971.

Boigk, H., & Porth, H.: Zur Frage der Erdölhöffigkeit des Außenschelfs, des Kontinentalabhanges und des Kontinentalanstieges. — Erdöl-Kohle-Erdgas-Petrochem., **23**, no. 3, 137—144, 1970.

Bonatti, E., Fisher, D. E., Joensuu, O., Rydell, H. S., & Beyth, M.: Iron-manganese-barium deposit from the northern Afar rift (Ethiopia). — Econ. Geology, 67, no. 7, 717—730, 1972.

Bostrom, K., & Peterson, M. N. A.: Precipitates from hydrothermal exhalations on the East Pacific Rise. — Economic Geology, **61**, 1258—1265, 1966.

BREWER, P. G., WILSON, T. R. S., MURRAY, J. W., MUNNS, R. G., & DENSMORE, C. D.: Hydrographic observations on the Red Sea brines indicate a marked increase in temperature. — Nature, 231, 37—38, 1971.

BURRETT, C. F.: Plate tectonics and the Hercynian orogeny. — Nature 239, 155—157, 1972.

BUTUSZOVA, G. Y.: Recent volcano-sedimentary process in Santorin Volcano caldera (Aegian Sea) and its effect on the geochemistry of sediments (in Russian). — Trudy Inst. Geol. Akad. Nauk SSSR, 194, 1969.

CALVERT, S. E., & PRICE, N. B.: Composition of manganese nodules and manganese carbonates from Loch Fyne, Scotland. — Contr. Min. & Petrol., 29, 215—233, 1970.

—: Recent sediments of the South West African shelf. — In: DELANY, F. M. (edit), ICSU/SCOR Working Party 31 Symposium, Cabridge 1970, Rep. no. 70/16, Inst. Geol. Sci., 1971.

CNEXO: Activité de l'industrie petrolière en mer. — In: Bulletin CNEXO, no. 39, 1972.

COULOMB, J.: Sea floor spreading and continental drift. — 184 p., Dordrecht (D.Reidel Publishing Company) 1972.

CRONAN, D. S., & THOMAS, R. L.: Geochemistry of ferromanganese oxide concretions and associated deposits in Lake Ontario. — Geol. Soc. Amer. Bull., 83, 1493—1502, 1972.

DEEP SEA DRILLING PROJECT: Scientific objectives and hoghlights. — University of California, San Diego, Scripps Institution of Oceanography, 20 p, 1972 a.

—: DSDP Arabian and Red Seas investigations reported. — DSDP Release 183, 1972 b.

DEGENS, E. T., & ROSS, D. A. (edit.): Hot brines and recent heavy metal deposits in the Red Sea. — 600 p., Berlin (Springer Verlag) 1969.

DEGENS, E. T.: Sea floor spreading: Lagerstättenkundliche Untersuchungen im Roten und im Schwarzen Meer. — Umschau no. 9, 268—274, 1970.

DEGENS, E.T., VON HERZEN, R. P., & WONG, H. K.: Lake Tanganyika: water chemistry, sediments, geological structure. — Naturwissenschaften, 58, no. 5, 1971.

DEGENS, E. T., OKADA, H., HONJO, S., & HATHAWAY, J. C.: Microcrystalline sphalerite in resin globules suspended in Lake Kivu, East Africa. — Mineral. Deposita, 7, no. 1, 1—12, 1972.

DIETZ, R. S.: Origin of continental slopes. — American Scientist, 52, 50—69, 1964.

DIETZ, R. S., & HOLDEN, J. C.: The breakup of Pangaea. — Scientific American, 223, no. 4, 30—41, 1970.

ELDERFIELD, H.: Effects of volcanism on water chemistry, Deception Island, Antarctica. — Mar. Geol., 13, no. 1, p. M 1—6, 1972.

ELDERS, W. A., REX, R. W., MEIDAV, T., ROBINSON, P. T., & BIEHLER, S.: Crustal spreading in southern California. — Science, 178, No. 4056, 15—24, 1972.

EL WAKEEL, S. K., & RILEY, J. P.: Chemical and mineralogical studies of fossil red clays from Timor. — Geochim. Cosmochim. Acta, 24, 260—265, 1961.

EMERY, K. O.: The sea off southern California: a modern habitat of petroleum. — 336 p., New York (John Wiley & Sons) 1960.

—: Continental rises and oil potential. — Oil Gas Journ., 67, no. 19, 231—240, 1969.

EMERY, K. O., UCHUPI, E., PHILLIPS, J. D., BOWIN, C. O., BUNGE, E. T., & KNOTT, S. T.: Continental rise off eastern North America. — Am. Assoc. Petroleum Geologists Bull., 54, no. 1, 44—108, 1970.

FALCON, N. L., GASS, I. G., GIRDLER, R. W., & LAUGHTON, A. S.: (organizer of:) A discussion on the structure and evolution of the Red Sea, Gulf of Aden and Ethiopia rift junction. — Phil. Trans. Roy. Soc. Lond. A. 267, 417 p., 1970.

FERGUSON, J., & LAMBERT, I. B.: Volcanic exhalations and metal enrichments at Matupi Harbor, New Britain, T.P.N.G. — Econ. Geol., 67, 25—37, 1972.

Finkelman, R. B.: Relationship between manganese nodules and cosmic spherules. — Mar. Techn. Soc. J., **6**/4, 34—39, 1972.

Frakes, L. A., & Kemp, E. M.: Generation of sedimentary facies on a spreading ocean ridge. — Nature, **236**, 114—117, 1972.

Garrison, R. E., & Fischer, A. G.: Deep-water limestones and radiolarites of the Alpine Jurassic. — In: G. M. Friedman, Depositional Environments in Carbonate Rocks. Soc. Econ. Paleont. Min., Spec. Publ., 14, 20—56, 1969.

Germann, K.: Mangan-Eisen-führende Knollen und Krusten in jurassischen Rotkalken der nördlichen Kalkalpen. — N. Jb. Geol. Pal., Monatsh., 133—156, 1971.

Girdler, R. W., & Sowerbutts, W. T. C.: Some recent geophysical studies of the rift system of East Africa. — Journ. Geomagnetism and Geoelectricity, **22**, no. 1—2, 153—163, 1970.

Glasby, G. P.: The mineralogy of manganese nodules from a range of marine environments. — Mar. Geol., **13**, 57—72, 1972.

Grasshoff, K.: Chemie der Manganknollen. — In: G. Dietrich, Erforschung des Meeres, 141—149, Frankfurt (Umschau Verl.) 1970.

Grill, E. V., Murray, J. W., & Macdonald, R. D.: Todorokite in manganese nodules from a British Columbia Fjord. — Nature, **219**, 358—359, 1968.

Grunwald, R. R.: The mineralogy and origin of manganese concretions in the Oacoma Zone of the Pierre Shale near Chamberlain, South Dakota. — Proc. S. D. Acad. Sci., **43**, 193—196, 1964.

Gümbel, W.: Über die im Stillen Ozean auf dem Meeresgrund vorkommenden Manganknollen. — Sitzungsber. Kgl. Bayr. Akad. Wissensch. München, Math.-Phys. Cl., **8**, 189—209, 1878.

Hammond, A. L.: Mars as an active planet. — Science, **175**, 286—287, 1972.

Hannak, W.: Die Blei-Silber-Zink-Lagerstätte von Bawdwin (Birma), ein Erzlager? — Nachr. Dtsch. Geol. Ges., no. 6, 93, 1972.

Harris, R. C., & Troup, A. G.: Chemistry and origin of freshwater ferromanganese concretions. — Limnol. Oceanogr., **15**/5, 702—712, 1970.

Hedberg, H. D.: Continental margins from the viewpoint of the petroleum geologist. — Am. Assoc. Petroleum Geologists Bull., **54**, no. 1, 3—43, 1970.

Heezen, B. C., Tharp, M., & Ewing, M.: The floors of the oceans. I. The North Atlantic. — Geol. Soc. America Spec. Paper 65, 122 p., 1959.

Heezen, B. C., & Hollister, C. D.: The face of the deep. — 659 p., New York (Oxford University Press) 1971.

Hockings, W. A., Rose, D. H., & Snelgrove, A. K.: Technological aspects of manganese nodules in Lake Michigan. — Skillings's Min. Rev., 7—9, Oct. 25, 1969.

Hollister, C. D., Ewing, J. I., Habib, D., Hathaway, J. C., Lancelot, Y., Luterbacher, H., Paulus, F. J., Poag, G. W., Wilcoxon, J. A., & Worstell, P.: 6. Site 105 — lower continental rise hills. — Initial reports of the Deep Sea Drilling Projects, **XI**, 219—312, Washington 1972.

Honnorez, J.: La formation actuelle d'un gisement sous-marin de sulfures fumerolliens a Vulcano (mer tyrrhénienne). — Mineral. Deposita (Berl.), **4**, 114—131, 1969.

Horn, D. R., Horn, B. M., & Delach, M. N.: Ferromanganese deposits of the North Pacific. — Techn. report no. 1, NSF GX-3361, 78 p., Washington 1972.

Horn, D. R., Ewing, J. I., & Ewing, M.: Graded-bed sequence emplaced by turbidity currents north of 20° N in the Pacific, Atlantic and Mediterranean. — Sedimentology, **18**, 247—275, 1972.

Horowitz, A.: The distribution of Pb, Ag, Sn, Tl, and Zn in sediments on active oceanic ridges. — Marine Geol., **9**, 241—259, 1970.

Illies, H.: Die großen Gräben: Harmonische Strukturen in einer disharmonisch struierten Erdkruste. — Geol. Rdsch., **59**, no. 2, 528—552, 1970.

838   Aufsätze

JENKYNS, H. C.: Fossil manganese nodules from the West Sicilian Jurassic. — Eclog. Geol. Helv., **63/3**, 741—774, 1970.

KEITH, M. L.: Ocean-floor convergence: A contrary view of global tectonics. — Journ. Geology, **80**, no. 3, 249—276.

KLEMME, H. D.: Heat influences size of oil giants. — The Oil and Gas Jounal, **70**, no. 29, 30, 1972.

KRAUS, E. C.: Die Entwicklungsgeschichte der Kontinente und Ozeane. — 429 p., Berlin (Akademie-Verlag) 1971.

LANCELOT, Y., HATHAWAY, J. C., & HOLLISTER, C. D.: 31. Lithology of sediments from the western North Atlantic Leg 11 Deep Sea Drilling Project. — Initial reports of the Deep Sea Drilling Project, **XI**, 901—949, Washington 1972.

LORBACH, M.: Die Erdölgewinnungstechnik zur Gewinnung fester Mineralien, insbesondere in Offshore-Gebieten. — Erdoel-Erdgas-Zeitschrift, **88**, 361—370, 1972.

LOWELL, J. D., & GENIK, G. J.: Sea-floor spreading and structural evolution of southern Red Sea. — Am. Assoc. Petroleum Geologists Bull., **56**, no. 2, 247—259, 1972.

MANHEIM, F. T.: Manganese-iron accumulations in the shallow marine environment. — Narragansett Mar. Lab., Univ. Rhode Island, Occ. Publ. 3, 217—275, 1965.

MANHEIM, F. T., & SAYLES, F. L.: Brines and interstitial brackish water in drill cores from the deep Gulf of Mexico. — Science, **170**, 57—61, 1970.

MAUCHER, A.: Die Lagerstätten des Urans. — 162 p., Braunschweig (F. Vieweg & Sohn) 1962.

MCDOWELL, A. N.: Practical application of continental-drift concept may find giants. — The Oil and Gas Journ., **69**, no. 26, 114—116, 1971.

MCIVER, R. D.: Cores show oil migration in Western Pacific. — Ocean Industry, **7**, no. 3, 18—19, 1972.

MENARD, H. W.: Marine geology of the Pacific. — 271 p., New York/London (McGraw-Hill) 1964.

MERO, J. L.: The mineral resources of the sea. — 312 p., Amsterdam (Elsevier Publ. Company) 1965.

MEYERHOFF, A. A., & MEYERHOFF, H. A.: "The new global tectonics": Major inconsistencies. — Am. Assoc. Geologists Bull., **56**, no. 2, 269—336, 1972.

MOLENGRAAFF, G. A. F.: On manganese nodules in Mesozoic deep-sea deposits of Dutch Timor. — Proc. Roy. Acad. Sci. Amsterdam, **23**, 997—1012, 1922.

MORGAN, W. J.: Deep mantle convection plumes and plate motions. — Am. Assoc. Petroleum Geologists Bull., **56**, no. 2, 203—213, 1972.

MURRAY, J., & RENARD, A. F.: Report on the scientific results of the voyage of H.M.S. "Challenger" during the years 1873—76. — Deep-sea deposits. 525 p., Edinburgh 1891.

NELSON, T. M., & TEMPLE, P. G.: Mainstream mantle convection: A geologic analysis of plate motion. — Am. Assoc. Petroleum Geologists Bull., **56**, no. 2, 226—246, 1972.

OXBURGH, E. R.: Flake tectonics and continental collision. — Nature, **239**, 202—204, 1972.

PAUTOT, G., AUZENDE, J. M., & LE PICHON, X.: Continuous deep sea salt layer along North Atlantic margins related to early phase of rifting. — Nature, **227**, 351—354, 1970.

PETRASCHECK, W. E.: Kontinentalverschiebung: Zerteilung und Neuschaffung von Erzprovinzen. — Umschau, **72**, no. 21, 677—680, 1972.

RAMBERG, I. B., & SMITHSON, S. B.: Gravity interpretation of the southern Oslo Graben and adjacent precambrian rocks. Norway. — Tectonophysics, **11**, 419—431, 1971.

REEVES, C. V.: Rifting in the Kalahari? — Nature, **237**, 95—96, 1972.

RONA, P. A.: Comparison of continental margins of eastern North America at Cape Hatteras and north-western Africa at Cap Blanc. — Am. Assoc. Petroleum Geologists Bull., **54**, no. 1, 129—157, 1970.

—: The continental margin between the Canary and Cape Verde Islands and symmetries with eastern North America. — In: DELANY, F. M. (editor), ICSU/SCOR Working Party 31 Symposium, Cambridge 1970: The geology of the East Atlantic continental margin. 2. Africa Rep. no. 70/16, Inst. geol. Sci., 38—42, 1971.

ROSSMANN, R., CALLENDER, E., & BOWSER, C. J.: Interelement geochemistry of Lake Michigan ferro-manganese nodules. — 24th Int. Geol. Congr., 10, 336—341, Montreal 1972.

RUNNELS, D. D.: Continental drift and economic minerals in Antarctica. — Earth and Plan. Sci. Letters, 8, 400—402, 1970.

SCHNEIDER, E. D.: Evolution of rifted continental margins and possible long-term economic implications. — Colloque International sur l'Exploitation des Océans, Thème IV, Tome II, Bordeaux 1971.

SCHNEIDER, E. D., & JOHNSON, G. L.: Deep-ocean diapir occurrences. — Am. Assoc. Petroleum Geologists Bull., 54, no. 11, 2151—2169, 1970.

SCHOETTLE, M., & FRIEDMAN, G. M.: Fresh water iron-manganese nodules in Lake George, New York. — Geol. Soc. Amer. Bull., 82, 101—110, 1971.

SCHOTT, W.: Gedanken zur Geologie des äußeren Schelfs und des Kontinentalabhanges. — Erdöl Kohle-Erdgas-Petrochem., 23, no. 4, 197—205, 1970.

SCHTERENBERG, L. Ye.: Some aspects of the genesis of iron-manganese concretions in the Gulf of Riga. — Dokl. Akad. Nauk SSSR, 201/2, 457—460, 1971.

SCHUILING, R. D.: Tin belts on the continents around the Atlantic Ocean. — Econ. Geol., 62, 540—550, 1967.

SCHULZ, O.: Unterwasserbeobachtungen im sublitoralen Solfatarenfeld von Vulcano (Äolische Inseln, Italien). — Mineral. Deposits (Berl.), 5, 315—319, 1970.

SEIBOLD, E.: Der Meeresboden als Rohstoffquelle und die Konzentrierungsverfahren der Natur. — Chemie-Ing.-Techn., 42, no. 23, A 2091—2102, 1970.

SEVASTJANOV, V. F., & VOLKOV, I. I.: Redistribution of chemical elements in the oxidated layer of the sediments in the process of iron-manganese nodules formation in the Black Sea (Russ.). — Trudy Inst. Okeanol., 83, 135—152, 1967.

SILLITOE, R. H.: A plate tectonic model for the origin of porphyry copper desposits. — Econ. Geol., 67, 184—197, 1972 a.

—: Formation of certain massive sulphide deposits at sites of sea-floor spreading. Inst. Minig Metall. — Transact., 81 B, Bull. no. 789, 141—148, 1972 b.

SMITH, P. A.: Underwater mining — insight into current U.S. thinking. — Mining Magazine, July 1972.

SORENSON, P. E., & MEAD, W. J.: A new economic appraisal of marine phosphorite deposits off the California coast, in the decade ahead, 1970—1980. — Transactions of the Annual MST Conf. and Exhibit: Marine Technology Soc., 491—500, 1969.

STANLEY, D. J. (edit.): The new concepts of continental margin sedimentation. — Washington, D.C. (American Geol. Institute), 1969.

TARLING, D. H.: Another Gondwanaland. — Nature, 238, 92—93, 1972.

TAYLOR, P. T., BRENNAN, J. A., & O'NEILL, N. J.: Variable seafloor spreading off Baja California. — Nature, 229, 396—399, 1971.

TERASMAE, J.: Notes on lacustrine manganese-iron concretions. — Geol. Surv. Canada, Dept. Energy, Min. & Res., Pap. 70—69, 12 p., 1971.

UNITED NATIONS: Mineral resources of the sea. ST/ECA/125. — 49 p., New York 1970.

van ANDEL, Tj. H., & KOMAR, P. D.: Ponded sediments of the Mid-Atlantic ridge between 22° and 23° North latitude. — Geol. America Bull., 80, 1163—1190, 1969.

VARENTSOV, I. M.: On the main aspects of formation of ferromanganese ores in recent basins. — 24th Int. Geol. Congr., Sect. 4, 395—403, Montreal 1972.

VINE, F. J., & HESS, H. H.: 1. Sea-floor spreading. — In: "The sea", 4, Part 11, 587—622, 1970.

von HERZEN, R. P., & VACQUIER, V.: Terrestrial heat flow in Lake Malawi, Africa. — J. Geophys. Res., 72, 4221—6, 1967.

WEGENER, A.: Die Entstehung der Kontinente. — Geol. Rundschau, 3, 276—292, 1912.

WENK, jr., E.: The physical resources of the ocean. — (In: The ocean. A scientific American book), 81—91, San Francisco (W. H. Freeman & Coy.), 1969.

WILSON, J. T.: 2. Continental drift, transcurrent, and transform faulting. — In: "The sea", 4, part II, 623—644, 1970.

ZELENOV, K. K.: Iron and manganese in exhalations of the submarine Banu Wuhu Volcano (Indonesia). — Doklady Akad. Nauk SSSR, 155, no. 6, 1317—1320, 1964.

38

BAUXITE FORMATION AND PLATE TECTONICS

By

Gy. Bárdossy

GEOCHEMICAL RESEARCH LABORATORY, HUNGARIAN ACADEMY OF SCIENCES, BUDAPEST

On the basis of paleomagnetic measurements and the well known reconstructions offered by the theory of continental drift, the author analyzes and explains the geographical distribution of bauxites.

He has found a close relationship between the tectonic position of the karstic bauxites on the one hand and their type of deposit, petrographic and mineralogic compositions and genesis on the other.

He has shown a chronological relationship between bauxitization and the tectonic evolution of the orogenic belts. The quantity of karstic bauxites has been calculated for all epochs of geological history. This too has been controlled by the tectonic evolution.

It is a matter of common knowledge that all recent laterites and lateritic bauxites occur in the tropical and subtropical climatic belts, roughly between the 30° of southern and northern latitudes. The deposits of Neogene age also fall in this range, while the Paleogene ones are situated already between the 35° and 50° of northern latitude. The Paleocene and Cretaceous bauxite deposits of Siberia occur yet farther north, between 50° and 60° N. The Paleozoic deposits of the Ural and Tyman' Mountains lie between the 50° and 68° N; moreover, the bauxites of the recently explored Pripolyarny Urals reach up to the Arctic Circle. In the last years bauxite deposits were found on the Taymir peninsula between the 74° and 77° N. Four possible explanations may account for the above facts:

i. It may be supposed that in the Mesozoic and Paleozoic eras the bauxites might have been formed under colder climates than today. This hypothesis is contradicted by the entire geochemical mechanism of bauxitization requiring both warm and humide climate. In addition, it is also refuted by the tropical and subtropical fauna and flora of the immediate hanging and foot-walls of these bauxite deposits.

ii. Another hypothesis may be that in certain geological epochs the climate of the Earth was more equilibrated, i.e. more uniformly warm, than today. This is how some workers explain the formation of the afore-mentioned North Urals bauxites. Paleoclimatological evidence, however, does not support these hypotheses. What can be taken to be a proven fact is that climatic zones similar to those existing today must have existed in the Paleozoic too (SCHWARZ-BACH, 1961). The most recent meteorological and geophysical studies agree with this opinion. Accordingly, this hypothesis has to be rejected, too.

iii. It may also be supposed that the entire earth crust has been displaced with regard to the poles. In this case the climatic zones would gradually shift away, whilst the distribution of the continents would have remained unchanged. This hypothesis was suggested by GORETSKY (1960), who supposed that in Devonian time the equatorial zone extended across the North Urals, roughly parallel to the present-day equator. If the equator is plotted on paleogeographical maps according to GORETSKY's hypothesis, considerable bauxite and coal basins will be placed elsewhere close to the arctic climatic zone. Anyhow the positions of the poles are changed, paleoclimatic evidence inconsistent with the given hypothesis will always be found. Briefly, the paleoclimatological pattern of the bauxites cannot be interpreted by the present-day distribution of the continents.

iv. A satisfactory explanation for all bauxite areas is only feasible if the theory of continental drift is adopted and this is how the author has arrived at using the laws of plate tectonics.

Several reconstructions of the relative positions of the continents in the different geological epochs have been proposed on the basis of plate tectonics. Unfortunately, these extend only to the Permian back in geological time. With the use of paleomagnetic results, however, one can locate the contemporaneous positions of Paleozoic bauxite areas as well. In doing so, the author has made use first of all of the data published by CREER (1968), HAMILTON (1970) and KHRAMOV and SHOLPO (1967). Since Paleozoic bauxitization had been confined to Eurasia a cartographic representation has been chosen in which the north pole is placed in the centre.

The oldest bauxite deposit known presently is of Lower Cambrian age: Boksonsk in the eastern Sayan Mountains. That time the Russian and Siberian continental plates were much farther away from each other than they are today. For their paleomagnetic poles are very distant apart (Fig. 1). The paleomagnetic north pole corresponding to the Russian plate is in the territory of the Pacific Ocean near the present-day equator; that corresponding to the Siberian plate falls south of present-day Australia in the southern hemisphere. Therefore the author could indicate only the position of the corresponding south pole in his figure. The equatorial circles corresponding to the two pole positions have been plotted. It can be readily seen that the bauxite area lies near to the paleomagnetic equator of the Siberian continental plate, hence the paleoclimatological conditions are granted. The tectonic position of the bauxite area: it lies in the Caledonian orogenic belt, close to the edge of the Siberian continental plate.

In Ordovician and Silurian times insignificant bauxitization took place

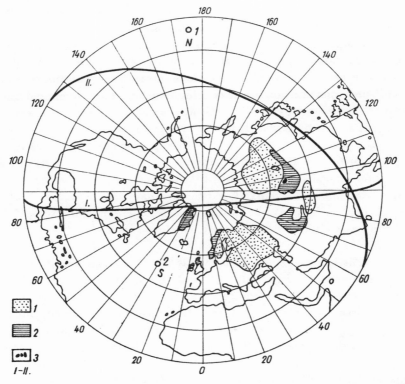

Fig. 1. Bauxitization and pole positions in the Cambrian. 1. continental plates; 2. Caledonian orogenic belt; ● bauxite area; Equator I. corresponds to the Russian continental plate, Equator II. to the Siberian one

in the Caledonian orogenic belt in Kazakhstan, USSR. This area also lies close to the contemporaneous equator.

It was in the Devonian that the first significant karstic bauxite deposits were formed: on the western and eastern margins of the Hercynian orogenic belt close to the Russian and Siberian continental plates (Fig. 2). The progressive shift of paleomagnetic pole positions suggests the Russian plate to have gradually approached the Siberian one. According to HAMILTON (1970), in the Early Devonian there was an island arc along a subduction zone in the line of the presentday Ural Mountains and it was reached by the Russian plate in the Lower Devonian. The bauxite deposits were formed later, in the Middle and Upper Devonian. Paleomagnetic pole positions also allow one to establish that both continental plates slowly rotated clockwise. As evident from Fig. 2, the bauxite deposits fall close to the equator, being situated nearly parallel to it. The bauxite area of Taymir Peninsula too gets much closer to the equator when considering the major transform fault recognized by HAMILTON (1970); a fault along which the Peninsula drifted southeastward for several hundred

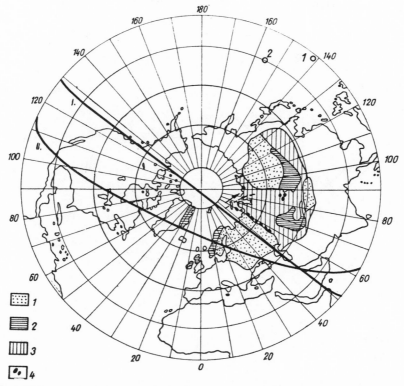

Fig. 2. Bauxitization and pole positions in the Devonian. 1. continental plates; 2. Caledonian orogenic belt; 3. Hercynian orogenic belt; ● bauxite area; Equator I. corresponds to the
Siberian continental plate, Equator II. to the Russian one

kilometres in the subsequent geological epochs. Even the other Devonian
bauxite area occurring near the Siberian plate does fall south of 30° N, the
climatic conditions being thus granted here too.

As deduced from the shifting of the paleomagnetic poles the clockwise
rotational motion of the two continental plates continued in Carboniferous
time and the plates came yet closer to each other (Fig. 3). The areas of bauxitization underwent considerable changes: over the Urals stretch of the Hercynian
orogenic zone the formation of bauxites came to an end, to be reinitiated in
its continuation, the Turkestan Mountains, over a length of about 1000 km.
The first important lateritic bauxite deposits appeared that time on the Russian
plate. As evident from the map, all these occurrences fall within the tropical
and subtropical climatic belts.

Bauxites began to be accumulated at a considerable rate on the microcontinental plates of North China and South China as well. These, however,
fall very far away from the equator plotted for the paleomagnetic pole of the

315

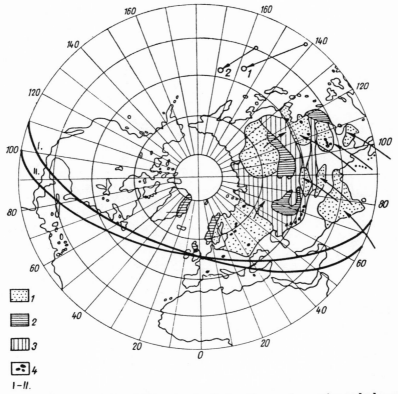

Fig. 3. Bauxitization and pole positions in the Carboniferous. 1. continental plates; 2. Caledonian orogenic belt; ● bauxite area; 3. Hercynian orogenic belt; Equator I. corresponds to the Siberian continental plate, Equator II. to the Russian one

Siberian plate, being situated in the temperate or even the boreal climatic zones. Obviously, no bauxite could be formed under climates like these. Unfortunately enough, the paleomagnetic data available on Southeast Asia are very scant and rather largely scattered. At any rate, it can be stated that they differ substantially from the pole positions corresponding to the Siberian plate. Therefore HAMILTON (1970) surmised that the microcontinental plates under consideration had only gradually welded with the vast Eurasian continent. He did not specify, however, how this process took place, because of the "structural chaos" characteristic of the area.

According to the present writer's opinion, the tectonic setting of this area, as related with bauxitization, can be readily interpreted by the aid of plate tectonics. Let us start from the assumption that the northeastward drift of the Indian subcontinental plate can be taken to be a proven fact now. It was welded with the Eurasian continent in the Tertiary, a process associated with the oroclinal folding of the Himalaya mountain system. The author believes that the five microcontinental plates of Central and Southeast Asia (Tarim, Tibet, North-China, South-China, Indochina) drifted in one and the same direction and that they were successively welded with the Siberian continental plate during the Late Paleozoic and Mesozoic. The supposed direction of motion is shown in Fig. 3. The force of thrusting was increased by the clockwise rotation of the Russian and Siberian continental plates.

This process can be traced well on the "Tectonic Map of Eurasia" published in 1966 in Moscow: A Baikalian orogenic belt was the first to be piled up on the southern margin of the Siberian plate. This is followed farther to the south by a Caledonian and then by a Hercynian orogenic belt. The latter indicates already the collision of the North-China and Tarim microcontinents. Between the microcontinental plates of North and South-China on the one hand and those of Tarim and Tibet on the other, a Hercynian orogenic belt can be found. In accordance with the direction of piling up, the strike of folding is east-west. The folding between the microcontinents of South-China and Indochina occurred during the Mesozoic era which is in good agreement with the gradular collision of the microocntinental plates. Finally, on the southern and western sides, respectively, of the Tibet and Indochina microcontinents it is again first a continuous, large Mesozoic orogenic belt that follows and only to the south of it can be found the vast Tertiary mountain range of the Alpine-Himalaya orogenic belt with the Indian continental plate on its southern side.

The tectonic mechanism of motion of this kind will bring a solution to the paleoclimatological problems of bauxite formation as well. Should a drift of this kind of the microcontinental plates of Southeast Asia be accepted, these will then be sited in the proximity of the Carboniferous equator and the formation of bauxites on them will become paleoclimatologically plausible. This holds true otherwise for the Carboniferous coal basins of China, too.

Naturally, the afor-outlined movements still continued during the Permian. The pole positions corresponding to the Russian and Siberian plates came close to each other that the welding of the two continental plates might be considered to have been completed by that time (Fig. 4). In the Permian the rate for bauxite accumulation decreased markedly and was going on in East Asia only. Accepting the above-postulated continental drift these bauxites too have been formed under subtropical to tropical climates. Finally, the Permo-Triassic bauxites of Turkey fall exactly in the equatorial zone.

The reconstruction of the earliest pattern of the Globe's continents so far available is that of the Upper Permian given by DIETZ and HOLDEN (1970, Fig. 5). The author has indicated on this map the afore-listed bauxite areas, and the positions of the three East Asian microcontinental plates and their direction of drifting as supposed by him. It can be seen well that the bauxite areas have thus been placed in the tropical and subtropical zones. Remarkably enough, it is in respect of Southeast Asia that DIETS-HOLDEN's model is most

317

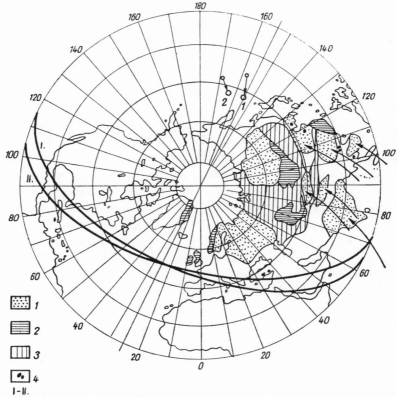

Fig. 4. Bauxitization and pole positions in the Permian. 1. continental plates; 2. Caledonian orogenic belt; 3. Hercynian orogenic belt; ● bauxite area; Equator I. corresponds to the Siberian continental plate, Equator II. to the Russian one

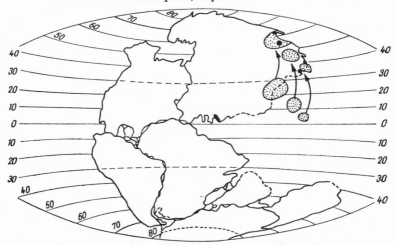

Fig. 5. Locations of the continents in the Permian according to DIETZ and HOLDEN (1970) ● bauxite area; 1. displacement of the microcontinental plates of East Asia as supposed by the author

318

uncertain, the pattern of the continental margin being indicated only by a dash-line of entirely schematic design.

Having been formed by the end of the Paleozoic, "Pangea" was disrupted at the beginning of the Mesozoic era, as shown by DIETZ-HOLDEN's model of the Upper Triassic. The direction and size of displacement of the individual continents are shown by arrows (Fig. 6). All bauxite deposits occur on the northern border of the Tethys, in the tropical and subtropical zones.

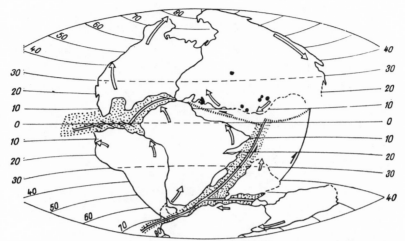

Fig. 6. Locations of the continents in the Upper Triassic according to DIETZ and HOLDEN (1970). ● bauxite area

In the Jurassic period the continental plates drifted farther away from one another (Fig. 7). The bauxite areas were still confined to the northern shore of the Tethyan sea. By the way, this was the period witnessing the minimum of bauxite formation with regard to the entire Mesozoic era. Territorially, they fall, all, in the tropical to subtropical zone.

The formation of karstic bauxites, which gradually increased in the Lower Cretaceous, attained its maximum in Upper Cretaceous time. Throughout the northern coastline of the Tethys, from Portugal as far as Afghanistan, bauxite deposits were formed (Fig. 8). This entire zone fell into the subtropical climatic range. According to a number of bauxite specialists, large-scale lateritic bauxite formation on the Indian and African plates began at the end of the Cretaceous, to attain its full-scale development during the Tertiary. According to the DIETZ-HOLDEN model, this is well conceivable paleoclimatologically.

Unfortunately, no model of this kind has been designed for the Tertiary. Using an interpolation, however, the author could ascertain that the karstic and lateritic bauxites of Tertiary age developed all in the tropical and subtropical zones, too.

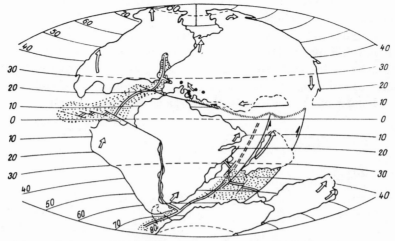

Fig. 7. Locations of the continents in the Upper Jurassic according to DIETZ and HOLDEN (1970). ● bauxite area

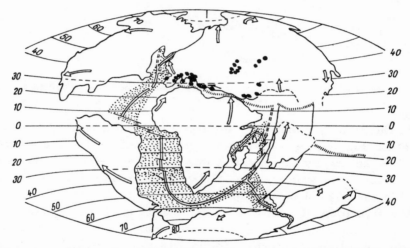

Fig. 8. Locations of the continents in the Upper Cretaceous according to DIETZ and HOLDEN (1970). ● bauxite area

Summing up the above, let us conclude that the theory of continental drift has allowed to give a satisfactory paleoclimatological interpretation of all bauxites formed during the geological history. The general framework of bauxite formation must have been provided for by the climatic zones controlled by the continental drift. However, bauxite did not develop everywhere in the favourable climatic zones either, and the amounts of bauxites formed in the various geological epochs were extremely different. According to the author's opinion, the Globe's tectonic evolution must have been largely responsible for this phenomenon as well.

According to the author's calculations, the total of the presently known bauxites of the Globe is 27 billion tons in round figure. 70% of this quantity are lateritic bauxites, 28% karstic and 2% are of the so-called "sedimentary" type. The author determined the major tectonic units comprising the bauxite deposits and has calculated the percentage distribution of the bauxite reserves:

	karstic bauxites	lateritic bauxites
1. in orogenic belts	85.6%	5%
2. on continental plates	14.4%	90%
3. on oceanic plates	—	5%

It can be seen that the overwhelming majority of the karstic bauxites occur on the territory of the orogenic belts, while the lateritic bauxites are concentrated on the continental plates. The author believes that this difference in their tectonic position is the main reason for the differences of their geological setting, their composition and origin.

If the orogenic belts have provided the optimum for karstic bauxite formation, then it will be logical that their formation was primarily restricted to the orogenic belts within the favourable climatic zones. However, the distribution of karstic bauxites is not identical even within the orogenic belts:

1. geosynclinal belts	70.4%
2. median masses	7.9%
3. intracontinental orogenic belts	7.3%
4. fore-deeps	—
Total	85.6%

Within the geosynclinal belts about 60% of the bauxites belong to eu-, 40% to miogeosynclinal areas.

A question intriguing students for a long time is to explain: why are some orogenic belts, for instance the so-called Mediterranean zone of the Alpine orogeny, extremely abundant in bauxites, as a contrast to the lack of bauxites in other belts such as the so-called circum-Pacific belt of the Alpine orogeny. The author has studied the tectonic evolution of the principal orogenic belts from the point of view of bauxite formation in the light of recent literature on plate tectonics. He has come to the conclusion that bauxite deposits could be formed mainly in areas, where continental plates had collided with intervention of microcontinental plates or oceanic microplates. Zones of this kind were in the Paleozoic the Uralides-Turkestanides range, while during the Alpine orogeny the Mediterranean and the Caribbean territories must have played the same role. It was this very complex tectogenetical evolution that provided conditions favourable for bauxitization: island ranges, coastal pene-

plain zones, alternating uplifts and transgressions and wide extension of carbonate rocks.

In zone of frontal collision of vast continental plates with large oceanic
ones, for instance the Andes and the Cordilleras, however, local phases of
emergence are characterized by deposition of continental coarse-detrital sequences. In these areas the role of the carbonate rocks is subordinate and quiet,
erosion-free phases of emergence are almost totally absent. This scheme of
development was unfavourable for bauxite formation.

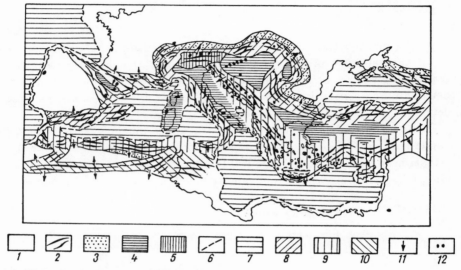

Fig. 9. Tectonic sketch of the Mediterranean region according to AUBOUIN and DURAND-
DELGA (1971). 1. Continental plate; 2. Orogenic belt; 3. Fore-deep; 4. Median mass; 5. Mass
of the Eastern Alps; 6. Limit of sea deeper than 1000 m; 7. Sea basins; 8. Northern geosynclinal zone; 9. Southern geosynclinal zone; 10. Intracontinental orogenic belt; 11. Main directions of tectonic movement; 12. Bauxite areas

Even though found to have been favourable for the accumulation of
bauxite, the Mediterranean belt itself includes bauxite-free areas. In fig. 9 the
tectonic map of this region, reproduced after AUBOUIN et DURAND-DELGA
(1971) and supplemented with information on bauxites, is shown. As evident
from it, the so-called northern arc system contains few bauxites, most of the
deposits being concentrated in the middle segment of the southern arc system.
The rotation of microcontinental plates and of certain orogenic arcs, island
ranges along subduction zones, brought about a most advantageous topography
and it was in these areas that both erosion and detrital sedimentation were
most efficiently handicapped in the zones of bauxite formation. On the contrary, where simple frontal collision took place, as in the areas of the Maghre-

bides, bauxite failed again to be deposited. In the author's opinion, this should account also for the absence of bauxite in the Himalayan range.

In addition to geographic limitations, the bauxite formation in the orogenic belts shows some *chronological regularities* as well. No bauxite was formed in the initial stage of geosynclinal evolution, a period characterized by lasting subsidence and pelagic sedimentation. In the next stage of development bauxite deposits were formed throughout the geosyncline when the long-spanned subsidence was followed by minor uplifts. This stage may differ in duration in dependence of the local characteristics of development of the given orogenic belt. In the Alpine orogenic belt it was the longest in the Dinarides and Hellenides, ranging from the Middle Trias up to the Middle Eocene with a series of bauxitization periods. This is one of the reasons why the greatest bauxite resources of the Mediterranean area occur here.

During the main orogenic period bauxitization discontinue even if the area has remained in the favourable climatical zone. The cause seems to consist in the substantial increase of relief as a result of orogenic movements and, consequently, in the resulting increase of erosion. Even if some bauxite may have been locally accumulated, it would be eroded and mixed up with coarse-detrital molasse-like sediments which were accumulated in large masses. For the same reason, no bauxite deposit could be formed in the so-called tardigeosynclinal phase which followed the orogenic phase either.

It was not until the so-called post-orogenic stage, after the completion of the orogeny, that bauxite formation could reappear if the area remained in the favourable climatic zone. The ancient orogenic belt had been consolidated in the meantime and its tectonical behaviour had become similar to that of the continental plates. Bauxite bodies appeared at the base of the post-orogenic sedimentary sequence of epirogenic character, with lateritic bauxites on the surface of alumosilicate-bearing rocks and karstic bauxites on carbonate sediments. However, these are always of Kazakhstan type!

The above chronological regularity is particularly well manifested in the Alpine orogenic belt. This is why in the Mediterranean area the maximum of bauxitization is in the Upper Cretaceous and its end in post-Middle Eocene time. The main orogenic phase of tectonic development and morphogenesis here falls in the Oligocene and Miocene (Pyrenean, Sava, Styrian, Attic phases), stopping bauxite formation in spite of the preservation of favourable climate. At the Western extremity of the Alpine-Tethyan orogenic belt, the Caribbean region, the initial stage of orogeny lasted longer on, the orogenic phase set in during the Pliocene and did not probably come to completion by the end of this epoch. Accordingly, deposits much younger (Miocene-Pliocene) than the Mediterranean bauxites can be found there (Jamaica, Haiti, etc.). At the eastern end of the Alpine-Tethyan belt, at its linkage with the "Circum-Pacific" belt, we are still in an earlier stage of orogenic development, for the orogenic phase proper has not even begun yet. Here we are in the period of bauxitization in the tectonic respect. And really, it is alone here that karstic bauxites are still being formed today (Rennel, Loyauté islands, Lau islands, Nioué).

Inasmuch as tectonic evolution would have set bauxitization such strict limits, the process of bauxite formation ought to have changed cyclically rather than linearly in the course of geological history. The author has calculated the amounts of karstic bauxites formed during the geological epochs:

Quaternary-Holocene	1.2%	Tertiary+Quaternary	
Miocene-Pliocene	24.4	total	39.6%
Oligocene	0.6		
Paleocene-Eocene	13.3		
Upper Cretaceous	20.0		
Lower Cretaceous	8.1		
Jurassic	3.2	Mesozoic total	35.3%
Trias	4.0		
Permian	3.1		
Carboniferous	15.9		
Devonian	5.9	Paleozoic total	25.1%
Silurian	—		
Ordovician	0.1		
Cambrian	0.2		
	100.0%		100.0%

Accordingly, the variation is really of cyclical pattern, the single maxima within the individual orogenic cycles correspond to the period preceding the orogenic phase. Each maximum is higher than the one preceding it. In other words, the rate of bauxitization increases with time. This is particularly obvious, if the summarized percentage values are considered. Taking into consideration that the Paleozoic era lasted for 345 M. Y., the Mesozoic for 160 M. Y., and the Cenozoic for only 65 M. Y., this increase will prove even more remarkable. In addition to the tectonic and paleoclimatologic causes, it is the general evolution of sedimentation on Earth that we have to do with here, as pointed out at several instances by STRAKHOV (1960) and E. SZÁDECZKY-KARDOSS.

Summarizing, let us conclude that in the light of the new model of plate tectonics a reasonable explanation for the spatial and chronological distribution of the bauxites can be given.

REFERENCES

1. AUBOUIN, J.—DURAND-DELGA, M.: Mediterranée. Encyclopedia Universalis, X, 738—745, 1971.
2. CREER, K. M.: Nature, 219, 41—44, 1968.
3. CREER, K. M.: Nature, 219, 246—250, 1968.
4. DIETZ, R. S.—HOLDEN, J. C.: Journ. Geophys. Research, 75, 26, 4939—4956, 1970.

5. HAMILTON, W.: Geol. Soc. Amer. Bull., **81**, 9, 2553—2576, 1970.
6. SCHWARZBACH, M.: Das Klima der Vorzeit. Stuttgart, 1—275, 1961.
7. SZÁDECZKY-KARDOSS, E.: A föld szerkezete és fejlődése. (Structure and Evolution of the Earth.) Akadémiai Kiadó, Budapest, 1—340, 1968.
8. SZÁDECZKY-KARDOSS, E.: Geonómia és Bányászat, **4**, 1—89, 1971.
9. SZÁDECZKY-KARDOSS, E.: MTA X. Osztályának Közleményei, 113—122, 1972.
10. GORETSKIY, YU. K.: Trudy VIMS, vyp. **5**, nov. ser. Moscow, 1—257, 1960.
11. STRAKHOV, N. M.: Izd. AN SSSR, Moscow, **I. II**, 1960.

39

Reprinted from *Geology* 3:613–616 (1975)

Atlantic evaporites formed by evaporation of water spilled from Pacific, Tethyan, and Southern oceans

Kevin Burke
Department of Geological Sciences
State University of New York at Albany
Albany, New York 12222

ABSTRACT

The distribution of salt diapirs around the Atlantic Ocean, both onshore and offshore, when plotted on a closed Atlantic indicates that there are four main areas without salt and three with salt. The occurrence of massive salt deposits formed at the opening of an ocean appears to depend on (1) proximity to saline ocean waters, (2) graben structure and related occurrences of hot-spot volcanism, and (3) climatic zonation, because there are no salt deposits at equatorial or high paleolatitudes. The salt is interpreted to have formed by evaporation of intermittent spills into sub–sea-level graben over periods of a few million years.

EXTENT OF SALT DEPOSITION INFERRED FROM DIAPIR DISTRIBUTION

Discussion of the origin of the evaporite deposits that fringe the Atlantic and formed in early opening has been confused because of uncertainty about their extent (see, for example, papers in Ion [1965], and Burk and Drake [1974]) and because of questionable identifications of igneous (Rona, 1969) and shale (Mascle and others, 1973) structures as salt diapirs. The basic assumption of this paper is that substantial thicknesses of Atlantic salt were deposited only in areas where numerous salt diapirs have been mapped. This simplifying assumption is considered justified because the many kilometres of sediment deposited in Atlantic miogeoclinal wedges are sufficient (Trusheim, 1960) to have induced general diapirism of underlying salt. Using the distribution of salt diapirs, we mapped seven provinces in the opening Atlantic, four without salt (1, 3, 5, and 7) and three with salt (2, 4, and 6); their distribution is sketched on the familiar continental reconstruction of Figure 1.

Inspection of this map leads to the new suggestion that the salt of each saline province formed by spillover from a different ocean. The salt deposits are interpreted as products of evaporation of oceanic waters repeatedly spilled over structural sills into sub–sea-level graben formed in the early stages of continental rupture. This mechanism can produce salt deposits with a maximum thickness comparable to graben relief ($\sim$1 to 2 km, Burke and Whiteman, 1973, Table 2) and typically will produce salt deposits with both sebkha and diagenetic replacement features in repeated horizontal and vertical cycles of increasing salinity (see Wardlaw and Nicholls, 1972, for descriptions of examples from Sergipe on the east coast of Brazil and Gabon on the west coast of Africa). A Jurassic episode of repeated spillover is considered responsible for province 4 evaporites in the Gulf of Mexico, which were contiguous with those of the Old Bahama Channel and those now preserved off the mouth of the Casamance River in Senegal (Aymé, 1965). Pacific Ocean waters are here suggested to have been the source of these salt deposits. The Tethys is a less likely source, because no saline basins of appropriate age are known between Morocco and Senegal or the Gulf of Maine and the Old Bahama Channel.

The Tethys is thought to have provided the water from which a second group of evaporites, those of province 6, formed in Late Triassic and Early Jurassic time by spillover into graben now preserved in Morocco (Faure-Muret and Choubert, 1971) and offshore Nova Scotia and Newfoundland (Pautot and others, 1970; Amoco and Imperial Oil Companies, 1973; McIver, 1972). The third and youngest group of evaporites, those of province 2, formed when waters of the Southern Ocean spilled over the sill formed by the youthful Walvis–Rio Grande ridge to flood graben preserved offshore and on land in West Africa, Brazil, and possibly also in the Benue Trough (Burke and others, 1971).

STRUCTURAL CONTROL OF SALT DISTRIBUTION

It has been suggested elsewhere (Dewey and Burke, 1974) that continental rupture commonly develops from the linking of rifts formed over hot-spot domal uplifts. The lateral boundaries of the thick peri-Atlantic salt bodies are generally the rift walls of members of the complex of more than 100 linked rift valleys that developed during opening of the Atlantic. Four of the terminal boundaries of the salt deposits are barriers caused by hot-spot volcanism and associated high topography marking the sites of former domal uplifts. These features controlled the flow of spilled ocean water through the graben system and restricted the distribution of evaporites. In the South Atlantic the salt deposits of province 2 are restricted to an area between the

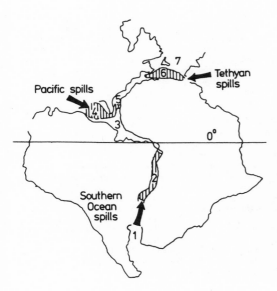

Figure 1. Africa, the Americas, and Iberia in their relative positions prior to the opening of the Atlantic (based mainly on Dewey and others, 1973; Le Pichon and Fox, 1971; and Bullard and others, 1965). Abundant diapirs in the shaded provinces (2, 4, and 6) indicate the extent of massive salt deposition. Diapirs are rare in provinces 1, 3, 5, and 7, and this is interpreted to mean an absence of massive salt. Approximate position of the Jurassic and Cretaceous paleoequator for Africa and South America indicates that provinces 1, 3, and 7 were in moist latitudes during rifting, whereas provinces 2, 4, 5, and 6 were in latitudes where evaporation was likely to exceed precipitation. The most important of many sources reviewed in compiling the distribution of diapirs were as follows: province 1, Emery and others (1975), Urrien and Zambrano (1973); province 2, Brognon and Verrier (1966), Belmonte and others (1965), Pautot and others (1973), Asmus and Ponte (1973); province 3, Asmus and Ponte (1973), Campos and others (1974), Delteil and others (1974); province 4, Antoine and others (1974), Meyerhoff and Hatten (1974), Aymé (1965); province 5, Sheridan (1974); province 6, Pautot and others (1973), Montadert and others (1974); province 7, Roberts (1974), Montadert and others (1974). Ewing and others (1974) was a major source of information for the entire Atlantic.

Tristan hot-spot trace (Walvis–Rio Grande ridge) and the site of igneous activity associated with the development of the triple-rift junction at the Niger delta (Cabo de San Agostinho granite; Vandoros, 1966). The southern termination of the salt bodies of province 6 off Morocco and in the Gulf of Maine corresponds to the zone of high topography associated with the White Mountain magmatism and its Jurassic trace in the westernmost Kelvin sea-mounts. Dietz (1973) pointed out that the occurrence of several kilometres of Mesozoic and Cenozoic carbonate in the Bahamas is best explained by considering the bank to overlie volcanic rocks produced at a hot spot associated with the opening of the Atlantic. Le Pichon and Fox (1971) recognized the concentricity of the Bahamas and the continental margin south of Newfoundland about an early Atlantic opening pole; Burke and others (in prep.) attributed this concentricity to the existence of a short-lived hot-spot trace below the Bahamas. This trace forms the northern boundary of the province 4 salt deposits in the Old Bahama Channel and at the mouth of the Casamance River on the African side of the Atlantic. Thus, four of the six terminal boundaries of the salt deposits are hot-spot boundaries. The remaining two, in Morocco and Mexico, are perhaps transform fault boundaries, because evidence of important transform motion in those areas at about the right time has been cited by Dewey and others (1973) and Silver and Anderson (1974).

CLIMATIC CONTROL OF SALT DISTRIBUTION

A necessary condition for filling ocean-flooded graben with salt is that evaporation should exceed the supply of fresh water. This generally applies in the horse latitudes but not in the more humid climates of higher and lower latitudes. Some information about the latitudes of the Atlantic graben during flooding is available from paleomagnetic measurements (McElhinny, 1973), although inconsistent poles from the then-contiguous continents suggest that this information is of limited worth (Burke and others, 1973).

Application of paleomagnetic results to the distribution shown in Figure 1 indicates that the paleolatitudes of provinces 1 and 7 during flooding—Aptian time for province 1, Campanian to Paleocene time for province 7—appear to have been too high (>30°) to accommodate a climate likely to produce evaporites; this accounts for stratigraphy in which normal marine deposits immediately overlie continental, including lacustrine, sediments (Bolli and others, 1975). Province 3 had a paleolatitude of <10° during graben flooding in Albian time, and equatorial rain there was probably sufficient to prevent desiccation except very locally (Asmus and Ponte, 1973). Province 5 lay between lat 15°N and 25°N during flooding and would be expected to have been desert. There are two main possible reasons for the apparent absence of salt from province 5. Either the complex of graben between the bulge of Africa and the eastern coast of North America may have been too poorly linked to allow completion of the seaway until spreading was well underway and waters could circulate freely to maintain normal salinity, or large rivers draining the moist latitudes of North America may have maintained a lake

that did not precipitate salt, just as the Volga today keeps the Caspian Sea fresh. A combination of these mechanisms is likely. In general, major rivers do not drain into graben forming at continental rupture, because drainage from the elevated shoulders takes most of the water away from the rift (for example, the Congo River), but there is some indication in the Baltimore Canyon sediment lens that a major river entered the young ocean in the neighborhood of the Long Island triple-rift junction (Burke and Dewey, 1973) at about this time. This may have helped to prevent evaporite accumulation in province 5. The saline provinces (2, 4, and 6) were in latitudes between 10° and 20° during graben flooding, and the formation of salt in them is as would be expected.

DEEP- OR SHALLOW-WATER SALT?

Schmalz (1969) pointed out that precipitation of salt in deep-ocean waters is possible if the precipitating brines underlie waters of normal salinity, because brines are dense and may not mix. However, in a system such as that represented by the opening Atlantic, high bottom temperatures at spreading centers in the rifts would induce convection, and a density-stratified ocean hundreds of kilometres wide could not persist. Isolated brine pools restricted to the axis of spreading, such as exist today in the Red Sea, would not have been stable over a large enough area to have precipitated the peri-Atlantic salt. For this reason, Atlantic salt is here considered more likely to have formed in subaerial graben of oceanic depths. Sediments such as those observed below the salt horizon in the Gulf of Mexico by Ladd and others (1975) are to be interpreted as graben fill deposited on young oceanic crust. These sediments might closely resemble the sediments brought into the Ethiopian Afar by the Awash River today. The resemblance between the environments of deposition of the Messinian salts of the Mediterranean (Ryan and others, 1973) and those of the opening Atlantic is intriguing. Salt deposition consequent on restriction of oceanic circulation by barriers is apparently associated with both the early and late stages of the Wilson cycle of ocean opening and closing (Dewey and Burke, 1974).

CONCLUSIONS

The distribution of salt deposits around the opening Atlantic is seen as controlled by three main factors: (1) Proximity to an ocean from which saline waters may spill. Each of the three saline provinces appears to occupy a graben complex into which the waters of a different ocean have spilled and evaporated. (2) The structure of the linked rift complexes associated with ocean opening. Rift walls bound the salt deposits, and topographic highs associated with hot-spot volcanism at rift junctions have restricted salt distribution. (3) Salt deposits are confined to latitudes in which evaporation rates are likely to exceed rainfall.

Drilling in the next few years may establish whether or not there is a substantial volume of salt off the east coast of the United States. Present evidence of the absence of diapirs indicates that there is no such salt.

REFERENCES CITED

Amoco and Imperial Oil Companies, 1973, Regional geology of the Grand Banks: Bull. Canadian Petroleum Geology, v. 21, p. 479-503.

Antoine, J. W., Martin, R., Pyle, T., and Bryant, W. R., 1974, Continental margins of the Gulf of Mexico, *in* Burk, C. A., and Drake, C. L., Geology of continental margins: New York, Springer-Verlag, p. 683-694.

Asmus, H. E., and Ponte, F. C., 1973, The Brazilian marginal basins, *in* Nairn, A.E.M., and Stehli, F. G., eds., The ocean basins and margins, Vol. 1, The South Atlantic: New York, London, Plenum Pub. Corp., p. 87-132.

Aymé, J. M., 1965, The Senegal salt basin, *in* Ion, D. C., ed., Salt basins around Africa: London, Inst. Petroleum, p. 83-90.

Belmonte, Y., Hirtz, P., and Wenger, R., 1965, The salt basins of the Gabon and the Congo, *in* Ion, D. C., ed., Salt basins around Africa: London, Inst. Petroleum, p. 55-74.

Bolli, H. M., Ryan, W.B.F., and scientific party, 1975, Basins and margins of the eastern South Atlantic: Geotimes, v. 20, p. 22-24.

Brognon, G. P., and Verrier, G. B., 1966, Oil and geology in Cuanza basin of Angola: Am. Assoc. Petroleum Geologists Bull., v. 50, p. 108-158.

Bullard, E. C., Everett, J. E., and Smith, A. G., 1965, The fit of the continents around the Atlantic, *in* Symposium on continental drift: Royal Soc. London Philos. Trans, v. 258A, p. 41-51.

Burk, C. A., and Drake, C. L., eds., 1974, The geology of continental margins: New York, Springer-Verlag, 1009 p.

Burke, K., and Dewey, J. F., 1973, Plume generated triple junctions: Key indicators in applying plate tectonics to old rocks: Jour. Geology, v. 81, p. 406-433.

Burke, K., and Whiteman, A. J., 1973, Uplift rifting and the breakup of Africa, *in* Tarling, D. H., and Runcorn, S. K., eds., Implications of continental drift to the Earth sciences, Vol. 2: London, Academic Press, p. 735-755.

Burke, K., Dessauvagie, T.F.J., and Whiteman, A. J., 1971, Opening of the Gulf of Guinea and geological history of the Benue depression and Niger delta: Nature Phys. Sci., v. 233, p. 51-55.

Burke, K., Kidd, W.S.F., and Wilson, J. T., 1973, Relative and latitudinal motion of Atlantic hot-spots: Nature, v. 245, p. 133-137.

Campos, C.W.M., Ponte, F. C., and Miura, K., 1974, Geology of the Brazilian continental margins, *in* Burk, C. A., and Drake, C. L., eds., The geology of continental margins: New York, Springer-Verlag.

Delteil, J.-R., Valery, P., Montadert, L., Fondeur, C., Patriat, P., and Mascle, J., 1974, Continental margin in the northern part of the Gulf of Guinea, *in* Burk, C. A., and Drake, C. L., eds., The geology of continental margins: New York, Springer-Verlag, p. 297-313.

Dewey, J. F., and Burke, K., 1974, Hot spots and continental breakup: Implications for collisional orogeny: Geology, v. 2, p. 57-60.

Dewey, J. F., Pitman, C. C., Ryan, W.B.F., and Bonnin, J., 1973, Plate tectonics and the evolution of the Alpine system: Geol. Soc. America Bull., v. 84, p. 3137-3180.

Dietz, R., 1973, Morphologic fits of North America/Africa and Gondwana: A review, *in* Tarling, D. H., and Runcorn, S. K., eds., Implications of continental drift to the Earth sciences, Vol. 2: London, Academic Press, p. 865-872.

Emery, K. O ,Uchupi, E., Bowin, C. O., Phillips, J., and Simpson, E.S.W., 1975, Continental margin off western Africa: Cape Francis to Walvis Ridge: Am. Assoc. Petroleum Geologists Bull., v. 59, p. 3-59.

Ewing, J., Ewing, M., Windisch, C., and Aitken, T., 1974, Underway marine geophysical data in the North Atlantic, June 1961-January 1971, Pt. F, *in* Talwani, M., ed., Seismic reflection profiles in Lamont-Doherty survey of the world ocean: Palisades, N.Y., Lamont-Doherty Geological Observatory.

Faure-Muret, A., and Choubert, G., 1971, Tectonics of Africa: Paris, UNESCO Earth Science Ser., 602 p.

Ion, D. C., 1965, Salt basins around Africa: London, Inst. Petroleum, 125 p.

Ladd, J. W., Buffler, R. T., Watkins, J. S., and Worzel, J. L., 1975, Multichannel seismic reflection results from the Gulf of Mexico: EOS (Am. Geophys. Union Trans.), v. 56, p. 381-382.

Le Pichon, X., and Fox, P. J., 1971, Marginal offsets, fracture zones and the early opening of the North Atlantic: Jour. Geophys. Research, v. 76, p. 6294-6308.

Mascle, J., Bornhold, B., and Renard, V., 1973, Diapiric structures off Niger delta: Am. Assoc. Petroleum Geologists Bull., v. 57, p. 1672-1678.

McElhinny, M. W., 1973, Palaeomagnetism and plate tectonics: New York, Cambridge Univ. Press, 358 p.

McIver, N. L., 1972, Cenozoic and Mesozoic stratigraphy of the Nova Scotia shelf: Canadian Jour. Earth Sci., v. 9, p. 54-70.

Meyerhoff, A. A., and Hatten, C. W., 1974, Bahamas salient of North America, *in* Burk, C. A., and Drake, C. L., eds., The geology of continental margins: New York, Springer-Verlag, p. 429-446.

Montadert, L., Winnock, E., Delteil, J.-R., and Grau, G., 1974, Continental margins of Galicia-Portugal and Bay of Biscay, *in* Burk, C. A., and Drake, C. L., eds., The geology of continental margins: New York, Springer-Verlag, p. 323-342.

Pautot, G., Auzende, J. M., and Le Pichon, X., 1970, Continuous deep sea salt layer along North Atlantic margins related to early phase of rifting: Nature, v. 227, p. 351-354.

Pautot, G., Renard, V., Daniel, J., and DuPont, J., 1973, Morphology, limits, origin, and age of salt layer along the South Atlantic African margin: Am. Assoc. Petroleum Geologists Bull., v. 57, p. 1658-1671.

Roberts, D. G., 1974, Structural development of the British Isles, the continental margin and the Rockall Plateau, *in* Burk, C. A., and Drake, C. L., eds., The geology of continental margins: New York, Springer-Verlag, p. 343-360.

Rona, P., 1969, Possible salt domes in the deep Atlantic off northwest Africa: Nature, v. 224, p. 141.

Ryan, W.B.F., Hsü, K. J., and others, 1973, Initial reports of the Deep Sea Drilling Project, Vol. 13: Washington, D.C., U.S. Govt. Printing Office, 514 p.

Schmalz, R. F., 1969, Deep water evaporite deposition: A genetic model: Am. Assoc. Petroleum Geologists Bull., v. 53, p. 798-803.

Sheridan, R. E., 1974, Atlantic continental margin of North America, *in* Burk, C. A., and Drake, C. L., eds., The geology of Continental margins: New York, Springer-Verlag, p. 391-408.

Silver, L. T., and Anderson, T. H., 1974, Possible left-lateral early to middle Mesozoic disruption of the south-western North American craton margin: Geol. Soc. America Abs. with Programs, v. 6, no. 7, p. 955-956.

Trusheim, F., 1960, Mechanism of salt migration in northern Germany: Am. Assoc. Petroleum Geologists Bull., v. 44, p. 1519-1540.

Urrien, C. M., and Zambrano, J. J., 1973, Geology of the basins of the Argentine continental margin and Malvinas Plateau, *in* Nairn, A.E.M., and Stehli, F. G., eds., The ocean basins and margins, Vol. 1, The South Atlantic: New York, London, Plenum Pub. Corp., p. 135-166.

Vandoros, P., 1966, Idades absolutas das rochas igneas da Regiao do Cabo, Pernambuco: XX Congr. Bras. Geol. Publicacao, no. 1, p. 65-66.

Wardlaw, N. C., and Nicholls, G. D., 1972, Cretaceous evaporites of Brazil and West Africa: Internat. Geol. Cong., 24th, sec. 6, p. 43-58.

ACKNOWLEDGMENTS

I thank Jim Hays for stimulating my interest in Atlantic evaporites and Bert Bally, John Dewey, Jeff Fox, and Warren Hamilton for very helpful reviews.

MANUSCRIPT RECEIVED JULY 18, 1975

MANUSCRIPT ACCEPTED AUGUST 20, 1975

40

Reprinted from *Implications of Continental Drift to the Earth Sciences*,
vol. 2, D. H. Tarling and S. K. Runcorn, eds., Academic Press,
London, 1973, pp. 1057–1068

Plate Tectonics and Evolution of
Precambrian Crust

A. M. GOODWIN

*Department of Geology,
University of Toronto,
Toronto, Ontario, Canada*

Iron Formation

Introduction

Precambrian sedimentary iron formations are common to all continents. The principal type which is particularly common in early Proterozoic rocks, is classified as Lake

Superior type. As described by Gross (1970a p. 20) this class of iron formation is a characteristically thin-banded cherty rock with iron-rich layers corresponding to various sedimentary facies. The rocks are particularly free of clastic material. The textures and sedimentary features are remarkably alike in detail wherever examined. The sequence of dolomite, quartzite, red and black ferruginous shale, iron formation, black shale and argillite, in that order from bottom to top, is so common on all continents as to be considered almost invariable. Volcanic rocks are nearly always present somewhere in the succession.

Continuous belts of Lake Superior type iron formation typically extend for hundreds or thousands of miles along the margins of geosynclines and basins. They apparently formed in fairly shallow water on continental shelves or margins. As stressed by Gross, it is still uncertain whether the iron and silica contained in the iron formation was derived from a land or a marine source.

This type of cherty iron formation is the host rock for the rich hematite–geothite ore bodies of Australia, Africa, Brazil and Venezuela, Quebec–Labrador and Lake Superior region in North America, Orissa and Bihar states in India, and Krivoy Rog and Kursk areas in USSR (Gross, 1970a p. 20). They include by far the largest and

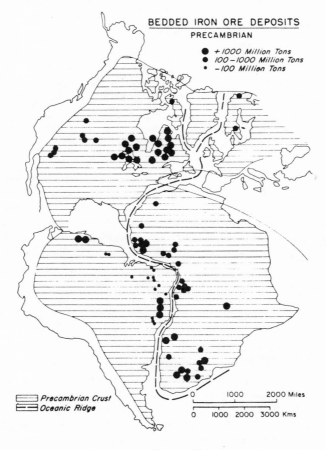

FIGURE 9(a)

BEDDED IRON ORE DEPOSITS   PRECAMBRIAN

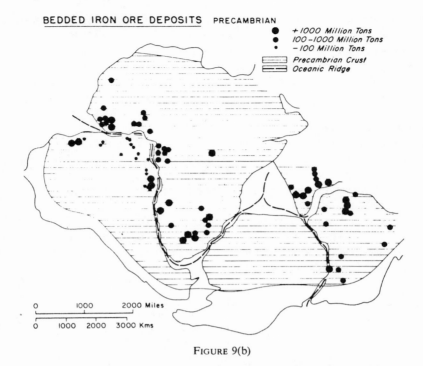

FIGURE 9(b)

FIGURE 9   Plot of bedded iron ore deposits in Precambrian iron formation: (*a*) Atlantic reconstruction (Bullard *et al.*, 1965), (*b*) Gondwanaland reconstruction (Veevers *et al.*, 1971). Iron deposits are plotted in three size ranges. All iron ore data is taken from United Nations Survey of World Iron Ore Resources (1970). The approximate distribution of Precambrian crust including those parts with Phanerozoic cover are illustrated. Oceanic ridges are Cenozoic in age.

economically most valuable iron formations in the world. As discussed below, the largest proportion of Lake Superior type iron formation appears to have been deposited in early Proterozoic time approximately 1900–2100 m.y. ago.

A second class of bedded iron formation, called Algoma type, is present in nearly all early Precambrian belts of volcanic and sedimentary rocks in many shields (Gross, 1970a p. 20). Characteristically, they are thin-banded or laminated with interbands of ferruginous grey or jasper chert, hematite and magnetite. Carbonate and sulphide facies are locally prominently developed. Usually a number of lenses of iron formation are present in a volcanic belt. They are intimately associated with various volcanic rocks and fine grained clastic sediments. As stressed by Gross, the associated rocks indicate a eugeosynclinal environment for their formation and a close relationship in time and space with volcanic activity.

Much Algoma type iron formation is present in older Precambrian rocks particularly Archean basement terrain upon which much of the younger, predominantly Lake Superior type iron formations have been deposited together with their associated supracrustal assemblages.

A possible third class of oolitic bedded iron formation is prominently developed in the Pretoria Series of the Transvaal system in South Africa. It resembles Minette type iron formation (Gross, 1970a p. 20–21) and has been so classified in the Survey of World Iron Ore Resources (U.N., 1970). Although not normally developed in Precambrian rocks, the arenaceous oolitic ironstone of the Pretoria Series is reported to constitute an extraordinarily persistent bed from 1·5 to 9 m thick which can be traced for hundreds of kilometers. It is reported by P. A. Wagner to contain 'one of the greatest iron deposits in the world, if not the greatest' (Haughton, 1969 p. 144). Normally Minette type iron formations are most abundant in Mesozoic and Tertiary rocks, particularly in Europe.

Precambrian Distribution

The distribution of Precambrian bedded iron ore deposits of the world taken from Survey of World Iron Ore Resources (U.N., 1970) is illustrated in Figs. 9a, b and 10c. All types of Precambrian bedded iron deposits are included. Massive Precambrian iron deposits (Bilbao, Magnitnaya, Kiruna and Taberg types) are excluded. The deposits are shown in three size categories: over 1,000 million tons, 100 million to 1000 million tons, and less than 100 million tons. The total listed reserves of Precambrian bedded iron deposits amounts to 469,409 million metric tons. This constitutes 60% of the listed total world reserves of 782,500 million metric tons.

The greatest concentrations by far of Precambrian bedded iron ore deposits are present in western Australia, India, South Africa, western Africa, Brazil and Venezuela in South America, Lake Superior and Quebec–Labrador regions of North America, and Krivoy Rog and Kursk regions in western USSR. In each of these regions iron formations of different Precambrian ages may be present as for example in Australia, India, Africa, and Lake Superior regions. Commonly iron formations of different ages in a particular region are separated by major unconformities demonstrating that conditions favourable for iron deposition recurred over major intervals of Precambrian time.

Figure 9a illustrates the distribution of Precambrian bedded iron ore deposits in a pre-drift Atlantic reconstruction after Bullard *et al.* (1965). Particularly notable are the iron concentrations in southern Africa, Brazil and Venezuela, and Lake Superior and Quebec–Labrador regions in North America. Figure 9b illustrates corresponding distributions in India and Australia in a pre-drift Gondwanaland reconstruction after Veevers *et al.* (1971). In Fig. 10c the main reserves present in the Krivoy Rog and Kursk regions with lesser reserves in Siberia, China and North Korea are illustrated in a Laurasian reconstruction after Hurley & Rand (1969).

Early Proterozoic Distribution

In Figs. 10a, b, c only ore reserves in iron formations of established or interpreted early Proterozoic age (c. 1800–2100 m.y.) are shown.

Total ore reserves in iron formations of Lower Proterozoic age are calculated at approximately 372,807 million metric tons or 48% of total listed world reserves. Immense tonnages as listed in the Survey of World Iron Ore Resources (tables accompanying regional appraisals) are present in western Australia, India, South Africa, western Africa, Brazil and Venezuela in South America, Lake Superior and Ungava regions of North America and Krivoy Rog and Kursk areas of USSR. Even though some iron formations may be incorrectly dated the fact is clear that early Proterozoic iron formations of the world contain vast iron ore reserves totalling approximately 50% of the total listed world reserves.

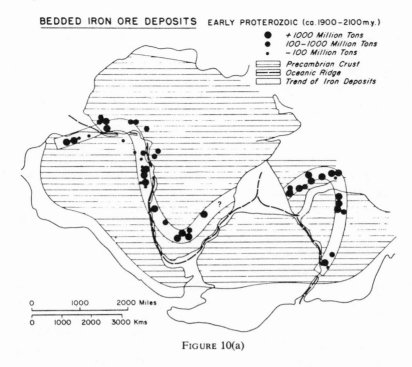

BEDDED IRON ORE DEPOSITS EARLY PROTEROZOIC (ca.1900-2100 m.y.)

- +1000 Million Tons
- 100-1000 Million Tons
- -100 Million Tons

Precambrian Crust
Oceanic Ridge
Trend of Iron Deposits

0   1000   2000 Miles

0   1000   2000   3000 Kms

FIGURE 10(a)

In Western Australia, the Hamersley Group which contains the Brockman Iron Formation, includes the Woongarra Volcanics dated at 2000 ± 100 m.y. (Trendall, 1970 p. 5). In South Australia the Middleback Group of iron formation lies at the base of more than 9000 m of metasediments; granulites from the Gneiss Complex yield an age of 1780 ± 120 m.y. which sets a minimum depositional age for this iron formation (Trendall, 1970 p. 6).

In India, iron ores of the Singhbhum Group with an age of 2000–1700 m.y. are to be correlated with the Iron Ore Stage of Dunn and the Iron Ore Series of Krishnan. These iron formations would then correlate with the Australian iron formations of the Hamersley Group in Western Australia (Crawford, 1969 p. 383).

Principal iron ores in South Africa which include some of the largest concentrations of iron in the world are contained in the Dolomite and Pretoria Series and equivalents of the Transvaal System dated between 2300 m.y. and 1950 m.y. (Haughton, 1969). Hematite-magnetite deposits of similar age are present in Namibia (Southwest Africa) within the Kaokoveld of Pre-Damara age as indicated by the age of the Huabian tectonic episode estimated at approximately 1760 m.y. (Haughton, 1969 pp. 286, 479).

Large iron reserves are present in western Africa particularly in Angola, Congo, Gabon, Ivory Coast, Liberia, Sierra Leone and Guinea. In some localities, the exact age of the iron formation is in dispute. Various workers report either an Archean or a Lower Proterozoic age depending upon whether they consider the iron formation to be an integral part of the Archean basement or to belong to young Lower Proterozoic infolds in the basement complex. Thus the Kambui schists of Sierra Leone show whole rock Rb–Sr ages of about 2700 m.y. But the pelitic and iron-bearing metamor-

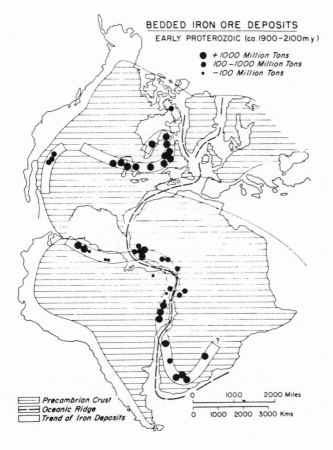

BEDDED IRON ORE DEPOSITS
EARLY PROTEROZOIC (ca.1900-2100 m.y.)

● +1000 Million Tons
● 100-1000 Million Tons
• -100 Million Tons

Precambrian Crust
Oceanic Ridge
Trend of Iron Deposits

0   1000   2000 Miles
0   1000   2000   3000 Kms

FIGURE 10(b)

phic rocks in the Marampa Formation of Sierra Leone and similar sections in the Nimba Range, Liberia, appear to be about 2200 m.y. old (Hurley *et al.*, 1971). Precise dating of the iron formation in adjoining countries is not yet available. Assessment of available literature indicates that a significant part, if not most of the itabirite, is of Lower Proterozoic age.

In South America, the main Precambrian iron formations contain immense reserves of iron ore particularly in Brazil and Venezuela. The Quadrilatero Ferrifero in Brazil contains one of the largest concentrations of iron ore in the world ranking in this respect with Western Australia. The Itabirite Group of the Minas Series has not been adequately dated. It is known to fall in the interval 2500–1350 m.y. (Dorr, 1969). It is commonly considered to be of Lower Proterozoic (i.e. Middle Precambrian) age. Exposures of itabirite are present northward to the Amazon River. Similar iron-bearing rock in Venezuela contain very large reserves of iron ore. There is controversy concerning the age of Venezuelan iron formation. Bucher (1952) places the iron-ore bearing formation above the older rocks of the Imataca Series. Lopez (1956) also considers the iron ferruginous quartzites to be part of a series which overlies the early

335

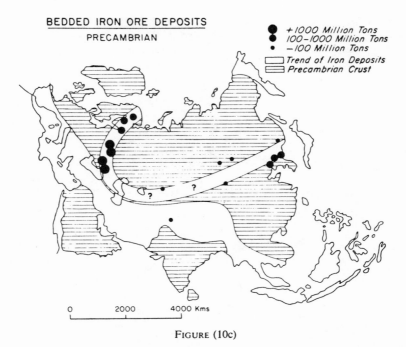

BEDDED IRON ORE DEPOSITS
PRECAMBRIAN

● +*1000 Million Tons*
● *100–1000 Million Tons*
· *–100 Million Tons*
▭ *Trend of Iron Deposits*
▤ *Precambrian Crust*

0   2000   4000 Kms

FIGURE (10c)

FIGURE 10   Plot of representative bedded iron ore deposits in early Proterozoic iron formations: (*a*) Gondwanaland reconstructions (Veevers *et al.*, 1971) (*b*) Atlantic reconstruction (Bullard *et al.*, 1965) (*c*) Laurasian reconstruction (Hurley, 1970). Approximate trends of early Proterozoic iron formations are delineated. The approximate distribution of Precambrian crust including those parts with Phanerozoic cover are illustrated. Oceanic ridges are Cenozoic in age. The distribution of iron deposits in southern Asia e.g. India, is not repeated in Fig. 10c.

Precambrian complex (Guyana System). Evidence supporting the younger age for the itabirite is presented by Morrison (in press). However, according to the Stratigraphic Lexicon of Venezuela (2nd ed., 1970), the ferruginous horizons are contained in the intensely metamorphosed and structurally complex sequence of Lower Precambrian; i.e. Archean age.

In North America, the main iron-bearing rocks in the Lake Superior and Quebec–Labrador regions have been comparatively well dated. Although the iron-bearing Animikian rocks of Lake Superior region are known to be younger than 2000 m.y. (Morey, 1970) there is still some uncertainty whether the minimum age is 1685 ± 24 m.y. (Faure & Kovach, 1969), 1750 ± 25 m.y. (Peterman, 1966) or 1900 ± 200 m.y. (Hurley *et al.*, 1961). A corresponding Lower Proterozoic age of 1900–2000 m.y. has been indicated by recent radiometric dating for the cirum-Ungava iron formations including Quebec–Labrador and Hudson Bay regions (Fryer, 1972). Similar ages have been suggested for the Mary's River hematite deposits of Baffin Island (Gross, pers. comm.).

In the eastern part of the Baltic Shield, the ferruginous quartzites of Kola Peninsula and Karelia are reported to be confined almost exclusively to the Lower Proterozoic age of 2600–2000 m.y. (Chernov, 1970) and to everywhere overlie the Archean basement (Goryainov, 1970). Ukrainian Shield data shows that cycles of iron-silica deposition occurred repeatedly in various stages ranging in age from 3500 m.y. to 1800–1700 m.y. (Semenenko, 1970 p. 16). However, the main development of iron deposits in the Krivorozhsko-Kremenchogsky syncline which is more than 200 km long, is found to be approximately 2000–1800 m.y. old (Semenenko, 1970 p. 11). Very large iron deposits in the Kursk region are considered to be Lower Proterozoic age (Plaksenko *et al.*, 1970). Indeed, the Kursk iron formations (Belgorod region) are considered to correspond to the Upper Krivoy Rog series of the Ukrainian Shield (Kalyaev, 1966 quoted in Zaitsev, 1970 p. 12). Iron formations of Kazakhstan have much in common with analogous formations of the world. The estimated absolute age of the formations is 2600–1900 m.y. (Novokhatsky, 1970). Data on other Siberian iron formations indicate that Lower Proterozoic ages are represented. Little information is available on Precambrian iron formations of China and North Korea other than their designated Proterozoic age.

Interpretation

The trends of Early Proterozoic iron formations as defined above are illustrated in Figs. 10a, b, c within pre-Cretaceous continental reconstructions. The reconstructions used for the Atlantic (Fig. 10a), Gondwanaland (Fig. 10b) and Laurasia (Fig. 10c) regions are as previously listed. The global reconstruction of Pangaea (Fig. 11) is after Dietz and Holden (1970).

Lower Proterozoic iron formations occupy distinct trends in the reconstructed land masses. Starting in southeastern Gondwanaland (Fig. 10a) the iron trend crosses Australia and India as shown. To the west of the gap in the Veever *et al.* reconstruction it loops around southern Africa and passes northward and westward alternately through western Africa and eastern South America with un-determined trend north of Venezuela. The North American iron trend (Fig. 10b) is uncertain in western United States, well-established in the Lake Superior and Ungava regions, including the Hudson Bay loop, and indefinite in Baffin Island. In northern Europe, USSR and Asia (Fig. 10c) the iron trend runs southerly through the Baltic Shield to the Kursk and Krivoy Rog regions. Scattered occurrences to the east culminating in those of North Korea–China suggest an easterly trend as illustrated.

The global iron trend of Lower Proterozoic iron formations is summarized in the Pangaela reconstruction (Fig. 11). Regardless of the precise age of specific iron formations, vast quantities of iron ore lie along this global trend. A salient feature of the global iron trend is that it lies mainly well within pre-drift continental reconstructions rather than towards the margins of Precambrian crust. The exceptions are in western United States and North Korea–China. Elsewhere the indicated iron trends are intracratonic. If we accept the shallow marine shelf environment (Gross, 1970b) and designated ages of these iron formations, then this particular environment extended intermittently along the proposed iron trend during early Proterozoic time.

Another salient feature of the global iron trend is its subparallelism to modern plate boundaries mainly those represented by oceanic ridges but including some fold belts. The degree of parallelism to oceanic ridges is moderately high in Gondwanaland including the loop in southern Africa (Fig. 10a); in the north Atlantic region the iron trend is subparallel to the mid-Atlantic ridge system; in USSR and Asia it is in part parallel to the Himalayan fold belt.

The relations suggest that some global plate pattern in early Proterozoic time

PANGAEA

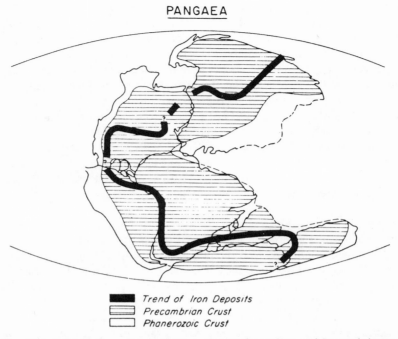

FIGURE 11   Global distribution of early Proterozoic iron formations and Precambrian crust shown on a reconstruction of Pangaea, the universal continent (Dietz & Holden, 1970). The zones of undetermined or questionable continuity are indicated by question marks. Approximately 50% of the total iron reserves of the world listed in the United Nations Survey of World Iron Ore Resources (1970) fall on or in the immediate vicinity of this global iron trend.

influenced the designated alignment of geosynclines and basins in which accumulated vast quantities of iron-silica sediments. Some unique feature of crustal evolution would be required to explain the scope and alignment of the deposits. The answers are, in all likelihood, to be found in first-order features of crustal evolution. In this regard it is noted that this time of massive iron-silica deposition shortly followed that all-important Archean-Proterozoic transition during which the main Precambrian shields of the world were stabilized. Unique consequences of this stabilization were (1) large-scale development of stable continental shelves, (2) global transition to predominantly oxidizing environment, and (3) accelerated biological evolution and activity. These interrelated developments may, in turn, have been a function of some significant stage in core-mantle-crust evolution. They may all have been causally related to development of the iron formations.

The indicated time-space restriction of major iron-silica deposition makes it unlikely that any single factor such as simple change in atmospheric environment is adequate explanation. Rather, it is suggested that explanation is to be sought within the context of crustal evolution in terms of (1) a global pattern of plate boundaries in early Proterozoic time; (2) introduction of large quantities of iron possibly from deep sources to the resulting basins and geosynclines and (3) deposition of iron formation in response to the prevailing oxidizing environment and accelerated organic activity. The exact nature of the early Proterozoic ocean environments is largely unknown.

Possibly the Labrador sea between Ungava and Greenland which is parallel in trend to the Labrador iron formations offers an example of the type of confined seaway in which early Proterozoic iron formations were deposited. Detailed stratigraphic studies of world iron formations are required to resolve these important problems.

The indicated distribution pattern of Lake Superior type iron formation is subject to test by paleomagnetic methods. Preliminary work on iron formation in the Lake Superior region by Symons (1966) has yielded an indicated Animikian (early Proterozoic) paleomagnetic pole position of 94° W, 28° N, site in western United States, with some evidence of pole reversals during deposition of the iron formation. Additional paleomagnetic studies of iron formation and associated rocks to test the proposed global pattern are clearly desirable.

Summary and Conclusions

1. The plate tectonics model, suitably modified in accordance with the great time interval involved, offers satisfactory explanation for several established and indicated features of Precambrian crust.
2. Archean volcanic belts contain within them the main petrogenetic suites and display the fundamental characteristics of Cenozoic island arcs. Accordingly, the Archean belts may have formed at sialic-oceanic interfaces in a similar manner to their modern counterparts.
3. The world-wide distribution pattern of Precambrian sedimentary iron formations in general and those of early Proterozoic age in particular indicate their deposition according to some global plate boundary pattern which is generally subparallel to that of modern oceanic ridges at accreting plate boundaries. The origin and pattern of the iron formations may be related to some type of plate motion that affected Proterozoic crust. If so, the apparent subparallelism in the distribution of early Proterozoic iron formations and modern plate boundaries supports Hurley's contentions of the stability of 'pre-drift' continental nuclei. This would impose some broad restrictions on the degree of pre-Cretaceous dispersion, disorientation and reassemblage of Precambrian crustal nuclei.
4. Additional widespread examination of the Precambrian record in the context of the plate tectonic model is warranted.

Acknowledgements

I have drawn freely from the United Nations Survey of World Iron Ore Resources 1970 as well as from unpublished papers presented at the International Symposium on the Geology and Genesis of Precambrian Iron/Manganese Formations and Ore Deposits, Kiev, 20–25 August 1970 which was organized jointly by Unesco and the International Association of Geochemistry and Cosmochemistry of the International Union of Geological Sciences in collaboration with the Ukrainian Academy of Sciences.

I have benefited greatly from discussion of iron formation problems with G. A. Gross, Geologic Survey of Canada. However, interpretative aspects including any shortcomings of this paper are my responsibility.

References

Bucher, W. H., 1952. Geologic structure and orogenic history of Venezuela, *Geol. Soc. Amer., Memoir* 49.

Bullard, E. C., Everett, J. E. and Smith, A. G., 1965. The fit of the continents around the Atlantic, *Phil. Trans. Roy. Soc. London, Ser. A*, **258**, 41.

Chernov, V. M., 1970. The ferruginous-siliceous formations of the eastern part of the Baltic Shield; Intl. Symposium on the geology and genesis of Precambrian Iron/Manganese Formations and Ore Deposits; Kiev.

Crawford, A. R., 1969. Indian, Ceylon and Pakistan: new age data and comparisons with Australia, *Nature*, **223**, 380–4.

Dietz, R. S. and Holden, J. C., 1970. Reconstruction of Pangaea: Break-up and dispersion of continents, Permian to Present, *J. Geophys. Res.*, **75**, 4939–4956.

Dorr, J. V. N., 1969. Physiographic, stratigraphic and structural development of the Quadrilatero Ferrifero, Minas Gerais, Brazil, *Geol. Surv. Professional Paper 641—A*.

Faure, G. and Kovach, J., 1969. The age of the Gunflint Iron Formation of the Animikie Series in Ontario, Canada, *Geol. Soc. America Bull.*, **80**, 1725–36.

Fryer, B. J., 1972. Age determinations in the Circum-Ungava geosyncline and the evolution of Precambrian banded iron-formations, *Can. J. Earth Sci.*, accepted for publication.

Goryainov, P. M., 1970. Structural and stratigraphic disposition of the iron-formation of the Baltic Shield and some concepts of the Lower Precambrian geology; Intl. Symposium on the geology and genesis of Precambrian Iron/Manganese Formations and Ore Deposits; Kiev.

Gross, G. A., 1970a. Nature and occurrence of iron ore deposits. *In:* Survey of World Iron Ore Resources. United Nations, New York.

Gross, G. A., 1970b. Continental drift and the depositional environments of principal types of Precambrian iron-formation; in Abstracts, Intl. Symposium on the geology and genesis of Precambrian Iron/Manganese Formations and Ore Deposits: Kiev.

Haughton, S. H., 1969. Geological History of Southern Africa, Geol. Soc. of South Africa, 535 pp.

Hurley, P. M., 1970, Distribution of age provinces in Laurasia, *Earth Planet. Sci. Letts.*, **8**, 189–96.

Hurley, P. M., Fairbairn, H. W., Pinson, W. H. and Hower, J., 1961. Unmetamorphosed minerals in the Gunflint Formation used to test the age of the Animikie, *J. Geol.*, **70**, 489–92.

Hurley, P. M., Leo, G. W., White, R. W. and Fairbairn, H. W., 1971. Liberian age province (about 2,700 m.y.) and adjacent provinces in Liberia and Sierra Leone, *Geol. Soc. America Bull.*, **82**, 3483–90.

Hurley, P. M. and Rand, J. R., 1969. Pre-drift continental nuclei, *Science*, **164**, 1229–42.

Lopez, V. M., 1956. Venezuelan Guiana. *In:* Handbook of South American Geology, edited by W. F. Jenks, *Geol. Soc. Amer., Memoir 65*.

Morey, G. B., 1970. Mesabi, Gunflint and Cuyuna Range, Minnesota; Intl. Symposium on the Geology and Genesis of Precambrian Iron/Manganese Formations and Ore Deposits; Kiev.

Novokhatsky, I. P., 1970. Pre-Cambrian ferruginous-siliceous formations of Kazakhstan; Intl. Symposium on the geology and genesis of Precambrian Iron/Manganese Formations and Ore Deposits; Kiev.

Peterman, Z. E., 1966. Rb-Sr dating of Middle Precambrian metasedimentary rocks of Minnesota, *Geol. Soc. America Bull.*, **77**, 1031–44.

Plaksenko, N. A., Koval, I. K. and Shchegolev, I. N., 1970. Ferruginous cherty formation of the Precambrian of the territory of the Kursk Magnetic Anomaly; Intl. Symposium on the geology and genesis of Precambrian Iron/Manganese Formations and Ore Deposits; Kiev, August, 1970.

Semenenko, N. P., 1970. The iron-chert formations of the Ukranian Shield; Intl. Symposium on the geology and genesis of Precambrian Iron/Manganese Formations and Ore Deposits; Kiev.

Stratigraphic Lexicon of Venezuela, 2nd ed., 1970. Boletin de Geologia, Publ. Esp. No. 4. Ministerio de Minas E Hidrocarburos, Caracas, Venezuela.

Symons, D. T. A., 1966. A palaeomagnetic study of the Gunflint, Mesabi and Cuyuna iron ranges in the Lake Superior region, *Econ. Geol.*, **61**, 1336–61.

Trendall, A. F., 1970. The iron formations of the Precambrian Hammersley Group, Western Australia, *Geol. Survey of Western Australia, Bulletin* **119**, 366 pp.

Veevers, J. J., Jones, J. G. and Talent, J. A., 1971. Indo-Australian stratigraphy and the configuration and dispersal of Gondwanalahd, *Nature*, **229**, 383–8.

Zaitsev, Yu. S., 1970. Geology of the Precambrian cherty-iron formations of the Belgorod iron-ore region; Intl. Symposium of the Geology and Genesis of Precambrian iron/manganese Formations and Ore Deposits; Kiev.

TO FIND A GIANT, FIND THE RIGHT BASIN

H. Douglas Klemme

Lewis G. Weeks Associates Ltd.

[*Editor's Note:* Figure 1 referred to in the first paragraph is reproduced here from the first part of the series and the Bibliography is taken from the third part of the series.]

BASINS may be classified as to their location and character, regional tectonics, and by their relation to continental and oceanic crust. This classification follows that of Weeks, 1952 (Fig. 1, Part 1 and Fig. 2).

Those basins located on the stable continental area have been termed cratonic and those basins located in the mobile zone, on the edge of the craton, between the continental crust and the oceanic crust, are termed intermediate. Oceanic basins underlain by oceanic crust exist. However, prospecting techniques are not sufficiently developed to permit significant commercial development (see Part 1 of this series, OGJ, Mar. 1, p. 85).

The simple basins (cratonic, interior—Type 1) and the less simple multicycle basins (cratonic, intracontinental, foreland shelf to intermontane-Type 2) developed most often as extensive Paleozoic embayments upon the stable cratonic platform, and result in nearly all the Paleozoic hydrocarbon reserves.

When orogenic Mesozoic sediments were deposited over Paleozoic platform sediments Type 2 or composite intracontinental basins were formed (see Fig. 2).

The relative field sizes within these basins appear to depend upon whether or not a significant regional arch was present and growing during the basins' development. If this large trap is present, the basins' largest field may contain 50-70% of the basins' total reserves (often as supergiant gas fields), whereas if the large structural trap is not provided, small structural and stratigraphic traps become the loci of accumulation, with

the largest field being less than 10% of the basins' total reserves.

Evaporites, if present, tend to increase the amount of hydrocarbons that are recoverable in the Type 2 basins. These basins, for the most part, contain high-gravity crudes. Generally high-sulfur crude occurs in Paleozoic carbonates, whereas mixed to low-sulfur crudes are found in clastic facies.

Over 75% of the world's gas reserves are found in clastic facies in Type 2 basins. Much of the structure in these basins and in the Type 3 basins is related to basement-controlled uplifts typical of cratonic areas.

Both Type 1 and Type 2 basins appear to be primarily related to the process of continental accretion, although when drifting starts, these basins may be literally split in two by sea-floor spreading (i.e. North Sea, Aquitaine, and Pechora).

The possible effect of sea-floor spreading during the development of a basin is first noted in the Type 3 cratonic, graben or half-graben basins. They may represent a rift developed on the stable craton during the initial stage of drift (Fig. 3).

It is believed that drift starts as a rift, and as the sea floor spreads, the rift opens. However, in many cases, the rift has remained dormant. The classic African Rift Valley appears to be an example, and many of the Type 3 basins in Africa are related to this trend.

These basins, at least those where sufficient data is available, are graben-like depressions which have dropped down in Mesozoic or Tertiary time

(post sea-floor spreading).

Where these basins developed close to the continental margin, or where they opened into Mesozoic or Tertiary seaways, evaporites and reefs sometimes developed. In some of these basins, development of evaporites which cap the source shales and reservoir rocks may enhance the generation, accumulation, and retention of hydrocarbons, resulting in high hydrocarbon yields.

The structural traps in Type 3 basins are developed as horsts or block uplifts with the largest field averaging 25% of the basins' reserves. In general, these basins produce low-sulfur, high-gravity, waxy crudes. Half of these basins and more than 80% of their reserves have been found since World War II.

Many of the Type 2 basins have a tendency to locally incorporate Tertiary-Cretaceous rift-like troughs resembling Type 3 basins (North Sea, Erg Oriental, Erg Occidental, Aquitaine) in either an incipient or pronounced manner. In one such basin (Norwegian Tertiary trough), which is superimposed on the North Sea/Netherlands-German basins, recent drilling at Ekofisk (OGJ, May 25, 1970, p. 23) has resulted in what appears to be a giant or supergiant field.

If the rift remains essentially dormant, the Type 3 cratonic, graben or half-graben basin may simply continue to subside. However, if for some unknown reason the rift begins to open due to the introduction of oceanic crust—for example the Red Sea—then a stable coastal (Weeks, 1952) or "pull-apart" basin is formed. The rift continues to open, and each half of

Basins—regional relation through time.

Fig. 3

DIAGRAMMATIC

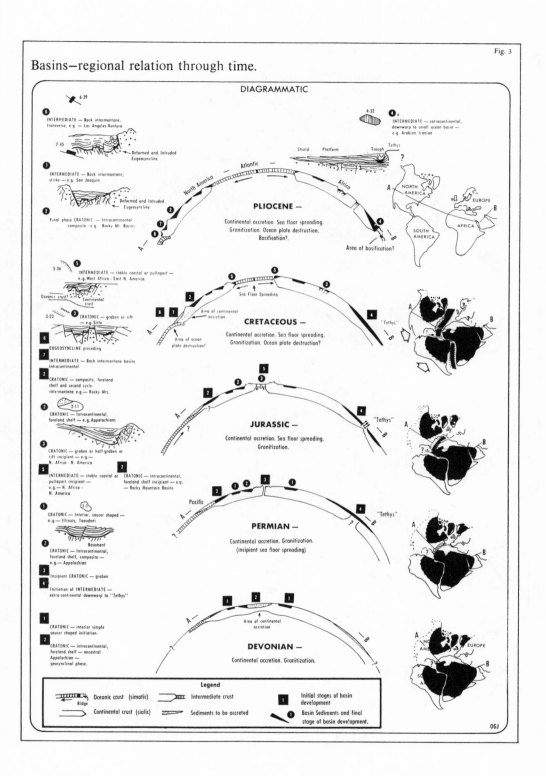

The giants and the supergiants, the geology of their location

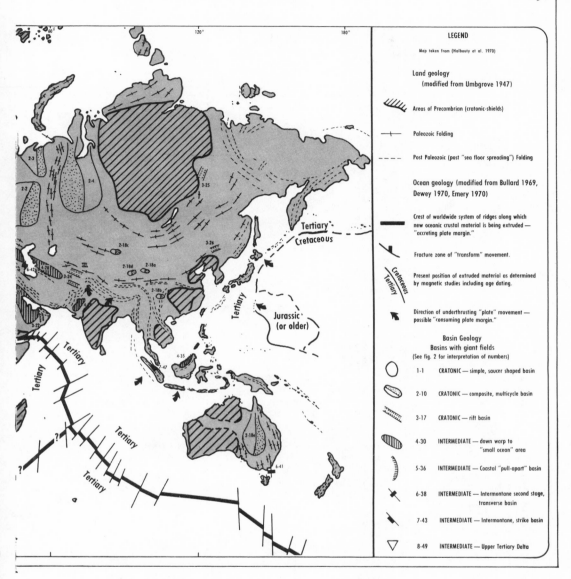

Fig. 1

LEGEND

Map taken from (Halbouty et al. 1970)

Land geology
(modified from Umbgrove 1947)

Areas of Precambrian (cratonic-shields)

Paleozoic Folding

Post Paleozoic (post "sea floor spreading") Folding

Ocean geology (modified from Bullard 1969,
Dewey 1970, Emery 1970)

Crest of worldwide system of ridges along which
new oceanic crustal material is being extruded —
"accreting plate margin."

Fracture zone of "transform" movement.

Present position of extruded material as determined
by magnetic studies including age dating.

Direction of underthrusting "plate" movement —
possible "consuming plate margin."

Basin Geology
Basins with giant fields
(See fig. 2 for interpretation of numbers)

1-1   CRATONIC — simple, saucer shaped basin

2-10   CRATONIC — composite, multicycle basin

3-17   CRATONIC — rift basin

4-30   INTERMEDIATE — down warp to
"small ocean" area

5-36   INTERMEDIATE — Coastal "pull-apart" basin

6-38   INTERMEDIATE — Intermontane second stage,
transverse basin

7-43   INTERMEDIATE — Intermontane, strike basin

8-49   INTERMEDIATE — Upper Tertiary Delta

Fig. 2

Some common characteristics of hydrocarbon-bearing areas

Crustal Types / Basin Types	CRATONIC					INTERMEDIATE		
	Type 1 Interior—simple, saucer shaped 	**Type 2** Intracontinental — including composite foreland shelf, remote interior to intermontane 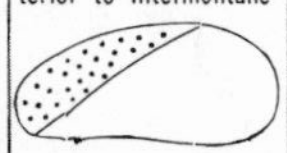 basin	**Type 3** Graben or half graben or rift 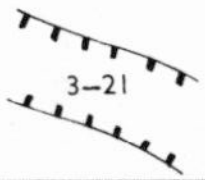	**Type 4** Extracontinental downwarp to small ocean basin 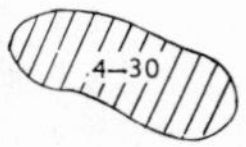	**Type 5** Stable coastal or pull-apart 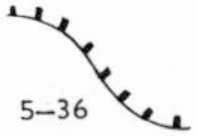	**Type 6** Intermontane (back)—Second stage, transverse 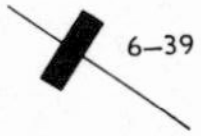	**Type 7** Intermontane (back)—second stage, strike 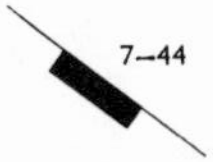	**Type 8** Upper Tertiary Delta 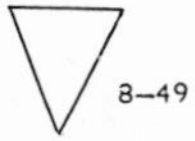
	1—1	2	3—21	4—30	5—36	6—39	7—44	8—49
Characteristics Description	Moderately large, flat, dish-like, single cycle basins—often with Paleozoic platform or embayment facies — with basement uplifts and sedimentary structures located on interior portion of craton.	Large (subcontinental) to small (intermontane) basins with two or more cycles of deposition—first cycle often platform or shelf sediments and second cycle orogenic clastics—located around margin of craton (Requires further subdivision)	Small to medium sized graben or rift faulted basins. Possible incipient or dormant "sea floor spreading"	Large to medium sized basins located along margin of small ocean basins. With development along "tethys" of 4a platform and trough 4b trough only — and development along small ocean basins 4c, downwarp into Mediterranean-like sea.	Linear Coastal basins with downfaulted Mesozoic and Tertiary located along cratonic margins—possibly separated by "sea floor spreading."	Small, second cycle Tertiary clastic basins lying transversely upon the deformed eugeosynclinal troughs which were developed parallel to certain coast lines possibly underthrusting of "oceanic plates"	Same as 6 with different orientation and greater linear extent.	Upper Miocene to recent birdfoot deltas in coastal areas where major continental drainage areas exist.
Number of basins (with number of fields) and % of world reserves with giant fields (0.5 billion +)	1 (1)	22 (82) 26%	8 (30) 8%	9 (84) 55%	1 (1)	6 (21) 7%	5 (12) 2%	2 (10) 1.5%
Number of producing basins without giant fields	4	16	6	14	6	2	10	1
Estimated* number of undeveloped - basins (various stages of exploration)	25+	23+	18±	15±	37±	14±	34±	19±
Estimated number of undeveloped basins lying offshore or in inland seas.	5 (inland seas)	3 (inland seas)	6	10	37	14	24	19 extending offshore
% of world reserves from all producing basins.	1%*	22%	5%	57%	0.6%	7.4%	2.5%	4.5%
Estimated % of world reserves by AGE 1. Tertiary 2. Mesozoic 3. Paleozoic	1. ...... 2. 2.2% 3. 3.98%	1. ...... 2. 2.50% 3. 3.50%	1. 1.38% 2. 2.47% 3. 3.15%	1. 1.23% 2. 2.75% 3. 2.%	1. 1.40% 2. 2.60% 3.	1. 1.98% 2. 2.2% 3. 3.1%	1. 1.95% 2. 2.5% 3.	1. 1.100% 2. 2.? 3. .....
Estimated % of world reserves from 1. sandstone 2. carbonate 3. other	1. 1.31% 2. 2.67% 3. 3.2%	1. 1.75% 2. 2.23% 3. 3.2%	1. 1.55% 2. 2.42% 3. 3.3%	1. 1.52% 2. 2.47% 3. 3.0.5%	1. 1.83% 2. 2.17% 3. 3....	1. 1.97% 2. 2.2% 3. 3.1%	1. 1.98% 2. 2.2% 3.	1. 1.100% 2. 2.? 3. .....
Estimated hydrocarbon recovery*1. Range*2. highest recovery lowest *2. giant field basins (Average)*3. All producing basins (Average)*3	1/10 Average Average to low	1/16 Above average Average	1/10 High High	1/80 Very High Low to High	1/10 Average to above average Low	1/25 Very high Very high	1/100 Very High Average to above average	1/1.2 Very High Very High
% of basin's reserves in largest field—average age of								

	all giant field basins: 16% / all producing basins: 13%	Variable / 25%	27% / 40%	22% / 20%	50% / 50%	Variable / 27%	27% / 15%	6% / 6%
STRUCTURAL-TRAP TYPES 1. Giant fields / 2. Other fields	1. Major basement arch 2. Small sedimentary traps	1. Major basement arch with combination structural-stratigraphic traps. 2. Mainly small stratigraphic-structural traps	1. Major uplifts, reefs 2. Tilted horst blocks	1. Large Anticlines 2. Small Anticlines	1. Combination structural-stratigraphic over salt swells and under salt 2. Diapirs	High basement relief 1. Large combination fault block and faulted anticline with stratigraphy change 2. Same + compression anticline 2. Diapirs	1. Large fault block anticlines 2. Same	1. Large flowage structures and "roll-over" anticlines 2. Same
TRAP SIZE	Generally small	Generally small with exceptions of giant fields on major arch	Both large and small	See above	Moderate	Moderate (Thick pay zone)	Moderate	Small to Moderate
Relation to evaporites	Evaporites do not appreciably affect hydrocarbon recovery	Evaporites, if extensive, appreciably affect recovery rate	Evaporites associated with increased hydrocarbon recovery rates	Evaporites appreciably affect recovery rate	Possibly related to higher recovery	Generally absent	Generally absent or limited	Flowage features developed
Relation to unconformities	Often related to accumulation	Often related to accumulation	Seldom	About half the accumulations related — 4 (giants related)	Often related	Giants often related	Often related	Deltaic deposition
Hydrocarbon character 1. Gravity / 2. Sulphur / 3. Gas	High (Paleozoic carbonates—mixed to low) Average	High High; sandstones Very high (70% of giant fields are gas) Generally low (some basins almost all gas)		Intermediate to high Intermediate to high (Hi 4a, lo 4b, mixed 4c Average (High in 4b) Low	Intermediate Low Low	Intermediate to low (Variable) Often High (Variable) Low	Variable Variable Average to low	Intermediate to High Low Extremely High
Basins with Giant Fields (and Super Giant fields [*6])	1-1 Illinois	2.2 Volga-Ural[°] 2.3 Pechora 2.4 West Siberian[°] 2.5 North Sea 2.6 Netherlands-German[°] 2.7 Aquitaine 2.8 Erg-Oriental[°] 2.9 Erg Occidental[°] 2.10 Fort Polignac 2.11 Alberta 2.12 Appalachian 2.13 Powder River 2.14 Uinta 2.15 San Juan 2.16 Anadarko-Amarillo[°] 2.17 Oklahoma-Mid Continent 2.18 Permian 2.18a Kansu 2.18b Szechuan? 2.18c Dzungaria? 2.18d Tsaidam? 2.18e Coopers Creek?	3.19 Sirte[°] 3.20 Suez[°] 3.21 Red Sea 3.22 Oman 3.23 Dnieper-Donetz 3.24 Bukhara-Tadzik 3.25 Ver Khayano Vilyuy 3.26 Heilungkiang	4a Arabian-Iranian[°] 4:27 4a Pre-Caucasus-Mangyshlak-Turkmen[°] 4:28 4b Indus 4:29 4c North Slope[°] 4:30 4c East Texas[°] 4:31 4c Gulf Coast 4:32 4c Tampico 4:33 4a East Venezuela 4:34 4b North Borneo 4:35	Cabinda 5-36	Ventura 6-37 Los Angeles 6-38 Maracaibo[°] 6-39 Pivura 6-40 Gippsland 6-41 Baku 6-42	San Joaquin 7-43 Sacramento 7-44 Cook Inlet 7-45 Magdelena 7-46 Central Sumatra[°] 7-47 Java Sea?	Mississippi 8-48 Niger 8-49
Basin without giant fields but production established	Williston Michigan Paris	North Texas Ahnet-Mouydir Pre-Caspean Surat Oriente Cuyo Chaco Neuquen Magellan Big Horn West Kansas Wind River Denver Green River Cantabrian	Cambay Rhine Graben Amadeus Bahia San Jorge Qattara	4a Gaziantep-Sirte 4b Carpathian foredeep 4b Po Valley 4b East Italy 4a Cyrenaica-N Egypt 4b East Borneo 4b Molasse 4b Assam 4b Barru Potwar 4b Rharb 4b East Pakistan 4b Rumania foredeep 4c Vera Cruz	Gabon Cuanza Carnarvon Serguipe-Alagoas Agadir Perth	Vienna Santa Barbara	South Sumatra North Sumatra Java Sea? Cuyama-Salinas Interior Iran Taiwan West Japan (Honshu) Taranaki Panonian Transylvania	Nile

*1 Excluding Antarctica. *2 Primary recovery of oil or gas and based upon the projected estimates of ultimate recovery for each producing basin. *3 Range—**Lowest** yield rate over multiple of lowest yield for **highest** yield rate. *3 bbl of oil (or thermal gas equivalent) per cubic mile of basin sediments; 0-25,000, low; 25-50,000, average; 50-100,000, above average; 100-150,000, High; 150,000+, Very High. *4 Less than 0.5%—Low. *5 Based on 35% gas to 65% oil for world's hydrocarbon reserves. *6 Basins with "Super Giant" fields—4 billion bbl +. G—gas~, O—oil~.

the Type 3 graben basin is separated by oceanic distances throughout time This results in the formation of the Type 5 intermediate stable coastal or pull-apart basins which are located along the coastal zones of the continental margin (i.e. West Africa, eastern South America, eastern North America, East Africa and West Australia).

Often a major ridge or uplift is developed down-dip and offshore on the outer edges of these basins. Recent evidence suggests that some of these ridges mark initial intrusions along the axis of the initial rift—for example, early Permian to Jurassic off the East Coast of U.S. (Emery 1970). These basins are most often filled with clastic sediments. Evaporites are often present.

So far, Type 5 basins have been tested only on their up-dip or landward side with the result that their relative recovery of hydrocarbons appears to be low.

The gamble being taken by the industry in West Africa, eastern South America and eastern Canada indicates that there may be reason to expect much higher recovery seaward in offshore areas which technology is now making available. Both "Cabinda B" and Emeraude Marine (OGJ, Mar. 16, 1970) are relatively recent discoveries found in the offshore or seaward side of this type of basin.

Small ocean basins (Menard, 1967) such as the Tethyan zone, Arctic Canada basin, Gulf of Mexico, Caribbean, Mediterranean and China seas have been referred to as "some of the most promising areas in the world for petroleum accumulation" where the younger age and richer character of sediments enchance the potential prospect (Weeks, 1965 and Hedberg, 1970, p. 28).

Intermediate, marginal extracontinental basins (Type 4) are located adjacent to and often extend into these small ocean basins. They are a group which can be further subdivided on the basis of their tectonic form (4a vs. 4b) and history of development (4a and 4b vs. 4c). This has been outlined in Fig. 2. They constitute 20% of the world's producing basins, and over 50% of the world's reserves.

Although these figures result from the presence of the unique Arabian/Iranian basin and its supergiant

fields, other basins within the group (around the Gulf of Mexico, Mangyshlak and possibly North Slope) have extremely high yields.

The potential for extremely high recovery rates in some of these basins is always present. However, the geologic parameters which qualify this category of basins' ability to yield hydrocarbons are so variable that within this class a wide range of recovery is encountered.

As more of the fundamental characteristics of small ocean basins are unravelled by the earth scientists and oceanographers, perhaps a greater predictability of hydrocarbon recovery will be possible within this class of basins.

Both basement-controlled structures and marginal mobile zone structures are present in these basins, with the result that cratonic trends often extend at angles into the fold trends developed along the margins of colliding oceanic and continental plates.

The hydrocarbons found within some of these basins include a greater amount of intermediate-gravity crude containing relatively high amounts of sulfur (4a). When these basins are limited to troughs or foredeeps (4b) and single cycle downwarps into small oceanic areas (4c) low-sulfur, waxy crudes and greater than average gas are found.

Where an oceanic plate appears to underthrust a continental plate (Circum-Pacific and possibly portions of eastern Tethys) or is overridden by the continental plate (western U.S.), small, relatively rich basins are often developed. These basins are located on intermediate crust and have been classified as back intermontane types (Weeks, 1952). They are second-stage Tertiary basins, developed either at right angles (transverse Type 6) or parallel (strike Type 7) to a deformed Mesozoic or older eugeosyncline which occupied the continental borderland. Their sediments consist of a thick series of alternating sands and shales, and production comes from multiple pay sands generally over a wide range of depths.

The recovery of hydrocarbons in these basins (Halbouty et al, 1970) is strongly affected by their stratigraphy, particularly the provenance of reservoir sands in relation to contamination by volcanic material. The

wide range of both hydrocarbon types and relative recovery rates within these basins may be due to their variable stratigraphy, geothermal gradient, and the intensity of deformation associated with the post-Paleozoic fold belts.

The worldwide character of Miocene to Recent deltas is relatively unique and they have been classified as a separate group—(Type 8) Intermediate-Upper-Tertiary delta. The unique character of these basins may be related to late Tertiary continental elevations and the rates of sea-floor spreading. Their hydrocarbon character is likewise unique with extremely high gas reserves and field sizes which seldom reach over 5% of the basins' total reserves.

Within each of the basin types, the range of recoverability seems to be related to such time-honored factors as:

1. Potential trap size and timing of development.

2. Potential reservoir including provenance, character, geometry, and hydrodynamics.

3. Source rock—character and relation to reservoirs.

4. Marine content of basin sediments (only about 5% of the world's reserves are found in nonmarine environments).

5. Intensity of deformation.

6. Recent evidence suggests (Halbouty et al, 1970) that the geothermal gradient or paleogeothermal gradients play a role in the magnitude of hydrocarbon recovery. What relation geothermal gradients and heat flow have to thick crustal areas, the relatively thinner intermediate crustal zones, and the process of sea floor spreading remains to be determined.

Although these factors affect the ultimate recovery in any basin, their emphasis and relative interassociation vary among the several classes or types of basins.

For example, if all factors above are favorable there is a chance that oil in normal quantities will be developed in most basins. However, if in addition to these factors 1) evaporites cover some of the types of basins, 2) there is a higher than normal geothermal gradient in some types of basins, and 3) there are significant unconformities developed in other types of basins—then giants and supergiants might result.

BIBLIOGRAPHY

Alexander, Tom "The Secret of the Spreading Ocean Floors" Fortune, p. 112, February 1969.

Bird, J. M. and Dewey, J. F. "Lithosphere Plate-Continental Margin Tectonics and the Evolution of the Appalachian Orogen" Geol. Soc. of Am. Bull, v. 81, p. 1,031-1060, 1970.

Dewey, J. F. and Bird, J. M., "Mountain Belts and the New Global Tectonics." Journal of Geophys. Research, May 10, 1970.

Dewey, J. F. and Horsfield, B. "Plate tectonics, orogeny and continental growth" Nature, Vol. 225, p. 521-525, Feb. 7, 1970.

Dickinson, William R. "The New Global Tectonics," Geotimes p. 18, April 1970.

Drake, C. L. and Kosminskaya, L. P. "The transition from Continental to Oceanic Crust", Tectonophysics 7 (5-6) p. 363-384, 1969.

Emery, K. O. et al., "Continental Rise off Eastern North America", Bull. AAPG, Vol. 54, No. 1, p. 89, 1970.

Halbouty, M. T., Dott, R. S., King, R. E.,

Klemme, H. D. Meyerhoff, A. A., Shabab, T. "Giant Oil and Gas Fields and Geologic Factors Affecting their Formation", AAPG, Mem. 14, 1970.

Hamilton, W. "Mesozoic California and The Underflow of Pacific Mantle" Geol. Soc. of Am. Bull., V. 80, p. 2,409-2,430, 1969.

Hedberg, H. D. "Continental margins from viewpoint of the petroleum geologist", Bull. AAPG, Vol. 54, No. 1, p. 3-43, 1970.

Knebel, G. M. and Rodriquez-Eraso, G. "Habitat of Some Oil" Bull. AAPG, Vol. 40, No. 4, p. 547-561, 1956.

Menard, H. W. "Transitional types of crust under small ocean basins," Journ. Geophysical Research, V. 72, p. 3,061-3,073, 1967.

Weeks, L. G. "Fractors of sedimentary basin development that control oil occurrence" Bull. AAPG Vol. 36, No. 11, p. 2,096, 1952.

Weeks, L. G. "World Offshore Petroleum Resources". Bull. AAPG, Vol. 49, No. 10, p. 1,680-1,693, 1965.

Continental Drift and Reserves of Oil and Natural Gas

D. H. TARLING

Department of Geophysics and Planetary Physics, University of Newcastle upon Tyne, Newcastle upon Tyne NE1 7RU

> **The necessary physiographic location of deltaic sources deposited during the past 200 million years requires that continental movements be taken into account.**

ANY new concept which is likely to assist in the location and assessment of the remaining reserves of oil and hydrocarbon gas is of obvious importance so it is opportune to use reviews[1,2] of the conditions of formation of the principal hydrocarbon sources and their subsequent migration and preservation in reservoirs to see how these are related to movements of the Earth's surface.

Deposition of Organic Substances

The ability of hydrocarbons to migrate[3] and their complex chemical nature complicates matters but it is generally agreed that they are organically derived[4,5], the oil coming chiefly from animal proteins and the gas including some vegetal content[1]. Sources some 10^7 yr old are almost entirely deltaic and occur independently of latitude, the organic materials being trapped within the rapidly accumulating sediments. The rich terrestrial vegetation of tropical areas means that there may be some latitudinal control on the percentage of gas to oil, but this is unlikely to be as significant as later factors. The location of deltas today is evident but somewhat older deltas exist which receive little or no drainage today and are therefore harder to distinguish physiographically as continental movements have separated the headwaters from deltas which now lie isolated and unexploited on other continents. The formation of mountain systems, along subduction zones or by continental collisions, also drastically changes the drainage pattern so that, for example, dispersed oil fields of the Andes may once have formed part of the proto-Amazon delta[6] when the Amazon drained from Africa. Deltaic areas older than 200 m.y., which covers the "Wegenerian" phase of continental drift, are almost impossible to locate physiographically but are of less economic importance because most productive deltaic areas are less than 200 m.y. old and geological identification is possible, irrespective of age.

Shales are generally considered to have the highest initial

organic content, followed by carbonates and finally sandstones—although sandstones are the principal rock type for major reservoirs[1] and can be a significant contributor to their own reservoirs. But the productivity of source rocks depends on the ease of migration of the hydrocarbons so that the most productive source rocks are the massive carbonates. although the productivity of all source rocks is increased by heating. Deep oceanic sediments accumulate slowly and are unable to trap significant protein matter before oxidation so that deep oceans are not potential oil or gas sources, except where shallower sediments have slumped into them, but much of the hydrocarbon is also likely to be lost at this time.

Because more than 90% of known oil and gas fields[1,7-9] are associated with evaporite deposits the conditions for evaporite accumulation are relevant to the source, migration and preservation by hydrocarbons. Most marine evaporites are formed of anhydrite, with some gypsum, and contain only small amounts of the more soluble halite. Even when halite forms a large part of the deposit, it only represents temporary conditions as it was laid down some 10^2 to 10^3 times faster than anhydrite[10]. Significant deposits cannot therefore have formed by prolonged, simple evaporation of seawater because halite would always predominate. Two depositional environments have been postulated to account for these features—the silled basin and the sabkha. The former[10-12] consists of a pool of water into which only surface waters can flow. As these surface waters evaporate, saline brines form and, being denser, sink to the bottom where the least soluble salts are eventually precipitated ; but the salinity can only rarely increase to the concentration appropriate to halite deposition. "Sabkha" regions[13,14] form from the occasional flooding of supratidal flats which create extensive areas of shallow marine waters. As these waters evaporate and retreat, the least soluble salts are left behind. These deposits are greatest where the flats are bordered by lagoons in which the salinity is already higher than normal seawater. Such extensive flat regions would be more common when orogeny was less active than during the past 20 m.y. and when sealevel was more constant than during the eustatic changes of the most recent Ice Age.

In both environments the depositional area must sink relative to the barrier and sealevel to account for the observed thicknesses of evaporites. Rapid subsidence in areas of low evaporation would cause persistent flooding and the solution of previous salts, as would also happen with rare rainstorms in areas of low evaporation and slow subsidence. So although both types of environment could exist temporarily at any latitude, thick deposits could only form in regions with a very high rate of evaporation, that is where hot, dry conditions are accompanied by strong diurnal on-shore and off-shore winds. Cyclothems and structures in associated sediments indicate intermittent subsidence so that the evaporation rates would have to match the maximum rate of subsidence. Such prolonged subsidence implies movements of the lower crust and upper mantle, as occurs in response to isostatic adjustments, for example during continental separation[15]. For continued deposition, the barriers of a silled basin and the shoreline of a sabkha must maintain their position relative to the subsiding depositional area. A fault controlled structure, where very localized tectonic movements happen to cancel out regional movements, could create this situation, but it is more likely that such persistent barriers or shorelines would result from sedimentary or biological processes. Sandbars could be maintained relative to sealevel if marine currents and a sediment supply were unaffected by the subsidence, but the most persistent type of barrier/shoreline would be formed by shallow reef structures. Reefs in low latitudes would easily maintain their position relative to an intermittently deepening sea as their growth is optimized in shallow, warm, clear waters. These low latitude conditions, necessary for optimum evaporite formation, are bordered by those for the optimum production and preservation of hydrocarbon source materials as the concentration of nutrients in warm, saline waters results in rich organic blooms of microorganisms, the salinity restricts the entrance of larger organisms which could feed on them and the nearby anaerobic bottom conditions are ideal for the preservation of the organic debris[16-18].

Migration

The gradual freeing of hydrocarbons from the original organic matter occurs continuously but is accelerated by small temperature rises[19,20] to which plate tectonic theories and continental drift have varying degrees of significance. Tectonic heating during collision between two continental blocks is likely to be locally severe and drive off all hydrocarbons, but the less disturbed areas and margins of collision zones may be areas of significant oil and gas migration. Such areas can be located from palaeomagnetic measurements of the angle and rate of collision but the shape of the original edges is also critical and this can only be detected by detailed geological and geophysical studies which, among other things, yield the information of sources, migration and reservoirs. Similarly, in subduction zones there is higher heat flow and igneous activity in a region up to 1,000 km wide, but the distribution of heat is not regular and the local controlling factors are only just becoming clear[21] and the estimation of heat flow in different areas, now and in the past, should become more precise as such interacting factors are considered further.

The igneous activity and high heat flow along spreading oceanic ridges are not of any importance within the oceans, where there are no source rocks, but become highly significant when these ridges intersect, or have intersected, continental plates. Local destruction of reserves is likely where significant igneous activity was involved, such as at previous plume centres, but these will also be rimmed by high gas pressures. In general, the high heat flow of the ridges will have been a major cause of migration as the hydrocarbons would move slowly from deep to shallow levels as the regional temperature increased. Thus the numerous, highly productive gas fields of north-west Siberia occur over the continental extension of the Arctic ridge ; similarly the productive oil fields around Los Angeles[22] are where the East Pacific Rise underlies continental rocks. So the identification of present and past intersecting regions is likely to be of great importance.

As the Atlantic and Indian Oceans originated from the break up of the Gondwanan and Laurasian continents, the fractured edges must have been subjected to oceanic ridge heating at their inception. This heating would coincide with the accumulation and migration of organic materials in these areas as the persistent subsidence afforded troughs into which organic material could be trapped and also allowed a thick accumulation of sediments to develop in them, giving rise to burial heating of older sediments. In general, direct heat flow is likely to have been most significant along the margin[23] and burial heating would be dominant within a few hundred kilometres of the margin. This subsidence was usually along pre-existing fractures where these paralleled the opening margins and the deepest subsidence took place on the continental side of fault blocks as they respond to upper mantle and lower crustal flow. Seafloor spreading also seems to be the cause of the formation[24] of marginal basins, such as the Japanese Sea, which also contain shallow sedimentary rocks. Thus the availability of source rocks and heating by burial and elevated heat flow implies that these regions, particularly in the West and Southwest Pacific, are likely to be areas of high hydrocarbon potential.

On the plate tectonic model igneous activity and high heat flow are absent away from plate margins, with the minor exception of kimberlite-type intrusions[21]. Deep burial heating can still occur away from plate margins as a result of upper mantle/lower crustal movements, but significant heating in such areas presumably arises mainly from exothermic chemical and biochemical reactions which may accompany diagenesis and subsequent compaction. Most of these reactions are independent of plate tectonic concepts, but some important reactions, such as between organic materials and anhydrite[13], will occur chiefly in sediments originally deposited at low latitudes.

The final composition of the reserves reflects the nature of the source materials, the temperature and organic activity in the source, during migration and after preservation in reservoirs, and the distance between source and the reservoir. So fatty, heavy grade oils are particularly associated with deltaic conditions[25] where migration has been restricted and the original organic material is often derived from brackish environments. Deltas also tend to have substantial gas associated with their high vegetal content. The variations of sulphur content within oils, or nitrogen within gas, are significant to extraction and refining techniques and prediction of the nature of the rocks through which the hydrocarbons have passed by determining their palaeolatitudinal history will lead to a closer understanding of natural refining processes and therefore to a prediction of the quality of hydrocarbon reserves. Sulphur content[26,27], for example, seems to correlate broadly with the presence of evaporites in the source rocks, but the final composition is probably determined largely by bacterial activity, particularly within the reservoir rocks.

Reservoirs

Various structures may trap oil and gas[5] and plate tectonic activity is involved in each, although some links are tenuous. Stratigraphic traps, for example, are related to the detailed local palaeogeography, but the number of traps in any one region is controlled by broader considerations. For example, structural traps, such as anticlines and domes, are directly related to tectonic processes along subduction and continental collision zones. Thus the gas fields marginal to the Pyrenees owe their origin to the relative movements between Iberia and France and the Argentine fields probably formed in response to movements of the Patagonian arc[6].

Evaporites are again significant because they lubricate tectonic movements, and afford an impervious seal for trapped hydrocarbons. Their ability to flow when the overburden exceeds some 600 m (ref. 28) means that they can create their own structural traps. The role of reefs and sand bars in evaporite formation also means that these offer potential reservoirs close to hydrocarbon sources. The margins of separating continents also provide structures, accompanying the faulting and folding movements, which may act as reservoirs.

Implications for Fuel Supplies

Concepts of continental drift and plate tectonics have a much more fundamental part to play in the evaluation of potential oil and gas reservoirs than in the mere matching of, for example, oil traps bordering the Atlantic. Physiographical location of deltaic sources deposited during the past 200 m.y. can be undertaken effectively only when continental movements are taken into account and similarly the structural evolution of an area should be related to the different distribution, movements, collisions and separations of the continents. The most striking feature is that palaeo-climatic consideration plays a vital part in the origin, migration and accumulation of hydrocarbons. Palaeomagnetic techniques, which offer a precise, economic measure of palaeolatitude, are therefore likely to be of increasing significance.

As technology advances, it should be possible to extract oil and gas from some of the extensive tar shales as well as from inhospitable yet potentially productive areas such as the Labrador Sea, Newfoundland Shelf, northern Greenland and the Canadian Arctic Islands. Nonetheless, the fact that low palaeolatitudes are so significant suggests that the southern continents which, as Gondwanaland, have mostly lain at high to intermediate palaeolatitudes during the past 400 m.y., are unlikely to be as productive as the northern hemisphere—although the size of Gondwanaland implies that some areas must have been at low latitudes for appreciable periods of time. So something like two-thirds of the world's total reserves are likely to be in the present northern hemisphere, which are already the best known areas and the regions of highest exploitation. Major reserves which are largely undetected or unexploited must still exist, for instance, off north-western Australia, the south-west and western Pacific, as well as in uptapped Mesozoic-Cainozoic deltas. It is also clear that the deep oceanic parts of the world are not likely to contribute to future needs for organic fuels. It should now be possible to make a realistic estimate of the world's hydrocarbon potential and thereby evaluate the optimum development of these resources for the benefit of mankind as a whole.

Received Novmeber 16, 1972; revised April 2, 1973.

[1] Halbouty, M. T., Meyerhoff, A. A., Dott, R., King, R. E., and Klemme, H., *Amer. Assoc. Petrol. Geol. Mem.*, **14**, 502 (1970).

[2] Radchenko, O. A., *Geochemical Regularities in the Distribution of the Oil-Bearing Regions of the World*, 312 (Israeli Prog. Sci. Trans., 1968).

[3] Silverman, S. R., *Amer. Assoc. Petrol. Geol. Mem.*, **4**, 53 (1965).

[4] Dott, R. H., and Reynolds, M. J., *Amer. Assoc. Petrol. Geol. Mem.*, **5**, 571 (1969).

[5] Levorsen, A. I., *Geology of Petroleum*, 724 (Freeman, San Francisco, 1967).

[6] McDowell, A. N., *Oil and Gas J.*, **69**, 114 (1971).

[7] Martinez, J. D., *Amer. Assoc. Petrol. Geol. Bull.*, **55**, 810 (1971).

[8] Peterson, J. A., and Hite, R. J., *Amer. Assoc. Petrol. Geol. Bull.*, **53**, 884 (1969).

[9] Stöcklin, J., *Geol. Soc. Amer. Spec. Paper*, **88**, 157 (1968).

[10] Brongersma-Sanders, M., *Marine Geol.*, **11**, 123 (1971).

[11] Woolnough, W. G., *Amer. Assoc. Petrol. Geol. Bull.*, **21**, 1101 (1937).

[12] Raup, O. B., *Amer. Assoc. Petrol. Geol. Bull.*, **54**, 2246 (1970).

[13] Bush, P. R., *Trans. Inst. Min. Metall.*, B79, 137 (1970).

[14] Kinsman, D. J. J., *Amer. Assoc. Petrol. Geol. Bull.*, **53**, 830 (1969).

[15] Bott, M. H. P., and Dean, D. S., *Nature Physical Science*, **235**, 23 (1972).

[16] Phleger, F. B., *Amer. Assoc. Petrol. Geol. Bull.*, **53**, 824 (1969).

[17] Fuller, J. G. C. M., and Porter, J. W., *Amer. Assoc. Petrol. Geol. Bull.*, **53**, 909 (1969).

[18] Kendall, C. G. St. C., and Skipworth, P. A. D. E., *Amer. Assoc. Petrol. Geol. Bull.*, **53**, 841 (1969).

[19] Philippi, H. W., *Geochim. Cosmochim. Acta*, **29**, 1021 (1965).

[20] Welte, D. H., *Amer. Assoc. Petrol. Geol. Bull.*, **30**, 1830 (1965).

[21] Tarling, D. H., *Nature*, **243**, 193 (1973).

[22] McFarland, L. C., and Greutert, R. H., *Oil and Gas J.*, **69**, 112 (1971).

[23] Sleep, N. H., *Geophys. J. Roy. Astron. Soc.*, **24**, 325 (1971).

[24] Karig, D. E., *J. Geophys. Res.*, **76**, 2542 (1971).

[25] Biederman, E. W., *Amer. Assoc. Petrol. Geol. Bull.*, **53**, 1500 (1969).

[26] Hood, A., and Gutjahr, C. C. M., *Abs. Geol. Soc. Amer.*, **4**, 542 (1972).

[27] Ho, T. Y., Drushel, H. V., Koons, C. B., and Rogers, M. A., *Abs. Geol. Soc. Amer.*, **4**, 539 (1972).

[28] Brunstrom, R. G. W., and Walmsley, P. J., *Amer. Assoc. Petrol. Geol. Bull.*, **53**, 870 (1969).

43

Reprinted from *Canadian Jour. Earth Sci.* 11(1):1–17 (1974)

Oil, Climate, and Tectonics[1]

E. IRVING

Earth Physics Branch, Department of Energy, Mines and Resources, Ottawa, Canada

F. K. NORTH

Department of Geology, Carleton University, Ottawa, Canada

AND

R. COUILLARD

Earth Physics Branch, Department of Energy, Mines and Resources, Ottawa, Canada[2]

Received July 25, 1973

Revision accepted for publication September 11, 1973

We identify four sets of factors governing oil occurrence—climate (especially temperature), mineral nutrients, tectonic factors controlling initial basin formation, and tectonic factors controlling preservation of the oil. We argue that all factors are themselves subject to the framework imposed by plate tectonics. If we are to consider all Phanerozoic oil deposits, the only factor capable of quantitative comparison for all the periods is the first one, in that it is partly a function of latitude.

A paleolatitude analysis has been made for both reservoir rocks and preferred source rocks for all petroliferous basins, with results weighted according to total reserves. No statistically satisfactory relationship was found between oil and paleolatitude that would embrace all Phanerozoic deposits. Most Paleozoic oil was formed in rocks deposited in low latitudes, but this may be an accident of preservation. The much larger Mesozoic deposits were similarly related to low paleolatitudes, but this result is heavily biased by the huge reserves of the Persian Gulf. If these are excluded, Mesozoic oil occurs with equal probability in high and in low paleolatitudes. Cenozoic oil is uniformly distributed with respect to paleolatitude.

The distribution of oil with time reveals that 72% of all known oil was probably formed in the late Mesozoic, most of it (60%) in the mid-Cretaceous. The first requirement in any general theory of oil occurrence, therefore, is to understand why so much oil was formed near the present Persian Gulf, and to a lesser extent in Middle America, during such a short interval of geological time. We attempt to show that all four controlling factors were optimized in these two places for this brief time-span. In the timetable of plate tectonics, two large marine embayments opened astride the equator in the late Mesozoic, and these may or may not have been connected through the western Mediterranean. One embayment contained the Persian Gulf, and the other, Middle America. The renewal of mantle convection at about −100 m.y. activated these embayments, abruptly increased the rate of sea-floor spreading, and enlarged the oceanic ridges, causing maximum development of warm, shallow seas and releasing, through igneous activity, greatly increased quantities of mineral nutrients.

The geometry of subsequent plate activity was such that the Persian Gulf was tectonically protected by the rapid northward movement of the Indian plate (which absorbed most of the impact with the Eurasian plate), and the Gulf of Mexico was protected by the northeastward movement of the Antillean arc.

Nous reconnaissons quatre séries de facteurs régissant l'occurrence de pétrole—le climat (principalement la température), les substances nutritives, les facteurs tectoniques contrôlant la formation initiale du bassin, et les facteurs tectoniques contrôlant la conservation du pétrole. Nous démontrons que tous

---

[1]Contribution from the Earth Physics Branch No. 472.

[2]Now with Chevron Standard Ltd., Calgary.

ces facteurs sont soumis aux limites imposées par la tectonique de plaque. Si nous devons considérer tous les gisements pétrolifères du Phanérozoique, le seul facteur capable de supporter une comparaison quantitative pour toutes ces périodes est le premier en ce sens qu'il est fonction de la latitude.

Une étude des paléolatitudes des roches sources et réservoirs a été faite pour tous les bassins pétrolifères, les résultats ont été pondérés selon les réserves. Nous n'avons trouvé aucun lien statistique satisfaisant entre le pétrole et la paléolatitude pour tous les gisements du Phanérozoique. La plupart des gisements du Paléozoique ont eu leur origine dans des roches de basses latitudes, ce qui peut être accidentel. Les gisements beaucoup plus considérables du Mésozoique sont reliés à de basses paléolatitudes, résultat qui peut être fortement influencé par les immenses réserves du Golfe Persique. Si on les exclut, le pétrole du Mésozoique se retrouve aussi bien aux paléolatitudes hautes que basses. Le pétrole du Cénozoique est distribué uniformément à cet égard.

La distribution du pétrole dans le temps révèle que 72% de tout le pétrole a été probablement produit au Mésozoique supérieur dont la plus grande partie (60%) au Crétacé moyen. La première exigence dans toute théorie générale sur l'occurrence du pétrole est de comprendre pourquoi il s'est formé autant de pétrole aux environs du Golf Persique et à un moindre échelle au centre de l'Amérique, durant un intervalle de temps aussi court. Nous essayons de démontrer que tous les quatre facteurs ont été à leur optimum à ces deux endroits durant cette courte période de temps. Au cours du mouvement de la tectonique de plaque, deux grandes baies marines se sont formées au niveau de l'équateur au Mésozoïque supérieur: elles peuvent ou non avoir été reliées par la Méditerrané occidentale. L'une de ces baies comprenait le Golfe Persique, l'autre comprenait le centre de l'Amérique. La reprise de la convection du manteau il y a environ 100×10^6 années a réactivé ces baies, modifié brusquement le taux d'expansion des océans et agrandi les dorsales océaniques, formant ains. des mers chaudes et peu profondes, et par l'activité ignée provoquant la formation de grandes quantités de substances nutritives.

Par la suite, l'activité de plaque a été telle que le Golfe Persique a été protégé tectoniquement par le mouvement rapide vers le nord de la plaque indienne (qui a absorbé presque tout le choc avec la plaque eurasienne), et le Golfe du Mexique a été protégé par le mouvement vers le nord-est de l'arc antillais.

[Traduit par le journal]

Introduction

Most petroleum geologists and geochemists believe that the initial step in oil genesis is the accumulation of planktonic remains. These remains must be transformed into oil, which must then be concentrated in reservoirs and preserved through subsequent time. We distinguish four sets of factors governing this process:

1. Climate, especially temperature.
2. Availability of nutrients.
3. Tectonic factors governing initial basin formation.
4. Tectonic factors governing subsequent preservation of the oil.

Hydrobiological studies suggest that the rate of accumulation of organic matter in sediments is a function of temperature. In warm climates and under clear skies, strong sunlight promotes photosynthesis, fertility increases, and life spans decrease. The greatest production of plankton occurs where warm climate is accompanied by high nutrient content of the seas, an abundance of phosphorus and nitrogen being especially necessary. Oceanic circulation concentrates nutrients in favored regions, but the global availability of most nutrients ultimately depends on the rate of their release from the mantle by volcanic processes. Hence

they may be expected to be most abundant at times of rapid plate tectonism, when volcanism at spreading ridges and island arcs was greatest.

Although the localization of regions of high organic production is influenced by both climate and nutrient supply, it is also affected by the form and orientation of the sea bottom. In the Pacific and Atlantic Oceans, rich plankton blooms occur off the southwest-facing coasts of southern California, Central America, Peru, South-West Africa, and northwest Africa, and off the northeast- and southeast-facing coasts of Argentina, Brazil, and Newfoundland. Upwelling nutrient-rich waters are guided to these localities by the configuration of the continental slope. The highest productivity of all is found along the northern margin of the Indian Ocean, where a south-facing, deeply embayed coastline lies immediately north of the equator. This configuration permits the southwesterly monsoon winds to drive nutrient-rich waters into the Arabian Sea, the Bay of Bengal, and the Indonesian seas. The influence of the shape of the water body is most clearly seen in the high productivity of most enclosed or semi-enclosed basins—Bering Sea, south Caspian Sea, Sea of Azov, Black Sea, Red Sea, and Lake Maracaibo.

Finally, oil must not be destroyed or lost. It is more likely to be preserved if it accumulates away from, rather than near to, plate margins. Hence long-lived oil, safely situated on stable cratons, is likely to have accumulated during times of widespread transgression. Oil accumulated near plate margins can, however, be preserved if it is very young, or if it has been tectonically shielded from destructive marginal plate interactions.

Among the three factors influencing the generation of oil (as distinct from its preservation), relative importances may have varied widely in both time and space. In an effort to evaluate them, we concentrated initially on the first of them, that of climate, because insofar as it is a function of latitude it is capable of quantitative analysis. The results of our paleolatitude study are then used as a springboard for a consideration of the roles played by the other factors in the large problem of oil occurrence.

Earlier Results

The first evaluation of oil paleolatitudes through paleomagnetic data was by Irving and Gaskell (1962). Their analysis showed a strong dependence of oil occurrence on the paleolatitude of the reservoir rock at the time of formation. This result was based mostly on Upper Paleozoic oilfields, because paleomagnetic evidence available at that time provided substantial control for those oilfields but comparatively little for younger fields. As paleomagnetic coverage improved, Deutsch (1965) and Irving (1964) showed that the Cenozoic and Mesozoic oil basins had a wider paleolatitude spread than those of the Paleozoic. In the last decade, oil reserves have increased by a factor of two, and paleomagnetic results by a factor of ten, so that a more complete analysis is now possible.

In these earlier studies, the paleolatitudes of the reservoir rocks only were calculated. Moreover, no account was taken of the relative volumes of oil in the fields; unit weight was accorded to each oilfield. In the present work we have also calculated the paleolatitude of the most probable source rock for each basin, and the data have been weighted proportionally to *total reserves*, so that it is the distribution of oil, rather than of oilfields, that is obtained.

Compilation of Oil Occurrences

The locations and ages of the sedimentary basins producing oil up to 1972 are listed in Tables 1 to 3, and plotted on Fig. 1. The ages of both source and reservoir rocks are given. Whereas the age of the reservoir rock is always known, the age of the source rock can never be conclusively established, and is commonly in dispute (*e.g.* in the North Slope and Cook Inlet basins of Alaska, the Maracaibo and Orinoco basins of Venezuela, the Polignac basin of Algeria, and the Vienna basin). The ages here assigned to source rocks reflect the opinions of the author (FKN) responsible for that aspect of this study. The geological evidence for these opinions has been given elsewhere (North 1971).

The 'total reserves' for each basin comprise cumulative production to 1972, plus the estimated, remaining, recoverable reserves. The estimates of reserves are compromises between various authoritative available figures, not all of them published. To a considerable extent, they again reflect the prejudices of one of the authors (*e.g.* Alberta basin, Gulf Coast basin of the United States, Uinta basin) are considered to have generated oil in more than one era. Such 'inherited' basins are entered in the tables for both eras, and their reserves apportioned between the two eras as accurately as the data allow.

Tar-sand deposits are not tabulated because of the possibly large uncertainty in their ages. For example, the age of the oil in the McMurray tar sands has been variously proposed to be Devonian, Mississippian, Jurassic, or Cretaceous. Every great tar-sand deposit, except that in the Botucatu sandstone of Brazil, is associated with a basin that contains orthodox pooled oil reserves of sufficient magnitude to appear in the tables in its own right.

Large oil-shale deposits are listed in Table 4. Their age is not in doubt. On the histograms, and in Fig. 1, the oil shales are distinguished by triangular symbols. Actual reserves of shale-oil are not included because it is only in a few well-studied cases that reliable estimates of reserves have been published. In most cases, the reserves are huge compared with associated

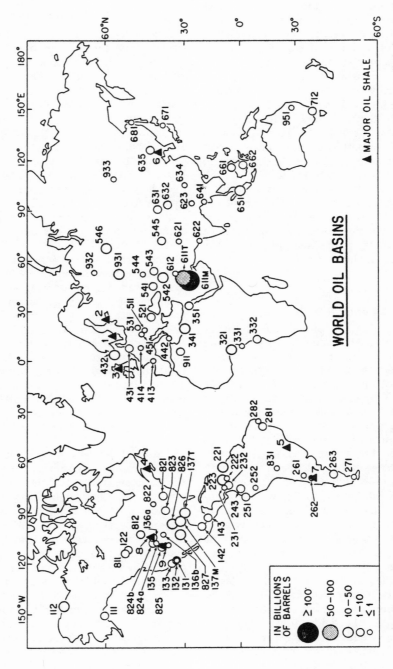

Fig. 1. Geographical distribution of oil. Symbol sizes proportional to reserves as listed in Tables 1 to 3. Triangles are oil-shale occurrences (Table 4), all are given the same size, as reserves are not accurately known.

TABLE 1. Oil basins wholly or largely Paleozoic

Ref. No.	Basin; Location	Total reserves (million bbls)	Period	Paleolatitudes Reservoir	Paleolatitudes Source	Ottawa List No.
112	North Slope; Alaska (+70, −145)	15 000	Carboniferous	+35	+35	860001
			Permian	+41	—	870001
			Triassic	+51	—	880001
811	Alberta (Paleozoic); Canada (+54, −114)	8000	Devonian	+05	+05	850001
812	Williston; U.S.A., Canada (+49, −103)	1800	Carboniferous	+07	+07	860001
821	Appalachian; Penn., W. Va., Ohio (+40, −080)	3200	Devonian	−18	−18	850001
822	Michigan; Michigan, Ontario (+44, −085)	600	Devonian	−13	−13	850001
			Carboniferous	−04	—	860001
823	Illinois; Illinois, Indiana (+39, −089)	3500	Devonian	—	−17	850001
			Carboniferous	−07	−07	860001
824a	Wind River; Wyoming (+43, −108)	400	Carboniferous	+03	—	860001
			Permian	+09	+09	870001
824b	Big Horn; Wyoming (+43, −108)	1300	Carboniferous	+04	—	860001
			Permian	+10	+10	870001
135	Uinta (Weber); Colorado (+40, −110)	700	Carboniferous	+02	—	860001
			Permian	—	+08	870001
825	Paradox; Utah (+38, −109)	600	Carboniferous	+01	+01	860001
826	Mid-Continent; Oklahoma, Kansas (+36, −097)	17 000	Ordovician	−28	—	830001
			Silurian	−20	—	840001
			Devonian	−16	−16	850001
			Carboniferous	−07	−07	860001
827	Permian; Texas, New Mexico (+32, −103)	18 000	Carboniferous	−08	−08	860001
			Permian	−02	−02	870001
831	Salta; Argentina, Bolivia (−20, −064)	650	Devonian	−65	−65	850004
			Carboniferous	−39	—	860004
911	Polignac; Algeria (+32, +006)	8850	Cambrian	−71	—	820002
			Ordovician	−64	—	830002
			Silurian	−68	−68	840002
			Devonian	*	*	*
			Carboniferous	−18	—	860003
931	Ural-Volga; U.S.S.R. (+56, +052)	34 000	Devonian	*	*	*
			Carboniferous	*	*	*
			Permian	+27	—	870006
932	Pechora; U.S.S.R. (+66, +053)	1000	Devonian	*	*	*
933	Irkutsk; Siberia (+58, +108)	500	Cambrian	+09	+09	820006
631	Dzungaria-Kansu; W. China (+43, +090)	2000	Carboniferous	—	+33	900006
			Permian	—	+51	870006
			Neogene	+40	—	910007
951	Surat; E. Australia (−28, +150)	100	Permian	—	−70	870005
			Jurassic	−68	—	890004

NOTES: The reference numbers are those of North (1971). Paleolatitude convention: positive north, negative south. A dash in either of the paleolatitude columns indicates that no reservoir (source) rock has been accorded that age for that basin. An asterisk indicates that, preferred ages notwithstanding, no paleolatitudes have been calculated for these reasons: 911, no suitable Devonian paleomagnetic data for Africa; 931 and 932, no maps for the Carboniferous and Devonian for northern Asia because data too scattered for calculation of mean pole. Sign convention in column two is as follows: for latitude north is positive, south negative; for longitude east is positive, west negative. See text for explanation of Ottawa list numbers.

orthodox pooled reserves, and their inclusion would seriously modify the distribution diagrams.

Estimation of Paleolatitude

Paleomagnetic observations are now available from most continental blocks for each Phanerozoic period. These have been compiled and average poles calculated for each block for each period. These average poles (indexed by six-figure numbers in the right-hand columns of Tables 1 to 4) may be found in the second issue of the Catalogue of Paleomagnetic Directions and Poles, to be published by Earth Physics Branch (Canada Department of Energy, Mines and Resources) in 1974. Using these average poles and Geuer's Terrascope (Geuer 1973), paleolatitude maps of each block were made photographically. Oil basins were then plotted on the maps corresponding

TABLE 2.   Oil basins wholly or largely Mesozoic

Ref. No.	Basin; Location	Total reserves (million bbls)	Period	Paleolatitudes Reservoir	Source	Ottawa List No.
111	Cook Inlet; Alaska (+61, −150)	1500	Jurassic	—	+62	890001
			Paleogene	+72	—	910005
122	Alberta Mesozoic; Canada (+53, −113)	3500	Jurassic	—	+61	890001
			Cretaceous	+57	+57	900001
136a	Powder River; Wyoming (+44, −105)	1250	Carboniferous	−03	—	860001
			Permian	+09	—	870001
			Cretaceous	+47	+47	900001
136b	Denver; Colorado, Nebraska (+40, −103)	300	Cretaceous	+43	+43	900001
137	Gulf Coast Mesozoic; Southern U.S. (+32, −095)	16 000	Jurassic	+29	—	890001
			Cretaceous	+33	+33	900001
142	Tampico-Nautla; Mexico (+21, −098)	6300	Cretaceous	+24	+24	900001
223	Maracaibo; Venezuela, Colombia (+10, −071)	38 000	Cretaceous	+17	+17	900004
			Tertiary	+08	—	910007
232	Magdalena; Colombia (+07, −074)	2100	Cretaceous	—	+14	900004
			Tertiary	+06	—	910007
243	Oriente; Ecuador (00, −076)	4000	Cretaceous	+08	+08	900004
252	Subandean; Peru (−08, −075)	150	Cretaceous	−01	−01	900004
261	Mendoza; Argentina (−33, −068)	1000	Triassic	−40	−40	880005
			Cretaceous	−36	—	900004
262	Neuquen; Argentina (−38, −070)	500	Jurassic	−38	−38	890003
263	San Jorge; Argentina (−45, −068)	2500	Cretaceous	−38	−38	900004
			Paleogene	—	−47	910007
271	Magallanes; Argentina, Chile (−53, −069)	315	Cretaceous	−45	−45	900004
281	Reconcavo; Brazil (−12, −039)	1250	Jurassic	−11	—	890003
			Cretaceous	−11	−11	900004
282	Sergipe-Alagoas; Brazil (−10, −036)	110	Cretaceous	−10	−10	900004
331	Gabon; West Africa (−01, +009)	600	Cretaceous	−10	−10	900003
			Paleogene	−07	−07	910008
332	Cabinda-Emeraude; Angola, Congo (−09, +013)	8400	Cretaceous	−19	−19	900003
			Paleogene	−15	−15	910008
341	Sirte; Libya (+30, +020)	33 000	Cretaceous	+15	+15	900003
			Paleogene	+25	+25	910008
413	Aquitaine; France (+44, 000)	380	Jurassic	+30	—	890005
			Cretaceous	+32	+32	900006
414	Rhine; France, Germany (+49, +008)	65	Jurassic	+36	+36*	890005
			Paleogene	+38	—	910014
431	Hannover; Germany, Holland (+53, +008)	1940	Jurassic	+40	+40	890005
442	Sicily; Italy (+37, +014)	450	Triassic	+02	+02	880007
			Jurassic	—	+26	890005
543	Mangyshlak; U.S.S.R. (+44, +054)	6000	Jurassic	+41	+41	890005
544	Emba; U.S.S.R. (+48, +052)	1000	Jurassic	+43	+43*	890005
546	Tyumen; U.S.S.R. (+60, +067)	35 000	Jurassic	+60	+60	890005
			Cretaceous	+52	+52	900006
611	Persian Gulf; Arabia, Kuwait, Iraq (+27, +048)	500 000	Jurassic	+03	+03	890002
			Cretaceous	+05	+05	900003
632	Tsai-Dam; China (+38, +093)	1500	Triassic	—	+46	880007
			Jurassic	—	+47	890005
			Paleogene	+45		910014
634	Szechuan; China (+30, +105)	1000	Permian	—	+43	870006
			Jurassic	+43	—	890005
635	Sungliao; China (+45, +126)	1500	Cretaceous	+49	+49	900006

*It is now the opinion of FKN that the preferred source of 414 should be Triassic, and of 544 Permian. These changes were made too late to be incorporated in the histograms, but do not materially affect them.

to the ages of the preferred source rocks and the reservoir rocks, and their paleolatitudes read off. This graphical method was chosen in preference to numerical methods because it is easier to deal in this way with basins of substantial areal spread, and because such maps, once produced, may be used for many other purposes. The accuracy is better than 1°,

TABLE 3. Oil basins wholly or largely Tertiary

Ref. No.	Basin; Location	Total reserves (million bbls)	Period	Paleolatitudes Reservoir	Source	Ottawa List No.
131	Los Angeles; California (+34, −118)	8000	Neogene	+34	+34	910002
132	Ventura; California (+34, −118)	3000	Neogene	+34	+34	910002
133	San Joaquin; California (+36, −120)	8200	Paleogene	+45	+45	910005
			Neogene	+36	+36	910002
			Pleistocene	+35	—	920001
135	Uinta (Green River); Utah, Colorado (+40, −110)	500	Paleogene	+48	+48	910005
137	Gulf Coast Tertiary; Louisiana, Texas (+30, −090)	30 000	Paleogene	+34	+34	910005
			Neogene	+29	—	910002
143	Isthmus; Mexico (+18, −093)	1500	Paleogene	—	+17	910005
			Neogene	+17	+17	910002
221	Orinoco; Venezuela, Trinidad (+10, −064)	10 500	Cretaceous	—	+16	900004
			Tertiary	+08	—	910007
			Pleistocene	—	+09	920007
222	Barinas; Venezuela (+08, −070)	500	Cretaceous	—	+14	900004
			Tertiary	+06	—	910007
231	North Bolivar; Colombia (+09, −075)	150	Cretaceous	—	+16	900004
			Tertiary	+07	—	910007
251	Talara-Sta. Elena; Ecuador, Peru (−03, −081)	1550	Cretaceous	—	−06	900004
			Tertiary	−04	−04	910007
321	Niger delta; Nigeria (+05, +007)	12 000	Cretaceous	—	−04	900003
			Neogene	+04	—	910006
351	Suez; U.A.R. (+28, +033)	8000	Neogene	+28	—	910006
432	North Sea; U.K., Norway (+58, +004)	16 000	Paleogene	+45	+45	910014
451	Vienna; Austria (+48, +017)	1000	Cretaceous	—	+36	900006
			Neogene	+44	—	910015
511	Pannonian; Hungary (+47, +019)	100	Neogene	+43	+43	910015
521	Carpathian; Romania (+45, +027)	4000	Paleogene	—	+34	910014
			Neogene	+41	+41	910015
531	Galician; Poland (+50, +020)	300	Cretaceous	—	+38	900006
			Neogene	+46	—	910015
541	North Caucasus; U.S.S.R. (+44, +045)	4400	Paleogene	—	+35	910014
			Neogene	+40	—	910015
542	South Caucasus; U.S.S.R. (+40, +050)	15 000	Paleogene	—	+32	910014
			Neogene	+46	+46	910015
545	Ferghana; U.S.S.R. (+41, +072)	1500	Cretaceous	—	+34	900006
			Paleogene	+36	+36	910014
611	Persian Gulf Tertiary; Iran, Iraq (+31, +050)	90 000	Jurassic	—	+07	890007
			Cretaceous	—	+09	900003
			Paleogene	+27	+27	910008
			Neogene	+29	—	910006
612	Qum; Iran (+35, 052)	1000	Paleogene	+27	—	910014
			Neogene	+31	—	910015
621	Potwar; Pakistan (+33, +052)	110	Paleogene	+37	+37	910014
622	Cambay; India (+22, +072)	440	Paleogene	+18	+18	910014
623	Upper Assam; India (+27, +094)	1000	Paleogene	+27	+27	910014
			Neogene	+24	—	910015
641	Irrawaddy; Burma (+20, +095)	600	Paleogene	+21	+21	910014
			Neogene	+18	—	910015
651	Sumatra-Java; Indonesia (00, +102)	10 800	Neogene	−02	−02	910015
661	Brunei; Borneo (+05, +115)	2400	Paleogene	—	+10	910014
			Neogene	+04	+04	910015
662	Balikpapan; Borneo (00, +117)	1600	Paleogene	—	+05	910014
			Neogene	−01	—	910015
671	Japan; Japan (+40, +140)	415	Neogene	+41	+41	910015
681	North Sakhalin; U.S.S.R. (+52, +142)	700	Neogene	+54	+54	910015
712	Gippsland; Victoria, Aust. (−38, +148)	3000	Cretaceous	—	−74	900005
			Paleogene	−14	−14	910009

NOTES: For 222 and 231 the reservoirs are Tertiary and the preferred source Upper Cretaceous. The paleolatitudes for reservoir and source are +06 and +14 for 222, and +07 and +16 for 231. These results were derived too late for inclusion in the main compilation, and their inclusion would not materially affect the histograms.

359

TABLE 4.   Important kerogen shale deposits

Deposit; Location	Period	Paleolatitude	Ottawa List No.
1. Kolm; Sweden (+58, +015)	Cambrian	−11	820006
2. Kukkersite; Esthonia (+60, +025)	Ordovician	+28	830005
3. Midlothian; Scotland (+56, −005)	Mississippian	*	*
4. Albert; New Brunswick (+46, −065)	Mississippian	−05	860001
5. Irati; Brazil (−25, −052)	Permian	−30	870004
6. Oil-shale series; Svalbard (+78, +015)	Triassic	*	*
7. Fushun; China (+42, +124)	Jurassic	+58	890005
8. Kashpirian; U.S.S.R. (+53, +048)	Jurassic	*	*
9. Neuquen; Argentina (−38, −070)	Jurassic	−38	890003
10. Frontier; Wyoming (+44, −105)	Cretaceous	+47	900001
11. Green River; Utah, Colorado, Wyoming (+40, −110)	Paleogene	+40	910005
12. Coal Measures; New Zealand (−38, +175)	Paleogene	*	*

NOTE: Numbers 6, 8 and 12 indicated by asterisks are omitted from the map and histograms.

which is negligible compared to the probable paleomagnetic errors (normally 5 to 10°). About sixty maps were produced, providing paleolatitudes for all but three oil basins (see footnote to Table 1).

The paleolatitude values were compiled into histograms, first for each period and then for each era (Figs. 2 to 6). The vertical scale in the histograms is proportional to total reserves. If the paleolatitude spread of an oil basin overlaps two latitude intervals, equal weight was given to each interval. Each histogram was compared to the distribution expected if oil was uniformly distributed, using the Kolmogorov-Smirnov test. An example is given in Fig. 2, showing the observed and expected frequencies of oil at various latitudes. The maximum absolute difference between the curves was then compared with the allowable difference at the 95% confidence level. If the difference is not significant then it is reasonable to say that oil is uniformly distributed with respect to latitude. The Paleozoic distribution is clearly non-uniform, the Mesozoic uniform (Fig. 2). The results of this test are indicated on each diagram as 'uniform' or 'non-uniform'.

This method of calculating paleolatitude assumes that the earth's field during the Phanerozoic has been on average that of an axial geocentric dipole (AGD); that is, the paleomagnetic pole corresponds to the paleogeographic pole. Although this assumption is almost certainly correct in a general way, its validity in an exact sense has been recently questioned on two grounds. Firstly, although

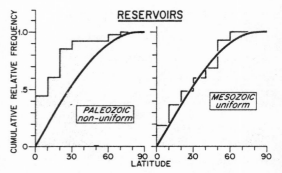

FIG. 2.   Cumulative frequency distribution of oil. The step curve is the observed relative frequency of oil; the smooth curve is the theoretical uniform distribution for 10° intervals for latitudes 0° to 90°.

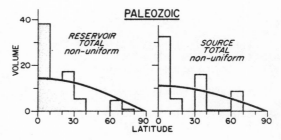

FIG. 3.   Volumes of oil reserves *versus* latitude for the Paleozoic. Histograms are plotted for both reservoir and source rock, without regard to latitude sign. The smooth curve represents the expected theoretical distribution containing the same volume as the histogram. Results of the Kolmogorov-Smirnov test are indicated.

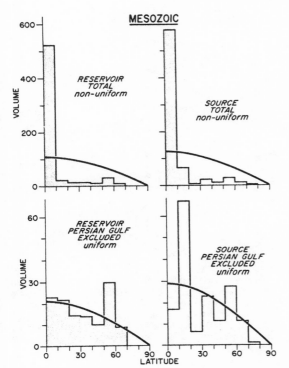

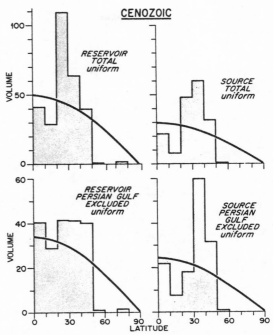

FIG. 4. Volumes of oil reserves *versus* latitude for the Mesozoic Era. Legend as for Fig. 3. The Persian Gulf is included in those plots designated TOTAL.

FIG. 5. Volumes of oil reserves *versus* latitude for the Cenozoic Era. Legend as for Fig. 4.

the mean paleomagnetic pole for the past 5 m.y. coincides with the present geographic pole, individual paleomagnetic poles are offset from it. The offsets vary systematically with respect to longitudes of sampling sites as if the axial dipole were displaced northwards by 285 ± 74 km (Wilson 1971). The error that this modification introduces is small (no more than 5°). The second line of argument is as follows. If the paleomagnetic poles from different continents are calculated, not with respect to the present frame but with respect to the Pangaea-type reconstructions of Wegener (1924) and later workers, the agreement between the poles, though improved, is still not as good as might be expected.

Thus paleolatitude analyses may be conducted in either of two ways. One may assume that Pangaea-type reconstructions are correct and that the AGD hypothesis, though true in general, requires modification. In this case one may use the maps of, say, Smith *et al.* (1973),

in which the positions of continents relative to one another are determined from geological evidence and the geographical grid from the paleomagnetic evidence. Discrepancies between the two procedures are then attributed to departures from a truly AGD field. Alternatively one may work directly from the paleomagnetic data assuming the AGD hypothesis to be correct, but without making any assumptions about the relative positions of continents. This is the procedure that we have used.

Paleolatitude Results

In Figs. 3 to 6, paleolatitudes of oil basins are compiled irrespective of sign, the values being plotted as if all were in one hemisphere. The Paleozoic distributions of both reservoir and proposed source are non-uniform, being predominantly in paleolatitudes of less than 30° (Fig. 3). If all Mesozoic results are plotted, the histograms of reservoir and source rock are again non-uniform, being strongly peaked between 0 and 10° (Fig. 4). Most of the world's reserves are in Mesozoic rocks immediately to the north and west of the Persian

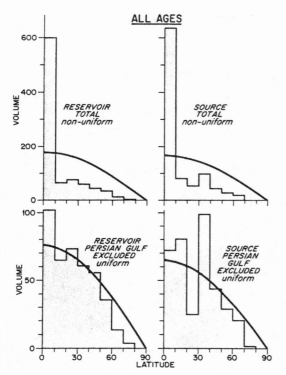

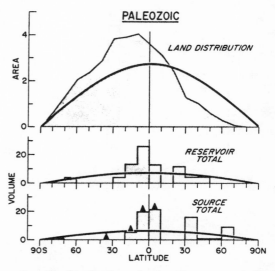

FIG. 7. Distribution of oil with respect to land in the Paleozoic Era, both distributions plotted with regard to sign. The smooth curves represent uniform distribution for an equal area of land and equal volume of oil, respectively. Small triangles indicate the paleolatitudes of important kerogen shale deposits (see Table 4).

FIG. 6. Volumes of oil reserves *versus* latitude for all ages. Legend as for Fig. 4.

Gulf, a region lying near the equator throughout the Mesozoic. If the reserves of the Middle East are excluded, the Mesozoic distributions are uniform. The paleolatitude distributions of both source and reservoir rocks for Cenozoic basins, with or without the Persian Gulf, are uniform (Fig. 5). Results for all oil basins, irrespective of age, are shown in Fig. 6. Excluding the Persian Gulf, the distributions are uniform, the reservoir beds in particular showing a remarkably uniform latitudinal distribution. If the Persian Gulf reserves are included they dominate the histograms, which then have a strong equatorial peak and are clearly nonuniform.

All known oil has formed in basins either wholly within continental crust or at continental margins. It is therefore possible that the distributions recorded in Figs. 2 to 6 simply reflect the latitude distribution of land. The latter has therefore been calculated and compared with oil distribution in Figs. 7 to 10. In these figures the distributions of reservoir and source rocks are plotted in their correct hemi-

spheres. Persian Gulf oil is excluded, because its inclusion would dominate the picture. Oil shale occurrences, however, have been added. Land distribution has been obtained by compiling the lengths of crust in small circles of latitude measured from the sixty paleolatitude maps already mentioned, using a latitude spacing of 10°. The continental crust was assumed to be bounded by the outer edges of the present continental shelves. Results are incorporated from all the main elements of continental crust with two exceptions: (1) Asia south and east of the Hindu Kush-Altai-Stanavoi ranges and north of the Himalayas for all Phanerozoic periods; and (2) northern Asia for the Devonian and Carboniferous (see footnote to Table 1). In the calculations, unit weight was given to each geological period (Tertiary treated as two, Paleogene and Neogene). This procedure gives too much weight to such short periods as the Silurian and the Neogene. A rigorous assessment would require equal divisions of time—a formidable undertaking we have not attempted. This inequality in the use of data is an outcome of the fact that most paleomagnetic results are keyed to the geological time scale and not to a numerical

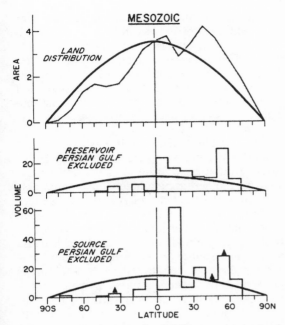

FIG. 8. Distribution of oil with respect to land in the Mesozoic Era. Legend as for Fig. 7.

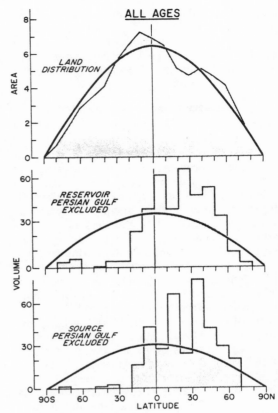

FIG. 10. Distribution of oil with respect to land for all ages. Legend as for Fig. 7.

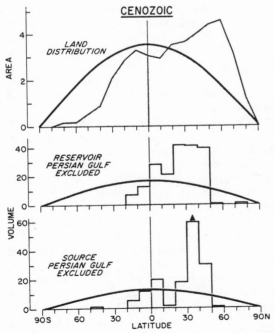

FIG. 9. Distribution of oil with respect to land in the Cenozoic Era. Legend as for Fig. 7.

scale of time. The effects of this inequality are probably random and should not prejudice the overall picture.

The distribution of land is biased towards the southern hemisphere in the Paleozoic and towards the northern hemisphere in the Mesozoic and Cenozoic. This is simply a confirmation in numerical terms of a general result obtained long ago by Köppen and Wegener (1924). Cumulatively for the entire Phanerozoic (Fig. 10, upper diagram), the distribution of continental crust is remarkably uniform. In the Mesozoic and Cenozoic there is a correlation between the distribution of land and the distribution of oil (Fig. 8). If Persian Gulf results were included this correlation would vanish. There is no correlation between land distribution and oil in the Paleozoic, nor in the overall result (Figs. 9 and 10). Its absence from the overall result may reflect the general

absence of Paleozoic oil in southern Gondwana, where sediments deposited were mainly
continental. The distribution of oil shales
mimics that of oil. They occur mostly in low
paleolatitudes in the Paleozoic, but have a
wider latitude spread at other times.

Oil Accumulation through Time

The interval of geological time witnessing
the greatest known generation of poolable oil
was that between the Nevadan (Columbian or
Cimmerian) orogeny, in the late Jurassic, and
the Laramide orogeny about at the end of the
Paleocene. Good cases can be made for assign-
ing to this interval the principal source rocks
for all the pooled oil in the Persian Gulf region, the Sirte basin of Libya, the northern
North Sea and circum-Atlantic basins, and the
Tyumen basin of western Siberia. On the other
side of the earth, the greatest concentration of
pooled, preserved oil lies in the region extending from northernmost South America to
the northern margin of the Gulf Coastal province of the United States. Many geologists
favor a Cretaceous origin for the greater part
of the oil of this 'Middle American' province.
It is conceivable that this very unequal distribution is a consequence of variable *preservation* of oil superimposed on a generally uniform
rate of *generation*, but this is unlikely. It appears rather that over 70% of all known pooled
oil was generated in the interval from -140
to -60 m.y., only one-seventh of Phanerozoic
time. For this interval the 'oil-incidence factor'
was more than five times the average.

About the middle of this time-interval, oil
generation was particularly rapid. Dunnington
(1967) argued that most of the Persian Gulf
oil was generated in the Middle Cretaceous
Epoch (Albian to Turonian, inclusively).
North (1971) proposed that the principal
source sediment in every 'Middle American'
oil basin was of Turonian age. If these arguments are accepted, then 60% of the world's
known oil was generated between about -110
and -80 m.y. (5% of Phanerozoic time), the
'oil-incidence factor' being then twelve times
the average.

Summary of Results and Discussion

(*1*) For the Phanerozoic as a whole, over
80% of all oil occurs in reservoir rocks whose

latitude at the time of deposition was less than
30°, and over 60% was in paleolatitudes less
than 10°. The same is true for preferred source
rocks. This result is due primarily to the low
latitude of the Persian Gulf region during the
Jurassic and Cretaceous. If the Persian Gulf
basin is excluded, the paleolatitude distributions of source and reservoir are uniform
(Fig. 6).

(*2*) As the Persian Gulf basin is unique
among conventional basins in its oil richness,
results from it are difficult to assess statistically.
The only other accumulations of comparable
size are in the tar sands of Alberta and eastern
Venezuela, and in the oil shales of the Green
River and Irati. Tar-sands oil cannot be realis-
tically incorporated into the analysis because
of the uncertainty in its age. Opinions concerning the origin of the oil in the McMurray
tar sands are dependent upon personal equations—geologist versus geochemist, migration-
ist versus immobolist, observer versus theoreti-
cian. One of us (FKN), aware of the
arguments in favor of a Mesozoic source, nonetheless considers the coincidence of its distribution with respect to the subcrop of certain
Upper Devonian formations to be good evidence for an ultimate source in the Upper
Devonian (Link 1951; Sproule 1955). Such
a source, if included in our analysis, would
reinforce the argument for latitudinal dependence, as Alberta lay in low latitudes in the
Devonian. Similarly, the large reserves of heavy
oil in tar sands in eastern Venezuela originated
in low latitude; their source can scarcely lie
outside the interval Turonian to Oligocene,
when the Orinoco basin lay in latitudes essentially similar to those of today.

(*3*) Cenozoic oil is uniformly distributed
with respect to paleolatitude, with or without
Persian Gulf oil. This essentially confirms the
results of Deutsch (1965).

(*4*) If Persian Gulf oil is excluded, Mesozoic oil is uniformly distributed, but if Persian
Gulf oil is included the distribution shows a
marked maximum at the equator.

(*5*) In the Paleozoic, most oil is within 30°
of the paleo-equator, as determined earlier
(Irving and Gaskell 1962). This distribution
is not simply an accident of land distribution,
but it may be an accident of another kind.

The contrasts between Paleozoic and Cenozoic oil basins, as they exist today, have been discussed by North (1971). It seems logical to believe that oil in Paleozoic basins (which are of enormously greater areal extent than productive Cenozoic basins) has not merely had more opportunity for distant migration, more opportunity for escape and loss during erosion, and for destruction during orogeny, than has oil in Cenozoic basins, but a great deal of it must also have suffered consumption into the mantle during the Mesozoic episode of continental redistribution. The Cenozoic basins should therefore reveal the fundamental controls of oil genesis and accumulation more faithfully than can the Paleozoic basins.

The implication is that oil in the Paleozoic basins was probably of uniform distribution, latitudinally, at the end of Paleozoic time, or at least uniformly distributed outside the non-generative Gondwana territory. This uniform distribution was then destroyed by the elimination of most Paleozoic continental margins, either by Paleozoic orogeny or by Mesozoic subduction. The only oil preserved on a large scale was that which had migrated on to one of three cratons—the North American, North African, and Russian cratons—each of which happened to lie in low latitudes during the late Paleozoic (when the northern hemisphere was the 'water hemisphere').

(6) During the Paleozoic, oil and land distributions show no correlation. In the Mesozoic and Cenozoic there is a good correlation, but only if the oil of the Persian Gulf is excluded from the analysis.

(7) Most of the world's oil accumulated in the later Mesozoic, the maximum accumulation probably taking place in the comparatively short interval from the Albian to the Turonian (approximately −110 to −80 m.y.).

Late Mesozoic Oil

Basic Assumptions

Insofar as present data allow, there are no general rules governing the paleolatitudinal distribution of oil applicable to all of Phanerozoic time. The results depend on whether the Persian Gulf data are included or not. Is the Persian Gulf therefore simply a gigantic exception, *sui generis*, or is it the archetype of all basins, inasmuch as all the factors favoring oil accumulation were optimal in it? We adopt the

latter viewpoint, maintaining that any theory of oil distribution must first and foremost explain the facts that two-thirds of the world's pooled reserves accumulated in one relatively small region in the last quarter of Mesozoic time and in equatorial paleolatitudes, and a good proportion of the remainder accumulated in another relatively small region (Middle America) at much the same time and also in low paleolatitudes.

We further consider that these occurrences reflect large-scale processes. We therefore set out below the major features of terrestrial history, first for the longer interval with high oil-incidence factor (five times average), from late Jurassic to Paleocene, and then for the shorter mid-Cretaceous interval when the oil-incidence factor was the maximum (twelve times average).

Interval from −140 to −60 m.y.

(1) Most of the sea-floor spreading responsible for the present ocean-continent distribution, and for the principal extensional oil basins of the world, took place during this interval. The interval began with important items in the timetable of continental drift—initiation of rapid spreading in the north Atlantic between North America and Africa, and in the southwestern Indian Ocean between Africa and Antarctica, as well as the opening and the beginning of reclosing of the new Tethys, between the Caribbean and Indonesia (Vine 1971; Valentine and Moores 1972).

(2) Rapid plate-margin subduction caused semi-continuous orogeny in the young mobile belts of the earth, and the creation of numerous compressional basins. In particular, the Persian Gulf and Maracaibo basins and the Mexican foredeep were created or accentuated during this time.

(3) Above the subduction zones, or marginal to them, the greatest bodies of synorogenic felsic magma and the largest volume of 'Alpine' ultramafic rocks in the Phanerozoic record were emplaced.

(4) There was an abrupt onset of deep-water sedimentation, beyond new, sharply demarcated continental margins lacking constructional shelves. Very widespread radiolarites and other ophiolite-associated sediments; ammonite-bearing shales lacking benthonic faunas;

aptychus marls and limestones; chalks; marine phosphorites (Mediterranean, Moscow, and northern Mexico phosphogenic provinces). Exceedingly widespread marine evaporites, especially salt, commonly as initial deposits of new basins (Middle America, circum-African basins, Arabian foreland, and the Qara Qum platform, all of them Oxfordian-to-Aptian).

(5) A rapid rise in the number of families of all groups of organisms (House 1971), and particularly a great increase in the number and variety of planktonic forms (radiolaria, pelagic foraminifera, tintinnids, dinoflagellates, calcareous nannoplankton).

Interval from −110 to −80 m.y.

(1) This interval, Aptian to Coniacian inclusively, began with the opening of the rift between Africa and South America (Vine 1971).

(2) Very rapid sea-floor spreading in the Atlantic and Pacific Oceans coincided with this interval, after which it reverted to rates comparable with those at present (Larson and Pitman 1972).

(3) The combination of fast-spreading ridges and the growth of new continental shelves yielded a stage in basinal development permitting widespread bathyal sedimentation to co-exist with the evaporite minimum. The consequence was the greatest of all post-Devonian marine transgressions (Sloss 1972), as ridges displaced oceanic waters on to the continental platforms (Pitman and Hayes 1973).

(4) Ocean temperatures reached possibly their highest level in the last 200 m.y., if oxygen isotope ratios in calcareous fossils are a reliable guide (Emiliani 1966). The highest temperatures (about 21 °C or 70 °F) are found for the Albian and Coniacian, separated by lower temperatures (16 °C or 60 °F) in the Cenomanian, all representing a level unattained since. (The nearest approach to it was a peak recorded for the Oligo-Miocene, a probable Cenozoic optimum for oil generation in the Gulf Coast, California, the Caucasus, and Indonesia.)

(5) There was a distribution of the continents about the narrowing Tethys such that both the present Persian Gulf and present Middle America were situated in huge marine embayments within 30° of the equator (Köppen and Wegener 1924; Smith *et al.* 1973).

(6) Major changes were brought about in the biogeography of important fossil groups (particularly foraminifera and mollusca), the cosmopolitan aspects characteristic of the late Paleozoic and early Mesozoic being replaced by greater provinciality as the Atlantic and Indian Oceans opened and Tethys began to break up.

(7) A Phanerozoic peak was reached in the emplacement of carbonatites, kimberlites, and related alkalic and ultrabasic rocks. This peak may be accorded an age of about 100 ± 20 m.y. (Macintyre 1971; 1973). If, as many believe, these intrusions were of very deep origin and provided the driving mechanism for continental plate tectonics (Morgan 1972), then a major renewal of mantle-wide convection occurred at about −100 m.y. This renewal was presumably responsible for items (1) and (2). This idea derives some support from the apparent absence of reversals of the geomagnetic field between −112 and −83 m.y. (Helsley and Steiner 1967; Larson and Pitman 1972; Irving and Couillard 1973). This major geomagnetic change may reflect changes in core convection, in turn caused by changes in the temperature distribution at the core-mantle interface, an expected consequence of the renewal of mantle-wide convection (Couillard and Irving 1973). Another expected consequence is that there would be an increase in the release of mineral nutrients into the oceans, roughly in proportion to the increase in rate of sea-floor spreading—a factor of two at −112 m.y. according to Larson and Pitman (1972). Life elements, such as sulfur, are ubiquitous products of volcanism, and phosphorus is especially common in alkalic rocks and carbonatites of which there was an abundance at this time. Numerous authorities have argued for a volcanic source for the great nitrate deposits of Chile (Lindgren 1933), but the nitrogen in the oceans may well be derived from the atmosphere.

Synthesis

Cretaceous palaeographic maps are of two main types—those that depict two land-masses (Laurasia and Gondwana) joined in the western Mediterranean region (*e.g.* Smith *et al.* 1973), and those that depict them separated at that point, with the Caribbean, Atlantic, and Tethys forming one continuous seaway

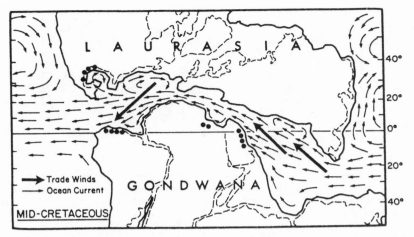

FIG. 11. Possible relationship of the continents during period of rapid drift, in the middle Cretaceous, with continuous Tethys-Atlantic seaway. Modified from Smith *et al.* (1973). Dots are major oil basins of the Persian Gulf, Libya, and Middle America. Ocean current directions are taken from Luyendyk *et al.* (1972).

(*e.g.* Phillips and Forsyth 1972). In fact the maps of Phillips and Forsyth show the separation to be a maximum in the mid-Cretaceous. The Cretaceous sediments of the Betic cordillera are of pelagic limestone facies, suggesting that there was an open seaway in the western Mediterranean.

Luyendyk *et al.* (1972) have simulated experimentally the ocean currents for the second configuration (Fig. 11). They suggest that the currents then, as now, were dominated by winds, and that there was a forceful westerly flow from the Tethys through the Atlantic to the Pacific, motivated by the southeast trade winds in Tethys, and the northeast trade winds in the Atlantic. Because of the configuration of the seaway and the trade winds the principal impact of this Tethyan current would be against the northern margins of Gondwana, including what is now the Persian Gulf and Middle America (Fig. 11).

If, on the other hand, the western Mediterranean was closed, it might be expected that the circulation would have been less vigorous but with currents, as before, impinging on the northern coasts of Gondwana (Fig. 12).

Recalling now that the general level of available nutrients may have been higher than normal (item 7), and that there were unusually wide expanses of very warm, shallow seas (items 3 and 4), then we suggest that rich oil generation in these two regions was a natural consequence. This is likely to be true for both configurations, but especially if there was an open seaway connecting Tethys and Middle America, with its warm, vigorous, equatorial or tropical Tethyan current (Fig. 11).

Most late Mesozoic oil, however, is now pooled in clearly demarcated basins, each relatively small, and not spread uniformly over the Tethys – Middle America region. We suggest that this is largely a function of preservation; petroliferous basins were originally more widespread and the producing basins of today are simply those that have escaped destruction by plate-margin interactions. The northern margin of the Tethyan ranges eventually impinged directly against their forelands, preserving a foredeep only along the front of the Caucasus and Kopet Dag (where it is petroliferous). The southern margin of Tethys was deformed out of recognition by the northward-driving 'prongs' of Assam, the Punjab, and the Adriatic (North 1965), except for the stretches between the Oman and Turkey (now the Persian Gulf basin in front of the Zagros Mountains), and between Tunisia and Agadir (in front of the Atlas Mountains). In Middle America, comparable basin-obliteration took place between Guatemala and Puerto Rico, and along the southeast front of the eastern cordillera of Colombia (North 1965). In contrast basin preservation occurred in the Gulf of Mexico, which was protected by the north-

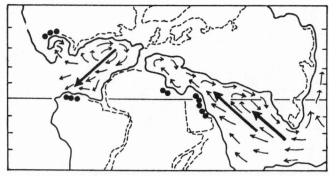

Fig. 12. Possible relationships of the continents during interval of rapid drift in the middle Cretaceous, with Tethys and Atlantic separated. (From Smith *et al.* (1973)). Symbols as for Fig. 11. The ocean currents are surmized by the authors.

eastward movement of the Antillean arc, and in the Maracaibo basin, which became in intermontane fragment when the Merida Andes rose to the southeast. But the clearest example of all is the Persian Gulf basin, which was protected from destructive plate-margin interaction by the rapid northward drive of India.

Our argument, in brief, is therefore this. Oil *generation* is a function of planktonic productivity, which in turn is a function of water-body geometry, of climate, and of availability of nutrients—all ultimately controlled by plate tectonics. *Pooling* of oil is largely a function of tectonism. *Preservation* of oil is a function of tectonism and little else.

Older Oil, Younger Oil

The preceding section offers a preliminary evaluation of the factors that controlled the generation of some 72% of the known, orthodox, pooled oil in the world, based on a plate-tectonic model. To narrow down still further the relative significances of these factors, and their variation in time, it is necessary to discuss their operation during the generation of the remaining 28% of the world's oil.

This 28% may be subdivided as follows, assuming that the source-age assignments proposed in this paper are correct. About 12% of known oil was generated subsequent to the Laramide orogeny (-60 m.y.); about 5% subsequent to the Hercynian orogeny (-300 m.y.) but prior to the Nevadan orogeny (-135 m.y.); about 9% subsequent to the Caledonian orogeny (-375 m.y.) but prior to the Her-

cynian orogeny; and less than 2% prior to the Caledonian orogeny.

A consideration of post-Laramide oil should be possible in our terms, though we have not yet attempted it. A consideration of pre-Nevadan oil will not be possible in comparable terms until more accurate plate-tectonic maps are available for older episodes of crustal displacement.

Acknowledgments

The authors are grateful to the Department of Energy, Mines and Resources for Research Grant No. 195/72, awarded to North and making possible Couillard's participation in the project—participation without which the project would never have been finished. The paper itself has materially benefited from suggestions and criticisms by W. A. Robertson, R. W. Yole, and the reviewers.

COUILLARD, R. and IRVING, E. 1973. Paleolatitudes and reversals: evidence from the Cretaceous. Geol. Assoc. Canada, Spec. Publ. (in press).
DEUTSCH, E. R. 1965. The paleolatitude of Tertiary oil fields. J. Geophys. Res., **70**, pp. 5193–5203.
DUNNINGTON, H. V. 1967. Stratigraphical distribution of oilfields in the Iraq-Iran-Arabia basin. J. Inst. Petrol., **53**, pp. 129–161.
EMILIANI, C. 1966. Isotopic paleotemperatures. Science, **154**, pp. 851–857.
GEUER, J. W. 1973. The Terrascope. Can. J. Earth Sci., **10**, pp. 1164–1169.
HELSLEY, C. E. and STEINER, M. B. 1969. Evidence for long interval of normal polarity during the Cretaceous Period. Earth Plan. Sci. Lett., **5**, pp. 325–332.
HOUSE, M. R. 1971. Evolution and the fossil record. *In* Understanding the Earth (Gass *et al.* Eds.). Mass. Inst. Technol. Press, Cambridge, Mass., pp. 193–211.

IRVING, E. 1964. Paleomagnetism. John Wiley and Sons, Ltd., New York, pp. 287–289.

IRVING, E. and COUILLARD, R. 1973. The Cretaceous normal interval. Nature (in press).

IRVING, E. and GASKELL, T. F. 1962. The palaeogeographic latitude of oil fields. Geophys. J. Roy. Astron. Soc., 7, pp. 54–63.

KÖPPEN, W. and WEGENER, A. 1924. Die Klimate der geologischen Vorzeit. Bornträger, Berlin, 255 p.

LARSON, R. L. and PITMAN, W. C. 1972. Mesozoic sea-floor spreading. Bull. Geol. Soc. Am., 83, pp. 3645–3662.

LINDGREN, W. 1933. Mineral deposits. McGraw-Hill Book Co., New York, pp. 323–325.

LINK, T. A. 1951. Source of oil in "tar sands" of Athabaska River, Alberta, Canada. Bull. Am. Ass. Petrol. Geol., 35, pp. 854–864.

LUYENDYK, B. P., FORSYTH, D., and PHILLIPS, J. D. 1972. Experimental approach to the paleocirculation of oceanic waters. Bull. Geol. Soc. Am., 83, pp. 2649–2664.

MACINTYRE, R. 1971. Apparent periodicity of carbonatite emplacement in Canada. Nature, 230, pp. 79–81.

—— 1973. Possible periodic pluming. Eos, 54, p. 239.

MORGAN, W. J. 1972. Deep mantle convection plumes. Bull. Am. Ass. Petrol. Geol., 56, pp. 203–213.

NORTH, F. K. 1965. The curvature of the Antilles. Geol. Mijnb., 44, pp. 73–86.

—— 1971. Characteristics of oil provinces. Bull. Can. Petrol. Geol., 19, pp. 601–658.

PHILLIPS, J. D. and FORSYTH, D. 1972. Plate tectonics, paleomagnetism, and the opening of Atlantic. Bull. Geol. Soc. Am., 83, pp. 1579–1600.

PITMAN, W. C. and HAYES, J. D. 1973. Upper Cretaceous spreading rates and the great transgression. Eos, 54, p. 240.

SLOSS, L. L. 1972. Synchrony of Phanerozoic sedimentary-tectonic events of the North American craton and the Russian platform. 24th Int. Geol. Cong., 6, pp. 24–32.

SMITH, A. G., BRIDEN, J. C., and DREWRY, G. E. 1973. Phanerozoic world maps. Systematics Ass., Publ. 9, pp. 1–42.

SPROULE, J. C. 1955. Discussion of "in situ origin" of McMurray oil. Bull. Am. Ass. Petrol. Geol., 39, pp. 1632–1636.

VALENTINE, J. W. and MOORES, E. M. 1972. Global tectonics and the fossil record. J. Geol., 80, pp. 167–184.

VINE, F. J. 1971. Sea-floor spreading. In Understanding the Earth (Gass et al. Eds.). Mass. Inst. Technol., Press, Cambridge, Mass. pp. 233–250.

WEGENER, A. 1924. The origin of continents and oceans. Translated by J. G. A. Skerl, Methuen, London. 212 p.

WILSON, R. L. 1971. Dipole offset—the time average paleomagnetic field over the past 25 million years. Geophys. J. Roy. Astron. Soc., 22, pp. 491–504.

44

Reprinted from *Tectonophysics* 26:40, 41, 42, 52, 53, 54 (1975)

NORTH SEA TROUGHS AND PLATE TECTONICS

A. Whiteman, D. Naylor, R. Pegrum, and G. Rees

[*Editor's Note:* In the original, material precedes this excerpt.]

CONCLUSIONS

We propose that most of the trilete structural trough pattern which characterises the North Sea and adjacent areas (Fig. 1) was initiated in Late Carboniferous and Early Permian times. Plume-generated crestal uplifts (Burke and Whiteman, 1973) developed in an area extending from the Rockall Bank to the Skagerrak. Plate movement then carried the crestal uplifts off plume so preventing the rrr systems from developing by lithospheric dyke injection into RRR spreading systems. The positions of the Mainz and Indefatigable systems in this scheme are not clear.

Crustal thinning developed beneath the Northern North Sea Trough in (?) Late Carboniferous and Permian times (Fig. 3). The Northern North Sea Trough was infilled with predominantly continental and evaporitic sediments in Permo-Triassic times and with predominantly paralic-marine sequences in Jurassic—Cretaceous times.

The Shetland, Forties and Skagerrak Systems (Fig. 1, Systems 1, 2 and 5) and the Rockall—Hatton and Rockall Troughs began to develop as major depocentres in Early Mesozoic time with individual "failed" arms marked by troughs, following different rates and patterns of structural and sedimentational development (Fig. 2).

A large intracratonic basin was established along Hercynian lines in the Southern North Sea Basin in Permian times. It is not clear from the data available whether the Indefatigable Trough System (Fig. 1, System 3) was initiated in Permian or Triassic times but it was in existence in the Jurassic and appears to be superimposed on the Permian Basin.

Several periods of epeirogenic uplift and block faulting affected the North Sea trough systems between from Late Triassic and Cenozoic times. These movements were due to adjustments beneath the continental margin consequent on new crust being generated in the Atlantic, as well as to failed-arm development and modification. The structures resulting from these movements have proved to be important for the migration and entrapment of oil and gas in the North Sea. Marine incursions entered the area from the spreading Atlantic and from "Tethys". Most of the North Sea troughs became inactive in Late Cretaceous times, when the widespread marine Chalk Sea incursion occurred. Some differential movement persisted into the Paleocene, especially association with the Odin High (Hopkinson and Nysaether, in press).

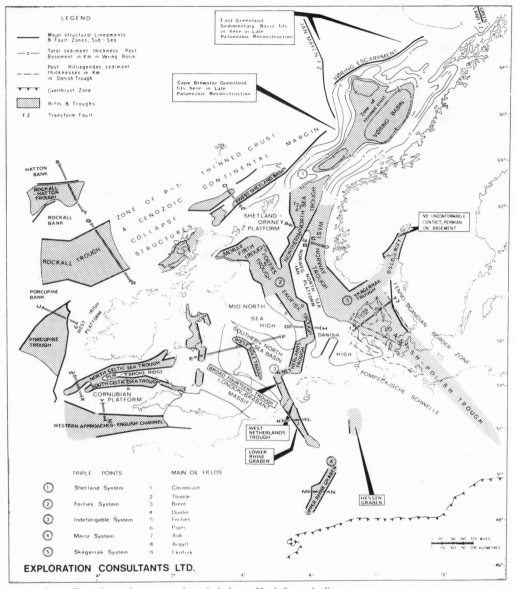

Fig. 1. General map showing troughs and platforms. North Sea and adjacent areas.

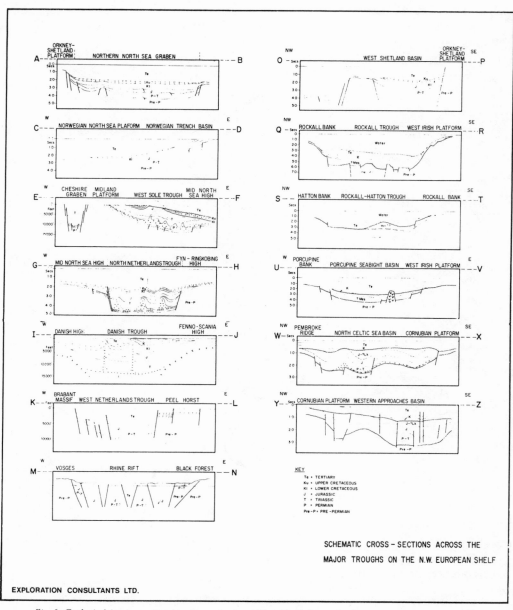

Fig. 2. Geological interpretations based on seismic profiles, North Sea and adjacent areas.

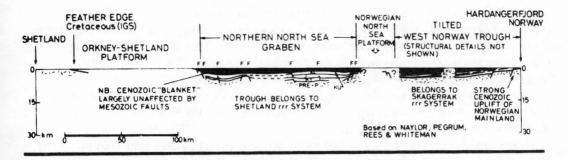

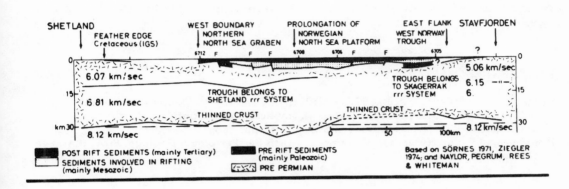

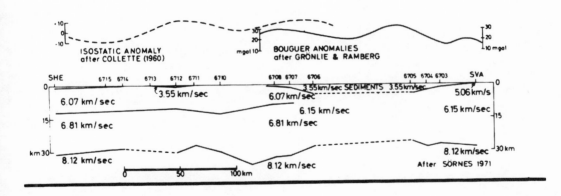

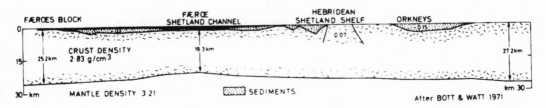

Fig. 3. Generalized sections showing crustal thinning beneath the Northern North Sea Trough Faeroe—Shetland Channel.

373

The Rhine—Hessen Grabens developed in Late Mesozoic—Early Tertiary time are regarded as part of the same fragmentation sequence, although there may be evidence of Permian or older fault activity. The West Netherlands Trough appears to be currently subsiding and contains the Rhine Delta Complex, and the Rhine Graben is probably still active.

The North Sea area subsided regionally during most of the Cenozoic and this has been coupled with uplift of the adjacent land masses of Norway, Scotland, England and Wales, etc.

There are close relationships between oilfield distribution and trough patterns especially in the Northern North Sea (Fig. 1). These will become much clearer as more details are published. The generation of hydrocarbons, geothermal history and the formation of these trough systems are closely linked.

REFERENCES

Burke, K. and Whiteman, A.J., 1973. Uplift, rifting and the break-up of Africa. In: D.H. Tarling and S.K. Runcorn (Editors), Implications of Continental Drift to the Earth Sciences, 2. NATO Advanced Study Inst., pp. 735—755.

Hopkinson, J.P. and Nysaether, E., in press. North Sea Petroleum Geology. Proc. Stavanger Oil Conf., 1972.

Hydrocarbon potential of marginal basins bounded by an island arc

ABSTRACT

The Japan Basin is an example of a western Pacific marginal basin that is bounded by an island arc. Such basins were produced by an extensional process that involves formation of new, hot sea floor below basins that are opening as newly created arc systems migrate away from the continental borderland or older arcs. Rapid deposition of sediment on the hot basin floors allows development of steep geothermal gradients in these sedimentary prisms. In these sedimentary basins, the relationship between temperature and strata age is such that the kerogen-hydrocarbon transformation occurs in young, shallow strata. The sealing effect of turbidite deposits and the presence of numerous gravity faults ensures entrapment conditions. Hydrocarbon potential of these basins can be evaluated by considering the age of basin opening, sedimentation rates adjusted for compaction, and heat-flow data.

S. O. Schlanger*
Department of Earth Sciences
University of California
Riverside, California 92502
Jim Combs
Institute for Geosciences
University of Texas at Dallas
Richardson, Texas 75080

INTRODUCTION

Geologic and geophysical investigations of island arcs and the marginal basins that separate them from continental borderlands suggest that these marginal basins were formed by an extensional mechanism that produces new, hot, basaltic sea floor upon which the basinal sedimentary prism is deposited. Heat-flow values measured within these basins are commonly higher than normal oceanic values and indicate that steep geothermal gradients exist within the sedimentary section.

Field studies of producing oil and gas basins (Phillippi, 1965) and laboratory research on the geochemistry and kinetics of the kerogen-hydrocarbon transformation (Welte, 1972) indicate that the rate of oil and gas generation, together with accumulation and degradation, is functionally dependent on relationships between the age of the hydrocarbon-bearing strata and the temperature regime through time to which these strata have been exposed. The conclusion has been drawn that knowledge of the present, as well as past, geothermal gradients within a basin is important to the evaluation of the hydrocarbon potential of that basin. Klemme (1972a, 1972b) discussed the influence of geothermal gradients on both the size and depth distribution of oil fields of various ages and tectonic settings.

In this paper, using the Japan Basin as an example, major conclusions are integrated from the geochemical and field studies in order to outline an approach to the evaluation of hydrocarbon potential in marginal basins. The approach uses data on heat flow, thermal conductivity, and basin geometry to provide a picture of the thermal structure within the sedimentary fill of the basin. The relationship of temperature versus strata age that governs hydrocarbon generation is considered within the framework of the thermal structure to determine if the sediment has been subjected to the "correct" temperature regime for the "proper" period of time in order for significant oil and gas generation to have occurred.

*Present address: Sedimentology Group, Geological Institute, University of Leiden, The Netherlands.

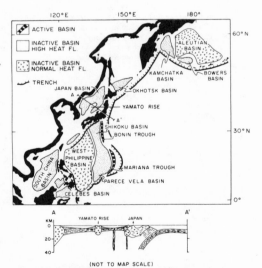

Figure 1. Marginal basins of northwest Pacific and cross section through Japan Basin (after Karig, 1971, 1974).

DISCUSSION

In a series of papers, Karig (1971, 1974) developed the hypothesis that the marginal basins of the western Pacific (Fig. 1) were formed essentially by extension. The kinematics postulated are simply that the island arc migrates relatively away from the mainland toward the oceanic plate after the breaking up of the continental margin. New sea floor is created between the migrating arc and the mainland. This new basin floor becomes the site of sediment accumulation. According to Karig's scheme, pulses of extension may produce complex marginal basins characterized by remnant arcs, such as the Yamato Rise in the Japan Basin (Fig. 1).

Heat-flow measurements within these marginal basins (Fig. 2) indicate that, given representative thermal conductivity values within their respective sedimentary covers, the basaltic basements of the marginal basins are considerably hotter than the upper surface of the plate of oceanic basalt seaward of the trench. Whatever the heat-producing and heat-transporting mechanism is beneath these marginal seas, it is sufficient for the purposes of this paper to demonstrate that steep geothermal gradients exist and most likely existed in the past in the sediment that is accumulating on the new sea floor.

On the basis of study of approximately 45,000 km of seismic reflection and bathymetric records, the heat-flow data of Yasui and others (1968), and Karig's model, Hilde and Wageman (1973) postulated that the Japan Basin first opened during a Cretaceous extensional pulse along a northeast-trending axis between the Yamato Rise and the Chinese borderland. A second extensional pulse began about 22 m.y. B.P. and further opened the Japan Sea along another northeast-trending axis between the Japanese islands and the Yamato Rise. Sediment thicknesses of as much as 2.2 and 2.6 km were found by Hilde and Wageman (1973) along the eastern and western margins of the basin, respectively. Figure 2 shows a generalized interpretation of the structure of the Japan Basin along a line from central Honshu to the Chinese borderland. The basin is characterized by high heat-flow values of from 2.0 to 2.5 $\mu cal/cm^2/sec$ over areas corresponding to the extensional axes and by lower values of 1.5 to 2.0 $\mu cal/cm^2/sec$ over the Yamato Rise. Other pertinent characteristics include high sedimentation rates due to intense volcanic activity, relatively high biological productivity of the surface waters in the Sea of Japan, the proximity of large land masses drained by major rivers, and the occurrence of turbidity currents. Sedimentation rates as great as 180 m/m.y. were noted at DSDP site 301 on the northwest flank of the Yamato Rise (Ingle and others, 1973) where Miocene diatomite deposits are overlain by Pliocene-Pleistocene distal turbidite deposits.

The temperature structure within the sedimentary prism in the Japan Basin can be approximated (after making suitable assumptions) by a heat-flow modeling program (D. Langenkamp and J. Combs, in prep.). A finite difference approximation was made to the heat-conduction equations, and the resulting set of simultaneous equations was solved by the method of successive overrelaxation (Young, 1954). The temperature distribution in

the subsurface was calculated from the input of an arbitrary two-dimensional thermal conductivity structure and specified boundary conditions. The boundary conditions can consist of heat-flow values, heat-source distributions, and (or) fixed temperatures.

For the case study of the Japan Basin from the Yamato Rise to the Chinese borderland, Hilde and Wagemen's (1973) isopach data on the geometry of the sedimentary prism were used to fix the thermal conductivity structure. The values along the heat-flow profile, used as heat-sink inputs at the sediment-water interface, were taken from Yasui and others (1968). Temperatures of 4°C at the sediment-water interface and of 800°C for the intrusive body (Fig. 3) were assumed. All other temperature nodes were unspecified. The raw data from the computer program were temperatures printed out on a grid representing nodes 500 m apart vertically and 10 km apart horizontally. Figure 3 presents the hand-contoured output of the computer program—the subsurface isothermal distribution and heat-flow (Q) profile over the northwestern part of the Japan abyssal plain. Geothermal gradients were calculated; their steepness is considered critical to the depth at which formation of hydrocarbons takes place.

Phillippi (1965) studied the relationship between geothermal gradients and the depths at which oil generation occurred in the Los Angeles and Ventura basins. He concluded that the hydrocarbon content of a basin, given suitable source beds and trapping situations, was largely dependent on the length of time the sedimentary deposits had been exposed to temperatures higher than those needed to initiate the kerogen-hydrocarbon transformation. Since Phillippi's (1965) study, the geochemistry of the kerogen-hydrocarbon transformation has been examined in considerable detail. Tissot and others (1974) summarized the chemical changes involved. These changes include the decrease of the carbon-atom number of the n-alkanes, the ring number of the cycloalkanes, and the carbon-atom number of the aromatics. The chemical pathway followed during the transformation of the original organic material (biopolymers) through kerogen to oil and gas (geomonomers) takes place in several diagenetic realms and involves a complex series of reactions as shown in Table 1.

Figure 4 is a plot of various data on the relationship of temperature versus strata age that prevails during the kerogen-hydrocarbon transformation in oil- and gas-producing basins. In this paper, the kerogen-hydrocarbon transformation is discussed in a general sense, although it is realized that different kerogen types (woody and [or] coaly versus sapropelic) will yield oil over different time periods and temperature ranges. Connan

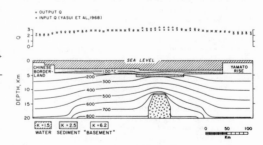

Figure 2. Generalized cross section from Chinese borderland through Japan Trench showing pertinent geologic, sedimentologic, and geophysical features (based on data from Hilde and Wageman, 1973; Yasui and others, 1968; Kitamura and Onuki, 1973; Ingle and others, 1973).

Figure 3. Computer-modeled, hand-contoured plot of subsurface isothermal distribution beneath Japan abyssal plain.

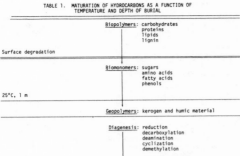

TABLE 1. MATURATION OF HYDROCARBONS AS A FUNCTION OF TEMPERATURE AND DEPTH OF BURIAL

	Biopolymers: carbohydrates proteins lipids lignin
Surface degradation	
	Biomonomers: sugars amino acids fatty acids phenols
25°C, 1 m	
	Geopolymers: kerogen and humic material
	Diagenesis: reduction decarboxylation deamination cyclization demethylation thermal alteration cracking
50°C, 1,000 m	
	Geomonomers: low-molecular-weight hydrocarbons other organic compounds
175°C, 6,000 m	

Note: After I. R. Kaplan, unpub. data.

(1974), using data similar to those plotted in Figure 4 plus his experimental data, applied kinetic laws for first-order reactions to the kerogen-hydrocarbon transformation to arrive at the equation

$$\log t \ (10^6 \ \mathrm{yr}) = 3.014/T \ (°\mathrm{K}) - 6.498.$$

Analysis of the field data shown in Figure 4 indicates the complex relationship of temperature versus strata age. Perhaps in nature, the kerogen-hydrocarbon transformation is not taking place isothermally as is the assumption in analysis of first-order reactions.

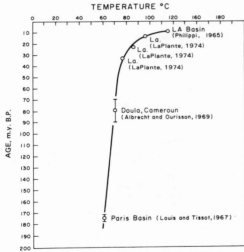

Figure 4. Plot of temperature at which initiation of hydrocarbon generation has been observed versus age of containing strata in various oil- and gas-producing basins.

Application of these studies to the geology and geophysics of marginal basins and the geochemistry of the kerogen-hydrocarbon transformation can be discussed with Figure 5 as a reference. The columns on the right in Figure 5 represent the stratigraphic sections that would result from sediment accumulating at rates of 150 and 100 m/m.y. with regular compaction of clay-rich sediment. As shown in the columns, the postcompaction boundaries of each million years' accumulation of sediment can be determined by using the method of compaction evaluation of Perrier and Quiblier (1974), thereby establishing absolute time scales within the strata.

Values used in the computer model in Figure 3 are from curve A, which represents the geothermal gradient generated by assuming a heat flow Q of 2.5 μcal/cm^2/sec and constant thermal conductivity K of 2.5 mcal/cm·sec·°C throughout the section. Actually, because of compaction and consequent increase in bulk density, the value of K generally increases with depth. Curve B was plotted by using various values for K: 2.4 for the 0- to 500-m interval, 3.0 for the 500- to 1,000-m interval, and 3.4 for depths greater than 1,000 m. These K values were given by Marshall and Erickson (1974) for sediment cored at DSDP sites 241 to 249. The effect on the shape of the geothermal gradient of upward-percolating water that is expelled from the compacting sediment (Panda, 1973) is not considered in our analysis. According to von Herzen 1973), the effect in rapidly accumulating sections would be quite small.

Curves C and C' are the data on age of strata versus temperature from Figure 4 replotted using the time scales in the columns on the right in Figure 5. Curve C is plotted against the 150 m/m.y. section. Curve C' is plotted against the 100 m/m.y. section. The geothermal gradient is assumed to be established in a steady-state condition as the sediment accumulates and compacts (Panda, 1973). Hydrocarbon generation from the contained kerogen would commence in strata as young as 10 to 11 m.y. old in the section deposited at 150 m/m.y. (as shown by the intersections of curves A and B with curve C) and in strata 13.5 to 14.5 m.y. old in the section deposited at 100 m/m.y. (as shown by the intersections of curves A and B with curve C').

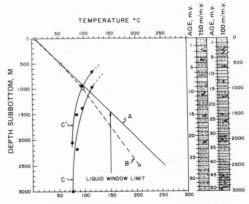

Figure 5. Geothermal gradients (curves A and B); age and thickness of sedimentary fill; and time, depth, and temperature conditions (curves C and C') of initiation of hydrocarbon generation in a marginal basin. Circles are data points (see text for sources of data).

If Pusey's (1973) end-limit temperature of 150°C is accepted for the purpose of illustration, maturation of these hydrocarbons would continue until temperatures reached the upper temperature limit of the "liquid window" at 150°C. At this temperature and above, the destruction of liquid hydrocarbons dominates, and thermal gas is the end product of the kerogen-hydrocarbon transformation (Tissot and others, 1974). Thus, strata older than 14.5 m.y. (intersection of 150°C line with curve A) or 17.5 m.y. (intersection of 150°C line with curve B) in the section deposited at 150 m/m.y. would contain thermal gas. At a sedimentation rate of 100 m/m.y., thermal gas would be present in strata older than 22.5 m.y. (curve A) or 27 m.y. (curve B). These ages are well within the age range of strata in the Japan Basin. That such a marginal basin contains suitable stratigraphic traps is indicated by the fact that flows of ethane gas hampered drilling (Ingle and others, 1973) at DSDP site 299 between Japan and the Yamato Rise and at DSDP site 301 between the Yamato Rise and the Chinese borderland. Both gas occurrences at subbottom depths of less than 500 m were at the boundary between diatomite deposits of late Miocene age and overlying turbidite deposits of Pliocene-Pleistocene age. The impermeable turbidite section evidently had prevented the upward migration of gas.

CONCLUSIONS

The oil and gas potential of a marginal basin bounded by an island arc (for example, the Japan Basin) depends on the age of the extensional pulses that created it and the rate and duration of the extension. The basin geometry is determined by the age of opening and rate of extension, whereas the geometry of the sedimentary prism is determined by the sedimentation rate adjusted for compaction. The extensional pulse history also plays a role in the distribution and abundance of horst-and-graben or gravity-fault structures that can provide hydrocarbon-migration traps. As is the case in the Tertiary basins of Sumatra (Wennekers, 1958) and in the Tertiary and Mesozoic strata of Iraq (Dunnington, 1958), pelagic sediment within the sedimentary prism can serve as source beds. Similarly, river-borne sediment drained from large adjacent land masses can contain abundant source material as does the modern sediment of the Yellow Sea (Wageman and others, 1970). Permeability seals are provided by intercalated turbidite deposits. The texture, thickness, and frequency of occurrence of turbidite deposits in the section are a function of the geology, rate of erosion, and areal extent of the bounding land masses. Volcanic activity can contribute abundant montmorillonite, which enhances the kerogen-hydrocarbon transformation because of its catalytic effect (Andreev and others, 1968; Johns and Shimoyama, 1972). The insulating effect of the rapidly accumulating sedimentary prism on the new, hot basin floor results in the development of steep geothermal gradients, hence thin "liquid windows," and probably in very shallow production of hydrocarbons.

REFERENCES CITED

Albrecht, P., and Ourisson, G., 1969, Diagénèse des hydrocarbures saturés dans une série sedimentaire épaisse (Doula, Cameroun): Geochim. et Cosmochim. Acta, v. 33, p. 138–142.

Andreev, P. F., Bogomolou, A. I., Dobryanskii, A. F., and Kartsev, A. A., 1968, Transformation of petroleum in nature: New York, Pergamon Press, 461 p. (English edition translated by R. B. Gaul and B. C. Metzner and edited by E. Barghoorn and S. Silverman).

Connan, J., 1974, Time-temperature relation in oil genesis: Am. Assoc. Petroleum Geologists Bull., v. 58, p. 2516–2521.

Dunnington, H. V., 1958, Generation, migration, accumulation, and dissipation of oil in northern Iraq, in Weeks, L. G., ed., Habitat of oil: Tulsa, Okla., Am. Assoc. Petroleum Geologists, p. 1194–1251.

Hilde, T.W.C., and Wageman, J. M., 1973, Structure and origin of the Japan Sea, in Coleman, P. J., ed., The western Pacific, island arcs, marginal seas, geochemistry: New York, Crane, Russak & Co., p. 415–434.

Ingle, J. C., Jr., and others, 1973, Western Pacific floor: Leg 31, Deep Sea Drilling Project: Geotimes, v. 18, no. 9, p. 22–25.

Johns, W. D., and Shimoyama, A., 1972, Clay minerals and petroleum-forming reactions during burial and diagenesis: Am. Assoc. Petroleum Geologists Bull., v. 56, p. 2160–2167.

Karig, D. E., 1971, Structural history of the Mariana arc system: Geol. Soc. America Bull., v. 82, p. 323–344.

———1974, Evolution of arc systems in the western Pacific, in Donath, F., ed., Annual review of Earth and planetary sciences, Vol. 2: Palo Alto, Calif., Annual Reviews, Inc., p. 51–75.

Kitamura, N., and Onuki, Y., 1973, Geological and crustal sections of the A-zone, northeast Japan, in The crust and upper mantle of the Japanese area, Pt. II: Japanese Natl. Comm. for the Upper Mantle Project, Geol. Survey of Japan, 163 p.

Klemme, H. D., 1972a, Heat influences size of oil giants, Pt. 1: Oil and Gas Jour., v. 70, no. 29, p. 136–144.

——— 1972b, Heat influences size of oil giants, Pt. 2: Oil and Gas Jour., v. 70, no. 30, p. 76–78.

LaPlante, R. E., 1974, Hydrocarbon generation in Gulf Coast Tertiary sediments: Am. Assoc. Petroleum Geologists Bull., v. 58, p. 1281–1289.

Louis, M. C., and Tissot, B. P., 1967, Influence de la température et de la pression sur la formation des hydrocarbures dans les argiles à kérogène: World Petroleum Cong., 7th, Proc., v. 2, p. 47–60.

Marshall, B. V., and Erickson, A. J., 1974, Heat flow and thermal conductivity measurements, Leg 25, Deep Sea Drilling Project, in Simpson, E. S., Schlich, R., and others, Initial reports of the Deep Sea Drilling Project: Washington, D.C., U.S. Govt. Printing Office, v. 25, p. 349–355.

Panda, P. K., 1973, The transient and steady-state influence of sediment compaction and interstitial water expulsion on the temperature distribution in oceanic sediments: Marine Geophys. Researches, v. 2, p. 37–49.

Perrier, R., and Quiblier, J., 1974, Thickness changes in sedimentary layers during compaction history, methods for quantitative evaluation: Am. Assoc. Petroleum Geologists Bull., v. 58, p. 507–520.

Phillippi, G. T., 1965, On the depth, time and mechanism of petroleum generation: Geochim. et Cosmochim. Acta, v. 29, p. 1021–1049.

Pusey, W. C., III, 1973, Paleotemperatures in the Gulf Coast using the ESR-kerogen method: Gulf Coast Assoc. Geol. Socs. Trans., v. 23, p. 195–202.

Tissot, B., Durand, J., Espitalié, J., and Comboz, A., 1974, Influence of nature and diagenesis of organic matter in formation of petroleum: Am. Assoc. Petroleum Geologists Bull., v. 58, p. 499–506.

von Herzen, R. P., 1973, Geothermal measurements, Leg 21, in Burns, R. E., Andrews, J. E., and others, Initial reports of the Deep Sea Drilling Project: Washington, D.C., U.S. Govt. Printing Office, v. 21, p. 443–457.

Wageman, J. M., Hilde, T.W.C., and Emery, K. O., 1970, Structural framework of East China Sea and Yellow Sea: Am. Assoc. Petroleum Geologists Bull., v. 54, p. 1611–1643.

Welte, D. H., 1972, Petroleum exploration and organic geochemistry: Jour. Geochem. Exploration, v. 1, p. 117–136.

Wennekers, J.H.L., 1958, South Sumatra basinal area, in Weeks, L. G., ed., Habitat of oil: Tulsa, Okla., Am. Assoc. Petroleum Geologists, p. 1347–1348.

Yasui, M., Kishii, T., Watanabe, T., and Uyeda, S., 1968, Heat flow in the Sea of Japan, in Knopoff, L., Drake, C. L., and Hart, P. J., eds., The crust and upper mantle of the Pacific area: Am. Geophys. Union Geophys. Mon. 12, p. 3–16.

Young, D., 1954, Iterative methods for solving partial difference equations of elliptic type: Trans. Am. Math. Soc., v. 76, p. 92–111.

ACKNOWLEDGMENTS

Reviewed by Leigh C. Price, Paul Morgan, and Dewey A. Baker. Schlanger supported by the University of California Intramural Research Fund; Combs supported by a research grant to the University of Texas at Dallas from Gulf Mineral Research Company.

We thank Lewis H. Cohen and Tien-Chang Lee for helpful discussions.

Contribution number 283, Institute for Geosciences, University of Texas at Dallas, P.O. Box 688, Richardson, Texas 75080.

MANUSCRIPT RECEIVED APRIL 10, 1975

MANUSCRIPT ACCEPTED MAY 6, 1975

POSTSCRIPT–LATER DEVELOPMENTS

This compilation includes most papers on the topic up to the end of 1975, with one major omission: in mid-1974, there was a joint Geological Association of Canada/Mineralogical Association of Canada meeting, during which several papers on Metallogeny and Plate Tectonics were presented. These have been compiled into a special volume (Strong, 1976), which contains twenty-six papers and six abstracts, most of them covering much the same ground as those in this volume—indeed, several are by authors already represented here.

The most promising recent researchers appear to be those involving attempts to "fingerprint" the plate tectonic setting of ore deposits using trace-element and minor geochemistry (Pearce and Gale, in press; Cronan et al., in press); and use of REE and isotopes to investigate the role of sea water in volcanic-hydrothermal mineralization at spreading centres (Robertson and Fleet, in press; Spooner and Heaton, in press).

It is now clear that the generalization phase is coming rapidly to an end, and from now on we can expect the emphasis to be increasingly on specific ore deposits. While these studies will undoubtedly increase our understanding of how mineral deposits have formed, it may still be some time before they can be used in a predictive way, as an aid to exploration.

REFERENCES

Anderson, R. N., and Halunen, A. J., 1974, Implications of heat flow for metallogenesis in the Bauer Deep: Nature, v. 251, p. 473–475.

Anguita, F., and Hernan, F., 1975, A propagating fracture model versus a hot spot origin for the Canary Islands: Earth and Planetary Sci. Letters, v. 27, p. 11–19.

Arculus, R. J., and Smith, F. W., 1972, Contributed remarks to paper by C. J. V. Wheatley: Inst. Mining and Metallurgy Trans., B, v. 81, B109–B110.

Badham, J. P. N., and Halls, C., 1975, Microplate tectonics, oblique collisions, and evolution of the Hercynian orogenic systems: Geology, v. 3, p. 373–376.

Bailey, D. K., 1974, Continental rifting and alkaline magmatism, in Sorensen, H., ed., The Alkaline Rocks, Wiley, New York, p. 148–159.

Bailey, D. K., 1975, Lithosphere control of continental rift magmatism (abstract only): Geol. Soc. London Newsletter, v. 14, no. 5, p. 12.

Barnes, J. W., 1973, Contributed remarks to paper by D. L. Searle: Inst. Mining and Metallurgy Trans., B, v. 82, no. 798, p. B75.

Beck, R. H., and Lehner, P., 1974, Oceans, new frontiers in exploration: Am. Assoc. Petroleum Geologists Bull., v. 58, p. 376–395.

Beloussov, V. V., 1970, Against the hypothesis of sea-floor spreading: Tectonophysics, v. 9, no. 6, p. 489–511.

Bernard, A. J., and Soler, E., 1971, Sur la localisation géo-tectonique des amas pyriteux massifs du type Rio Tinto: C. R. Acad. Sci. Paris, v. 273, p. 1087–1090.

Blackett, P. M. S., Bullard, E., and Runcorn, S. K., eds., 1965, A symposium on continental drift: Royal Soc. London, 339 p.

Boström, K., and Fisher, D. E., 1969, Distribution of mercury in East Pacific sediments: Geochim, et Cosmochim. Acta, v. 33, p. 743–745.

Bullard, E. C., 1974, Minerals from the deep sea: Endeavour, v. 33, p. 80–85.

Burk, C. A., 1972, Global tectonics and world resources: Am. Assoc. Petroleum Geologists Bull., v. 56, no. 2, p. 196–202.

Burke, K., and Dewey, J. F., 1973, Plume-generated triple junctions: key indicators in applying plate tectonics to old rocks: Jour. Geology, v. 81, p. 406–433.

Carozzi, A. V., 1970, New historical data on the origin of the theory of continental drift: Geol. Soc. America Bull., v. 81, p. 283–286.

References

Carr, R. A., Jones, M. M., and Russ, E. R. 1974, Anomalous mercury in near-bottom water of a mid-Atlantic Rift valley: Nature, v. 251, p. 489–490.

Carvalho, D. de, 1972, The metallogenetic consequences of plate tectonics and the Upper Palaeozoic evolution of southern Portugal: Estud. Not. Trab. Serv. Fom. Min., v. 20, p. 297–320.

Clifford, T. N., 1966, Tectono-metallogenic units and metallogenic provinces in Africa: Earth and Planetary Sci. Letters, v. 1, p. 421–434.

Cloud, P., 1973, Palaeoecological significance of the banded iron-formations: Econ. Geology, v. 68, p. 1135–1143.

Cox, A., 1973, Plate tectonics: W. H. Freeman, San Francisco, 702, p.

Crawford, A. R. 1973, A displaced Tibetan massif as a possible source of some Malaysian rocks: Geol. Mag., v. 109, p. 463–469.

Crawford, A. R., 1974, The Indus suture line, the Himalaya, Tibet and Gondwanaland: Geol. Mag., v. 111, p. 369–383. *See also* A Greater Gondwanaland: Science, 1974, v. 184, p. 1179–1181.

Crockett, R. N., and Mason, R., 1968, Foci of mantle disturbance in southern Africa and their economic significance: Econ. Geol., v. 63, p. 532–540.

Cronan, D. A., Smith, P. A., and Bignell, R. D., in press, Modern submarine hydrothermal mineralization: examples from Santorini and the Red Sea: Inst. Mining and Metallurgy Special Publication.

Curray, J. R., 1975, Marine Sediments, geosynclines and orogeny, *in* Fischer, A. G., and Judson, S., eds., Petroleum and global tectonics, Princeton University Press, Princeton, p. 157–224.

Degens, E. T., and Ross, D. A., eds., 1969, Hot brines and recent heavy metal deposits in the Red Sea: Springer-Verlag, Berlin, 600 p.

Degens, E. T., and Kulbicki, G., 1973, Hydrothermal origin of metals in some East African rift lakes: Mineralium Deposita, v. 8, p. 388–404.

Dixon, C. J., and Pereira, J., 1974, Plate tectonics and mineralization in the Tethyan region: Mineralium Deposita, v. 9, p. 185–198.

Dmitriev, L., Barsukov, V., and Udintsev, G., 1971, Rift-zones of the ocean and the problem of ore-formation, *in* Takeuchi, Y., ed., Proceedings of the IMA-IAGOD meetings: Spec. issue no. 3, Society of Mining Geologists of Japan, Tokyo, p. 65–69.

Drewry, G. E., Ramsay, A. T. S., and Smith, A. G., 1974, Climatically controlled sediments, the geomagnetic field and trade wind belts in Phanerozoic time: Jour. Geology, v. 82, p. 531–554.

Dunham, K. C., 1972, Basic and applied geochemists in search of ore: Inst. Mining and Metallurgy Trans., B, v. 81, p. B44–B49.

Dunham, K. C., 1973, Geological controls of metallogenic provinces: Bur. Min. Res. Geol. Geophys. Bull. (Canberra), v. 141, p. 1–12.

du Toit, A., 1937, Our wandering continents: Oliver and Boyd, Edinburg, 366 p.

Embleton, B. J. J., and McElhinny, M. W., 1975, The palaeoposition of Madagascar: palaeomagnetic evidence from the Isalo Group: Earth and Planetary Sci. Letters, v. 27, p. 329–341.

Eugster, H. P., and Chou, I. M., 1973, The depositional environment of Precambrian banded iron-formations: Econ. Geol., v. 68, p. 1144–1168.

Evans, A. M., 1975, Mineralization in geosynclines—the Alpine enigma: Mineralium Deposita, v. 10, p. 254–260.

Fischer, A. G., 1975, Origin and growth of basins, *in* Fischer, A. G., and Judson, S., eds., Petroleum and global tectonics: Princeton University Press, Princeton, p. 47–82.

Fisher, D. E., and Boström, K., 1969, Uranium-rich sediments on the East Pacific Rise: Nature, v. 224, p. 64–65.

Flower, M. J. F., and Strong, D. F., 1969, The significance of sandstone inclusions in lavas of the Comores Archipelago: Earth and Planetary Sci. Letters, v. 7, p. 47–50.

Garrels, R. M., Perry, E. A., and MacKenzie, F. T., 1973, Genesis of Precambrian iron-formations and the development of atmospheric oxygen: Econ. Geol., v. 68, p. 1173–1179.

Gass, I. G., Smith, P. J., and Wilson, R. C. L., eds., 1972, Understanding the Earth: Artemis Press, Sussex, 355 p.

Goodwin, A. M., 1973, Archaean iron-formations and tectonic basins of the Canadian shield: Econ. Geol. v. 68, p. 915–933.

Goossens, P., 1972, Metallogeny in Equadorian Andes: Econ. Geol., v. 67, p. 458–468.

Goossens, P. J., and Hollister, V. F., 1973, Structural control and hydrothermal alteration patterns of Chaucha porphyry copper, Ecuador: Mineralium Deposita, v. 8, p. 321–331.

Gordon, W. A., 1975, Distribution by latitude of Phanerozoic evaporite deposits: Jour. Geology, v. 83, p. 671–684.

Green, A. G., 1972, Sea floor spreading in the Mozambique channel: Nature Phys. Sci., v. 236, p. 19–21.

Gross, G. A., 1973, The depositional environment of principal types of Precambrian iron-formations, *in* Genesis of Precambrian iron and manganese deposits: Proceedings of the Kiev Symposium 1970, Earth Sciences 9, UNESCO, New York, p. 15–21.

Guilbert, J. M., 1971, Known interactions of tectonics and ore deposits in the context of new global tectonics: Soc. Mining Engineers Preprint no. 71-S-91, 15 p.

Guild, P. W., 1974, Distribution of metallogenic provinces in relation to major Earth features, *in* Petrascheck, W. E., ed., Metallogenetic and geochemical provinces, Symposium Leoben, November 1972, Springer-Verlag, Berlin, p. 10–24.

Guild, P. W., 1977, Application of global tectonic theory to metallogenic studies, *in* B. Bogdanov, ed., *IAGOD Report 4th Symposium*, v. 2, Varna, Bulgaria, (in press).

Haddock, M. H., 1936., The wandering of the continents: Discovery, v. 17, p. 67–70.

Hallam, A., 1973, A revolution in the Earth Sciences: Oxford University Press, London, p. ix, 127.

Hammond, A. L., 1975, Minerals and plate tectonics: a conceptual revolution: Science, v. 189, p. 779–781.

Hammond, A. L., 1975, Minerals and plate tectonics, II: sea water and ore formation: Science, v. 189, p. 868–869, 915–917.

Hastings, D. A., 1974, Proposed origin for Guianan diamonds: Geology, v. 2, p. 475–476.

Hedberg, H. D., 1970, Continental margins from the point of view of the petroleum geologist: Am. Assoc. Petroleum Geologists Bull., v. 54, p. 3–43.

Henderson, G., 1973, The implications of continental drift for the petroleum prospects of West Greenland, *in* Tarling, D. H., and Runcorn, S. K., eds., Implications of Continental Drift to the Earth Sciences, Vol. 2: Academic Press, New York, p. 599–608.

Hodder, R. W., and Hollister, V. F., 1972, Structural features of porphyry copper deposits and the tectonic evolution of continents: Canadian Inst. Mining and Metallurgy Trans., v. 75, p. 23–27.

Holland, H. D., 1973, The oceans: a possible source of iron in iron-formations: Econ. Geol., v. 68, p. 1169–1172.

Holland, J. G., and Lambert, R. St. J., 1975, The chemistry and origin of the

Lewisian gneisses of the Scottish mainland: the Scourie and Inver assemblages and subcrustal accretion: Precambrian Res., v. 2, p. 161–188.

Hollister, V. F., 1974, Regional characteristics of porphyry copper deposits of South America: Soc. Mining Engineers Trans., v. 225, p. 45–53.

Hollister, V. F., and Sirvas, E. B., 1974, The Michiquillay porphyry copper deposit: Mineralium Deposita, v. 9, p. 261–269.

Holmes, A., 1931, Radioactivity and earth movements: Geol. Soc. Glasgow Trans., v. 18, pt. 3, p. 559–606.

Holmes, A., 1944, Principles of physical geology, 2nd ed.: Thomas Nelson, London, p. xii, 532.

Holmes, A., 1953, The South Atlantic: land bridges or continental drift: Nature, v. 171, p. 669–671.

Holmes, A., 1965, Principles of physical geology, 2nd ed.: Thomas Nelson, London, 1303 p.

Hutchinson, R. W., and Hodder, R. W., 1972, Possible tectonic and metallogenic relationships between porphyry copper and massive sulphide deposits: Canadian Inst. Mining and Metallurgy Trans., v. 75, p. 16–22.

James, T. C., 1972, Concepts in mineral exploration: Inst. Mining and Metallurgy Trans., B, v. 81, p. B138–B140.

Jankovic, S., 1972. The origin of base-metal mineralization on the mid-Atlantic ridge: 24 Int. Geol. Congr., sec. 4, p. 326–334.

Jeffreys, H., 1970a, The Earth, 5th ed.: Cambridge University Press, London, 537 p.

Jeffreys, H., 1970b, Imperfections of elasticity and continental drift: Nature, v. 225, p. 1007–1008.

Jensen, M. L., 1971, Provenance of cordilleran intrusives and associated metals: Econ. Geol., v. 66, p. 34–42.

Kanamori, H., Takeuchi, H., and Uyeda, S., 1967, Debate about the Earth: An approach to geophysics through analysis of continental drift: W. H. Freeman & Co., Reading, England, 253 p.

Kasbeer, T., 1973, Bibliography of continental drift and plate tectonics: Geol. Soc. America Spec. Paper 142, 96 p.

King, L. C., 1953, Necessity for continental drift: Am. Assoc. Petroleum Geologists Bull., v. 37, p. 2163–2177.

Kinsman, D. J. J., 1975, Rift valley basins and sedimentary history of trailing continental margins, *in* Fischer, A. G., and Judson, S., eds., Petroleum and global tectonics: Princeton University Press, Princeton, p. 83–128.

Klemme, H. D., 1972, Heat influences size of oil giants: Oil and Gas Jour., v. 70, no. 29, p. 136–144; and v. 70, no. 30, p. 76–78.

Klemme, H. D., 1975a, Giant oil fields related to their geologic setting: a possible guide to exploration: Canadian Petroleum Geologists Bull., v. 23, p. 30–66.

Klemme, H. D., 1975b, Geothermal gradients, heat flow, and hydrocarbon recovery, *in* Fischer, A. G., and Judson, S., eds., Petroleum and global tectonics: Princeton University Press, Princeton, p. 251–306.

Krauskopf, K. B., 1971, The source of ore metals: Geochim. et Cosmochim. Acta, v. 35, p. 613–659.

Kutina, J., 1969, Hydrothermal ore deposits in the western United States: a new concept of structural control of distribution: Science, v. 165, p. 1113–1119.

Kutina, J., 1972, Regularities in the distribution of hypogene mineralization along rift structures: 24 Int. Geol. Congr., sec. 4, p. 65–73.

Landwehr, W. R., 1968, The genesis and distribution of major mineralization in western United States: Econ. Geol., v. 63, p. 967–970.

Lange, I. M., and Cheney, E. S., 1971, Sulfur isotopic reconnaissance of Butte, Montana: Econ. Geol., v. 66, p. 63–74.

Laznicka, P., and Wilson, H. D. S., 1972, The significance of a copper-lead line in metallogeny: 24 Int. Geol. Congr., sec. 4, p. 25–36.

Longwell, C. R., 1944, Some thoughts on the evidence for continental drift, Am. Jour. Sci., v. 242, p. 218–321.

Lowell, J. D., 1974, Regional characteristics of porphyry copper deposits of the southwest: Econ. Geol., v. 69, p. 601–617.

McDowell, A. N., 1971, Practical application of continental-drift concept may find giants: Oil and Gas Jour., v. 69, pt. 26, p. 114–116.

Mero, J. L., 1965, The mineral resources of the sea: Elsevier, Amsterdam, 312 p.

Meyerhoff, A. A., and Teichert, C., 1971, Continental drift III: Late Palaeozoic glacial centres, and Devonian-Eocene coal distribution: Jour. Geology, v. 79, p. 285–321.

Meyerhoff, A. A., Meyerhoff, H. A., and Briggs, R. S., 1972, Continental drift V: proposed hypothesis of earth tectonics: Jour. Geology, v. 80, p. 663–692.

Meyerhoff, A. A., and Meyerhoff, H. A., 1972a, The new global tectonics: major inconsistencies: Am. Assoc. Petroleum Geologists Bull., v. 56, no. 2, p. 269–336.

Meyerhoff, A. A., and Meyerhoff, H. A., 1972b, The new global tectonics: age of linear magnetic anomalies of ocean basins: Am. Assoc. Petroleum Geologists Bull., v. 56, no. 2, p. 337–359. *See also* discussion of the Meyerhoffs' papers, with replies, by Mill, D. E., and MacKenzie, D. B., 1972, Am. Assoc. Petroleum Geologists Bull., v. 56, no. 11, p. 2290–2295; and by Facer, R. A., and Page, N. J., Am. Assoc. Petroleum Geologists Bull., v. 57, no. 6, p. 1134–1142.

Morgan, W. J., 1971, Convection plumes in the lower mantle: Nature, v. 230, p. 42–43.

Morgan, W. J., 1972, Deep mantle convection plumes and plate motions: Am. Assoc. Petroleum Geologists Bull., v. 56, p. 203–213.

Morrissey, C. J., Davis, G. R., and Steed, G. M., 1971, Mineralization in the Lower Carboniferous of Central Ireland: Inst. Mining and Metallurgy Trans., B, v. 80, no. 777, p. B174–B185.

Noble, J. A., 1970, Metal provinces of the western United States: Geol. Soc. America Bull., v. 81, p. 1607–1624.

O'Hara, M. J., 1973, Non-primary magmas and dubious mantle plume beneath Iceland: Nature, v. 243, p. 507–508.

Pearce, J. A., and Gale, G. H., in press, Identification of ore-depositional environment from trace-element geochemistry: Inst. Mining and Metallurgy Special Publication.

Pereira, J., and Dixon, C. J., 1965, Evolutionary trends in ore deposition, Inst. Mining and Metallurgy Trans., B, v. 74, p. 505–527.

Petrascheck, W. E., 1965, Typical features of metallogenetic provinces: Econ. Geol., v. 60, no. 8, p. 1620–1634.

Petrascheck, W. E., 1969, Ore metals from the crust or mantle: Econ. Geol., v. 64, p. 576–578.

Petrascheck, W. E., 1972, Kontinentalverschiebung: Zerteilung und Neuschaffung von Erzprovinzen: Umschau, v. 72, p. 677–680.

Petrascheck, W. E., 1973a, Orogene und kratogene Metallogenese: Geol. Rundschau, v. 62, p. 617–626.

Petrascheck, W. E., 1973b, Some aspects of the relations between continental drift and metallogenic provinces, *in* Tarling, D. H., and Runcorn, S. K., eds., Implications of continental drift to the earth sciences, Vol. 1: Academic Press, New York, p. 563–568.

Petrascheck, W. E., 1974, Die Herkunft der Erzmetalle, *in* Metallogenetic and geochemical provinces, Symposium Leoben, November 1972, Springer-Verlag,

Berlin, p. 174–183. *See also* English summary in Mineralium Deposita, 1973, v. 8, p. 96.

Piper, D. Z., 1973, Origin of metalliferous sediments from the East Pacific Rise: Earth and Planetary Sci. Letters, v. 19, p. 75–82.

Rea, W. J., 1972, Igneous activity and metallogenesis in the Ordovician in North Wales: a review, *in* Wood, A., and Davies, W., eds., Mineral exploitation and economic geology: University College of Wales, Aberystwyth, 80 p.

Reid, A. R., 1974, Proposed origin for Guianan diamonds: Geology, v. 2, p. 67–68.

Ridge, J. D., 1973, Volcanic exhalations and ore deposition in the vicinity of the sea floor: Mineralium Deposita, v. 8, p. 332–348.

Robertson, A. H. F., and Fleet, A. J., in press, Rare-earth element evidence for the genesis of the metalliferous sediments of Troodos, Cyprus: Inst. Mining and Metallurgy Special Publication.

Routhier, P., et al., 1973, Some major concepts of metallogeny: Mineralium Deposita, v. 8, p. 237–258.

Rupke, N. A., 1970, Continental drift before 1900: Nature, v. 227, p. 349–350.

Russell, M. J., 1969, Structural controls of base metal mineralization in relation to continental drift: Inst. Mining and Metallurgy Trans., B, v. 78, pp. 44–52, 127–131.

Russell, M. J., 1973, Base metal mineralization in Ireland and Scotland and the formation of Rockall Trough, *in* Tarling, D. H., and Runcorn, S. K., eds., Implications of Continental Drift to the Earth Sciences, Vol. 1: Academic Press, New York, p. 577–594.

Schermerhorn, L. J. G., 1974, Variscan specialists meet in Rennes: Geotimes, v. 19, p. 23–25.

Schermerhorn, L. J. G., 1975, Spilites, regional metamorphism and subduction on the Iberian pyrite belt: some comments: Geologie en Mijnbouw, v. 54, p. 23–36.

Schilling, J. G., 1973a, Iceland mantle plume: geochemical study of Reykjanes Ridge: Nature, v. 242, p. 575–571.

Schilling, J. G., 1973b, Iceland mantle plume: Nature, v. 246, p. 141–143.

Schoell, M., 1975, Previous and current ideas on the origin of the Red Sea brines: Oceanology Internat., v. 75, p. 115–117.

Schönberger, H., and Roever, E. W. F. de, 1974, Proposed origin for Guianan diamonds: Geology, v. 2, no. 10, p. 474–475.

Schopf, J. M., 1973, Coal, climate, and global tectonics, *in* Tarling, D. H., and Runcorn, S. K., eds., Implications of Continental Drift to the Earth Sciences, Vol. 1: Academic Press, New York, p. 609–622.

Shin, B. W., 1974, Geotectonic movements and metal ore deposits in South Korea: J. Korean Institute of Mining Geol., v. 7, no. 1, p. 1–21.

Smith, A. G., and Hallam, A., 1970, The fit of the southern continents: Nature, v. 225, p. 139.

Snider, A., 1858, La Création et ses Mystères dévoites, A. Frank and E. Dentu, Paris, 487 p.

Spooner, E. T. C., and Fyfe, W. S., 1973, Sub-sea floor metamorphism, heat and mass transfer: Contr. Mineralogy and Petrology, v. 42, p. 287–304.

Spooner, E. T. C., and Heaton, T. H. E., in press, Isotopic evidence for the origin of mineralization associated with the ophiolitic rocks of the Troodos Massif, Cyprus: Inst. Mining and Metallurgy Special Publication.

Strong, D. F., ed., 1976, Metallogeny and plate tectonics: Geol. Assoc. Canada Spec. Pub. No. 14.

Szadeczky-Kardoss, E., 1974, Metallogenesis and distribution of elements in the zones of subduction, *in* Petrascheck, W. E., ed., Metallogenetic and geochemical provinces, Symposium Leoben, November 1972, Springer-Verlag, Berlin. *See also* Abstract published in Mineralium Deposita, 1973, v. 8, p. 94.

Tarling, D. H., 1973, Metallic ore deposits and continental drift: Nature, v. 243, p. 193–196.

Taylor, D., 1974, The liberation of minor elements from rocks during plutonic igneous cycles and their subsequent concentration to form workable ores, with particular reference to copper and tin: Geol. Soc. Malaysia Bull., v. 7, p. 1–16.

Taylor, F. B., 1910, Bearing of the Tertiary mountain belt on the origin of the Earth's plan: Geol. Soc. America Bull, v. 21, p. 179–226.

Taylor, G. R., 1974, Volcanogenic mineralization in the islands of the Florida group test: Inst. Mining and Metallurgy Trans., B, v. 83, p. 120–130.

Thonis, M., and Burns, R. G., 1975, Manganese ore deposits and plate tectonics: Nature, v. 253, p. 614–616.

Tooms, J. S., 1970, Review of knowledge of metalliferous brines and related deposits: Inst. Mining and Metallurgy Trans., B, v. 79, p. 116–126.

Turneaure, F. S., 1955, Metallogenetic provinces and epochs: Econ. Geol., v. 1, p. 38–98.

Turneaure, F. S., 1971, The Bolivian tin-silver province: Econ. Geol., v. 66, p. 215–225.

Vening Meinesz, F. A., 1964, The Earth's crust and mantle: Elsevier, Amsterdam, 133 p.

Vokes, F. M., 1973, Metallogeny possibly related to continental break-up in southwest Scandinavia, *in* Tarling, D. H., and Runcorn, S. K., eds., Implications of Continental Drift to the Earth Sciences, Vol. 1: Academic Press, New York, p. 569–576.

Wegener, A., 1912, Die Entstehung der Kontinente: Geol. Rundschau, v. 3, p. 276–292.

Wegener, A., 1929, The origin of continents and oceans, 4th ed.: Dover, New York, 1966, 278 p.

Wheatley, C. J. V., 1971, Aspects of metallogenesis within the Southern Caledonides of Britain and Ireland: Inst. Mining and Metallurgy Trans., B, v. 80, p. 211–223.

White, D. E., 1968, Environments of generation of some base-metal ore deposits: Econ. Geol., v. 63, p. 301–335.

Whiteman, A. J., Rees, G., Naylor, D., and Pegrum, R. M., 1975, North Sea troughs and plate tectonics: Norges Geol. Undersökelse, v. 316, p. 137–161.

Williams, D., Stanton, R. L., and Rambaud, F., 1975, The planes-San Antonio pyritic deposit of Rio Tinto, Spain: its nature, environment and genesis: Inst. Mining and Metallurgy Trans., B, v. 84, p. B73–B82.

Willis, B., 1944, Continental drift, ein Märchen: Am. Jour. Sci., v. 242, p. 509–513.

Wilson, J. T., 1965, A new class of faults and their bearing on continental drift: Nature, v. 207, p. 343–347.

Wright, J. B., 1970, Controls of mineralization in the older and younger tin fields of Nigeria: Econ. Geol., v. 65, p. 945–951.

Wright, J. B., 1973, Continental drift, magmatic provinces and mantle plumes: Nature, v. 244, p. 565–567.

Wright, J. B., and McCurry, P., 1970, Comments on Flower and Strong Earth and Planetary Sci. Letters, v. 8, p. 267–268.

Wright, J. B., and McCurry, P., 1973a, Magmas, mineralization, and sea-floor spreading: Geol. Rundschau, v. 62, no. 1, p. 116–125.

Wright, J. B., and McCurry, P., 1973b, Sea-floor spreading and continental ore deposits, *in* Tarling, D. H., and Runcorn, S. K., eds., Implications of continental drift to the Earth Sciences Vol. 1: Academic Press, New York, P. 559–562.

AUTHOR CITATION INDEX

389

SUBJECT INDEX

About the Editor

J. B. WRIGHT is Reader in Earth Sciences at The Open University, Milton Keynes, England, where he plays a major role in the preparation of course texts and related television and radio programs. Mr. Wright received his M.A. in Geology in 1955 and a B.Sc. for research on Iron-Titanium Oxide Minerals in 1957, from Oxford University. He has served as a government geologist for the Geological Survey of Kenya, lecturer in Geology at the University of Otago, New Zealand, and was founder and first head of the Geology Department at Ahmadu Bello University in Nigeria.

Mr. Wright is a Fellow of the Geological Society of London, and a member of the Mineralogical Society, for which he acts as an assistant editor, and he has published more than fifty papers and articles in scientific journals and books. He also holds a flying license.